ANAEROBES AND ANAEROBIC PROCESSES

ANAEROBES AND ANAEROBIC PROCESSES

Om Prakash PhD Microbiology
Scientist
National Centre for Microbial Resource
National Centre for Cell Science, Pune, Maharashtra, India

and

Dilip R. Ranade PhD Microbiology
Ex-Officiating Director and Scientist-G (Retd.)
MACS-Agharkar Research Institute
Pune, Maharashtra, India

NEW INDIA PUBLISHING AGENCY
New Delhi – 110 034

Distribution Information: *Co-Published with CRC Press, UK. Meant for Sales Only in

India, Pakistan, Nepal, Sri Lanka and Bhutan.*

NEW INDIA PUBLISHING AGENCY

101, Vikas Surya Plaza, CU Block, LSC Market
Pitam Pura, New Delhi – 110 034, India
Email: info@nipabooks.com
Web: www.nipabooks.com

For customer assistance, please contact
Phone: + 91-11-27 34 17 17 Fax: + 91-11- 27 34 16 16
E-Mail: feedbacks@nipabooks.com

ISBN: 978-93-90175-06-2

Composed and Designed by NIPA.

Prof. T. Satyanarayana
Professor Emeritus
Department of Biological Sciences & Engineering
Netaji Subhas University of Technology
New Delhi.

Foreword

The basic premise of geobiochemistry is that life emerged on Earth where there were opportunities for catalysis to expedite the release of chemical energy in water–rock–organic systems. In this framework, life is a planetary response to the dilemma that cooling decreases the rates of abiotic processes to the point that chemical energy becomes trapped, and catalysis via metabolism releases the trapped energy, and life benefits by capturing some of the energy released.

Microbes that thrive in the absence of oxygen are known as anaerobes. These are distinguished from the aerobes in their ability to use alternative electron acceptors other than oxygen to complete their energy/carbon cycle (anaerobic respiration / fermentation). In case of anaerobic respiration, terminal electron acceptors are mainly reduced compounds such as nitrate (NO^{3-}), ferric (Fe^{3+}), sulphate (SO_4^{2-}), carbonate (CO_3^{2-}) and occasionally certain organic compounds. In case of fermentative way of energy generation, organic compounds serve as electron donors and acceptors both. Fermentations are accompanied by the production of more or fewer reduced compounds such as alcohols, organic acids, ammonia, hydrogen (depending on the regeneration of the cofactor) and carbon dioxide as the oxidized product. However, via this strategy, anaerobes obtain less net energy gain. This fact has implications for their metabolism, with the result that they are considered as slow growers as compared to their aerobic counterparts.

Anaerobes can be differentiated into aerotolerant anaerobes, which are microorganisms unable to respire oxygen (O_2) but they do not grow in the presence of oxygen while obligate (or strict) anaerobes die with momentary exposure with O_2. The alternative energy strategy of anaerobes has allowed them to colonize diverse environments on earth where oxygen is not present or has been depleted by aerobic organisms. Obligate anaerobes are found in all three domains of life. The eukaryotes are represented by anaerobic fungi, ciliates and flagellates, and the Archaea by the methanogens.

The energy generation system of anaerobic bacteria is not as efficient as in their aerobic counterparts. For example, anaerobic fermentation of 1mole glucose produce only 2 molecules of ATP for the cell in comparison to 36-38 ATP molecules formed by aerobic respiration. However, in terms of biotechnological process application, this provides the advantage of less carbon being used for cell biomass generation and maintenance, and more carbon directed towards product formation, which leads to a high product yield. Another point of interest for biotechnology is the wide range of products formed and substrates fermented. However, when the metabolic routes are known for the microorganism, shifting the production towards one (or more) specific products can be achieved by process engineering and/or by metabolic engineering.

Opting for process engineering adaptations brings the advantage of maintaining all the routes still active in the microorganism. It also offers the benefit of using the same strain to produce different products, while obtaining high productivities by shifting the process operational conditions. Most anaerobic bacteria have a substrate versatility and flexibility that is lacking in their aerobic counterparts, for example adaptability for the uptake of hexose and pentose sugars. Moreover, they not only ferment sugars but also have the machinery to degrade complex structural biomass (e.g., cellulose, hemicellulose). This offers the possibility to potentially hydrolyse and ferment plant biomass into an efficient and integrated process, termed consolidated bio-processing. The simplification of several processing steps could provide many operational and economic advantages on an industrial scale.

Anaerobic microorganisms, and especially bacteria, have played an outstanding role in the rise and expansion of industrial biotechnology. The first industrial fermentation process for the production of chemicals and fuels was the acetone-butanol-ethanol process (ABE) by *Clostridium acetobutylicum*. The process was discovered and patented by Chaim and Weizmann in the United Kingdom and was soon upgraded to an industrial scale, initially to supply acetone to the British artillery during World War I (Goldstein, 1995). Subsequently, the solvents were used in the chemical industry and as fuel for the expanding automobile industry, resulting in the construction of industrial plants. With the expansion of the petrochemical industry, the production of chemicals and fuels by fermentation was mostly replaced by oil refining. In certain places (e.g. South Africa), the ABE process, however, continued in operation till 1980. After oil crises at the beginning of the 21st century, the concern for replacing oil-derived products by renewable resources has once again focused attention on biological processes and is now one of the main targets of the chemical industry. Genetically modified strains with anaerobic microbial metabolic routes are the core for the production of such compounds by industry nowadays. One example is the butanol production by Butamax TM. Another successful example of chemicals in large scale production that can be produced by an anaerobic bacterium, but commercialized from a genetically modified strain, is 1,3-propanediol (Bio-PDOTM) by DuPont.

Anaerobic microorganisms have been used in fermentation processes for food production for thousands of years. Food items (such as beer, cheese and vinager) that are produced by anaerobic fermentation by yeast or bacteria, get their characteristics by the formation of compounds such as propionic acid, lactic acid, ethanol and acetic acid. These compounds are also of interest for the chemical industry nowadays. To use traditional food-fermenting microorganisms for the chemical industry is an approach in the study of chemicals and fuels produced from renewable resources. A commercially exploited example is lactic acid from *Lactobacillus bulgaricus* for the use in formulations for the cosmetic and pharmaceutical industry and polymers. However, wild anaerobic microorganisms encountered in natural selective processes (such as cellulose/hemicellulose rich environments) are a potential source of microorganisms or genes for reaching next generation processes for the conversion of biomass into value added products.

Depletion of easily accessible fossil energy resources, threat of climate change and political priority to achieve energy self-sufficiency and sustainable solutions prioritize a conscious and smart use of renewable resources to generate a bio-based economy. Bio-based compounds can replace chemicals and fuels that are now mainly produced from crude oil. Efficient processes for the conversion of plant biomass into compounds of interest to the biorefinery industry occur naturally in anaerobic environments such as in the forestomach of herbivores. Exploration of anaerobic microorganisms for industrial biotechnological applications creates the possibility to convey efficient and flexible processes, with lower implementation and running costs, making it also applicable to developing and emerging economies.

The Editors of the book Dr. D.R. Ranade and Dr. Om Prakash deserve appreciation for bringing out this very useful book with contributions from the experts working on various aspects of anaerobic microbes. I sincerely hope and wish that the book will be useful to students, scholars, teachers and scientists working in the disciplines of microbiology, biotechnology and life sciences.

Prof. T. Satyanarayana

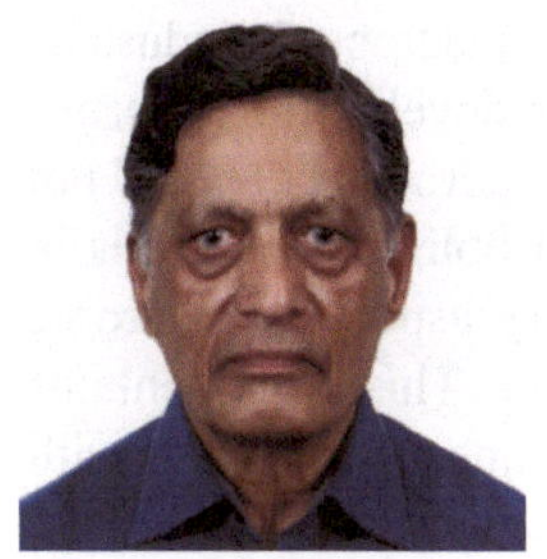

Prof. Rup Lal
FNA, FNASc, FNAAS
NASI Senior Scientist Platinum Jubilee Fellow
The Energy and Resources Institute
Darbari Seth Block, IHC Complex, Lodhi Road
New Delhi-110003, India

Foreword

I am delighted to write about the Editors and Contents of current book entitled "***Anaerobes and Anaerobic Processes***" Edited by Dr. Om Prakash from National Centre for Microbial Resource (NCMR), National Centre for Cell Science(NCCS) Pune and Dr. D.R. Ranade (Ex-Officiating-Director, ARI Pune). It is also a matter of great delight that one of the Editors Dr. Om Prakash Sharma was my PhD student who worked on bacterial degradation of polycyclic aromatic hydrocarbons (PAHs). Dr Om Prakash is a very keen and devoted researcher and impressed everyone with dedication towards to research in microbiology.

After attaining his PhD from the University of Delhi, Dr. Prakash has moved to USA and received his postdoctoral training from Florida State University and Georgia Institute of Technology. During his stay in USA he developed expertise of handling of obligate anaerobes like metal reducers and methanogenic archaea under the supervision of a renowned scientist Prof. Joel E Kostka. After Postdoctoral training he Joined NCMR-NCCS and established the anaerobic facility and also handling cultivation, preservation and study of obligate anaerobes. He has more than 13-years of experience in anaerobic microbiology.

Dr. D. R. Ranade, also has longstanding experience in methanogenic archaea and obligate anaerobes and has published several papers on these aspects. The designing and selection of the contents for this book is the outcome of in-depth understanding of the subject by the editors . Needless to say that this book was really required for the scientific community and students.

Although anaerobes are less explored in comparison to aerobes due to technical limitations and tough nature of cultivation and maintenance. They are the valuable components of human-gut microbiota, anaerobic digestion of waste, wastewater treatment, global climate change, anaerobic probiotics and anaerobic infection. Human-gut is totally anoxic and anaerobic based gut research is getting extra attention in microbiology. Considering the importance of anaerobes in industrial and ecosystem services, I always have a fascination to develop the expertise and collaboration with microbiologists working with anaerobes but could not achieve this due to very little expertise of handling of obligate anaerobes in India. This book is unique as it covers each and every aspect of anaerobic microbiology to impart basic and advance knowledge. This book contains chapters on cultivation and preservation of obligate anaerobes along with description of different anaerobic groups like, methanogenic Archaea, anaerobic methanotrops, alternative electron accepting process like denitrification and sulfate reduction, anaerobic composting and solid waste management. It incorporated anaerobic process of bio-hydrogen, bio-butanol, production along with wastewater treatment. From human perspective it included human gut ecology, anaerobic probiotics and anaerobic infection which is quite relevant in current scenario. In addition, this book also covers anaerobic fungi and photoautotroph and microbiology of animal rumen. Biotechnological and industrial applications of anaerobes have also been covered.

I feel the contents of the book are very appropriate and will be of much interest to undergraduate, graduate and PhD students of the Microbiology, Environment science, Biochemistry and Biotechnology background. I must congratulate the Editors for innovative design, selection of experts and content for this book.

Professor Rup Lal

Dr. Hemant J Purohit
Head Environmental Biotechnology & Genomics Division
National Environmental Engineering Research Institute, CSSRI-NEERI
Nehru Marg, Nagpur 440020 India

Foreword

The book entitled "***Anaerobes and Anaerobic Processes***" is an extraordinary piece of compilation of selected expertise in area of anaerobe microbiology; and are working with different tools towards common goal of understanding the anaerobic system. The compilation takes to next level of knowledge by discussing the various applications like chemicals, bioenergy, food and other industrial application. Studying anaerobe biology is important aspect for understanding the ecological function of primitive organism like anoxygenic phototrophs to the recently mined microbiome of human gut. Physiologically, occurrence of microbes across the contrasting niches like anoxygenic marine environment to indigenous the gut of human and ruminants and finally its interaction and survival as pathogens have been nicely discussed.

The content of the book is selectively planned keeping in the mind the reader's interest, understanding and aptitude while challenging their curiosities. The chapters are arranged systematically beginning with explanation about the "technical considerations for cultivation and purification of Anaerobic Microorganisms". With details about what and how to handle anaerobic culture for cultivation and purification, the chapters provides the comprehensive overview about various technical considerations. Once the culture is purified and is cultivated for performing various studies, there come as stage of preserving such culture for future use. In this line chapter 2 deals with the Past Present and Future of Preservation methods for anaerobes.

Tracing back the origin of life, one must be aware of the anaerobic atmosphere present at that time. The editors have compiled few chapters which provide details about the power and prowess of anaerobes in that environment and their sustenance till this age. In this line, chap 3 provided insights into the diversity of anoxygenic phototrophic bacteria and its application. Likewise, methanogens (chap 4) has served as the major carbon sink/source since origin of life form since beginning. Certain physiological variations like anaerobic methanotrophy (chap 5) is also discussed. Sulfur has been an elemental in making and shaping the functions of life forms which is predominated by the microbial world as discussed in Microbial Mediated Anaerobic respiration of sulfur compounds (chap 6). Additionally, there is separate section for past present and future of Anaerobic Fungi (chap 7)

In another section the focus is on using anaerobic system for bio-hydrogen production and its use as green energy (chap 8), Biobutanol production and its use as biofuel (chap 9). Larger role of anaerobes in ecosystem functioning is also discussed, as in what are the aerobic principles behind the solid-waste management and Bio-composting (chap 10) and what is the role of anaerobes in wastewater treatment and nitrogen cycling (chap 11).The book passes through the basics about the anaerobic Process of Animal Rumen and its role in ruminant metabolism (chap 12) and discuss the trend and role of anaerobes as an Emerging Probiotics (chap 13) and how human Gut is an Ideal habitat for anaerobes (chap 14). Since anaerobe co-exists at the interface of human and animals, there are possible health threats too which are discussed in following chapter (15) "emerging threat of anaerobic infection". Last but not least, microbiology has always served the human need ranging from development of drugs like antimicrobials to food through fermentation of cereals to chemicals like butanol and others. The book compiles the biotechnological application of anaerobes in its last serving (chap 16) to the readers.

The structure of the books shows meticulous planning by The editor towards compilation of such selected topics. Cooking such a feast for the readers is definitely an expert chef's task and I am sure the readers would relish the content of this book and it would serve a quick reference guide for aspiring professionals for implementation of anaerobic studies in their research work or teaching curriculum. I congratulate the editors for putting all aspects of anaerobic microbiology and bring out the relevance in most organized way, such an interesting compilation; my complements to authors for sharing their expertise. I wish readers to use the "grab and go-through" content provided in this book, which will definitely force them to dive deep in plethora of knowledge presented in this book.

Dr Hemant J Purohit

Preface

Microorganisms are interesting live forms not only for morphological types but also for its unique physiological traits and biotechnological applications. In comparison to aerobes; anaerobic Bacteria, Fungi and Archaea are understudied. Earlier studies on anaerobic bacteria were mostly confined to anaerobic pathogen like *Clostridium*. Development of improved handling protocols and specialized techniques expanded the study of anaerobes like fastidious anaerobic bacteria, methanogenic archaea, anaerobic fungi and human gut- resident. Parallel to these developments, we have realized that anaerobic microorganisms including bacteria fungi and archaea are the key component of clean-energy, global climate change, waste to energy generation, solid waste management, waste water treatment, bio-toilets, Human-gut-microbiome, anaerobic probiotics, fecal-microbiota-transplantation (FMT), fecal microbiome banking and emerging threat of anaerobic infection. Therefore; cultivation, bio-banking and In-depth-understanding of functionality and physiology of anaerobes are imperative to exploit them for biotechnological processes.

Due to fastidious nature and need of special setup for cultivation, purification, characterization, handling and bio-banking in comparison to aerobic microorganisms very less work has been done on anaerobic bacteria, archaea and fungi. In addition, despite ecologically, industrially and clinically important group very little description is available on anaerobes and anaerobic process in most of the textbooks of microbiology. If we go through with most of the good textbooks available in microbiology we find that only 2-4 pages are dedicated to this important group of microbes. Most of the textbooks of microbiology only provide partial information about anaerobes and anaerobic process and these hidden heroes of the nature. We realized that a textbook in simple languages with ability to provide extensive knowledge about cultivation, purification, handling and bio-banking of different classes of anaerobes including the obligate, facultative, aero-tolerant and micro-aerophillic is very much essential. The aim of preparation of current text book is to provide the state-of-the-art knowledge about anaerobes and anaerobic process to undergraduate, graduate and PhD students. During the preparation of book we tried to incorporate contents from very basic like cultivation, handling, preservation and various important process related to anaerobes like biomethanation, anaerobic methanotrophy, solid waste management, anaerobic biohydrogen and biobutanol production, anaerobic nitrogen treatment, anaerobic solid waste management. In addition considering the importance of anaerobes in human health and disease we tried to dedicate a complete section like anaerobic probiotic, anaerobes and human gut microbiology and anaerobes and human infection. Finally we tried to incorporate a chapter on role of anaerobes in animal rumen and food digestion. We believe that current book will serve its aim to satisfy the hunger of knowledge about anaerobes and anaerobic process for the UG, PG and PhD students of Microbiology, Biotechnology, Environmental Science and Biochemistry background.

Editors

Contents

List of Contributors

1. **Abhijit Kulkarni**
   The YSS lab, National Centre for Microbial Resource, National Centre for Cell Science Pune, India

2. **Akshay Gaike**
   The YSS lab, National Centre for Microbial Resource, National Centre for Cell Science Pune, India

3. **Anshuman A. Khardenavis**
   Academy of Scientific and Innovative Research (AcSIR), Ghaziabad 201002, India
   Environmental Biotechnology and Genomics Division (EBGD), CSIR–National Environmental Engineering Research Institute (NEERI), Nehru Marg
   Nagpur-440 020, Maharashtra, India

4. **Anurag Kumar Bari**
   Dept of Pathology & Microbiology, Breach Candy Hospital Trust, Mumbai Maharashtra, India

5. **Aruna Poojary**
   Dept of Pathology & Microbiology, Breach Candy Hospital Trust, Mumbai Maharashtra, India

6. **Ashish Kumar Singh**
   Academy of Scientific and Innovative Research (AcSIR), Ghaziabad 201002, India
   Environmental Biotechnology and Genomics Division (EBGD), CSIR–National Environmental Engineering Research Institute (NEERI), Nehru Marg
   Nagpur 440 020, Maharashtra, India

7. **Ashish Polkade**
   Bharat Eco Solutions and Technologies, A19, Meera Classics, Aundh Ravet BRTS Road, Kalewadi, Pune, Maharashtra, India

8. **Atul R. Chavan**
   Academy of Scientific and Innovative Research (AcSIR), Ghaziabad 201002, India
   Environmental Biotechnology and Genomics Division (EBGD), CSIR–National Environmental Engineering Research Institute (NEERI), Nehru Marg
   Nagpur 440 020, Maharashtra, India

9. **Charu Dogra Rawat**
   Department of Zoology, Ramjas College, University of Delhi, Delhi

10. **Dilip R. Ranade**
MACS-Agharkar Research Institute, Pune, Maharashtra

11. **Geetika Sharma**
Amity Institute of Biotechnology, Amity University Uttar Pradesh, India

12. **Hemant J. Purohit**
Environmental Biotechnology and Genomics Division (EBGD), CSIR–National Environmental Engineering Research Institute (NEERI), Nehru Marg Nagpur-440 020, Maharashtra, India

13. **Jyoti A. Mohite**
C2, Bioenergy group, MACS Agharkar Research Institute, G.G. Agarkar Road, Pune 411004, Maharashtra, India
Savitribai Phule Pune University, Ganeshkhind Road, Pune 411007, Maharashtra India

14. **Kajal Gopal Singh**
A2-991, Saarrthi Souvenir, Mhalunge, Pune-411045, Maharashtra, India

15. **Kasturi Deore**
Bioenergy Group, MACS-Agharkar Research Institute, Pune, Maharashtra, India

16. **Krishna K Yadav**
National centre for Microbial Resource, National Centre for Cell Science, Pune, India

17. **Kumal Khatri**
Bioenergy Group, MACS-Agharkar Research Institute, Pune, Maharashtra, India

18. **Leena Kamlaskar-Kulkarni**
MACS - Agharkar Research Institute,G G Agarkar Road, Pune-411004, Maharashtra India

19. **Mahesh S. Sonawane**
National Centre for Microbial Resource, National Centre for Cell Science, Pune, India

20. **Manasi P. Tukdeo**
Department of Microbiology, Modern College of Arts, Science and Commerce Ganeshkhind, Pune, Maharashtra, India

21. **Manisha Arora Pandit**
Department of Zoology, Kalindi College, University of Delhi, Delhi, Indai

22. **Monali C. Rahalkar**
C2, Bioenergy group, MACS Agharkar Research Institute, G.G. Agarkar Road, Pune 411004, Maharashtra, India

23. **Nancy Garg**
Amity Institute of Biotechnology, Amity University Uttar Pradesh, India

24. **Om Prakash**
National Centre for Microbial Resource, National Centre for Cell Science, Pune, India

25. **Prachi Karodi**
National centre for Microbial Resource, National Centre for Cell Science, Pune Maharashtra, India

26. **Pranitha S. Pandit**
C2, Bioenergy group, MACS Agharkar Research Institute, G.G. Agarkar Road Pune-411004, Maharashtra, India
Savitribai Phule Pune University, Ganeshkhind Road, Pune 411007, Maharashtra India

27. **Prashant K Dhakephalkar**
Bioenergy Group, MACS-Agharkar Research Institute, Pune, Maharashtra, India

28. **Rahul A. Bahulikar**
BAIF Development Research Foundation, Central Research Station, Urulikanchan Pune 412202

29. **Rahul Bodkhe**
The YSS lab, National Centre for Microbial Resource, National Centre for Cell Science Pune, India

30. **Rajesh K. Srivastava**
Departmant of Biotechnology, GITAM Institute of Technology (GIT), Gandhi Nagar Rushikonda, Visakhapatnam-530045, Andhra Pradesh, India

31. **Rakesh Kumar Gupta**
Academy of Scientific and Innovative Research (AcSIR), Ghaziabad 201002, India

32. **Rohit Sharma**
National centre for Microbial Resource, National Centre for Cell Science, Pune, India

33. **Sarah Fatima**
Centre of Excellence, Department of Microbiology, Dr. Ram Manohar Lohia Avadh University, Faizabad (U.P.) 224001

34. **Shailendra Kumar**
Centre of Excellence, Department of Microbiology, Dr. Ram Manohar Lohia Avadh University, Faizabad (U.P.) 224001

35. **Sheetal Shirodkar**
Amity Institute of Biotechnology, Amity University, Uttar Pradesh, India

36. **Shraddha Shaligram Vajjhala**
National centre for Microbial Resource, National Centre for Cell Science, Pune Maharashtra, India

37. **Shrikant P. Pawar**
National centre for Microbial Resource, National Centre for Cell Science, Pune Maharashtra, India

38. **Shubhanjali Singh**
National centre for Microbial Resource, National Centre for Cell Science, Pune, India

39. **Sruthy Vineed Nedungadi**
Departmant of Biotechnology, GITAM Institute of Technology (GIT), Gandhi Nagar, Rushikonda, Visakhapatnam-530045, Andhra Pradesh, India

40. **Suraj P. Nakhate**
Academy of Scientific and Innovative Research (AcSIR), Ghaziabad 201002, India
Environmental Biotechnology and Genomics Division (EBGD), CSIR–National Environmental Engineering Research Institute (NEERI), Nehru Marg, Nagpur-440020 Maharashtra, India

41. **Tushar Lodha**
National centre for Microbial Resource, National Centre for Cell Science, Pune, India

42. **Vaidehi Mirashi**
National centre for Microbial Resource, National Centre for Cell Science, Pune, India

43. **Vikas C. Ghattargi**
National centre for Microbial Resource, National Centre for Cell Science, Pune, India

44. **Vikas Patil**
National centre for Microbial Resource, National Centre for Cell Science, Pune, India

45. **Vikram B. Lanjekar**
Bioenergy Group, MACS-Agharkar Research Institute, Pune, Maharashtra, India

46. **Yogesh Nimonkar**
National centre for Microbial Resource, National Centre for Cell Science, Pune, India

47. **Yogesh S. Shouche**
National centre for Microbial Resource, National Centre for Cell Science, Pune, India

1

Technical Considerations for Cultivation and Purification of Anaerobic Microorganisms

Om Prakash and Dilip R. Ranade*

National Centre for Microbial Resource, National Centre for Cell Science Pune-411007, Maharashtra, India

Abstract

Based on oxygen requirement, microorganisms have been classified into different groups, like facultative anaerobes, micro-aerophilic, aerotolerant and strict/obligate anaerobes. Obligate or strict anaerobes are the key components of anoxic environments and provide valuable ecosystem services like methanogenesis, sulphate reduction, metal reduction, degradation of pollutants etc. Apart from being valuable for the ecosystem, they are causative agents of human infections. They are the key players of human-gut microbiology, and can be used in bio-toilets, solid waste management, anaerobic digestion of waste, wastewater treatment processes and generation of waste to energy etc. Only few anaerobic microorganisms are pathogenic in nature. A mechanistic understanding about physiology and metabolism of anaerobic microorganisms is imperative for their industrial applications. Cultivation of these organisms is the first and very important step for the basic and applied study of anaerobes. Unlike cultivation of aerobic microorganisms anaerobic cultivation needs special care and facilities for successful handling of anaerobes. In the current chapter, We tried to incorporate the components and technicalities of anaerobic pre-reduced media preparation, cultivation and successful isolation of strict anaerobes. In addition, we also tried to discuss the different tools used for cultivation of anaerobes.

**Corresponding Author: omprakash@nccs.res.in,prakas1974@gmail.com*

Background

According to the definition, anaerobes are those microorganisms which show survival and growth potential under anoxic conditions or in complete absence of oxygen. They mainly constitute members of domain Bacteria and Archaea (prokaryotes) and some Fungi (Eukarya) (Cheng et al. 2018, Doddema & Vogels 1978, Edwards et al. 2013, Haitjema et al. 2014). In brief, based on oxygen relationships microbes are classified under different categories.

Strict or obligate aerobes: Only grow in the presence of oxygen

Facultative anaerobes: Prefer oxygen for growth but can grow in the absence of oxygen.

Aerotolerant: They can survive in aerobic conditions although their growth rate is slow.

Obligate or strict anaerobes: Molecular oxygen is lethal for them and can survive and grow only in the complete absence of O_2.

Microaerophilic: They cannot grow normally at atmospheric O_2 concentration and need reduced oxygen levels for their growth.

The extended description about these can be found from several reviews and topics published in this area (Balch et al. 1979, Wolf 2011, Speers et al. 2009, Stieglmeier et al. 2009). To survive and grow in anaerobic environment, anaerobic organisms use an alternative way of energy generation (Hong & Gu, 2009). It includes fermentation and anaerobic respiration types of metabolisms. Fermentative type of metabolism is widespread among anaerobes. Similar to respiration, initially it begins with glycolysis but pyruvate produced at the end of glycolysis does not enter in the citric acid cycle. Electron transport chain (ETC) does not operate and ATP generation occurs via substrate level phosphorylation. Fermentation is energetically less favourable because in comparison to respiration, only two ATP molecules are produced from glucose (Wolfe 1999). In addition, anaerobic respiration is similar to aerobic respiration wherein, released or extracted electron from electron donors travels through electron transport system for generation of ATP using oxidative phosphorylation mechanisms, but instead of using the O_2 as terminal electron acceptor, it uses several other alternative electron acceptors for this process due to absence of oxygen. Nitrate, sulphate, some metals and CO_2 etc. are used as alternative electron acceptors in anaerobic respiration (Hong & Gu, 2009, Prakash et al. 2010, Green et al. 2010). Furthermore, anaerobic respiration is energetically less efficient than aerobic respiration and generates less amount of ATP depending on the redox potential of respective alternative electron acceptors involved in process.

Lack of superoxide dismutase, catalase or peroxidase in anaerobic microorganisms is mainly responsible for lethal effects of oxygen. In aerobic microorganisms these enzymes detoxify the system from reactive oxygen species (ROS) like superoxide anion (O_2^-), hydroxyl radicals, singlet oxygen and hydrogen peroxide produced during growth and metabolism of organisms in the presence of oxygen (Jab³oñska & Tawfik 2019). The lethal effects of ROS also depends on several other factors including the location of detoxifying enzymes, rate of formation and accumulation of ROS in cells and susceptibility of cellular components like DNA, proteins and lipids with produced ROS. Due to use of low redox potential flavoproteins in anaerobic respiration they probably generate excess superoxide and hydrogen peroxide when exposed with molecular oxygen. In addition, obligate anaerobes also use different classes of dioxygen-sensitive enzymes which is not produced in aerobic organisms (Imlay 2002, Zhao & Drlica 2014).

Importance of anaerobes in ecosystem services

Anaerobic life is generally adopted for resource limited and harsh environmental conditions and show immense biotechnological, ecological and environmental significance (Franzmann et al. 1997, Bräuer et al. 2006, Brauer et al. 2011, Ze et al. 2013, Lowe et al. 1993, Podolsky et al.2019, Hatti-Kaul and Mattiasson 2016, Cadillo-Quiroz et al. 2009, Kurr et al. 1991). It is considered that anaerobes are the most primitive type of organisms and first originated during anoxic condition of the earth. They have the ability to grow in harsh conditions of the environment like extremely high concentration of salt, acidic or alkaline pH, very high or very low temperatures, and metal toxicity. Anaerobes play crucial role in biogeochemical cycling of carbon, nitrogen, sulphur, and iron etc. They also have great potential of biotechnological and industrial importance. For example, anaerobes used in degradation of starch and cellulose, work as source and sink of methane by the process of methanogenesis and methanotrophy. Anaerobes also play a role in global climate change, bioenergy generation, CO_2 reduction, biomass valorization, nitrate, sulphate, metal and CO_2 reduction, ethanolagenesis and acetogenesis. Wastewater treatment and remediation of waste contaminated sites are associated with activities of anaerobic microbes in oxygen limited conditions. Anaerobic fungi from digestive tract can be a good source of industrial enzymes of biotechnological importance (Cheng et al. 2018). Fastidious nature, strict growth requirement, limited strain availability, difficult scale-up process are some of the issues that need to be addressed for better exploitation of anaerobes in future. In order to provide a more detailed description of anaerobic microorganisms in environmental and biotechnological process, a complete chapter is included on this theme in the current book.

Importance of Cultivation

Due to continuous development and refinement of culture independent OMICS approaches, cultivation is a lagging art compared to OMICS. Very few laboratories, researchers and Scientists have a fascination about cultivation, purification and bio-banking of the organisms for future generations. Development of high-throughput next generation sequencing platforms, generation of culture independent molecular data has become affordable and less labour intensive in comparison to cultivation. Although it has been realized that inference about microbial diversity and functionality achieved by mainly culture independent metagenomic approach is incomplete and premature and it needs additional support from cultivation-based approaches (Prakash et al. 2013). In fact, for study of physiological behaviour in changing environmental conditions and effects of different stressors, exploitation of microbes for antibiotics, enzymes, vitamins and drugs of industrial and clinical significance, and discovery of novel pathways and biomolecules, availability of pure culture is essential (Prakash et al. 2013, Borner et al. 2013, Brady et al. 2009). Importance of pure culture study can be understood by putting gut-microbiology research as an example. Recently, it has been realized that gut microbiota is involved in gut health and several diseases. Several hypotheses about involvement of gut-bacteria and archaea have been proposed in different conditions of disease and health based on metagenomic data but due to lack of pure culture, conclusive evidence is not established (Lagier et al. 2018, Goodman et al. 2011). Similarly, in the area of diagnosis we can detect the pathogens using culture independent approaches, but for the study of antibiotic sensitivity (AST) and resistance, pure culture is a must. Furthermore, prediction of functionality of organisms based on metagenomic data also is not very realistic in most of the conditions and needs actual organism to prove the concept. It has been discovered that keystone species of any ecosystem, which are mainly responsible for ecosystem functionality, are not always the abundant population but the ones belonging to sub abundant group in the community. Metagenomic studies are mainly biased for more abundant species of any ecological system but detection of keystone species is difficult and most of the time is neglected (Rafrali et al. 2013, Banerjee et al. 2019). Therefore, for accurate study of functionality of the ecological system and to commercially exploit the species, their cultivation is a must.

Cultivation of not-yet-cultured microorganisms

Culture independent metagenomic approaches have unveiled the fact that immense microbial diversity of clinical, industrial and environmental potentials is not-yet-cultured and substantiated the truth of great plate count anomaly (Prakash et al. 2013, Bollman et al. 2007, Borner et al. 2013). It has been mentioned in almost all articles published on this concept that despite all the efforts, 85-99% microbial diversity, specially bacteria and archaea from different ecological systems are still not available in the form of isolated pure culture. These not-yet-cultured and unstudied microbial communities are known as "Microbial Dark Matter" (Hedlund et al. 2014). They could be the dominant player of the ecosystem and valuable source of novel biomolecules and pathways. Several different factors have been discovered for non-culturability of the microbes on common laboratory medium. The detailed description about them is given elsewhere (Prakash et al. 2013). In brief, toxicity of medium component, overgrowth of fast growers r-strategic organism, nutritional shock during oligotrophic to copiotrophic transition, lack of social communication, special nutrient requirement and fastidious nature, slow growth rate and longer generation time, syntrophism, requirement of specific host or partner are some of the key factors behind non-culturability of microbes (Prakash et al. 2013, Bollman et al. 2007, Aoi et al. 2009, Carbonero et al. 2010, Hamilton-Brehm et al. 2012,Horris 1983, Lagier et al. 2018, Nyonyo 2013, Tamaki et al. 2009, Thompson et al. 2015, Vilkova et al. 2015, Overmann et al. 2017). Considering all the above possibilities, efforts have been made from time to time to improve the culturability of microbes and to get the uncultured organisms in culture. Design of diffusion chamber and recent concept of isolation chip (I-Chip) to generate the simulated natural environment under laboratory set-up are the best examples and efforts in this direction (Nichols et al. 2010). In addition, design of low nutrient medium for oligotrophic and increased incubation time for slow growing organisms is another link in the same direction. Additional nutrient supplements, suppression of competition by cell separation, use of different gelling agents, mixing of agar after autoclaving, use of different range and combination of media to capture wide range of microbial diversity are some of the other aspects to cultivate the novel organisms (Overmann et al. 2017). Additionally, recently genomics-based approaches of cultivation are practiced by several research groups for successful cultivation of not-yet- cultured organisms (Garza & Dutilh 2015). In brief, using culture independent Metagenomic approach, whole genome sequence of not-yet- cultured organisms of interest can be achieved first. Subsequent to that, prior idea about metabolic pathways and nutritional requirement of the organisms should be generated based on the annotation of deduced complete genome. Information obtained from complete genome sequence, assists in media formulation and nutritional requirement of the organisms for cultivation of uncultured microorganisms. Several microorganisms have been cultivated successfully using genome guided approach, which were tough to culture (Berdy et al. 2017, Schwieterman et al. 2019).

Some example of successful cultivation using genomics

Survey of the published literature on microbial diversity indicates that more than 50% of described bacterial and archaeal phyla do not have cultured representatives and are only created based on the culture independent metagenomics, genomics and 16S-rRNA based data with minimum documentation. Phyla containing not-yet-cultured representative are known as candidate phyla. Now we can get the complete or almost complete genome sequence of not-yet-cultured microbes using culture independent shotgun metagenomic approaches. Detail description of about methodology and protocol about how to obtain whole genome sequence of not-yet-cultured organisms is not the subject matter of current chapter and can be obtained from somewhere else (Garza & Dutilh 2015, Handelsman, 2004). Construction of whole genome sequence from environmental DNA gave birth to genome assisted or genome guided cultivation of microorganisms. It provides information about metabolic potential, nutrient requirement, and interaction with other microbes, which can be mimicked during the time of cultivation to obtain the live culture of microorganisms. Cultivation story of SAR-11 clade of abundant marine bacterium is a good example of genome guided cultivation (Rappé et al. 2002). After getting the whole genome sequence of SAR-11 clade, using metagenomic approach it was discovered that the organism lacks assimilatory sulphate reduction gene. Supplementation of reduced sulphur from outside, during the cultivation, boosted the growth of the organisms and made it cultivable. Genome assisted cultivation is still not very well explored concept and needs to be explored for cultivation of not yet cultured microbes. For example, candidate phyla OP11, OD1, and BD1-5 are the widespread uncultured bacterial candidate phyla, with more than 15% representatives from bacterial domain (Garza & Dutilh 2015, Handelsman, 2004). Genomic and cell morphology data indicates that they have small genome size which lacks several biosynthetic pathways. Genome driven information can be used for the cultivation of these organisms.

Microbial culturomics: Although isolation, purification and preservation are considered as conventional art, due to the importance of pure cultures, to understand the physiology, functionality and biotechnological application of microorganisms, cultivation is re-emerging. In other words, it can be considered as rebirth of cultivation in the form of culturomics. Several modifications in traditional approach have been suggested in terms of media composition, incubation condition, use of gelling agent, selection criteria and technical and conceptual modifications for getting not-yet-cultured organisms in culture (Ze. et al. 2012, Ze et al. 2013, Zehnder & Wuhrmann 1976). Nakamura et al. (2011) developed a six well plate method for cultivation of obligate anaerobes with less labour and cost while Sakai et al. (2009) realized that high concentration of hydrogen can be inhibitory for the growth of certain group of methanogenic Archaea and tried to mimic the natural concentration of hydrogen during their cultivation and successfully cultivated novel methanogens with this modification.

Wolf and Metcalf (2010) introduced vacuum vortex method for preparation of pre-reduced medium for cultivation of obligate anaerobes while several other groups have used different gelling agents like phytagel, gelrite and others, to get growth of novel organisms (Harris 1985). Furthermore, several changes in media composition and incubation temperature have been introduced. For example, addition of mupirocin, acetic acid and norfloxacin has been suggested for selective cultivation and enrichment of *Bifidobacterium* (Vlková et al. 2015). Cultivation of anaerobic fungi from different clinical and environmental samples is also an emerging area of interest in anaerobic microbiology and a detailed description about it is provided in a separate chapter of this book. Interestingly, cultivation of anaerobes in aerobic atmosphere in presence of antioxidant has also been demonstrated by La-Scala et al. (2014). Kheliafia et al. 2013 designed improved medium for better growth of methanogens and Borner et al. 2013 developed small scale cultivation of obligate anaerobes by entrapping single cell in alginate beads. Brioukhanov and Netrusov (2012) tested positive effect of addition of exogenous hemin for cultivation of strict anaerobic archaeon *Methanobrevibacter arboriphilus* to protect the cells from oxidative stresses.

Tools and Techniques of Anaerobic Cultivation

Tools for Anaerobic Media preparation

Anaerobic chamber/glove box: Anaerobic chamber is also known as anaerobic glove box or anaerobic work station. Ralf Freter from University of Michigan Medical School for the first time suggested the idea of anaerobic chamber with a local engineer Dick Coy who, designed and built the same. Now Coy-type anaerobic chamber is very popular in anaerobic microbiology (Aranki and Freter1972). It is designed to create the anaerobic environment with controlled oxygen, temperature, humidity and CO_2. It constitutes of hydrogen and oxygen monitoring systems, airlock, incubator, desiccator and palladium sachets. Invention of anaerobic chamber opened the avenues of successful cultivation and purification of strict anaerobes on agar surface and contributed immensely to understand the anaerobic physiology, molecular biology biochemistry, cultivation of anaerobic pathogens and study of their antibiotic sensitivity/resistance. The inner environment of chamber contains a mixture of nitrogen, CO_2 and hydrogen. It works on very simple principal of reaction of hydrogen with O_2 to generate water in the presence of palladium as a catalyst. Water created by this reaction must be removed to maintain the anaerobicity. Desiccants like charcoal and charged silica gel are used for this purpose. Hydrogen sulphide generated during the cultivation of sulphate reducers hampers the activity of electronics and palladium catalyst. Therefore, it is mandatory to remove the generated hydrogen sulphide from inner environment of the chamber.

Coy Hydrogen Sulfide Removal Column (HSRC) is a good option as it requires less maintenance and is easy to handle. In addition, it needs charging of silica gel and palladium catalyst on regular basis for better activity (Fig.1)

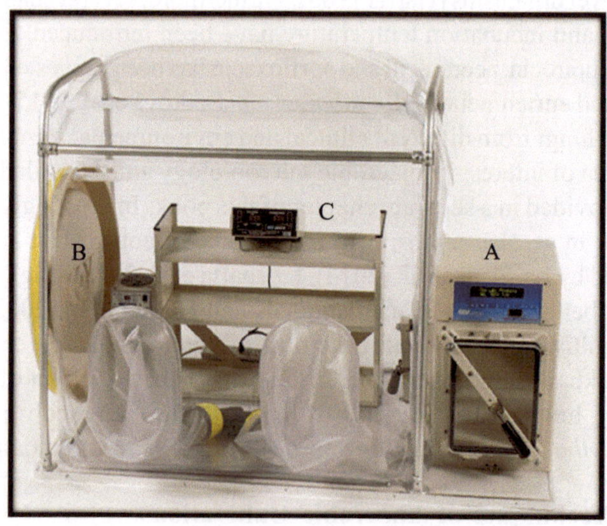

Fig. 1: Anaerobic chamber/Glove box (Coy Laboratory). Used for cultivation of strict anaerobes. **A)** Airlock for transfer of materials inside the chamber; **B)** port for transfer of small Instrument; **C)** Sensor for level of O_2

GasPak System/ Anaerobic Gas Jar: Anaerobic jar or GasPak (Fig. 2) is a very popular, easy to operate tool used for cultivation of anaerobes. Basically it is cylindrical jar made-up of metal or polycarbonate and like anaerobic chamber it works on the principle of scavenging the oxygen by the reaction with hydrogen in the presence of palladium catalyst. It needs a sachet (GasPak) with sodium bicarbonate, sodium borohydride and citric acid. It also include a strip impregnated with methylene blue as indicator of anaerobic conditions. For isolation of anaerobic bacteria, using GasPak system, the streaked plates are kept inside the jar and the reaction starts after the addition of a few drops of water inside the cut pouch. It generates the H_2 and CO_2. Generated H_2 reacts with the available O_2 of the Jar and produces H_2O while CO_2 assists in the growth. Other types of anaerobic Jars are

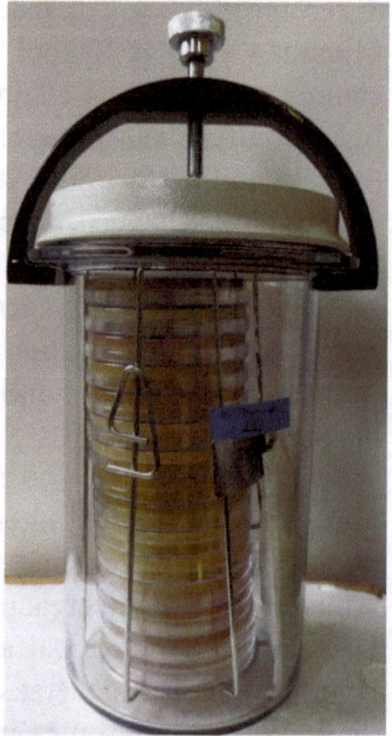

Fig. 2: Anaerobic Jar or GasPak System

also available. It is equipped with port and manometer for evacuation and filling of desired anoxic gases in the headspace of the Jar and to monitor the internal pressure respectively. Anaerobic jars are a very useful tool for incubation of the organism at different temperatures and can be used to transport the samples under anaerobic condition. However, it is suitable for cultivation of facultative anaerobes and aerotolerant type of microbes and does not work with strict/obligate anaerobes, as the exposure of O_2 cannot be avoided during plating, preparation of serial dilution etc. using this system. Candle jar or combustion jar can also be used.

Anoxomat: Anoxomat is a device based on the Mclntosh and Fildes system of evacuation and replacement for creating the anaerobic, capnophilic, microaerophilic and hypoxic conditions using microprocessor based automatic system (Fig.3). Along with the main instrument it includes an anaerobic jar and desired gas cylinder for creating the appropriate gas headspace conditions. It is quick, needs less instrumentation and is at a consumable cost. After evacuation and filling, 0.16% O_2 is left inside the jar which is further replaced by attached palladium catalyst. After completion of microaerophilic cycle 6% O_2 and 7% CO_2 is left inside the jar which is equivalent to conditions required to cultivate the members of the genus *Campylobacter*. It is equally applicable in industrial, clinical and environmental microbiology related research.

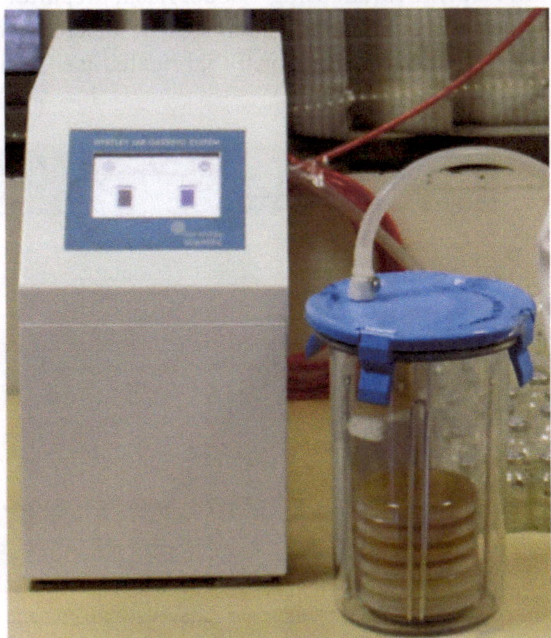

Fig. 3: Anoxomat, Don Whitley Scientific

Anaerobic lap system: Anaerobic lap system was developed by Frits Lap of Microbiology Department of Wageningen University, Netherlands and commercially marketed by GR instrument, Netherlands. It consists of a port for evacuation, gas supply line, pressure monitoring system and a port for filling of several tubes or serum vials at a time. As it is automatic in nature, it creates reproducible headspace in terms of desired gas and pressure and is considered ideal for study of anaerobic bacterial physiology. Detailed description about the instrument can be obtained from Plugge (2005).

Anaerobic gas station: Anaerobic gas station or anaerobic gas manifold is the simplest way of creating the anaerobic gas environment and preparation of pre-reduced anaerobic medium for cultivation and handling of anaerobes (Fig. 4). It is based on the principal of Hungate method (Hungate 1950) developed by Hungate and Nottingham (1969). A simplified method and protocol for fabrication of simplified gas station for cultivation of strict anaerobes like *Geobacter* has been discussed by O'Brian and Malavankar (2017). The main component of a gas station includes a gas-cylinder with mixed anaerobic gases, pressure regulator, gassing manifold, copper tubing, rubber tubing, air- lock fittings and 1ml capacity hypodermic syringe with port to attach different size cannulas. It is a very useful tool of anaerobic laboratory and can be used for different purposes like, transfer of culture, preparation of pre-reduced media, preparation of anaerobic medium using Widdel flask, preparation of anaerobic chemicals and stock. Any laboratory interested in anaerobic cultivation can fabricate it at a low cost. An indigenously fabricated gas station of my own laboratory is given in the Figure 4

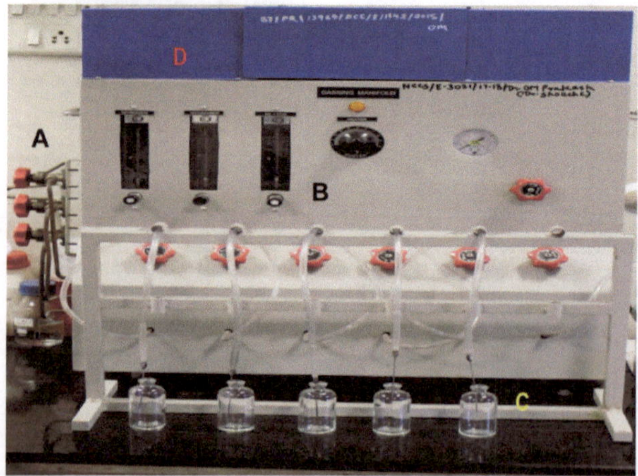

Fig. 4: View of indigenously fabricated anaerobic gas station used for preparation of pre-reduced anaerobic medium and flushing the tubes and vials with anoxic gas. **A)** Port for different anoxic gases; **B)** Pressure control knob; **C)** Serum vials with gas outlet port; **D)** Cover for heated copper scrubber

Widdel Flask: Widdel flask was designed by Friedrich Widdel from Max Planck Institute for Marine Microbiology, Bremen, Germany (Fig.5). He is well known for his work on anoxic microorganisms especially on microbiology and physiology of dissimilatory sulphate reducing bacteria (DSRB) (Widdel & Pfennig 1977). With respect to design it is similar to inverted conical flask with ports on two sides for releasing gas pressure and one central port with two tubes or two ports, one for supply of filter sterilized anoxic gas inside the flask, while the other port is used for dispensing the medium. Initially it was designed for preparation of pre-reduced cultivation medium using sodium sulphide as reducing agent for cultivation of DSRB. Due to inverted conical flask design, the surface area of the medium reduces gradually with pouring and there is minimized loss of sulphide gas from the surface. Our laboratory regularly use this device for preparation of pre-reduced medium using bicarbonate buffer system and found it to be more convenient. Bicarbonate buffer system is one of the best buffer systems although it requires saturation of medium with CO_2 to get the right pH and Widdel flask works very well for this purpose.

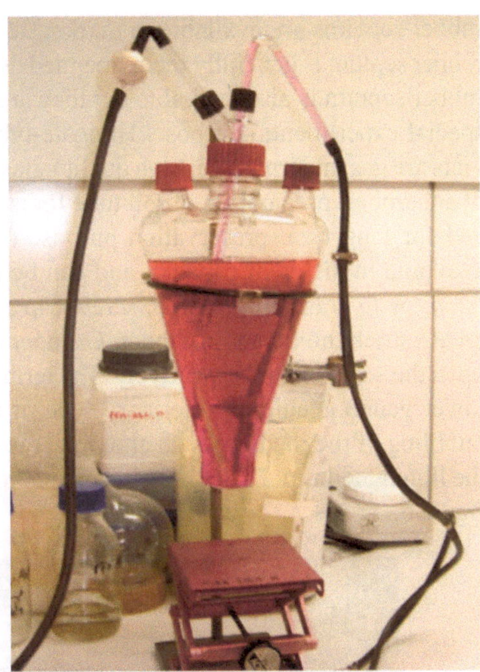

Fig. 5: View of Widdel Flask with resazurin added red coloured anaerobic medium

Small but important tools of anaerobic laboratory: Hypodermic needles and syringes with different lengths and widths, Hungate-tube, Balch-type tube, serum bottles, crimper, de-caper, gas cannulas, and aluminium crimps are the most frequently used tools of anaerobic laboratory and need their regular supply (Fig.6). Hypodermic needles and syringes are generally used to transfer the culture, inoculation of serum bottle, Hungate tube and Balch tube etc., It comes in different lengths and diameters. Eighteen, 20, 23, 25 gauge and 1 to 1.5 inch long needles are most common in practice but 1.0 inch long and 23-G needles are recommended for transfer of the culture as it creates a small pore size in the rubber septum and makes it less prone to entry of external O_2 by diffusion. Similar to needle and syringe, rubber septums are another essential tool. It comes in different shapes and colours. Blue butyl, grey butyl and black butyl rubber septums are available in market, sold by different vendors. Blue butyl rubber septum is generally recommended due to its better quality. Teflon coated rubber septum is also available but they are costly and are only used for some special experimental purpose. Hungate-tube, Balch-tube and serum bottle or vials are gas impermeable, high quality glass vessels used for liquid cultivation of anaerobes, preparation of roll tube for isolation of pure culture of anaerobes and for long term preservation purpose (Hungate 1969 figure 8). They are available in different capacity and can be purchased according to the needs. Along with aluminium crimp, screw cap and rubber septum they are used to create strict anoxic environment. Crimper and Decapper are used to seal and open the sealed vials respectively. The term Cannula is generally used for long lower gauge needle or veterinary type of needles. It is used for flushing or bubbling of the medium with anoxic gases and also for pouring or dispensing the liquid medium. Different tools of anaerobic laboratory are presented in the figure 6.

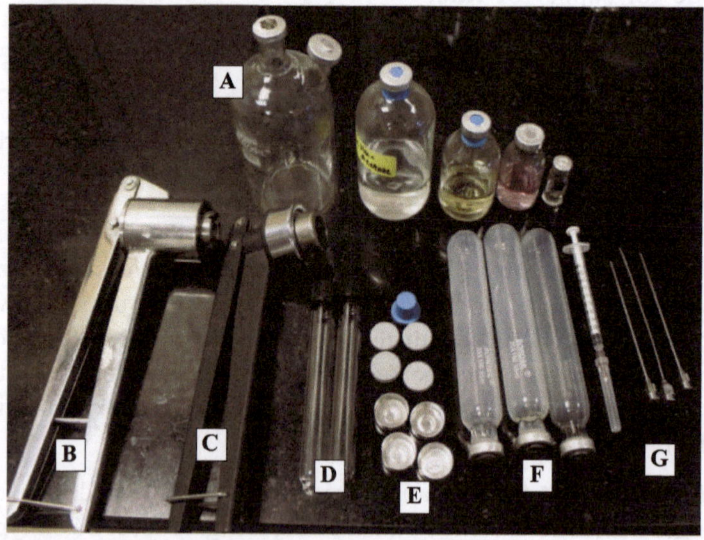

Fig. 6: Basic tools of anaerobic laboratory **A)** crimper; **B)** decaper; **C)** Hungate tube **D)** Septum and aluminum crimp **E)** Balch tube, **F)** hypodermic needle and syringes **G)** Different capacity serum vials.

Media Components

Redox indicator dye

Resazurin, also known as Alamar blue is commonly used redox indicator dye in anaerobic culture medium. In its oxidation state it is blue in colour and non-fluorescent in nature but in reduced stage it turns to resorufin which is pink in colour, and highly fluorescent and irreversible in nature. Resorufin is further reduced to dihydroresorufin through a reversible reaction which gives a colourless and non-fluorescent compound called hydroresorfin. Complete colourless condition of the dye in culture medium indicates that redox potential (Eh) of the culture medium has reached around -110 mV which indicates the right redox condition of the medium for inoculation because for most of the anaerobic bacteria only elimination of oxygen is not sufficient, they need low redox potential to start growing. Slightly pinkish colour indicates -51 mV Eh of the culture medium. For more detail description see the technical note published by DSMZ on anaerobic cultivation (https://www.dsmz.de/fileadmin/Bereiche/Microbiology/Dateien/Kultivierungshinweise/englAnaerob.pdf). Sometimes even boiling or prolonged bubbling of the medium with anoxic gases does not give colourless medium. Several factors like impurities of residual O_2 in anoxic gas, pH of the medium and presence of some intermediate like NO_2, also prevents the medium from turning colourless. Methylene blue is another indicator of the anaerobic condition. It gives blue colour in the oxidized state and turns colourless in anaerobic condition. Resazurin and methylene blue impregnated strips are generally used to indicate the anoxic conditions of the Glovebox and GasPak environment. The liquid solution of both can also be used as an indicator but for better results, it should be bubbled with gas phase of the inner environment of chamber. In addition, hydrogen and oxygen gas analyser is a good alternative, it detects *parts per million* (*ppm)* levels of gases but is expensive.

Reducing Agents

Reducing agents or reductants are crucial components of pre-reduced media formulation in anaerobic microbiology for cultivation of anaerobes. They quickly react with residual oxygen available in culture medium and further reduce the Eh of the medium. As mentioned earliar, the exclusion of O_2 is not enough as certain organisms require redox potential below -300 mV for growth, especially for cultivation of methanogenic archaea. Such low oxidation reduction potential can only be achieved with the help of reducing agents. Common reducing agents used in anaerobic culture media are cysteine hydrochloride, sodium sulphide (Na_2S), iron sulphide (FeS), sodium thioglycolate, dithiotheritol, sodium dithionate, ascorbate, $FeCl_2$ etc. (Prakash et al. 2010, Harmsen et al. 1998, Imachi et al. 2008) Sodium thioglycolate is more robust type of reductant and commonly used in commercially available medium as it does not need special care in handling. Cysteine hydrochloride and a combination of cysteine hydrochloride and sodium sulphide works better to achieve desired redox potential in most of the strict anoxic culture medium. In addition, sodium thioglycolate only provides -100 mV Eh to the medium, which is not suitable for some group of strict anaerobic microbes. As they are quickly oxidized, special care should be taken during the preparation of stock solution of reducing agents. It is recommended that one should weigh the salt inside the anaerobic chamber and quickly add it to the deoxygenated water to make the stock solution. One should prepare the anoxic water by boiling and bubbling it with nitrogen, and then add the reducing agents. It is advised not to bubble the contents after adding the reducing agents for better results. For preparation of stock solution of sodium dithionite, it should be dissolve it in anoxic water, filter sterilized it and to be stored in a dark place not more than-2-3 weeks. It is also recommended that before adding the sodium sulphide, one should, wash the crystal with water and, quick dry with tissue paper after which it should be immediately added to the anoxic medium. Cysteine hydrochloride can be added alone or in combination with sodium sulphide. Reducing agents can be added either during the pre-reduced media preparation or after the sterilization of the medium or during the time of inoculation. For immediate inoculation (within one week) addition of reducing agent at the time of media preparation is fine but if the medium needs to be stored for a longer period without inoculation, then it is better to add the reducing agent 24 h prior to inoculation for best results. It is also found that different reducing agents work differently and are not equally applicable for all sorts of cultivation practices. Addition and choice of reducing agent depends on the nature of organisms and required levels of Eh in culture medium.

Buffers

Bacteria generally produce organic acids or alkali (ammonia) during their growth and the gradual accumulation of these metabolites challenges the buffering capacity of the medium, sometimes eventually leading to a change in the pH. Hence, the uses of pH buffers which neutralize the change in the pH of the growth medium are recommended during the cultivation and physiological experiments. Buffering of the media, according to the nature of sample, to mimic the pH of the sampling sites is one of the good strategies to promote the cultivation of uncultured anaerobes. Although different kinds of buffers are recommended for cultivation of anaerobes, bicarbonate buffer system with N_2 + CO_2 in the headspace is considered ideal. Inorganic buffers are generally not recommended for cultivation and physiological studies due to their less compatibility with live cells. Sometimes inorganic buffers tend to be detrimental for cell growth and inhibit cellular cultivability. In addition, organic buffers, biological buffers or Goods buffers are generally recommending for cultivation of live cells. A list of organic buffers with their pKa and pH range is given in table. 1 and can be used based on the required pH of cultivation medium. It has been found that media with high load of organic carbon provides better protection and buffering than minimal or oligotrophic media. Therefore, during the cultivation of oligotrophic bacteria or use of minimal media with low carbon concentration, selection of appropriate organic buffer gives better results with a more diversity coverage.

Table: List of Biological buffers with their pH and pKa range used for buffering of media

Buffer	pH Range	pKa (25 ºC)	Ion binding	Uses
HEPES	6.8 - 8.2	7.45 - 7.65	Negligible metal ion binding	Buffer mammalian cell cultures Buffer for in vitro fertilization and embryo culture
MOPS	6.5 - 7.9	7.0 - 7.4	Strong interaction only with Fe	Use in cell culture media for bacteria Used in cell culture media for yeast and mammalian cells
MES	5.5 - 6.7	5.9 - 6.3	Strong interaction only with Fe	Used in plant, bacteria and mammalian cell culture media
BES	6.4 - 7.8	6.9 - 7.3	Strong interaction only with Cu	Used in culture media for bacteriophage adsorption
MOPSO	6.2 - 7.6	6.7 - 7.1	Strong interaction only with Fe	Used as a buffer component of charcoal yeast extract medium
ACES	6.1 - 7.5	6.6 - 7.0	Strong interaction with Cu and Mg	Used in cell culture media for hairy roots Used as a buffer component of charcoal yeast extract medium
TAPS	7.7 - 9.1	8.25 - 8.65	Strong interaction with Cu, Cr and Fe	Used in cell culture media for experiments with dinoflagellates
Bicine (Good's buffer)	7.6 - 9.0	8.1 - 8.5	Strong interaction with Cu, Fe, Co, Mg, Ca, Ni, Zn and Cd	Used for the culture of ammonia fungi
Tricine	7.4 - 8.8	8.0 - 8.3	Strong interaction with Mg, Ca, Co, Cu, Ni and Zn	Used in culture media of bacteria to prevent pre cipitation of iron salts Used in animal tissue culture

Sampling, handling and transportation of anaerobic samples

Cultivation of aerobes, facultative anaerobes, aerotolerant and microaerophiles are easier than strict/obligate or extremely oxygen sensitive anaerobes. They can die even with momentary exposure to oxygen and need special care during sampling and transportation. Based on our own experience at National Centre for Microbial Resource (NCMR), National Centre for Cell Science (NCCS), Pune India; we have found that most of the time researchers do not make clear cut demarcation between facultative, aerotolerant and strict anaerobic bacteria and archaea. If a culture is growing inside the GasPak system, they do not bother to test whether they are facultative or aerotolerant and just give the name anaerobes. It is generally accepted that strict anoxic microorganisms need Glovebox for cultivation and purification. It is also found that most of the time the distance of the sampling site and cultivation laboratory is quite far and it takes hours to transport the sample for cultivation purpose. Generally, the samples are transported on ice to maintain the temperature but at the same time extended exposure of O_2 during transportation can be lethal for strict anaerobic organisms and as a result during cultivation we lose desired type of organism. For this purpose, maintaining the anoxia and temperature, both during sampling and transportation is imperative to get right type of anaerobic organisms (Wilkins et al. 1975, Hallander et al. 1980, Citron 1984).

Gut is totally anoxic and harbours several thousand unexplored species of strict anaerobic Bacteria and Archaea with implication in human health and diseases (Kelley et al. 2004, Quigley 2013, Zhang et al. 2015, Prakash et al. 2014, Prakash et al. 2015). Currently gut research is like hot-cakes and several efforts have been made for appropriate sampling of faecal material and its transportation for cultivation purpose. Several papers have been published on faecal sample collection, storage and transportation. Collection of clinical specimen for cultivation of anoxic microorganisms using clinical swabs or hypodermic needles and syringes and transportation of the specimen in commercially available container have been discussed in detail. But work on anaerobic sample collection and transportation in environmental microbiology is scanty. For collection of anaerobic samples of sludge, sediment or soil; coring is a very good option. Information about core and coring method can be obtained from anywhere else (https://www.epa.gov/sites/production/files/2015-06/documents/Sediment-Sampling.pdf). Immediately after collection intact core can be sealed and transported to the laboratory and must be opened in anoxic environment of the Glovebox. Alternatively, after collection any type of sample, it can be placed inside the GasPak and is made anoxic using the reaction of sodium borohydride and bicarbonate in the presence of palladium catalyst and whole unit can be transported to the laboratory on ice and the sample inside must be opened in the anoxic Glovebox. In addition, some publications recommend collection of anaerobic samples in pre-reduced phosphate buffered saline inside serum bottles or Hungate tubes. Phosphate buffered saline supplemented with 0.2% tween-80 and resazurin and reduced with a mixture of cysteine and sodium sulphide can be used for transportation of samples. After collection, put the sample inside the reduced PBS and transport to the laboratory. Oxidation of the medium can be guessed by colour of the resazurin and can be again reduced by adding the reducing agent from outside. Additionally, sampling vessel should be filled with 2/3 volume of sample, headspace should be flushed with sterile anoxic gas by placing a filter in gas line of a portable gas cylinder of 1-2 L capacity. Commercially available Gut Alive containers are also used for transportation of faecal samples for cultivation purpose and showed better results than sample collected using traditional methods (Martínez, et al. 2019).

Common step for pre-reduced media preparation: Successful cultivation of strict anaerobes depends on formulation of pre-reduced media. For cultivation of different groups of anaerobes, different media components, pH buffers, micronutrients, reducing agents, types of anoxic gases and vitamins are required but all follow the same steps. Preparation of pre-reduced media is based on the methods of Hungate which is based on two gas cannulas: One for flushing the anaerobic gas and another for pouring the deoxygenated media in the vessel. Tools used for preparation of pre-reduced media are given in the figure 7. Following basic steps are used for pre-reduced media preparation.

1. Boiling and bubbling of the solvent (water) is the best way to achieve anaerobic conditions quickly. Boiling expels dissolved oxygen from water while bubbling with anoxic gas prevents the entry of external O_2 from atmosphere and also assists in expelling of O_2.
2. Boiling the water first and then cooling it under the stream of nitrogen or a mixture of N_2+CO_2 is a good practice to get anaerobic condition.
3. Continuous heating of the medium in a round bottom flask at sub-boiling temperature using a heating mantle and flushing the headspace with anoxic gas is also a good practice of anaerobic media preparation.
4. Weigh and dissolve the media components in water, except reducing agent.
5. It is always better to add the media components (nutrients) one by one to avoid the precipitation. When one component is completely dissolved, then the other one should be added.
6. Addition of reducing agent before inoculation is a good practice, but it can also be added during the media preparation.

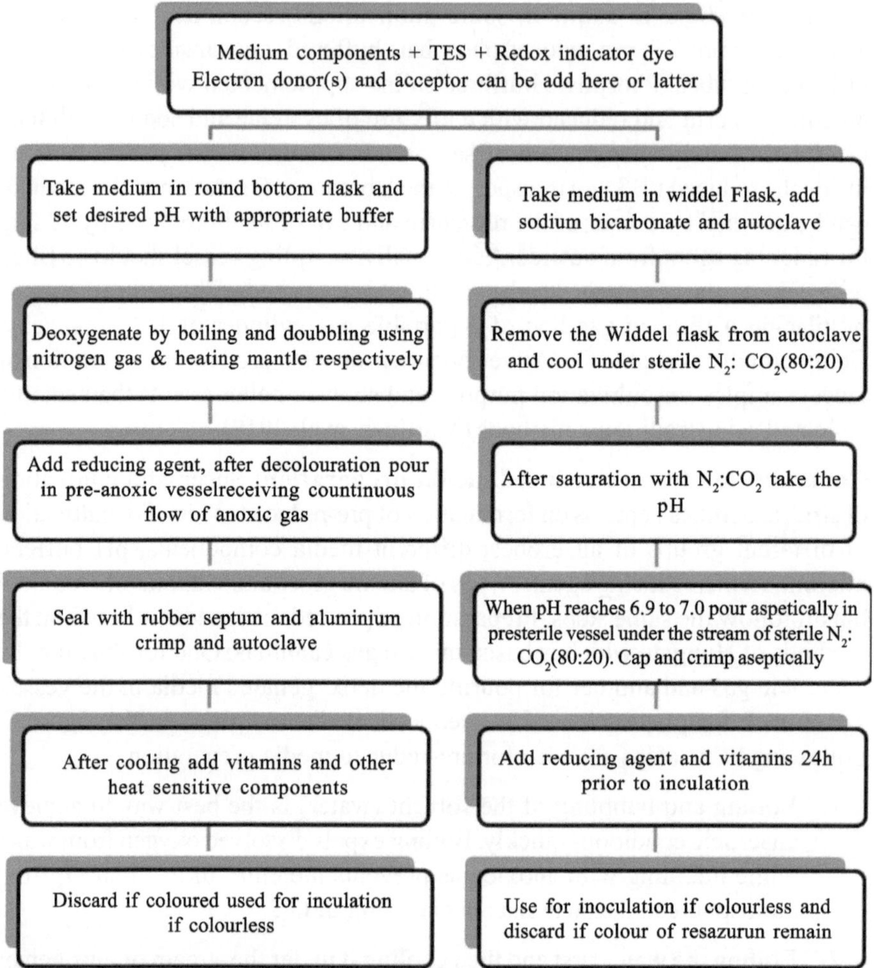

Fig. 7: Methods of pre-reduced media preparation using bicarbonate and other buffer systems for cultivation of strict anaerobes.
TES: trace element solution;
Anoxic Vessels: Hungate tube, Balch tube, Serum vials etc.

Reducing agent should be dissolved in deoxygenated water for better efficiency and to avoid exposure to oxygen.

Pre-reduced media preparation can also be done using Widdel flask. Preparation of medium with bicarbonate buffer system is briefly discussed below.

1. All media components, except the vitamins and heat sensitive components should be dissolved first.
2. The medium components pour in Widdel flask, loosely tighten the side port, to avoid explosion due to extra pressure, and is autoclaved.

3. After autoclaving, the pressure is released from side port and flask is transfered immediately to the nearest anaerobic gas station.
4. Then one of the ports is connected with sterile $N_{2:}$ CO_2 line and cooled the medium under the stream of sterile anoxic gas.
5. After cooling the medium; the heat sensitive components are added using hypodermic needle and syringe. It is better to add bicarbonate at this stage to avoid precipitation.
6. The medium is left as such for being saturated with CO_2 and checked the pH periodically. After getting the right pH of the medium adjust the inside pressure using the side port to pour the medium.
7. The medium is poured in pre-sterilized Hungate-tube or serum vials using aseptic anoxic techniques.

Methods of handling of active cultures of anaerobes

Unlike aerobes; anaerobes are sensitive to oxygen and need special care for their handling and transfer. If anaerobic chamber is available in the laboratory and cultures grow easily on plates then it becomes easy to handle them inside the anoxic environment of the chamber using disposable loops or toothpick but handling of liquid culture broth or anoxic samples for inoculation and enrichment purpose needs special care. Inoculation and handling of anaerobic culture can also be done successfully using aseptic anoxic techniques and anaerobic gas station or gassing manifold as discussed below:

1) First, the septum of Hungate-tube or serum bottle should be sterilized by placing a drop of ethanol and flaming it.
2) Hypodermic needles and syringes made anoxic by displacing oxic space by filling and pumping the filter sterile anoxic gases from gas station.
3) It should be repeated six times with moderate speed to get satisfactory anaerobicity.
4) Sterilized anoxic gas can be obtained by adding a membrane filter in gas tubing of anaerobic gas station or manifold.
5) If vessel is over pressurized, one should vent it using anoxic hypodermic needle and syringe.
6) After that, the vessel need to be turned upside down and sterile anoxic gas should be injected in the headspace of the vessel. Equal amount of culture is removed and inject or inoculate in new vessel.
7) Need to repeat this practice every time.

Cultivation and isolation of strict/ obligate anaerobes: Unlike aerobes, cultivation, isolation and purification of obligate anaerobes require utmost care. Only Elimination or exclusion of oxygen from growth medium is not sufficient because certain microorganisms require specific low redox potential for their growth (Miller and Wolin 1974, Prakash et al. 2010, Tian et al. 2010) . Even momentary exposure to oxygen can kill them or stop their growth. Several different methods including the conventional methods of removal of oxygen to more advanced instrumentation like anaerobic Glovebox, GasPak and Anoxomat are used to cultivate them. Initial cultures of anaerobes were mostly mixed but development of Hungate methods in 1950 improved anaerobic cultivation and purification (Sowers & Noll 1995, Speers et al. 2009, Stieglmeier et al. 2009). Cultivation of facultative anaerobes and aerotolerant anaerobes is reasonably easy, but fastidious and slow growing organism which need gaseous and volatile electron donor like hydrogen, methanol, ethanol, etc. need special care. Drying of the plate is another problem with slow growing organism as volatile electron donors escape from the medium in the form of gaseous fumes. Bottle plate, gelrite shake/agar shake and roll tube are the some of the methods of choice for isolation of fastidious and slow growing organisms like methanogens. The methods of preparation of roll tube have been discussed in detail by Hungate (1969). The growth of the colonies in roll tube is given in figure-8. It includes the following basic steps.

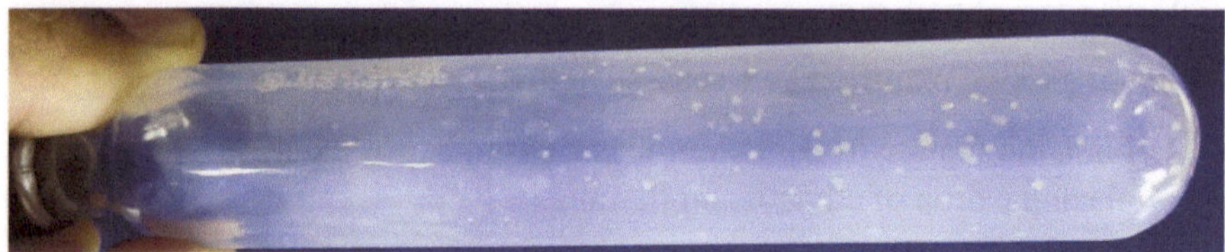

Fig. 8: Image of roll tubes prepared using Balch-tube. Bacterial colonies are visible like white patches on thin layer of agar

1. Pre-reduced medium is prepared in any one of the suitable anaerobic vessels like Hungate tube, Balch tube or serum bottle.

2. After autoclave maintain temperature at 45-50 °C to keep the agar in molten stage.

3. Serial dilution of sample is made using aseptic anoxic techniques.

4. From different dilutions inoculum is injected in molten agar and, mixed quickly. The tubes are rolled on bed of ice to form a thin layer of molten agar with bacterial cells.

5. Fill the space of the tube with appropriate gases like 80:20 H_2: CO_2 for methanogens.

6. Incubate at appropriate temperature for colony formation.
7. Isolated colonies can now be removed using bent anoxic pasture pipette and can be inoculate in fresh medium.
8. Check the purity of culture and use it for further characterization.
9. Take care of water of incineration. Remove it frequently using hypodermic needles and syringes.
10. Isolation using the Gelrite or agar shake is similar to Roll tube in principal but in this case the tubes are kept intact without rolling and making thin layer.

Precaution to be followed in anaerobic laboratory

1. Although, most of the gases used in anaerobic laboratory are not toxic, they are expensive. Therefore, try to switch off the gas cylinder immediately after work to avoid their misuse.
2. Take extra care while handling hydrogen. It is explosive in nature; concentrations above 5% can be explosive.
3. H_2S is toxic and corrosive. Laboratory working on dissimilatory sulphate reducers need to take care and avoid exposure.
4. Autoclaving of closed vessel like gas tight serum bottles and Hungate tubes generate overpressure inside the vessel and creates a possibility of explosion. They need to be sterilized in metal or plastic cage to avoid injury by flying glasses and always handle them after cooling.
5. Avoid exposure of O_2 to reducing agents as, it quickly reacts with reducing agents and oxidizes itself making them less efficient in the medium.
6. In bicarbonate buffer system, to get desired pH, sufficient saturation of headspace and medium with CO_2 is necessary. It may take little time and patience.
7. Always pressurize the serum vials behind shield to avoid the injury by explosion and flying glass.
8. Always transfer the gas cylinders on gas trolley and fix them tightly with chain and holder on gas station or in laboratory to prevent them from falling.
9. Hypodermic needles and syringes are the common tools for handling and transfer of live anaerobic culture. Take utmost care during handling. Poking the hand with contaminated needle will not only cause pain but will also create the possibility of health risk.

10. Puncturing of the neoprene gloves with needles and syringes is a very common mistake in an anaerobic laboratory.
11. All anaerobic cultures grow in gas tight, closed vessel systems. Active growth of culture, inside the vessel, creates a possibility of overpressure and explosion. Venting the culture vessel at daily basis should be done to avoid explosions.
12. Always overpressure the serum vials behind protective shield to avoid injury by flying glasses
13. Do not use over punctured septum, as they may become permeable to O_2 and make it unsuitable to maintain the anaerobicity.

For more detail information consult: Methanogens (Archaea : A Laboratory Manual) by K.R. Sowers and F.T. Robb (1995)

Future Prospects

It is clear from the above discussion that the strict anaerobic Bacteria and Archaea are the important players of various processes and are environmentally, clinically and industrially significant. Due to difficult nature of cultivation, purification and characterization of strict anaerobes, database of their cultivable representatives is not very rich and needs more work in this area. Culture independent molecular approach of diversity study from diverse anaerobic niches has revealed immense diversity of anaerobic microbes, which needs cultivation and characterization for their future exploitation on physiological, functional and commercial purposes. Despite immense importance, cultivation seems a traditional art to students and microbiologists, and no serious researchers show interest and inclination in cultivation. Furthermore, in comparison to molecular approach, the publication and data generation is difficult in cultivation and physiology, which also hampers the interest of microbiologists in this area. However, considering the values and application of pure culture and calculated potentials of uncultured organisms, several innovative approaches have been designed, and culturomics is gradually emerging again. Furthermore, cultivation is not as simple as researchers think, although for cultivation of not -yet- cultured organisms researchers need in-depth understanding of microbial nutrition, art of media formulation, growth kinetics and modern art of molecular biology like metagenomics and genomics. Many stories of successful cultivation of not-yet-cultured organisms have been published using genome centric approaches. It includes generation of data about microbial community structure from significant ecological niches, followed by prediction of their functional role in specific habitat. Subsequently, focus is turned to important but not yet cultured organisms and constructing whole genome sequence from metagenomic data. For unsuccessful cultivation, metabolic pathway and nutrient requirements must be studied and compared with the medium used. Supply the essential nutrient and try to culture it again.

The important aspect of anaerobic cultivation which needs to be addressed is the proper anaerobic set-up for cultivation of strict anaerobic archaea, bacteria and fungi (Moench and Zeikus 1983, Sowers and Noll 1995, Cheng et al. 2018). Adequate training of anaerobic cultivation, handling and transportation of anaerobic samples is another important aspect for successful cultivation of anaerobes. Certain strict anaerobic organisms start dying even due to momentary exposure to oxygen and need very smart handling during cultivation and transportation of the samples. Patent is the most important attribute of researchers involved in anaerobic microbiology because unlike aerobic organisms, anaerobes take longer time to grow and make visible colonies and it is more difficult with fastidious type of anaerobes. It creates frustration among the researchers and they terminate their work without culturing. Thus, in conclusion, considering the importance of anaerobic culture, their cultivation database needs to be expanded for future study and application purposes. It needs sampling of different anaerobic niches, and it should then focus on their cultivation. Niche wise database creation would be an important aspect. Pure culture database of methanogenic archaea, gut-archaea, Gut-bacteria, anaerobes from landfill leachate, subsurface groundwater, could work in favour of their biotechnological application in future.

Acknowledgement

This work was supported from the Grant no. BT/PR13969/BCE/8/1142/2015, Department of Biotechnology (DBT), Govt. of India. We are obliged to Dr. Tapan Chakarwarti from NCMR NCCS and Eden Jacques and Manali Vaijanapukar (Symbiosis Pune) for critical editing of the chapter.

References

Aoi Y, Kinoshita T, Hata T, Ohta H, Obokata H, Tsuneda S. 2009. Hollow-fiber membranse chamber as a device for in situ environmental cultivation. *Appl. Environ. Microbiol* 75: 3826–3833.

Aranki A, R. Freter. 1972. Use of anaerobic glove boxes for the cultivation of strictly anaerobic bacteria. *Am. J. Clin. Nutr* 25: 1329–1334.

Balch WE, Fox GE, Magrum LJ. Woese CR, Wolfe RS. 1979. Methanogens – re-evaluation of a unique biological group. *Microbiol Rev* 43: 260–296.

Banerjee S, Schlaeppi K, van der Heijden MG. 2019. Reply to 'Can we predict microbial keystones?' *Nat Rev Microbiol* 17(3): 194.

Berdy B, Spoering AL, Ling LL, Epstein SS. 2017. *In situ* cultivation of previously uncultivable microorganisms using the ichip. *Nat. protoco* 12(10): 2232.

Bollmann A, Lewis K, Epstein SS. 2007. Incubation of environmental samples in a diffusion chamber increases the diversity of recovered isolates. *Appl. Environ. Microbiol* 73: 6386–6390.

Borner RA, Aliaga MTA, Mattiasson B. 2013. Microcultivation of anaerobic bacteria single cells entrapped in alginate microbeads. *Biotechnol. Lett* 35: 397–405.

Brauer SL, Cadillo-Quiroz H, Ward RJ. Yavitt JB, Zinder SH. 2011. *Methanoregulaboonei* gen. nov., sp. nov., an acidiphilic methanogen isolated from an acidic peat bog. *Int. J. Syst. Evol. Microbiol* 61: 45–52.

Bräuer SL, Yashiro E, Ueno NG, Yavitt JB, Zinder SH. 2006. Characterization of acid-tolerant H_2/CO_2-utilizing methanogenic enrichment cultures from an acidic peat bog in New York State. *FEMS Microbiol Letts* 57: 206–216.

Brioukhanov AL, Netrusov AI. 2012. The positive effect of exogenous hemin on a resistance of strict anaerobic archaeon *Methanobrevibacterarboriphilus* to oxidative stresses. *Curr Microbiol* 65: 375–383.

Cadillo-Quiroz H, Yavitt JB, Zinder SH. 2009. *Methanosphaerulapalustris* gen. nov., sp. nov., a hydrogenotrophic methanogen isolated from a minerotrophic fen peatland. *Int. J. Syst. Evol. Microbiol* 59:928–935.

Canion A, Prakash O, Green SJ, Jahnke L, Kuypers MM, Kostka JE. 2013. Isolation and physiological characterization of psychrophilic denitrifying bacteria from permanently cold Arctic fjord sediments (S valbard, N orway). *Appl. Environ. Microbiol* 15(5): 1606-1618.

Carbonero F, Oakley BB, Purdy KJ. 2010. Improving the isolation of anaerobes on solid media: The example of the fastidious Methanosaeta. *J Microbiol Methods* 80: 203–220.

Cheng Y, Shi Q, Sun R, Liang D, Li Y, Li Y, Jin W, Zhu W. 2018. The biotechnological potential of anaerobic fungi on fiber degradation and methane production. *World J. Microbiol. Biotechnol* 34(10): 155.

Citron DM 1984. Specimen collection and transport, anaerobic culture techniques, and identification of anaerobes. *Rev Infect Dis 6*(Supplement_1), S51-S58.

Doddema HJ, Vogels GD. 1978. Improved Identification of methanogenic bacteria by fluorescence microscopy. *Appl. Environ. Microbiol* 36: 752–754.

Edwards AN, Suarez JM, McBride SM. 2013. Culturing and maintaining *Clostridium difficile* in an anaerobic environment. *J Vis Exp* 79: e50787.

Franzmann PD, Liu YT, Balkwill DL, Aldrich HC, Macario EC, Boone DR. 1997. *Methanogenium frigidum* sp. nov., a psychrophilic, H_2-using methanogen from Ace Lake, Antarctica. *Int. J. Syst. Evol. Microbiol* 47: 1068–1072.

Garza DR, Dutilh BE. 2015. From cultured to uncultured genome sequences: metagenomics and modeling microbial ecosystems. *Cell. Mol. Life Sci 72*(22): 4287-4308.

Goodman AL, Kallstrom G, Faith JJ, Reyes A, Moore A, Dantas G, Gordon JI. 2011. Extensive personal human gut microbiota culture collections characterized and manipulated in gnotobiotic mice. *Proc. Natl. Acad. Sci. U.S.A.* 108: 6252–6257.

Green SJ, Prakash O, Gihring TM, Akob DM, Jasrotia P, Jardine PM, Kostka JE. 2010. Denitrifying bacteria isolated from terrestrial subsurface sediments exposed to mixed-waste contamination. *Appl. Environ. Microbiol 76*(10): 3244-3254.

Haitjema CH, Solomon KV, Henske JK, Theodorou MK, O'Malley MA. 2014. Anaerobic gut fungi: advances in isolation, culture, and cellulolytic enzyme discovery for biofuel production. *Biotechnol and Bioeng* 111: 1471–1482.

Hallander HO, Flodström A, Åberg C 1980. Collection and transport of specimens for anaerobic culture. Infection, 8: S147-S150.

Hamilton-Brehm SD, Vishnivetskaya TA, Allman SL, Mielenz JR, Elkins JG. 2012. Anaerobic high-throughput cultivation method for isolation of thermophiles using biomass-derived substrates. *Methods Mol. Biol* 908: 153–168.

Handelsman J. 2004. Metagenomics: application of genomics to uncultured microorganisms. *Microbiol Rev* 68(4): 669-685.

Harmsen HJ, van Kuijk BL, Plugge CM. Akkermans AD, De Vos WM, Stams AJ. 1998. *Syntrophobacter fumaroxidans* sp. nov., a syntrophic propionate-degrading sulfate-reducing bacterium. *Int. J. Syst. Evol. Microbiol* 48: 1383–1387.

Harris JE. 1985. GELRITE as an agar substitute for the cultivation of mesophilic *Methanobacterium* and *Methanobrevibacter* species. *Appl. Environ. Microbiol* 50: 1107–1109.

Hatti-Kaul R, Mattiasson B. 2016. Anaerobes in Industrial- and Environmental Biotechnology. *Adv. Biochem. Eng. Biotechnol* 156: 1-33.

Hong Y, Gu JD. 2009. Bacterial anaerobic respiration and electron transfer relevant to the biotransformation of pollutants. *Int. Biodeterior. Biodegradation* 63(8): 973-980.

Hungate RE. 1950. The anaerobic mesophilic cellulolytic bacteria. *Bacteriol Rev* 14: 1–49.

Hungate RE. 1969. Chapter IV:A roll tube method for cultivation of strict anaerobes. *Methods Microbiol* 3B:117–132.

Imachi H, Sakai S, Sekiguchi Y, Hanada S, Ohashi A, Harada H, Kamagata Y. 2008. *Methanolineatarda* gen. nov., spnov., a methane-producing archaeon isolated from a methanogenic digester sludge. *Int. J. Syst. Evol. Microbiol* 58: 294–301.

Imlay JA. 2002. How oxygen damages microbes: oxygen tolerance and obligate anaerobiosis *Adv. Microb. Physiol* 46: 111-53.

Jab³ońska J, Tawfik DS. 2019. The number and type of O_2-utilizing enzymes indicates aerobic vs. anaerobic phenotype. *Free Radic. Biol. Med* 140: 84-92.

Kelly D, Campbell J, King TP, Grant G, Jansson EA, Coutts AG, Pettersson S, Conway S. 2004. Commensal anaerobic gut bacteria attenuate inflammation by regulating nuclear-cytoplasmic shuttling of PPAR-ã and RelA. *Nat Immunol* **5:** 104–112 doi:10.1038/ni101

Khelaifia S, Raoult D, Drancourt M. 2013. A versatile medium for cultivating methanogenic archaea. *PLoS One.*

Kurr M, Huber R, Konig H, Jannasch HW, Fricke H, Trincone A, Kristjansson JK, Stetter KO. 1991. *Methanopyruskandleri*, gen. and sp. nov., represents a novel group of hyperthermophilic methanogens, growing at 110°C. Arch *Microbiol* 156: 239–247.

La Scola B, Khelaifia S, Lagier JC, Raoult D. 2014. Aerobic culture of anaerobic bacteria using antioxidants: a preliminary report. *Clin. Microbiol. Infect* 33: 1781–1783.

Lagier JC, Edouard S, Pagnier I, Mediannikov O, Drancourt M, Raoult D. 2015. Current and past strategies for bacterial culture in clinical microbiology. *Clin. Microbiol* Rev 28: 208–236.

Lagier, JC, Dubourg G, Million M, Cadoret F, Bilen M, Fenollar F, Levasseur A, Rolain JM, Fournier PE, Raoult D. 2018.Culturing the human microbiota and culturomics. *Nat Rev Microbiol* 16: 540–550.

Martínez N, Hidalgo-Cantabrana C, Delgado S, Margolles A, Sanchez B. 2019. Filling the gap between collection, transport and storage of the human gut microbiota. *Sci rep* 9(1): 8327.

Miller TL. Wolin M 1974. A serum bottle modification of the Hungate technique for cultivating obligate anaerobes. Appl Microbiol 27: 985

Moench TT, Zeikus JG. 1983. An improved preparation method for a titanium (III) media reductant. *J. Microbiol Method* 1: 199–202.

Nakamura K, Tamaki H, Kang MS, Mochimaru H, Lee ST, Nakamura K, Kamagata Y. 2011. A six-well plate method: less laborious and effective method for cultivation of obligate anaerobic microorganisms. *Environ Microbio.* 26: 301–306.

Nichols D, Cahoon N, Trakhtenberg EM, Pham L, Mehta A, Belanger A, Epstein SS. 2010. Use of ichip for high-throughput in situ cultivation of "uncultivable" microbial species. *Appl.Environ Microbiol* 76(8): 2445-2450.

Nyonyo T, Shinkai T, Tajima A, Mitsumori M. 2013. Effect of media composition, including gelling agents, on isolation of previously uncultured rumen bacteria. *J Appl. Microbiol* 56: 63–70.

Overmann J, Abt B, Sikorski J. 2017. Present and future of culturing bacteria. *Ann rev microbiol* 71: 711-730.

Plugge C. M. 2005. Anoxic media design, preparation, and considerations. *Meth enzymol* 397: 3-16.

Podolsky IA, Seppälä S, Lankiewicz TS, Brown JL, Swift CL, O'Malley MA. 2019. Harnessing Nature's Anaerobes for Biotechnology and Bioprocessing. *Annual Review of Chemical and Biomolecular Engineering* 10:105-128.

Prakash OM, Gihring TM, Dalton DD, Chin KJ, Green SJ, Akob DM & Kostka JE. 2010. Geobacterdaltonii sp. nov., an Fe (III)-and uranium (VI)-reducing bacterium isolated from a shallow subsurface exposed to mixed heavy metal and hydrocarbon contamination. *Int. J. Syst. Evol. Microbiol 60*(3): 546-553.

Prakash O, Nimonkar Y, Vaishampayan A, Mishra M, Kumbhare S, Josef N, Shouche YS (2015). Pantoea intestinalis sp. nov., isolated from the human gut. *Int J Syst Evol Microbiol.* 65:3352-8. doi: 10.1099/ijsem.0.000419

Prakash O, Munot H, Nimonkar Y, Sharma M, Kumbhare S, Shouche YS (2014) Description of Pelistega indica sp. nov., isolated from human gut. *Int J Syst Evol Microbiol.* 64:1389-94. doi: 10.1099/ijs.0.059782-0.

Quigley EM 2013. Gut bacteria in health and disease. Gastroenterology & hepatology, 9:, 560.

Rafrafi Y, Trably E, Hamelin J, Latrille E, Meynial-Salles I, Benomar S, Giudici-Orticoni MT, Steyer JP. 2013. Sub-dominant bacteria as keystone species in microbial communities producing bio-hydrogen. *Int J Hydrogen Energy* 22: 4975-85.

Rappe MS, Connon SA, Vergin KL, Giovannoni SJ. 2002. Cultivation of the ubiquitous SAR11 marine bacterioplankton clade. *Nature* 418(6898): 630.

Lowe SE, Jain MK, Zeikus JG. 1993. Biology, ecology, and biotechnological applications of anaerobicbacteria adapted to environmental stresses in temperature, pH, salinity, or substrates. *Microbiol Rev* 57(2): 451–509.

Sakai S, Imachi H, Sekiguchi, Tseng IC, Ohashi A, Harada H, Kamagata Y. 2009. Cultivation of methanogens under low-hydrogen conditions by using the coculture method. *App. Environ. Microbiol* 75: 4892–4896.

Sowers KR, Noll KM. 1995. Techniques for anaerobic growth, methanogenic archaea. In: Robb FT, Plac AR, Sowers KR, Schreier HJ, DarSarma S, Fleischmann EM (eds) Archaea, a laboratory manual. CSHL 15–47.

Speers AM, Cologgi DL, Reguera G. 2009. Anaerobic cell culture. *Curr Protoc Microbiol* A4:A 4F

Stieglmeier M, Wirth R, Kminek G, Moissl-Eichinger C. 2009. Cultivation of anaerobic and facultatively anaerobic bacteria from spacecraft-associated clean rooms. *App. Environ. Microbiol* 75(11): 3484-3491.

Tamaki H, Hanada S, Sekiguchi Y, Tanaka Y, Kamagata Y. 2009. Effect of gelling agent on colony formation in solid cultivation of microbial community in lake sediment. *Environ Microbiol* 11: 1827–1834.

Thompson H, Rybalka A, Moazzez R, Dewhirst FE, Wade WG. 2015. In vitro culture of previously uncultured oral bacterial phylotypes. *App. Environ. Microbiol* 81: 8307–8314.

Tian JQ, Wang YF, Dong XZ. 2010. *Methanoculleushydrogenitrophicus* sp. nov., a methanogenic archaeon isolated from wetland soil. *Int. J. Syst. Evol. Microbiol* 60: 2165–2169.

Vlková E, Salmonová H, Bunešová V, Geigerová M, Rada V, Musilová Š. 2015. A new medium containing mupirocin, acetic acid, and norfloxacin for the selective cultivation of bifidobacteria. *Anaerobe* 34:27–33

Widdel F, Pfennig N.1977. A new anaerobic, sporing, acetate-oxidizing, sulfate-reducing bacterium, *Desulfotomaculum* (emend.) *acetoxidans*. *Arc. Microbiol* 112: 119–122.

Wilkins TD, Jimenez-Ulate FRANKLIN 1975. Anaerobic specimen transport device. J Clinic Microbiol 2: 441-447.

Wolfe RS. 2011. Techniques for cultivating methanogens. *Met. Enzymol* 494: 1–22.

Wolfe RS, Metcalf WW. 2010. A vacuum-vortex technique for preparation of anoxic solutions or liquid culture media in small volumes for cultivating methanogens or other strict anaerobes. *Anaerobe* 16: 216–219.

Wolfe R. S. 1999. Anaerobic life- a centennial view. *J Bacterial Res* 181(11): 3317-3320.

Ze X, Duncan SH, Louis P, Flint HJ. 2012. *Ruminococcus bromii* is a keystone species for the degradation of resistant starch in the human colon. ISME J 6: 1535-1543.

Ze X, Le Mougen F, Duncan SH, Louis P, Flint HJ. 2013. Some are more equal than others: the role of "keystone" species in the degradation of recalcitrant substrates. *Gut Microbes* 4(3): 236-40.

Zehnder AJ, Wuhrmann K. 1976. Titanium (III) citrate as a nontoxic oxidation-reduction buffering system for the culture of obligate anaerobes. *Science* 194: 1165–1166.

Zhao X, Drlica K. 2014. Reactive oxygen species and the bacterial response to lethal stress. *Curr. Opin. Microbiol* 21: 1-6.

Zhang YJ, Li S, Gan RY, Zhou T, Xu, DP, Li HB 2015. Impacts of gut bacteria on human health and diseases. *Int J Mol Sci*, *16*: 7493-7519.

2

Preservation of Anaerobic Microbes Past, Present and Future

Krishna K Yadav, Yogesh Nimonkar, Shubhanjali Singh
*Vaidehi Mirashi, Om Prakash**

National Centre For Microbial Resources, National Centre for Cell Science
Pune-411007, Maharashtra, India

Abstract

Microbes are key ingredients of various ecosystems and backbone of modern biotechnology and industry. In addition to their cultivation, purification and characterization, preservation of microorganisms is equally important. Different preservation methods and protocols have been developed for long term preservation of microbes. Microbial Resource Centers (MRCs) are providing active services in the field of life sciences. National Centre for Microbial Resource (NCMR) at National center for Cell Science (NCCS), Pune is one of the MRCs. Unlike aerobes little work has been done in the area of cultivation and preservation of anaerobic microbes. In the previous chapter we have discussed the technicalities and problems of anaerobic cultivation. The aim of current chapter is to provide the recent overview of preservation of anaerobic microorganisms like methanogens, sulfate and metal reducers and the obligate anaerobes of clinical and industrial importance with fermentative mode of life style.

Background

Microbes are involved in various ecosystem services and are imperative for their accurate functioning. They are the pillars of modern Red, Blue, Green and White biotechnology and providers of industrially important, enzymes, vitamins peptides, hormones and novel drugs. They produce breathable oxygen decompose dead organic materials, breakdown pollutants and are future hope of novel antibiotics and drugs. Microbes are key components of sustainable agriculture

*Corresponding Author: omprakash@nccs.res.in,prakas1974@gmail.com

and source of plant growth promoting bacteria, biocontrol agents and several groups of bio-pesticides. Concept of biofuels like biogas, bio-butanol, bio-methane bio-ethanol and bio-hydrogen is based on the activities of these microorganisms. They are also responsible for global climate change and emission of greenhouse gases like CO_2, CH_4 and N_2O etc. Despite many beneficial activities, several groups of microbes act as pathogens of plants, animals and humans and are responsible for major socioeconomic loss. Recent development in culture independent approach has revealed immense untapped microbial diversity of industrial, agricultural, environmental and clinical significance is not-yet-cultured. Although culture independent approach can give the information about their presence in any ecosystem but for the study of their functional role, hypothesis generation, proof of concept, industrial exploitation of their valuable traits, their cultivation, purification, characterization and preservation of characterized pure culture is very important.

Isolation, identification, cultivation and preservation of microbes have become backbone of several biotech industries. As these microbes were found to be involved in many biological cycles and playing key roles in maintenance of environment, humanity is relying more and more on these microbes to fulfill their increasing needs. To fulfill the needs of rising population and seeking the solution of problems like pollution, extensive research is going on microbes. Till date we are able to culture and isolate hardly 1-10% of microorganism, rest 90-99% remain uncultured. Scientist and many research scholars are focusing on development of culturing techniques for these not-yet-cultured at the place of 99% microorganisms as their unknown abilities could be answer to various problems humanity is facing. Along with method of cultivation of microbes, especially bacteria and fungi, there is a need to develop methods for their preservation. Microbial preservation can be defined as an adequate and effective maintenance of pure cultures of microbes in viable condition for longer period of time without any changes in their necessary morphological, physiological and genetic characteristics (Prakash et al. 2012).

Several methods of preservation for different types of microorganisms have been developed; which include preservation by continuous sub culturing, keeping cultures at 4° (refrigeration), use of mineral oil (sterile liquid Paraffin) to overlay culture agar slant, lyophilization or freeze drying and cryopreservation. Among these lyophilization and cryopreservation are most commonly used and reliable methods for preservation of microbes for longer period of time (Prakash et al. 2012). Though these methods are well established for preservation of aerobic as well as facultative anaerobic microbes, there is very limited data available on preservation of strict anaerobic microbes (Hubalek 2003). As it is a well-established fact that bacteria from strict anaerobic conditions like landfills geothermal vents, mines, polluted water processing plants and depths of rivers, sea sediments and human gut are playing crucial role in sustaining life on earth by carrying out different metabolic processes in absence of oxygen, which maintain continuous flow of carbon, nitrogen and other organic as well as inorganic elements in environment (Wolfe 1999, Burrell et al. 2004, Nunoura et al. 2013, Pokorny et al. 2005, Naradsu et al. 2019 and van Vliet at al. 2019). Also these strict anaerobic bacteria are common cause of blood infections (bacteremia) as well as many of them colonize our gut in early stages of life and remain there in commensal association (Brook 2010 and Bircher et al. 2018). Recent studies on human gut microbiome have revealed a connection between different gastrointestinal disorders, diseases and gut microbes. Also possible use of these gut microbes has been suggested by different studies as a therapeutic method for treatment of intestinal as well as other metabolic disorders in near future (Bircher et al. 2018). Having so many possible implications and usefulness, isolation and identification of strict anaerobes has become priority and research scholars are working hard on it as it is difficult to isolate and purify strict anaerobes. Not only their isolation but also their preservation is very difficult. Conventional method like freeze drying and cryopreservation were not that effective for preservation of anaerobes because around 50-70% loss in viability of anaerobic bacteria was observed when preserved using conventional methods (Bircher et al. 2018). Therefore there is a need to develop effective methods for preservation of these strict as well as facultative anaerobic microbes having implications and applications far greater than our knowledge. Considering the importance of cultivation in previous chapter we have discussed about technicalities and glitches of anaerobic cultivation and have also mentioned why cultivation is very essential even in the era of OMICS. Several good articles about aerobic preservation have been published but data and protocols about anaerobic preservation is lacking. The aim of current chapter is to provide a recent overview of the current development in the preservation of anaerobes and how it is technically different from normal aerobic preservation.

Nature of cryoprotectants and their mechanism of action

Cryopreservation and freeze drying are two of the most effective and commonly used methods for long-term preservation of microbes. These methods are commonly used in microbial research centers (MRCs) and microbiological laboratories for preservation of microbes but the efficiency of these methods is not widely tested for preservation of intact microbiome or consortium of uncultured microbes (Malik 1992, Hauffe and Barelli 2019). To improve the efficiency of these two methods for preservation of anaerobes, different combinations of cryoprotectants such as glycerol, sucrose and inulin are applied which minimize the injuries to cells from drying and freezing. This has improved viability of preserved microbial cells up to 60-80% (Bircher et al. 2018). Glycerol is one of the most commonly used cryoprotectant in cryopreservation method which penetrates microbial cells and enters intracellular space where it prevents or minimizes ice crystal formation and shrinkage of cells due to osmotic changes caused by slow freezing and lower temperature (Meryman 2007 and Lovelock 1953). Due to highly viscous nature of glycerol it is less suitable cryoprotectant for lyophilization method (Abidias M. et al. 2001). In lyophilization different types of sugars (disaccharides) such as sucrose, maltose and trehalose are used as cryoprotectants. Sugars are effective penetrating cryoprotectants which prevent the formation of ice crystals and denaturation of intracellular proteins induced by osmotic dehydration of cells followed by freezing (Flower and Toner 2005). Other than sugars different types of Inulin, such as fructans (polymers of fructose molecules) are use as cryoprotectants in lyophilization. In contrast to sugars, fructans are non-penetrating cryoprotectants. They act as bulking agent and form protective matrix around the cells which prevents changes in outer membrane by directly interacting with membrane lipids and stabilizing them while drying and freezing process (Khan et al. 2014 and Hubalek 2003). We can use different combinations of these cryoprotectants for more effective preservation of aerobic as well as anaerobic microbes and to achieve up to 80-90% viability of preserved microbial cells (Hubalek 2003).

Methods for preservation of different type of anaerobic bacteria

Preservation method for anaerobic phototrophic bacteria

Malik (1984) developed new method for preservation of phototrophic bacteria under complete anaerobic condition by modifying conventional cryopreservation method. Instead of using glycerol (10%) as a cryoprotectant he preserved phototrophic bacteria in DMSO solution (5% v/v in H_2O). To maintain complete anoxic condition small screw capped glass ampoules of 2ml capacity (10x30 mm) with red oxygen impermeable butyl rubber septa were used. Bacteria were grown in these ampoules phototrophically. When culture was in logarithmic phase of growth, it was harvested by centrifuging ampoules at 4000g and supernatant was removed using sterile syringe. Cell pellets were used for preservation. Sterile ice cold DMSO solution was added in ampoules with the help of sterile syringes while ampoules were flushed continuously with sterile N_2 gas to maintain complete anaerobic condition. After 15 min of equilibration period in ice bath, ampoules were kept in aluminum canisters and frozen by immersing it in liquid nitrogen. After 2 years these ampoules were taken out and bacteria were revived in respective medium and percentage viability was calculated. Percentage viability or survival rate of different types of bacteria was within the range of 60-100%. Recovery of *Rhodospirillaceae* family of bacteria was ranged from 70-100% whereas strict anaerobic bacterial strains like *Chlorobiaceae* and *Chromatiaceae* showed maximum loss in viability by 1-2 $\log_{10}$ count.

Preservation method for methanogenic and Fe (II)-oxidizing phototrophic bacteria

In 1988 Huber and Lafferty showed that different methanogenic bacterial strains such as *Methanosarcina barkeri, Methanospirillum hungatei, Methanococcus vannielii* and *Methanobacterium formicicum*can can be preserved by using conventional cryopreservation method in liquid nitrogen and at -20 °C as well as -80 °C in deep- freezer. These methanogens remained in viable condition at these two temperatures. Bhattad et al. 2017 compared two methods of drying (spray or heat and freeze), to find out the efficiency of these methods to preserve methanogenic property of methanogenic culture obtained from anaerobic digester. Anaerobic methanogenic culture was concentrated from 300 ml to 20 ml by centrifugation and different aliquots were prepared. One such aliquot of methanogens was dried up by keeping it in a 25 ml porcelain crucible, in an oven at 104 °C for 12 hrs. Another aliquot was initially shell frozen in freeze drying flask by swirling it into liquid nitrogen for minutes. This shell frozen culture was then freeze dried under vacuum (13.3 Pa) by using bench top freeze dryer with condenser at -45 °C. When methanogenic activity of rehydrated culture was measured, it was observed that heat drying has more adverse effect on viability and activity of methanogens than freeze drying. Therefore, it was concluded by Bhattad that freeze drying method can preserve or maintain methanogens and their activity for longer period of time more efficiently than spray or heat drying method. In July 2010, Hegler and Kepler developed method for storage and maintenance of anoxygenic phototrophic Fe (II)-oxidizing bacteria. Due to highly sensitive photosynthetic membrane it was already very difficult to preserve photosynthetic bacteria. In case of phototrophic Fe (II)-oxidizing bacteria preservation becomes more challenging because of Fenton reaction which causes formation of radicals when bacteria are exposed to light and water. These radicals are responsible for genetic mutations in phototrophic Fe (II)-oxidizing bacteria. Hegler and Kepler cultivated phototrophic Fe (II)-oxidizing bacteria from different taxa including *Rhodobacter* sp., *Thiodictyon* sp. and *Chlorobium ferroxidans* in mineral medium containing 10 mM Fe (II) as an electron donor. After complete oxidization of Fe (II) in medium, cultures were transferred in small screw capped bottles with oxygen impermeable rubber septum. These bottles were already containing filter sterilized glycerol solution (10% and 20%) and anoxygenic condition was created by flushing bottles and syringes with mixture of N_2:CO_2 gas (90:10). This mixture of bacteria and glycerol was vortexed for 20 seconds and immediately kept in -

72 °C refrigerator. After 17 months, these cultures were revived on appropriate medium and it was found that bacterial strains of phototrophic Fe (II)-oxidizing bacteria i.e. *Rhodobacter* sp. and *Chlorobium ferroxidans* can be preserved and revived by using 20% glycerol whereas strain of *Thiodictyon* sp. requires 10% glycerol. It was also found that by using this method of preservation anoxygenic phototrophic Fe (II)-oxidizing bacteria can be preserved for at least 17 months with only 5-10% percent loss in viability of bacteria.

Preservation method for enteric bacteria and their relationship with type of preservative and storage condition used

Strasser et al. (2009) proved that method of preservation, type of cryoprotectants and storage conditions has great influence on survival and stability of lactic acid bacteria. Two strains of lactic acid bacteria *Enterococcus faecium* and *Lactobacillus plantarum* were grown anaerobically in batch fermenter. After reaching log phase of growth, both strains were harvested and preserved by using three different methods of preservation i.e. cryopreservation, lyophilization and fluidized bed bottom spray drying method. For preservation by cryopreservation and lyophilization methods three sugars glucose, sucrose and trehalose were used as cryoprotectant and their concentration was kept the same i.e. 32%. After 6 months cultures were revived and it was observed that the viability of both cultures was significantly lower in absence of cryoprotectants (39 and 11% respectively). From this study Strasser concluded that both bacterial strains react differently to different preservation methods. Lyophilization and cryopreservation methods of preservation are more significant and reliable than fluidized bed drying method of preservation for lactic acid bacteria. Use of disaccharide sugars trehalose and sucrose as cryoprotectant can significantly increase the viability of preserved lactic acid bacteria. Similar study was carried out in 2009 by Siaterlis and his colleagues; where they studied effects of culture medium and types of cryoprotectants on growth and survival of lactobacilli strains *Lactobacillus plantarum* and *Lactobacillus rhamnosus,* preserved by lyophilization method. Complex medium was developed by his research group for cultivation of lactobacilli strains with glucose, yeast extract and vegetable peptone as primary constituents. Ability of three disaccharides, sucrose, trehalose and sorbitol (5-10%) to protect cells during lyophilization was examined. His group observed that yeast extract and glucose are important growth factor and carbon limiting conditions hampers growth as well as viability of preserved cells of lactobacilli. Also trehalose was found to be offering much better protection to cells than sorbitol and glucose during lyophilization. From their observation, they concluded that there is a definite relation between medium used for growth of lactose fermenting bacteria and their survival during lyophilization method.

Preservation method for ammonia and nitrite oxidizing bacteria

In 2012, Kim Heylen and his colleagues developed preservation method for anaerobic ammonia oxidizing bacteria. DMSO was used as cryoprotectant instead of glycerol. Biomass of anammox bacteria was harvested from enrichment culture and was directly frozen in mineral medium and combination of two cryoprotectants and diluted growth medium (5% DMSO, 1% trehalose and 0.3% Trypticase soy broth). After 29 weeks, culture of '*Candidatus* Kuenenia' (mix culture of anammox bacteria) was revived and around 40% recovery of bacteria was observed without the use of cryoprotectants. Whereas, combination of cryoprotectants and only DMSO as cryoprotectant showed improved recovery 51% and 30% respectively. Recovery of anammox bacteria was increased by 11% when cryoprotectants were used. Similar experiment was carried out in 2013 by Hoefman and his research group proving that anammox bacteria can be preserved for longer period by using 5% DMSO and 1% trehalose as cryoprotectants in media containing tenfold diluted Trypticase broth. Another study was carried out by Bram Vekeman et al. (2013) to find out best applicable method for preservation of nitrite oxidizing bacteria. Their study demonstrated that different genera of nitrite oxidizing bacteria (*Nitrobacter, Nitrospina, Nitrococcus, Nitrotoga* and *Nitrospira*) can be preserved for longer period of time by modified cryopreservation method, developed by his group. They verified ability of different cryoprotectants and their combinations with carbon rich medium, such as 1%, 5% and 10% DMSO and, tenfold diluted Trypticase soy broth (TSB), tenfold TSB supplemented with 1% sucrose and 1% trehalose for preservation of nitrite oxidizing bacteria. Though it was observed from their study that different nitrite oxidizing bacterial strains require different optimal preservation conditions and marine isolates were found to be more sensitive to cryopreservation. They concluded that all strains of nitrite oxidizing bacteria can be preserved for longer period of time by using 10% DMSO with or without TSB as a preservation medium.

Preservation method for oil degrading bacterial consortium

In 2014 Humiko Kurachi and his colleagues developed method for preservation of turbine oil degrading bacteria (anaerobically degrades petroleum products). Before this, no method was available for preservation of turbine oil degrading bacterial consortium or isolates. Turbine oil degrading Tank-2 consortia was formulated in 2009 and preserved at -80 °C as glycerol stock. After 3 years bacterial consortium was revived and its ability to degrade different petroleum product was examined and degradation potential was compared with original consortium. No significant change in oil degrading potential of consortium was observed. Therefore, it was concluded by Kurachi and his group that turbine oil as well as other petroleum products degrading bacteria can be preserved at -

80° in glycerol stock without altering their capacity to degrade turbine oil. Also, in 2014 research group of Muhammad Ali developed simple and rapid method for preservation of active biomass of anammox bacterial consortium '*Candidatus Brocadia sinica'*. Bacterial consortium was harvested from continuous up-flow bioreactor and was homogenized. This consortium was preserved as granular as well as immobilized biomass. Before preservation of both granular and immobilized consortium biomass were washed with specially designed inorganic medium. Immobilized consortium biomass was preserved anaerobically in screw capped 10 ml glass serum vials containing 7 ml specially designed inorganic media with 5 mM NH_4^+, 5mM NO_2^- , 5.7 mM NO_3^- and 3mM sodium Molybdate, at 4 °C and ambient temperature for bacterial growth. Granular washed consortium biomass was stored as it is, at -80 °C without supplementing it with inorganic media. After 45, 90 and 150 days of preservation, preserved consortium of anammox bacteria was revived and its activity was measured. 96, 92 and 65 % ammonia oxidizing activity was observed after 45, 90 and 150 days of preservation. It was concluded from this study that active anammox bacterial consortium biomass can be effectively preserved by using nutrient medium containing 3 mM sodium molybdate supplied with NH_4^+ and NO_2^- periodically (45 days) at room temperature.

Preservation method for activated anammox sludge

In 2017 Viancelli et al. examined different cryoprotective agents and storage conditions for development of effective preservation method for preservation of activated anammox sludge. Different cryoprotective as wells as preservative agents (KNO_3, glycerol and Cow's skimmed milk) were used for preservation of '*Candidatus* Anammoxoglobus propionicus' and '*Candidatus* Jettenia asiatica' at different temperature (-20, -80, -200 and 4 °C). It was concluded from their study that anammox bacterial biomass activity can be revived by preserving them with glycerol and skimmed milk at -80 °C while there was complete loss in their ammonia oxidizing activity when preserved using other methods. Another technique called 'lyopreservation' was developed by Schulte et al. (2017) for preservation of industrially important carbon monoxide degrading bacterial strain *Clostridium ljungdahlii*. In his method *Clostridium ljungdahlii* culture was coated on 3mm chromatography paper along with culture medium and was dehydrated using convective drying method. As *C. ljungdahlii* is known to be desiccation sensitive bacteria, optimum drying conditions for stable coating of bacteria was investigated prior to actual experiment. These freshly prepared biocomposite strips were then placed in Balch tube and sealed with rubber butyl septum crimped with aluminum seal. These tubes were then stored at different temperature (4, 21, 26 and 37 °C). These desiccated bacterial biocomposite strips were then revived after 75 hrs of preservation by rehydrating them with fresh medium and were flushed with mixed gas (45% CO_2 + 45% H_2 + 10% N_2). Survival of bacteria was measured by its ability to degrade carbon monoxide. It was concluded carbon monoxide degrading bacteria *C. ljungdahlii* can be preserved in active state as desiccated biocomposite strips by exposing it to convective drying condition for 25 minutes and storing it at 4 °C without hampering its ability to degrade carbon monoxide.

Preservation method for gut microbes and gut microbiota

Extensive research on human-gut- microbiota is revealing importance of gut bacteria and their relationship with different diseases and metabolic disorders. There is an increasing interest in understanding gut bacteria due to their possible application as modern day 'probiotics' for treatment of different metabolic disorders. As it is a well-known fact that human gut has strict anoxic conditions and most of the gut microbes are obligate anaerobes, their preservation and maintenance in viable condition is very difficult. Therefore, to find an effective method for preservation of strict anaerobic gut microbes, Bircher et al. (2017) for the first time quantitatively compared cryopreservation and lyophilization methods of preservation. Both the methods were modified to improve viability of obligate anaerobic bacteria *Bacteroides thetaiotaomicron, Faecalibacterium prausnitzii, Roseburia intestinalis, Anaerostipes caccae, Eubacterium hallii* and *Blautiaobeum* isolated from human gut. 5% w/v sucrose and inulin were used as cryoprotectants in lyophilization method and these combined with 15% v/v of glycerol in cryopreservation method. Increase in viability of some microbes including *Anaerostipes caccae, Bacteroides thetaiotaomicron and Faecalibacterium prausnitzii* from 50% to 80% was observed, after addition of sucrose and inulin as cryoprotectants during lyophilization method. Also, this combination of cryoprotectants enhanced the percentage viability of gut microbes such as *Roseburia intestinalis, Eubacterium hallii* and *Blautiaobeum* from 0.03-2% to 11-37%, which are difficult to preserve and maintain. Bircher concluded from his study that efficiency of cryoprotectants is specific to method of preservation used and bacterial species. Although interest in fecal microbiota transplant is increasing as one of the promising methods for treatment of various intestinal diseases and disorders, safety issues associated with this method, pose as a major obstacle in its widespread use. Therefore, with the help of *in vitro* fermentation technology, artificial intestinal microbiota has been developed as a potential alternative to fecal microbiota, but a right method for preservation of artificial consortium of gut microbiota is still not developed. To solve this problem Bircher et al. (2018) developed a method for preservation of consortium of gut microbiota by modifying traditional cryopreservation method. Three different artificial microbiota developed by *in vitro* fermentation technology were preserved by using glycerol (15% v/v), inulin (5% w/v) and their combinations. Their ability to preserve different short chain fatty acid producing bacteria (specifically butyrate producing key gut bacteria) present in artificial microbiota at -80°C for three months was evaluated. It was observed after 3 months that butyrate, propionate and acetate producing ability generated consortium of artificial gut microbiota can be maintained using glycerol alone, however, as a result of using glycerol and inulin in combination the production rate increased. Also, it was observed that species belonging to *Roseburia* genus and *Eubacterium rectale* group can be preserve effectively using glycerol (15% v/v) as cryoprotectant, whereas better recovery of *Faecalibacterium prausnitzii* can be obtain by using inulin (5% w/v) as cryoprotectant. Therefore, Bircher et al. concluded that butyrate producing gut bacteria can be effectively preserved using glycerol and inulin as cryoprotectants for cryopreservation.

A Brief Overview of Anaerobic Preservation

Bryukhanov and Netrusov (2004) preserved Clostridia, acetogenic and sulfate-reducing bacteria and methanogenic archaea in 25% glycerol at -70 °C for three years. Results obtained showed that Clostridia and the acetogenic bacterium *S. sphaeroides*, *Acetobacterium* and sulfate reducing *D. kuznetsovii, M. arboriphilus* needed two to eight times more cells into a fresh media to get the initial growth rate. Other Obligate anaerobes *D. nigrificans* sub sp. *Salinus* and *Thermohydrogenium has* lost its viability in this preservation method.

Castro et al. (2001) tested methanogenic activity of the microbes from the anaerobic sludge in different preservation conditions such as storage at room temperature (20–25 °C), refrigeration (4 °C), freezing (–20 °C) with 10% (v/v) of glycerol or trehalose as cryoprotectants (Colleran *et al.* 1992) and freeze-drying using glycerol and trehalose. Results showed that storage at room temperature and refrigeration gives better viability and methanogenic activity than freezing and freeze-drying.

Anaerobic phototrophic and other sensitive bacteria like anaerobic purple non sulfur bacteria(*Chromatiaceae, Ectothiorhodospiraceae*) *and green sulfur bacteria, Campylobacter, Spirillum, Hydrogenophaga, Vibrios and Zymophilus* were successfully lyophilized by adding activated charcoal (5% w/v), skim milk (20% w/v) and raffinose (5% w/v) or meso-inositol (5% w/v) in medium.

Heylen et al. (2012) preserved Anaerobic Ammonium-Oxidizing Bacteria using different cryoprotectants. Cell mass was harvested by centrifugation and concentration cell mass was suspended in 3 ml of fresh medium. Samples were frozen in mineral medium without any cryoprotectant, with 5% (v/v) DMSO, or with a combination of 5% DMSO (v/v), 1% trehalose (w/v), and 0.3% TSB (w/v) as CPA or in mineral medium with 1% trehalose (w/v) for 2h under an anaerobic headspace at 30 °C and was then frozen with 5% DMSO (v/v). In all processes no special care was taken to avoid the contact with oxygen. After 29 weeks of preservation at -80 °C *Candidatus Kuenenia was* recovered well from preservation with and without any CPA in contrast to single-cell cultures.

Methanosarcina barkeri, Methanospirillum hungatei, Methanococcus vanelii and Methanobacterium formicicum were found viable after preservation at -20 °C or in liquid nitrogen with or without 10% DMSO (Huber and Lafferty year). Freeze drying with protective agents trehalose and glycerol resulted in preserving higher methanogenic activity as compared to those cultures which were freeze dried without any protective agents. Trehalose (10% v/v) proved more effective in preservation of methanogenic activity than glycerol (Castro 2002).

Methanol at 10% concentration proved an effective cryoprotectant for rapid cooling in liquid nitrogen of anaerobic bacteria *Chloroflexus* (Malik 1998). Long term preservation of anaerobic rumen fungus using ethylene glycol at 4 to 10% concentration is also an effective cryoprotectant as compared to dimethyl sulfoxide. Dimethyl sulfoxide, methanol, ethylene glycol, propylene glycol, and serum or serum albumin, are most successful cryoprotectants while glycerol, polyethylene glycol, PVP, and sucrose are less successful. Also, other sugars such as dextran, hydroxyethyl starch, sorbitol, and milk are the least effective. However, diols are toxic for many microbes. Dimethyl sulfoxide might be regarded as the most universally useful CPA (Zdenek Hubalek 2003).

Total 41 cultures, including anaerobic phototrophic bacteria and cultures sensitive to freeze drying were successfully preserved using the liquid drying method without use of anaerobic chamber. In this method vials used in gas chromatography (GC-vials) were used for liquid drying purpose. To form the dried cake, 0.1 ml 20% (w/v) skim milk containing 10% (w/v) neutral active charcoal, one of the protective agents (5% w/v), such as glutamate, mesoinositol, raffinose, trehalose or 10% (w/v) honey was used. Ampoules were sterilized at 115 °C for 13 min followed by freezing at -40 °C, then freeze dried to form cake in each vial. To prepare a cell suspension, thick cell mass was harvested by centrifugation and cell pellet was suspended in a solution of protective agent containing 5% (w/v) neutral activated charcoal in distilled water. In case of Halophilic strains or cells which do not form pellet, to the thick cell suspension equal amount of double concentrated protective solution was added on dried cake of each vial. 2-3 drops of equilibrated cell suspension were injected and used for vacuum drying. The results indicated that all the cultures showed good viability after L-drying and during long-term storage at -30 °C. Loss of viability was found in cultures stored at >9 °C for 2 to 3 years. Cultures sealed under vacuum and kept at > 0 °C, gave higher recoveries for long term storage of >5 years as compared to unsealed cultures and those which were not stored under vacuum. (Malik 1991)

Anaerobic Cultures Preservation Protocols of different Culture Collections

1. Leibniz Institute DSMZ German Collection of Microorganisms and Cell Cultures GmbH

DSMZ- Germany supplies many of the strict anaerobes in the active form. Cultures which are non-stringent anaerobes like *Clostridium* spp. and sulfate reducers are available in lyophilized form (not sensitive to short exposure of oxygen). The freeze-dried pellet of most anaerobic strains available from the DSMZ is protected against short exposure to oxygen by the use of amorphous ferrous sulfide (FeS) which confers a black color to the pellet. However, certain non-stringent or spore forming anaerobes are suspended in skimmed milk without addition of FeS prior to lyophilization, so that the ampoules display a white pellet.

In most of the cases freeze dried cultures of anaerobic strains exhibit a prolonged lag period upon rehydration and should be given at least twice the normal incubation time before reporting is more appropriate than regarding them as non-viable.

Common Access to Biological Resources and Information (CABRI)

a. Anaerobic microbes are grown in screw capped bottles like serum bottles Hungate or Balch tubes.

b. These grown cultures are transferred to the centrifuge tubes under anaerobic conditions and are harvested anaerobically to get the cell pellet. Alternately to avoid the transfer of the culture to the centrifuge tubes, anaerobic culture can also be grown in thick walled Hungate or Balch tube and can be harvest as such using the same tube.

c. Prepare anoxic long cannula (needle) by flushing it with anoxic gases used for the preparation of the medium or N_2:CO_2 instead of CO_2. Withdraw or remove maximum amount of supernatant using anoxically prepared 20 to 50 ml syringe with long needle

d. Prepare the cell suspension by adding 0.5 to 1.5 ml suspending medium with cryoprotectants in the cell pellet.

e. Add 25 microliter of the dense cell suspension to the anaerobically kept sterile capillaries and are sealed on a marked position by a fine hot gas flame.

f. Containers of the accessioned capillaries are kept in liquid nitrogen.

g. For revival the culture: thaw capillaries in the warm water (37 °C). Draw 1 ml of suspension, using sterile syringe or pasture pipette and inoculate in the 5 ml of pre-sterile medium.

h. For preservation of the anaerobic cultures, medium with 10% (v/v) glycerol or 5% (v/v) dimethyl sulphoxide (DMSO) is used as a suspending medium.

i. Suspending medium is prepared before use by adding 0.55 ml of sterile glycerol or 0.27 ml of sterile DMSO to sterile 5 ml aerobic or anaerobic medium in a culture tube. Cryoprotectant should be sterilized separately by autoclaving (DMSO: 10 min, 115C) aerobically or in anaerobic test tubes under nitrogen atmosphere.

American Type Culture Collection (ATCC)

Protocol used for preservation of anaerobic culture according to American Type Culture Collection (ATCC) is discussed in brief. After cultivation, concentrate the cell suspension and add it to pre reduced suspending fluid containing 10% glycerol. Transfer the cell suspension to the cryovials and place the vials in the liquid nitrogen or store at -80°C. Strict anoxic practice should be followed at all the steps including cell harvesting, preparation of stock of cryoprotectants and preparation of suspension for preservation.

Method of Preparation of Anoxic Stocks of Cryoprotectants

Dissolve 20% (v/v) glycerol and 20% (v/v) DMSO either in dH20 or PBS and aliquot in pre-cleaned serum vials (2/3 of Vessel Volume) and make the vials anoxic by flushing with nitrogen for 10-15 minute. Cap the bottles with butyl rubber septum and aluminum crimp and sterilize by autoclaving (15-min for Glycerol and 10min for DMSO). For better activity DMSO can be filter sterilized using Teflon syringe filter. During the time of preservation anoxically and aseptically remove the sterilized stock of cryoprotectants using long needle syringes and suspend the bacterial cells adjusting the final concentration of glycerol to 10% and DMSO to 5% and store the serum vials or cryovials in vapor phase of liquid nitrogen or deep freezer. Description is according to Khursheed A. Malik, DSMZ-Deutsche Sammlung von Mikroorganismen und Zellkulturen GmbH published by: UNESC/WFCC (World Federation for Culture Collections)-Education Committee 1989.

Revival of Preserved Anoxic Culture

a. Grow the culture in liquid broth or on the plate.

b. Harvest the cell biomass from late log phase using strict anoxic and aseptic techniques.

c. Mix harvested cell suspension with anoxically made cryoprotectant stock (50:50) to get the final cell concentration to 10^8 to 10^{10} and final concentration of glycerol to 10% and DMSO to 5%.

d. Serum vials with aluminum crimp and septum are the best option.

e. Allow the mixture to equilibrate with cryoprotectants (15 min for DMSO and 30 min for glycerol) on cold ice bath.

f. Transferred the vials to–30 °C freezer for 1h and then transfer to -80 °C (deep freezer) or vapor phase of Liquid nitrogen.

Preservation of Anoxic Culture

a. Remove the serum vials /cryovials from deep freezer or from liquid nitrogen and thaw at 37 °C in water bath.

b. Withdraw 50-100 ul of thawed cell suspension using sterile anoxic needles and syringes and inoculate into fresh anoxic medium in serum vials or hungate tubes.

c. Incubate at optimum temperature for growth.

d. In case of DMSO dilute the inoculum from 100-200 times to reduce the inhibitory effect of the compound.

e. Replace the serum vials at original position for future use.

Lyophilization or Freeze Drying

Lyophilization or freeze drying is another method for preservation of culture. It is better for aero-tolerant or spore forming cultures but proves tricky for obligate anaerobes. Concept of lyophilization of anaerobes is similar to aerobes but there is a need to take care of oxygen exposure during the preparation of sample and revival. Following steps are used for lyophilization of anaerobes.

a. Prepare cryoprotective and oxygen protective medium in oxygen free water or PBS.

b. Incubate the entire component inside anoxic chamber for at least 24h to make them free from the traces of oxygen.

c. Add 1.0 gl^{-1}Cysteine hydrochloride and 0.3gl^{-1} riboflavin in the solution to reduce the protective medium.

d. Cryoprotectants like Inulin or skimmed milk are considered better for lyophilization, but not glycerol due to its viscosity.

e. Prepare the anoxic solution of lyoprotectant as discussed above in anoxic PBS or dH2O supplemented with cysteine-HCl and riboflavin.

f. Fill the ampoules inside the chamber and freeze them inside and out on dry ice is more simple statement.

g. Put on the port of manifold of lyophilizer and start drying.

h. Seal it under vacuum and store at 4 C.

i. For revival rehydrate the freeze-dried ampoule in anoxic chamber for 1h and inoculate in fresh medium in Hungate tubes.

j. Pre and post preservation viability can be accessed through MPN using multiwall plate inside the chamber as discussed in Bricher et al. 2018

Post-preservation Viability

a. Add 100 μl of thawed cell suspension in 900 μl of normal saline or in fresh anoxic medium.

b. Prepare serial dilution inside the anoxic chamber or by using Hungate tube and anoxic gas station.

c. Plate the inoculum or do the drop plate method of enumeration using anoxic globe box /or anoxic jar.

d. Roll tube method can also be used for enumeration purpose.

e. Organisms showing difficult growth in or agar medium can be enumerated using MPN series in Hungate tube

f. Finally, calculate the viability by comparing the enumeration result from before and after preservation.

Future Prospect

Anaerobes are special group of organisms which reside in oxygen depleted environment and play crucial role in various ecosystem services. Annamox, Methanogens, sulfate reducers, metal reducers, clostridia, and anoxic gut-bacteria are some of the prominent group of anaerobes. Unlike aerobes, cultivation and preservation of obligate/strict anaerobes is tricky and labor intensive and it requires special care to protect them from lethal effect of oxygen. Therefore, little work has been done in the area of anaerobic microbiology as compared to aerobic microbiology. Preservation of facultative anaerobes and aerotolerant anaerobes is not difficult but obligate anaerobes require special care during handling. Furthermore, post-preservation viability or recovery of the microbes during revival is very important it depends on several factors such as phase of harvesting the culture, nature of cryoprotectants use during preservation, temperature of preservation, cell density, exposure to oxygen and rate of cooling etc. Researchers should take all the factors into consideration at

Table1: List of some anaerobes and methods used for their long term preservation

S.No.	Type of bacteria preserved	Method of preservation	Cryoprotectant used	Revival Period (months)	Reference
1	*Rhodospirillaceae*sp., *Chlorobiaceae*sp., *Chromatiaceae*	Cryopreservation	DMSO (5%)	24	Malik (1984)
2	*Methanosarcinabarkeri, Methanospirillumhungatei, Methanococcusvanielii, Methanobacteriumformicicum*	Cryopreservation	DMSO (5%) and glycerol (10%)	-	Huber and Lafferty (1988)
3	*Rhodobacter sp.,Thiodictyon sp., Chlorobiumferrooxidans*	Cryopreservation	Glycerol (10 and 20%)	17	Hegler and Kepler (2010)
4	*Enterococcus faecium, Lactobacillus plantarum*	Cryopreservation, Lyophilization, Fluidized bed bottom spray drying	Glucose, Sucrose and Trehalose (32%)	6	Strasser et al. (2009)
5	*Lactobacillus plantarum, Lactobacillus rhamnosus*	Lyophilization	Sucrose, Trehalose and Sorbitol (5-10 %)	-	Siaterlis et al. (2009)
6	*Candidatus* Kuenenia*stuttgartiensis*	Cryopreservation	DMSO (5%), Trehalose (1%) and Trypticase soy broth (0.3%)	7	Heylen (2012)
	*Candidatus*Brocadiasinica	Immobilization and Cryopreservation	Immobilization in 6% Polyvinyl alcohol and 2% Sodium alginate	5	Ali (2014)
7	*Nitrobacter*sp., *Nitrospina*sp., *Nitrococcus* sp., *Nitrotoga*sp., *Nitrospira*sp.	Cryopreservation	DMSO(1-10%), sucrose (1%), Trehalose(1%) and 1/10th Trypticase soy broth	-	Vekeman et al. (2013)
8	Bacterial consortium	Cryopreservation	Glycerol (10%)	36	Kurachi (2014)
9	*Neocallimastixpatriciarum*	Cryopreservation	DMSO	12	N. C. Yarlett et al. (1986)
10	*Clostridium acetobutylicum*	Cryopreservation	Glycerol	6	A. L. Bryukhanov& A. I. Netrusov (2004)

(Contd.)

11	Clostridium formicoaceticum	Cryopreservation	Glycerol	6	A. L. Bryukhanov& A. I. Netrusov (2004)
12	Clostridium butyricum	Cryopreservation	Glycerol	6	A. L. Bryukhanov& A. I. Netrusov (2004)
13	Bacteroidesthetaiotaomicron	Lyophilization and cryopreservation	sucrose and inulin (both 5% w/v) + Glycerol (15% v/v)	3	Lea Bircher et al. (2018)
14	Faecalibacteriumprausnitzii	Lyophilization and cryopreservation	sucrose and inulin (both 5% w/v) + Glycerol (15% v/v)	3	Lea Bircher et al. (2018)
15	Roseburiaintestinalis	Lyophilization and cryopreservation	sucrose and inulin (both 5% w/v) + Glycerol (15% v/v)	3	Lea Bircher et al. (2018)
16	Anaerostipescaccae	Lyophilization and cryopreservation	sucrose and inulin (both 5% w/v) + Glycerol (15% v/v)	3	Lea Bircher et al. (2018)
17	Eubacteriumhallii	Lyophilization and cryopreservation	sucrose and inulin (both 5% w/v) + Glycerol (15% v/v)	3	Lea Bircher et al. (2018)
18	Blautiaobeum	Lyophilization and cryopreservation	sucrose and inulin (both 5% w/v) + Glycerol (15% v/v)	3	Lea Bircher et al. (2018)
19	Lactococcuslactis	Cryopreservation	Skimmed milk (10%) + Mannitol		Berner and Viernstein (2006)
20	Vibrios	Lyophilization	Activated charcoal (5%) + skimmed milk (20%) + Raffinose (5%)		K. A. Malik (1990)
21	Zymophilus	Lyophilization	Activated charcoal (5%) + skimmed milk (20%) + Raffinose (5%)		K. A. Malik (1990)
22	Piromonascommunis	Cryopreservation	DMSO	12	N. C. Yarlett et al. (1986)
23	Candidatus Brocadia caroliniensis	Lyophilization	Skimmed milk		Michael J. Rothrock et al. (2011)
24	Caecomyces	Cryopreservation	DMSO	3	Ravinder Nagpal et al. (2012)

Cryopreservation: preservation at-80 degree C or at-196C in liquid nitrogen.

the time of preservation to get better result in terms of viability and maintaining certain abilities of the microbes. It has also been reported that certain microbes do not give better post-preservation recovery with commonly used cryoprotectants such as DMSO and glycerol. In such situations optimization of cryoprotectants is essential. Due to limited manpower and funding, most of the Microbial Resource Centers do not have time to optimize the preservation protocol of newly discovered group of microbes. Therefore, it is responsibility of researcher to optimize the preservation protocol which does not only preserve the cellular viability but also protects functionality of the microbes, to prevent their loss after isolation in terms of viability and functionality. In addition, there are certain group of microbes, which are tough to cultivate and purify, Annamox and methanogens are good examples. In such conditions, it is advisable to develop the protocols for intact microbiome, enrichment culture and mixed culture preservation which give better viability and functionality.

Acknowledgement

We highly acknowledge the Funding from Department of Biotechnology (DBT) Govt. Of India for establishment of National Centre for Microbial Resource by Grant no. BT/Coord.II/01/03/2016

References

Abidias M, Teixido N, Usall J, Benabarre A. and Vinas I. 2001. Viablity, efficacy and storage stability of Freeze dried bio-control agent Candida sake using different protective and rehydration media. *J. Food Prot* 64(6): 856-861.

Ali M, Oshiki M and Okabe S. 2014. Simple, rapid and effective preservation and reactivation of anaerobic ammonia oxidizing bacterium "Candidatus Brocadiasinica". *Water Res* 57: 215-222.

Bhattad U, Venkiteshwaran K, Cherukuri K, Maki JS and Zitomer DH. 2017. Activity of methanogenic biomass after heat and freeze drying in air. *Environmental Science: Water Research & Technology* 3: 462–471.

Bircher L, Geirnaert A, Hammes F, Lacroix C and Schwab C. 2017. Effect of cryopreservation and lyophilization on viability and growth of strict human gut microbes. *Microb. Biotechnol*: 721-733.

Bircher L, Schwab C, Geirnaert A and Lacroix C. 2018. Cryopreservation of artificial gut microbiota produced with in vitro fermentation technology. *Microb. Biotechnol* 11: 163-175.

Brook I. 2010. The role of anaerobic bacteria in bacteremia. *Anaerobe* 16: 183-189.

Bryukhanoy L and Netrusov AI. 2006. Long-Term Storage of Obligate Anaerobic Microorganisms in Glycerol. *Appl. Environ. Microbiol* 42 (2): 177–180.

Burrell PC, O'Sullivan C, Song H, Clarke WP and Blackall LL. 2004. Identification, detection and spatial resolution of Clostridium populations responsible for cellulose degradation in methanogenic landfill leachate bioreactor. *Appl. Environ. Microbiol.* 70(4): 2414-2419.

Castro H, Queirolo M, Quevedo M, Muxi L. 2002. Preservation methods for the storage of anaerobic sludges. *Biotechnol. Lett* 24(4): 329-333.

Fowler A and Toner M. 2005. Cryo injury and bio-preservation. *Ann. N. Y. Acad. Sci* 1066: 119–135.

Furushima Y, Yamamoto H, Imachi H, Takai K.Nunoura T, Hirai M, Miyazaki M, Kazama H, Makita H, Hirayama H, Furushima Y, Yamamoto H, Imachi H, Takai K. 2013. Isolation and characterization of a thermophilic, obligately anaerobic and heterotrophic marine Chloroflexi bacterium from a Chloroflexi-dominated microbial community associated with a Japanese shallow hydrothermal system, and proposal for Thermomarinilinea lacunofontalis gen. nov., sp. nov. ***Microbes Environ.*** 28(2): 228–235.

Ghurde VR, Wakode DD. 1981. A new report of Morchella from Central India. *Indian Journal of Mycology. Plant Pathol* 11: 314-315.

Hauffe HC and Barelli C. 2019. Conserve the germs: the gut microbiota and adaptive 549 potential. Conserv. Genet 20(1): 19–27.

Heggler F and Kappler A. 2010. Cryopreservation of anaoxygenic phototrophic Fe(II) -oxidizing bacteria. Cryobiology 61: 158-160.

Heylen K, Ettwig K, Hu Z, Jetten M and Kartal B. 2012. Rapid and Simple Cryopreservation of Anaerobic Ammonium-Oxidizing Bacteria. *Appl. Environ. Microbiol* 3010–3013.

Heylen K, Ettwig K, Hu Z, Jetten M and Kartal B. 2012. Rapid and simple method for cryopreservation of Anaerobic Ammonia-oxidizing bacteria. *Appl. Environ. Microbiol* 78: 3010-3013.

Hoefman S, Pommerening-Roser A, Samyn E, De Vos P and Heylen K. 2013. Efficient cryopreservation protocol enables accessibility of broad range of ammonia-oxidizing bacteria for the scientific community. *Microbiology Res* 164: 288-292.

Hubalek Z. 2003. Protectants used in the cryopreservation of microorganisms. *Cryobiology* 46: 205–229.

Huber K and Lafferty RM. 1988. Cultivation and preservation of Methanogenic bacteria. *Zentralbl Bakteriol B* 143: 149- 155.

Khan MT, van Dijl JM and Harmsen HJ. 2014. Antioxidants keep the potentially probiotic but highly oxygen sensitive human gut bacterium *Faecalibacterium prausnitzii* alive at ambient air. PLoS ONE 9: e96097.

Kurachi K, Hosokawa R, Takahashi M and Okuyama H. 2014. The potential of glycerol in freezing preservation of turbine oil degrading bacterial consortium and the ability of revised consortium to degrade petroleum wastes. *Int. Biodeterior. Biodegradation* 88: 77-82.

Lovelock JE. 1953. The mechanism of the protective action of glycerol against haemolysis by freezing and thawing. *Biochim. Biophys. Acta* 11: 28–36.

Malik KA. 1984. A new method for liquid nitrogen storage of phototrophic bacteria under anaerobic condition. *J. Microbiol. Methods* 2: 41-47.

Malik KA. 1990. Use of activated charcoal for the preservation of anaerobic phototrophic and other sensitive bacteria by freeze-drying. *J. Microbiol. Methods* 12: 117-124

Malik KA. 1992. Freeze drying of microorganisms using a simple apparatus. *World J. Microbiol. Biotechnol.* 8: 76–79.

Malik KA. 1998. Preservation of Chloroflexusby deep-freezing and liquid-drying methods. *J. Microbiol. Methods* 32: 73–77.

Meryman HT. 2007. Cryopreservation of living cells: principles and practice. *Transfusion* 47: 935–945.

Naradasu D, Miran W, Sakamoto M and Okamoto A. 2019. Isolation and Characterization of Human Gut Bacteria Capable of Extracellular Electron Transport by Electrochemical Techniques. *Front Microbiol* 9: 3267.

Pokorny R, Olejnikova P, Balog M, Zifcak P, Holker U, Janssen M, Bend J, Hofer M, Holiencin R, Hudecova D and Varecka L. 2005. Characterization of microorganisms isolated from lignite excavated from the Zahorie coal mine (southwestern Slovakia). *Microbiol Res* 156(9): 932-943.

Prakash O, Nimonkar Y and Shouche YS. 2012. Practice and prospects of microbial preservation. *FEMS Microbiology Letters* 339: 1-9.

Schulte MJ, Solocinski J, Wang M, Kovacs M, Kilgore R, Osgood Q, Underwood L, Flickinger MC and Chakraborty N. 2017. A technique for lyopreservation of *Clostridium ljungdahlii* in a biocomposite matrix for CO absorption. *PloS ONE* 12(7): e0180806.

Siaterlis S, Deepika G and Charalampopoulos D. 2009. Effect of culture medium and cryoprotectants on the growth and survival of probiotics lactobacilli during freeze drying *Lett. Appl. Microbiol.* 48: 295-301.

Strasser S, Neureiter M, Geppl M, Braun R and Danner H. 2009. Influence of lyophilization, fluidized bed drying, addition of protectants and storage on the viability of lactic acid bacteria. *J. Appl. Microbiol* 107: 167-177.

van Vliet DM, Palakawong Na Ayudthaya S, Diop S, Villanueva L, Stams AJM and Sánchez – Andrea I. 2019. Anaerobic degradation of sulfated polysaccharides by two novel *Kiritimatiellales* strains isolated from Black sea sediment. *Front Microbiol* 10: 253.

Vekeman B, Hoefman S, Vos PD, Spieck E and Heylen K. 2013. A generally applicable cryopreservation method for nitrite oxidizing bacteria. *Syst. Appl. Microbiol.* 36: 579-584.

Viancelli A, Pra MC, Scussiato LA, Cantao M, Ibelli AMG and Kunz A. 2017. Preservation and reactivation of Candidatus Jettenia asiatica and anammoxoglobus propionicus using different preservative agents. *Chemosphere* 186: 453-458.

Wolfe RS. 1999. Anaerobic life – a Centennial View. *J. Bacteriol. Res* 181(11): 3317-3320.

3

Insights Into the Diversity of Anoxygenic Phototrophic Bacteria and Its Application

Tushar Lodha[1*], *Prachi Karodi*[1]

[1]*National Centre for Microbial Resource, National Centre for Cell Science Pune, Maharashtra, India*

Abstract

Anoxygenic phototrophic bacteria (APB) are having the genetic potential of converting radiant energy into chemical energy useful for growth and metabolic activities without evolving oxygen. They are physiologically, metabolically and phylogenetically versatile group of organism spread in five distinct phyla; Proteobacteria *(purple bacteria),* Chlorobi *(green-sulfur bacteria),* Chloroflexi *(green non-sulfur bacteria),* Firmicutes *(heliobacteria) and* Acidobacteria. *Different group of APB harbors distinct photosynthetic pigments, carotenoids, and a photosynthetic reaction center. These group of bacteria succeeds in a natural environment with light availability and presence of anoxic conditions as the pigment, and the molecular oxygen inhibits bacteriochlorophyll synthesis whereas few APB can grow in hypooxic as well as in aerobic conditions. Calvin-Benson-Bassham (CBB) pathway, reverse tricarboxylic acid (rTCA) cycle, and hydroxypropionate cycle are reported pathways for carbon assimilation in APB to fix carbon dioxide. Photosynthesis occurs in either intracytoplasmic membrane (ICM) or membrane bound chlorosomes and in the cytoplasm for few APB. APB have a wide range of potential applications including removal of* H_2S *from wastewater and gas streams, hydrogen production, degradation of xenobiotic, toxic metals and radioisotopes, work as host for the production of challenging membrane proteins, source of carotenoids and terpenoids, single cell protein, extensive applications in agriculture and production of bioactive compounds with antimicrobial, antioxidant*

**Corresponding Author: passimtushar10@gmail.com*

and anticancer activity. The chapter is focused on the diversity of different types of APB, habitats, their growth requirements, photosynthetic machinery and biotechnological applications of APB.

Keywords: Anoxygenic photosynthesis, Purple bacteria, green bacteria, diversity CO_2 fixation, biotechnology

Introduction

Photosynthetic microorganisms are the oldest group of microorganisms on the earth, having the genetic potential of converting radiant energy into chemical energy, which is useful for growth and metabolic activities. All the photosynthetic organisms have machinery of light absorbing (photosynthetic) pigments and reaction centre, which makes them capable of trapping the light and translating it into food for the survival. There are two types of phototrophic microorganisms; one which evolves oxygen as a by-product of photosynthesis (oxygenic) and another carry out photosynthesis without the release of oxygen (anoxygenic). Oxygenic phototrophic microorganisms include cyanobacteria, prochlorophytes, and microalgae which have potential enzymes required for oxidizing water to molecular oxygen and photochemical type II reaction center receives electrons.

$$2H_2O \rightarrow 4H^+ + 4e^- + O_2\uparrow$$

Phototrophs which does not evolve oxygen after photosynthesis, whereas oxidises reduced sulphur compounds (sulfide, thiosulfate, sulfite, and tetrathionate), hydrogen, ferrous ions (Fe2+) or small organic compounds (acetate, succinate, pyruvate, malate) and electrons are donated to the photochemical reaction center are known as anoxygenic phototrophic bacteria (APB).

$$H_2S \rightarrow 2H^+ + 2e^- + S^\circ$$

APB encompasses a heterogeneous group of eubacteria, which are physiologically and phylogenetically diverse. APB are metabolically versatile and reported from diverse habitats including aquatic, terrestrial and extreme habitats, like hyper salinity, extreme pH, and high temperature, and light (Frigaard 2016). Versatile nature of APB is demonstrated by its diverse physiology, wide range of morphology, capacity to harbor several types of bacteriochlorophylls and carotenoids and widespread generic and species level phylogenetic relationship etc. APB can be differentiated based on the type of reaction center they harbour; green non-sulfur bacteria and purple bacteria contains type II reaction center which is analogous to photosystem II of cyanobacteria. Green sulfur bacteria and photosynthetic *Firmicutes* members have type-I reaction center.

APB have a potential for the application in removal of H_2S from wastewater and gas streams, hydrogen production, a host for the production of membrane proteins, source of carotenoids and terpenoids, wide applications in agriculture, production of bioactive compounds with antimicrobial, antioxidant, anticancer activity and degradation of xenobiotic, toxic metals and radioisotopes (Blankenship et al., 1995; Frigaard 2016; Sasikala and Ramana 1995). Though APB have the potential for various biotechnological applications yet not exploited as widely as chemotrophs. APB are widely distributed in marine, freshwater and terrestrial habitats with anaerobic/microaerobic conditions and sufficient light intensities. They are reported from diverse ecological niches such as freshwater lakes, lagoons, estuaries, ponds, sediments of ditches, moist soils, intertidal zones, hypersaline environments, alkaline environment, acid water, hot springs, Arctic and Antarctica (Glaeser and Overmann, 1999; Madigan and Jung, 2008; Martínez-Alonso et al., 2005; Ramana et al., 2013; Ramaprasad et al., 2016; Srinivas et al., 2014; Sucharita et al., 2010). APB are non-homogenously distributed into five distinct phyla; purple bacteria in phylum *Proteobacteria* (includes class α-, β -, γ-proteobacteria with interspersing chemotrophs), green sulfur bacteria in phylum *Chlorobi*, green non-sulfur bacteria in phylum *Chloroflexi*, *Heliobacteria* in the phylum *Firmicutes* and *Chloracidobacterium thermophilum* in the phylum *Acidobacteria* (Sattley and Madigan, 2014; Tank and Bryant, 2015; Trüper and Pfennig, 2013).

History of anoxygenic phototrophic bacteria

In early twentieth century, there were three conflicting theories on purple bacteria, wherein Engelmann's (1883) observation on purple bacteria's phototactic response in the presence of light was correlated with photosynthetic activity of green algae. And it was concluded that although purple bacteria photosynthesize but requires proper illumination. It was also noted that dark rays (rays in infrared region) which are not visible, was also absorbed by these bacteria. Meanwhile, Sergei Winogradsky (1887) was studying sulfur bacteria and observed that environmental sulfide can be converted to sulfur granules in the presence of light by purple sulfur bacteria. These sulfur granules after oxidation were getting accumulated in the cells of purple bacteria. As the sulfide content in the environment was getting exhausted the cells were also getting emptied off, these sulfur globules were further oxidized to sulfate and thus was released into the atmosphere as a non-toxic form. A constant fresh supply of sulfide was needed for the bacteria to survive. Above findings of phototactic movement of purple bacteria and hypothesis, about accumulation of sulfur globules in the cells was contradicted by Molisch in 1907 by proposing new findings. He concluded that purple bacterium does not assimilate CO_2 but assimilates organic substrate in both conditions of light and dark. All three works were later accepted because all three had some or the other parameter different from each other. Engelmann's work on colored bacteria, Winogradsky's work on metabolism of sulfur in sulfur bacteria and Molisch work on bacteria which did not require sulfide media to grow (i.e., purple non-sulfur bacteria) are essential to understand the ecology of APB (van Niel, 1941). The detail history of discoveries of anoxygenic phototrophic bacteria and their mecanisms of photosynthesis has been given in Table 1.

Types of anoxygenic phototrophic bacteria

Purple bacteria

Purple bacteria are the first group of phototrophic bacteria reported which do not evolve molecular oxygen but generate sulfur as a byproduct of photosynthesis. They are Gram-stain-negative, having bacteriochlorophy-ll *a* or *b or both* together with various carotenoids, and show different pigmentation ranging from purple, brown, orange and red. The significant carotenoids include spheroidene, spirilloxanthin, rhodopsin, lycopene, and their derivatives (Takaichi, 2006). The photosynthetic pigments are located within vesicular/lamellar intracytoplasmic membranes in the form of vesicles, bundled tubes, tubes, or in stacks. Purple bacteria belong to phylum *Proteobacteria* and harbors type- II photosynthetic reaction center with LH1 and LH2 as a light-harvesting antennae located in the cytoplasmic membrane or intracytoplasmic membrane. Most of the purple bacteria grow autotrophically and fix CO_2 by reductive pentose phosphate cycle (CBB pathway). All purple bacteria possess peptidoglycan in the cell wall and lipopolysaccharides in the outer cell membrane. These group of bacteria succeeds in natural environment, with sufficient light and anoxic conditions as molecular oxygen inhibits pigment and bacteriochlorophyll synthesis. Marine and coastal ecosystems (estuaries, mangroves, beaches and sub-surface sediments), freshwater ecosystems (rivers, ponds, paddy fields and lakes) and other aquatic environments harbor purple bacteria but the abundance and diversity is modulated by the concentration of sulfide, light quality, temperature and pH (Blankenship et al., 1995; Imhoff, 1995). Purple bacteria of the genera *Rhodopseudomonas, Rhodobacter, Rubrivivax* and *Rhodovastum* have also been reported from the terrestrial ecosystems such as paddy-fields (Harada et al., 2005; Hiraishi and Okamura, 2017; Okamura et al., 2009; Ramana et al., 2006; Vinay Kumar et al., 2013). There are reports of occurrence of halophilic, alkaliphilic, acidophilic, thermophilic and psychrophilic purple bacteria from extreme conditions (Blankenship et al., 1995). Purple bacteria can be readily grown in laboratory conditions in mineral media either photoautotrophic growth mode (media supplemented with sulfide and bicarbonate) or photoheterotrophic growth mode (media supplemented with an organic carbon like pyruvate, acetate succinate, malate, etc.).

Table 1: History of discoveries of anoxygenic photosynthetic bacteria and its photosynthesis

Year	Scientist or Research group	Contribution
1873-1876	Sir E. Ray Lankester	First observations of purple photosynthetic bacteria;Pigmented bacteria named as "*Bacterium rubescens*" and the pigment has been named as 'bacteriopurpurin".
1883	Theodon W. Engelmann	Dispersed spectrum accumulated at specific wavelengths in "Bacterium photometericum".
1884	Dr. Charles MacMunn	Discovery of cytochromes
1887	Erwin von Esmarch	Isolated *Spirillum rubrum* which was later renamed as *Rhodospirillum rubrum* by Molisch (1907)
1888	Sergei Winogradsky	Discovery of chemoautotrophy; Describe purple sulfur bacteria and *Chromatium;* Winogradsky column for the enrichment of phototrophic bacteria
1888	Theodon W. Engelmann	Bacteriopurpurin is a true chlorophyll
1906	Academician G.A. Nadson	First time coined the name green sulfur bacteria
1907	Hans Molisch	Pioneered in isolating and describing several species of nonsulfur bacteria in pure culture. Described *Rhodonostoc capsulatum* later renamed as *Rhodobacter capsulatus*; Assigned pigmented sulfur and nonsulfur bacteria to one order, *Rhodobacterales*
1912	G.A. Nadson	Describes green bacterium *Chlorobium limicola*
1931	Keita Shibata	Carbon and nitrogen assimilation; Shibata's theory of photosynthesis.
1932	Robert Emerson and William Arnold	Proposed the concept of the photosynthetic unit
1932	Van Neil	Photosynthetic carbon dioxide assimilation
1933	F.M. Muller	PSB can grow photosynthetically under anaerobic conditions
1937	Hans A. Krebs	Describes the citric acid cycle
1937	Pelsh	Distinguished *Ectothiorhodospira* and Endothiorhodospira
1944	C.B. van Neil	Physiology and characteristics of the photosynthetic pigments
1949	H. Gest and M.D. Kamen	Light dependent production of Hydrogen and nitrogen fixation by *Rhodospirillum rubrum*
1952	L.N.M Duysens	Discovery of absorption changes in bacteriochlorophyll
1953	Helge Larsen	Optimal conditions to grow *Chlorobium limicola* and *C. thiosulfatophilum*
1954	J. Postgate	Cytochrome in obligate anaerobic *Desulfovibrio desulfuricans*
1957	G. Cohen-Bazire, W.R. Sistrom and R.Y. Stanier	Regulatory mechanisms of photopigment production in purple bacteria

(Contd.)

1959	A.A. Krasnovsky	Pigments are organized as oligomeric complexes in green sulfur bacteria
1959	R. Clinton Fuller and Martin Gibbs	Reported Calvin-Benson-Bassham (CBB) cycle in purple bacteria
1959	Tuttle and Gest	Photopigments are localized in 'chromatophores'
1960	Clayton and Smith	Generated blue green mutant of *Rhodobacter sphaeroides* R-26
1961–1978	N. Pfennig	Successfully employed a new synthetic medium suited for the enrichment and isolation of virtually all anaerobic photosynthetic bacteria
1962	John Olson	Discovery of bacteriochlorophyll *an* antenna protein
1962	L.E. Mortenson	Describe low redox potential Fe protein
1963	C. Sybesma and J. Olson	A first quantitative study on the energy transfer efficiency in green sulfur bacteria
1963	C.F. Kettering Foundation	First international symposium on bacterial photosynthesis
1963	R.K. Clayton	The name 'reaction center' was proposed
1963	Wim Vredenberg and L.N.M. Duysens	Concept of 'puddle' and 'lake' organisation of antenna systems
1964	G. Cohen-Bazire	Observed 'chlorobium vesicles' (chlorosomes) of green sulfur bacteria under an electron microscope
1966	M.C.W. Evans, B.B. Buchanan and D.I. Arnon	Proposed reductive tricarboxylic acid cycle in photoautotrophic bacterium *Chlorobium*
1968	D.W. Reed and R.K. Clayton	Isolation of reaction center complex from *Rhodobacter sphaeroides* (Basonym as *Rhodopseudomonas sphaeroides*)
1970	A. Borisov and V. Godik	Determined life time of bacteriochlorphyll
1974	B.K. Pierson and R.W. Castenholz	Isolated thermophilic *Chloroflexus aurantiacus*
1974	Barry Marrs	Genetic recombination system in *Rhodobacter capsulatus*
1975	E. Broda	Proposed conversion hypothesis which states that non-photosynthetic bacteria descended from purple photosynthetic bacteria
1975	Melvin Okamura and George Feher	Established that the ubiquinone is the first stable electron acceptor in the reaction center
1975	R. Fenna and B. Matthews	Crystallographic structure of the FMO protein of antenna
1976	Barry Marr and cowerkers	Cloning of photosynthetic gene cluster
1978	K. Sato and K. Harashima	Reported the presence of Bchl *a* in aerobic anoxygenic photosynthetic bacteria
1978	V. Shuvalov	Identified bacteriochlorophyll as first electron acceptor
1979	T. Swarthoff and J. Amesz	Prepared a highly enriched reaction center particle from the green sulfur bacteria
1981	H. Zuber	The obtained sequence of an LH1 antenna protein
1983	F. Gest and J. Favinger	First report of anoxygenic phototrophic bacteria from phylum *Firmicute* (*Heliobacterium chlorum*)

(Contd.)

1984	Johann Deisenhofer and colleagues	Reported the electron density and chromophore structure of the reaction center complex
1984	Michael Madigan	Isolated thermophilic *Thermochromatium tepidum* containing novel LH1 photopigment complex that absorbs at 920nm
1985	R.C. Fuller and co-worker	Reported that bacteriochlorophyll *g* is the primary donor in *Heliobacterium chlorum*
1988	J. Deisenhofer, R. Huber and H. Michel	Awarded Nobel Prize in chemistry for their contribution in elucidating the structure of photosynthetic reaction center
1989	J. Trost and R. Blankenship	Reaction center in *Heliobacillus mobilis* is homodimer protein
1993	F. Widdel	Discovery of autotrophic growth of purple bacteria on CO_2 and Fe^{2+} ,which has significance for the hypothesis of early life on Earth.
1994	Carl Bauer	Phototactic behavior of *Rhodospirillum centenum*
1995	Richard Cogdell	Elucidated the crystallographic structure of LH2 antenna complex
2000	C. Bauer	Purple bacteria contains the most ancient Mg-tetrapyrrole biosynthesis genes
2000	Raymond	Establishes that the horizontal gene transfer has played a significant role in the evolution of photosynthesis

The table is prepared from an article entitles “Timeline of discoveries: anoxygenic bacterial photosynthesis” (Gest and Blankenship, 2004).

APB is ancient organisms harboring photosynthetic machinery which makes them important to understand the evolution of photosynthesis (Madigan and Jung, 2008). Due to simplest form of photosynthesis and easily accessible for cultivation, purple bacteria are used as model microorganism to study the physiology, biochemistry, and molecular event in the photosynthesis. In an anaerobic environment, purple bacteria contribute to the carbon cycle by fixing CO_2 (primary producer) and consuming organic carbon in the presence of light. Based on the utilization and tolerance of sulfide purple bacteria can be differentiated into two groups: Purple sulfur bacteria (PSB) and purple non-sulfur bacteria (PNSB). PSB and PNSB can be distinguished again based on the physiology as well as phylogenetic, relationship. It is also reported that PSB and PNSB are always intermingled with their non-photosynthetic neighbors. PSB belongs to the class γ-*Proteobacteria,* and PNSB are accommodated in class á-*Proteobacteria* and β-*Proteobacteria* (Imhoff, 1995). PSB can tolerate the millimolar levels of the sulfide, whereas PNSB does not. Hansen and van Gemerden carried out an experiment and suggested that the criteria of tolerance to sulfide is not absolute way to classify PSB and PNSB. PNSB isolated from sulfide rich habitat has shown that PNSB have the genetic potential to tolerate sulfide and can use as electron donor, but PNSB deposit elemental sulfur (S^0) outside the cell, unlike PSB where it is stored intracellularly (except for species of *Ectothiorhodospiraceae;* Hansen and van Gemerden, 1972). PSB and PNSB can be distinguished by microscopic observations of globules of S^0 formed.

Purple sulfur bacteria

PSB are anoxygenic photosynthetic bacteria, belongs to the order *Chromatiales* of the class γ-*Proteobacteria* (Imhoff, 1995). They can tolerate millimolar concentration of sulfide and oxidize it to sulfur globules (van Niel, 1944). Initially taxonomy of the purple phototrophic bacteria was completely based on phenotypic traits, but the later stage it was established that the formation of two families (*Chromatiaceae and Ectothiorhodospiraceae*) was supported by phylogenetic (16S rRNA gene sequence analysis) as well as physiological properties. Molecular analysis (DNA-DNA hybridisation, rRNA–DNA hybridisation, restriction digestion patterns of DNA) and chemotaxonomic characters like polar lipids, fatty acids, quinone, lipopolysaccharides etc. (Imhoff, 1995). Initially, all the sulfur metabolizing purple bacteria with capacity to deposits sulfur globules were accommodated in the family *Chromatiaceae*. The genus *Ectothiorhodospira* which deposit the sulfur globules outside the cell was also a member of family *Chromatiaceae* (Trüper and Pfennig, 2013). Later the family *Chromatiaceae* was divided into two; family *Ectothiorhodospiraceae* which encompasses purple sulfur bacteria which deposit sulfur (S^0) outside the cell whereas family *Chromatiaceae* (Endothiorhodospira) accommodates PSB which deposit sulfur globules inside the cells (Imhoff, 1995). Till May 2019, family *Chromatiaceae* encompasses 31 distinct validly published genera, whereas family *Ectothiorhodospiraceae* accommodates 19 validly published genera (http://www.bacterio.net/). The members of the family *Ectothiorhodospiraceae* are mainly alkaliphilic and halophilic in nature, and forms a different lineage which is phylogenetic descent of the family *Chromatiaceae*. The genus *Thiorhodospira* of the family *Ectothiorhodospiraceae* produces sulfur globules, which remain attached to the cells and deposited in the periplasmic space of the cell (Bryantseva et al., 2009).

Occurrence and diversity of PSB depend on intensity and availability of light, oxygen concentration and presence of sulfide. PSB are reported from different natural habitat and mostly recorded from the stagnant water bodies. Sulfate reducing bacteria produces sulfur in the sediments of aquatic ecosystems which diffuses upward creating the sulfur gradient. PSB utilize these sulfurs as electron donor and grow phototrophically. PSB are isolated from various habitats like salterns, marine aquaculture ponds, brackish water, river, and extremophilic niches like alkaline, acidic, cold, and hot (Imhoff, 1995). PSB can grow in bloom (large masses; either of the mixed PSB or single species of PSB), in association with green sulfur bacteria and purple non-sulfur bacteria as a mixed community. Sometimes lake water becomes pigmented (reddish brown, purple or red) because of the large cell density of PSB. They form a dense pigmented layer between the cyanobacterial layer and lower layers of the microbial mat and are commonly reported in freshwater microbial mats, marine microbial mats, hypersaline effluents of thermal spring. Members of the genus *Thiocapsa* and *Allochromatium* are metabolically versatile, which makes them to grow in variable conditions like different layers of marine microbial mats and dynamic marine habitats (Overmann, 2008a). There are reports of the extremophilic PSB like *Ectothiorhodospira, Halochromatium, Halorhodospira, Thiohalocapsa, and Marichromatium,* which are haloalkaliphilic in nature (Imhoff, 2006). Members of family *Chromatiaceae* are also reported from thermal spring and sea ice (Madigan, 1986; Petri and Imhoff, 2001). Thermophilic *Theromochromatium tepidum* isolated from thermal spring can perform photosynthesis at 57 °C (Madigan, 1986). PSB are capable of carryout photosynthesis in diverse environmental conditions like temperature ranging from 0 °C to 57 °C, NaCl concentration ranging from 0 to ~32 % (w/v; saturation limit), pH ranging from 3 to 11 (Imhoff, 1995).

Unlike other bacterial genera members of PSB are not abundant in natural habitats therefore, to isolate PSB in laboratory condition needs specialized skill and set-up are mandatory. There is a need for enrichment of PSB in the samples to increase the population of these sulfur oxidizing bacteria for cultivation purpose. Initially in 1988, Russian microbiologist Sergei Winogradsky used glass column also known as Winogradsky column to selectively enrich the phototrophic bacteria which was modified subsequently by Pfennig and van Neil. Winogradsky column a self-contained mixed community system obtaining energy from light is the best model to study the independent role of different microorganisms. Glass column is filled by sediment sample from the environment supplemented with cellulose (newspaper), calcium carbonate and sodium sulfate, water from the source of sediment and incubated in light (Fig. 1).

Fig. 1: Winogradsky column designed to enrich the phototrophic bacteria from the sediments collected from the Mutha river, Pune, India at National Centre for Microbial Resource, Pune. **(1)** Column prepared with autoclaved sediment sample incubated under illumination; **(2)** Column prepared with sediment sample and incubated in dark; **(3, 4 and 5)** Column prepared with sediment sample and incubated in light and showing the green and brown growth of phototrophic organisms (**3** and **4** are the sides facing towards light and **5** is facing away from light). **(6)** Isolates of purple non-sulfur bacteria purified from the Winogradsky column.

In 1944, van Neil proposed the physiological basis of enrichment of purple non-sulfur bacteria in the Winogradsky column (van Niel, 1944). Later van Neil has designed a defined media to enrich and cultivate purple bacteria specifically. The enrichment of the phototrophic purple bacteria depends on the specific minerals, carbon source (inorganic or organic), vitamins, sulfide source, sulfide concentration and incubation conditions (light intensity, pH, anoxic conditions and temperature; Pfennig and Trüper, 1981). The growth of PSB is inhibited by high sulfide concentration; therefore, the sulfide concentration in the enrichment media should be optimum so that it should allow the growth of a sensitive form of purple bacteria also. Generally, 1-2 mM concentration of sulfide is optimum, which support the growth of sensitive microorganisms and in some, cases a continuous feeding is essential to maintain the sulfur concentration in the medium.

The salinity of the enrichment media should be same as that of the environment from where samples are collected. For the samples collected from the marine habitat 2-3% of NaCl gives better enrichment. But not always the environmental salinity is the optimum for the culture to enrich so varying different saline concentration may help to enrich purple bacteria form marine habitats.

Temperature is one of the crucial parameter which decide the growth of purple bacteria. When enrichments are incubated at 25 – 35 °C fast growing bacteria grow at the higher rate and suppress the growth of other slow growers, and biased for enrichment of single type of cultures. On the other hand, when enrichments are incubated at 15-22 °C, enrichment grows slowly, but a different type of phototrophs grow simultaneously. For enriching moderately thermophilic purple sulfur bacteria enrichment should be incubated at 40°C (*Thermochromatium tepidum* enriched at 40 °C; Madigan, 1986). By altering the media composition and incubation conditions, different types of purple bacteria can be enriched and isolated in laboratory. A general procedure to enrich purple sulfur bacteria is; inoculate approximately 1 g or 1 ml of sample (sediment or water) in completely filled screw cap tubes (20 ml) or bottles (50 ml) with mineral media and additional supplements. After inoculation tubes or bottle should be incubate at desired temperature under the illumination of photosynthetically active radiation (PAR) of 100-200 μmol m^{-2} s^{-1} photon (light).

Phototrophs selectively absorb the light of different intensity or wavelength and this property can be exploited to enrich different group of phototrophs. Generally, a continuous high-intensity light (1,000 to 2,000 lux) at 30 °C is ideal for growing members of *Chromatiaceae* (*Thiocystis minor, T. violascens, T. violacea, Allochromatium vinosum, Allochromatium minutissimum, Thiocapsa roseopersicina* and *Marichromatium gracile*; Imhoff, 1995). Strains of *Thiococcus pfennigii* and *Thioflavicoccus mobilis* can be enriched by modulating the light using IR filters (above 900-1000 nm). Members of the genus *Ectothiorhodospira* can be enriched by incubating in incandescent light (1000-10000 lux) and at the desired temperature. Flagellated members of the family *Chromatiaceae* can be enriched where initially non-motile forms enrich at the bottom then motile members having gas vesicles harboring members gets enrich at the surface.

Purple non-sulfur bacteria

Purple non-sulfur bacteria (PNSB) are a non-taxonomic group of physiologically and are energetically versatile organisms capable of photolithotrophic photoorganotrophic, and chemoorganotrophs growth. PNSB are the most diverse anoxygenic phototrophic bacteria varying in morphology, photosynthetic pigments utilization of electron donors, carbon sources and nature of internal structures.

They can switch from one growth mode to another (e.g., photoautotrophy to photoheterotrophy) depending on external conditions available (Imhoff, 1995). PNSB can grow photoautotrophically by using hydrogen or low concentration of sulfide as electron donor, and some species even can utilize $S_2O_3^{2-}$ or Fe^{2+} as an electron donor. To grow PNSB, organic acids, fatty acids, carbohydrates, tricarboxylic acid (TCA) cycle intermediates, amino acids, yeast extract as a source of vitamin B_{12}, nicotinic acid, thiamine, p-aminobenzoic acid and biotin can be added to media. PNSB also grow in dark anoxic conditions using anaerobic respiration and fermentation type of metabolism. They also grow diazotrophically, in nitrogen limited conditions. This group of bacteria can be found in anoxic zones of water including fresh, marine, and polluted water bodies with high activity of sulfate reduction, microaerophilic conditions and availability of hydrogen sulfide, stagnant water bodies like lakes, ponds, lagoons, and other aquatic environments with soluble organic matter and low oxygen tension considerd good habitats for PNSB. As these bacteria are sensitive to sulfide, no colored bloom is formed. PNSB are also found in the water of eutrophic ponds, lakes, mud, and ditches which have low sulfide tolerance and less abundance in pelagic zone. It is also reported from marine, hypersaline environment, water bodies, sediments of salt marshes, intertidal zones and polluted harbor basins but not reported from the open ocean. Marine isolates can tolerate high sulfide and use thiosulfate as electron donors. *Rhodopseudomonas palustris, Rhodomicrobium vannielii,* and *Rhodobacter maris* are isolated from the marine habitat but does not have a salt requirement for the growth (Gandham et al., 2018; Ramana et al., 2013; Vinay Kumar et al., 2013). *Rhodovulum euryhalinum, Rhodobium orientis,* and *Rhodobacter maris* are also confined to only marine habitat (Imhoff, 1995). PNSB are also reported from the extremophilic conditions like high salt concentration (*Rhodothalassium salexigens, Rhodovibrio salinarum, Roseospira sp.* and *Rhodovibrio sodomensis*), acidic conditions (*Rhodoblastis acidophilus*), psychrophilic condition (*Rhodoferax antarcticus*). *Rhodovibrio salinarum* forms a colored layer on salt deposits in evaporated seawater pools and salterns. Dead sea sediments, thermal springs and cold polar environments also harbor purple non-sulfur bacteria (Imhoff, 1995).

PNSB can be enriched in a media with low concentration of sulfate (to avoid the growth of sulfate reducing bacteria) or sulfate free media supplemented with trace elements with nitrogen and organic carbon source. PNSB can be enriched by using mineral media supplemented with common organic substrate and incubation in anoxic illumination conditions. During enrichment, fermentative processes generate acetate, lactate, butyrate, and propionate, which enhances the growth of PNSB. The concentration of nutrients, mineral salt composition, and incubation parameters (light intensity, range of light, temperature and pH) are crucial factors for the enrichment of PNSB. Some PNSB (*Rhodovulum sulfidophilum, Rhodovulum adriaticum,* and *Blastochloris viridis*) can tolerate and/or require sulfur source as an electron donor in photosynthesis (Hiraishi and Ueda, 2009). For enrichment of isolates from marine habitat, salinity was adjusted to 3% (w/v) NaCl whereas for freshwater isolates media should be without NaCl. Biebl and Pfennig have developed anaerobic infusion method to selectively enrich *Pararhodospirillum photometricum* (Biebl and Pfennig, 1981). Enrichment of the species of genus *Rhodospirillum,* amino acids and methanol bicarbonate to enrich *Rhodoblastis acidophilus* can be used. Strains of *Rhodothalassium* spp., *Rhodovibrio salinarum,* and *Rhodovibrio sodomensis* can be enriched by maintaining salt concentration (NaCl) more than 10% (w/v; Frigaard 2016). *Phaeospirillum fulvum and Rhodopseudomonas palustris* can be enriched by exploiting their ability to metabolize benzoate. *Blastochloris sulfoviridis* have capability to use toluene as carbon source, whereas higher fatty acids at low concentration can be used to enrich *Phaeospirillum fulvum* and *Phaeospirillum molischianum. Blastochloris viridis* and *Blastochloris sulfoviridis* have Bchl *b,* which needs long radiation (1015 to 1055 nm) to grow and these radiations can be exploited to enrich such group of bacteria (Frigaard 2016). PNSB belongs to the class α-*Proteobacteria* and β-*Proteobacteria* of the phylum *Proteobacteria* (Imhoff, 1995). Phylogenetically they didn't form a distinct cluster but interspersed in between the variety of chemoheterotrophic (non-photosynthetic) members. They are a model system to study the anaerobic photosynthesis, especially members of genus *Rhodobacter* and *Rhodopseudomonas* are considered as ideal model system to study photosynthesis and India stands first globally in the description of novel species of PNSB.

Green sulfur bacteria

Green sulfur bacteria (GSB) are a distinct group of bacteria, which have a unique type of photosynthetic apparatus, presence of chlorosomes, and use sulfide as a photosynthetic electron donor (Imhoff, 2014a). Green phototrophic bacteria have pigment system dominated by Bchl *c* and *d* with small amount of BChl *a*. GSB found at the lowermost part of the stratified environment as they harbor a unique property of carrying out photosynthesis even at minute amount of light. They are obligate anaerobic, obligate phototrophic and have sulfur metabolisms capabilities, which provoke researchers to study the physiology, genomics and ecology of these organisms. GSB are in syntrophic association with the sulfur/sulfate reducing bacteria, which results in the recycling of sulfur. Norbert Pfennig had isolated different GSB from the freshwater habitat and had a collection of them. Later, Hans George, Truper, and Pfennig established the taxonomy of GSB in the family *Chlorobiaceae* (Pfennig and Truper, 1971). GSB are phylogenetically not related to other bacteria and belong to the family *Chlorobiaceae* of the phylum *Chlorobi*. The few genera of GSB are *Chlorobium, Chlorobaculum, Prosthecochloris,* and *Chloroherpenton*. GSB are metabolic specialist, which can oxidize sulfide with deposition of sulfur outside the cell and have unique CO_2 fixation pathway (reductive tricarboxylic acid cycle. The unique metabolic potential of these bacteria shows that they perform district role in the environment.

GSB perform photolithoautotrophy in anoxic condition, and occurred in habitat with low intensity of light, inorganic electron donors and utilize CO_2 as carbon source. In 1966 Evans et al. has reported a reductive tricarboxylic acid cycle in GSB (Evans et al., 1966). GSB can photoassimilate few organic compounds (acetate, propionate, and pyruvate) in the presence of bicarbonate and sulfide (Imhoff, 2014a). The presence of vitamin B_{12} has a direct correlation with the synthesis of bacteriochlorophyll: therefore, it is crucial growth factor to grow GSB in laboratory conditions. All GSB can use sulfur compounds as electron donor like sulfide by all GSB, elemental sulfur by most and thiosulfate by few, which gets oxidized to sulfate as a final product. Some GSB oxidizes thiosulfate to sulfate without the formation of intermediate elemental sulfur globules, whereas *Chlorobaculum parvum* forms elemental sulfur as intermediate from thiosulfate and sometimes there is a formation of other intermediates also (Imhoff, 2014a).

A single subunit enzyme sulfide:quinone oxidoreductase (SQR) present in cell membrane oxidize sulfide to polysulfide have many homologs. Dissimilatory sulfite reductase (DSR) oxidize polysulfide to sulfite is absent in early diversing *Chloroherpenton thalassium* which shows that it might have acquired by horizontal exchange (Imhoff, 2014a). Genomes of GSB has revealed that many genes involved in sulfur oxidation were not present in the common ancestor, which indicates that the possibility of horizontal exchange with other organisms. A repertoire of the enzyme in GSB varies between the species which, shows that they have a different evolutionary background. Formation of elemental sulfur externally makes them associated syntrophically with sulfate reducing bacteria like *Desulfuromonas acetoxidans* (Imhoff, 2014b). The co-culture of *D. acetoxidans* and *Prosthecochloris aesturii* was reported at low intensity of light (5-10 lux).

The members of the family *Chlorobiaceae* occurred in the lowermost part of aquatic ecosystems where sulfide (photosynthetic electron donor) and low amount of light is available. As GSB are obligately anaerobic, they are susceptible to even low amount of oxygen and can tolerate high sulfide concentration. The important factors which decide the GSB occurrence are light, oxygen, and sulfide concentration in the stratified environment. GSB are observed as a separate colored layer under the purple bacterial growth, as the presence of chlorosomes and efficient light- harvesting system helps to carry out photosynthesis in low intensity light as well. There is a positive correlation with the abundance of green and brown colored GSB and depth of zones. Green colored GSB are found in the environment lacking chemocline shallow depth of 2 m - 4 m and brown colored at below 9 m, whereas in between there is mixed occurrence of green and brown GSB. Brown colored members of *Chlorobiaceae* have special carotenoids (isorenieratene/â-isorenieratene), which effectively capture light (450 to 550nm) at anoxic depth (Imhoff, 2014b; Takaichi, 2006).

GSB associated with purple sulfur bacteria can be found in anoxic habitat of all types of stagnant water bodies like a freshwater lake, marine lagoons, brackish water and sediments. GSB are the major primary producer and carry out massive development in the deep anoxic layer of the lake. They are also reported from the hypersaline habitat like Black sea. In the marine sediments, the microbial mats are consisting of a distinct layer of GSB and red colored once are purple sulfur bacteria (Imhoff, 2014a). Bright bloom of *Chlorobium limicola, Chlorobium luteolum,* and *Ancalochloris perfilievii* is observed on oxygen/sulfide borderline with low intensities of light in the stratified lakes. Flagella is absent in green sulfur bacteria which corresponds to lack of motility, but some members (*Chloroherpenton thalassium*) shows the gliding type of motility. GSB have gas vesicles which give buoyancy to the cells and helping them to position in low light intensity environment. With the help of symbiotic associate GSB are flexible to move with the help of central motile bacterium which is attached with few cells (6-12) of GSB. The phototactic and metabolic response of the community helps them to position in favorable conditions. Sulfur spring ecosystem is a favorable habitat for GSB due to continuous sulfur flow in the ecosystem. *Chlorobaculum tepidum* is reported from the sulfur spring, which can grow at 55 °C (optimum is 48 °C; Imhoff, 2003). GSB are abundant in the coastal sediments having reduced sulfate and exposed to light. There is massive bloom formation in the marine habitat such as closed bays, shallow estuaries water, and lagoons. *Chloroherpenton thalassium* and *Prosthecochloris* spp. are typically found in the marine and hypersaline environment whereas some GSB are adapted to low salt concentration as well (Imhoff, 2014a).

Green non-sulfur bacteria

Green non-sulfur bacteria (GNSB) are a group of filamentous gliding bacteria carrying out the anoxygenic photosynthesis. First GNSB was described by Pierson and Castenholz in 1974, which was isolated from the microbial mats of the hot spring (Castenholz and Pierson, 2006). Phylogeny of 16S rRNA gene showed that GNSB forms separate branched lineage forming phylum *Chloroflexi* which also harbor few chemotrophic bacterial members. They are green or orange colored Gram-stain-negative filamentous bacteria having gliding motility and photoheterotrophic metabolisms, which separate them from other anoxygenic phototrophic bacteria. Cultivated members of the order *Chloroflexales* are obligate or facultative anoxygenic phototrophs and have versatile physiology which helps them to adapt to the changing environment by the tactic response and gliding motility. Metagenomics studies have shown their presence in the oxic pelagic region of the ocean, freshwater, sediments, lakes, aphotic zones of the deep sea, deep biosphere, endolithic communities, periodontal microbiota, marine sponges, tubeworms and wastewater treatment plant (Hanada and Pierson, 2006). The members of the genus *Chloroflexus, Oscillochloris Heliothrix, Roseiflexus,* and *Chloronema* are filamentous in nature; therefore it is also known as filamentous anoxygenic phototrophic (FAP) bacteria. *Oscillochloris, Chloronema,* and *Chloroflexus* possess the light harvesting organelle called chlorosomes, which is a distinct membrane-bound ovid structure attached to the cytoplasmic membrane (Overmann, 2008b).

The cells of *Chloroflexi* are Gram-stain-negative in nature except for *Oscillochloris chrysea,* which shows Gram-stain-positive character (Overmann, 2008b). These bacteria may lose the gliding motility with the subsequent subculturing. Cells are arranged in a single row, undergoes division by fission and filaments are not branched. There is similarity between the cell wall of *Chloroflexi* and Gram-stain positive bacteria (Sutcliffe, 2010). Although the cell wall is composed of outer membrane and lipoproteins, GNSB has the peptidoglycan layer made up of L-ornithine and cross linking to polysaccharides (via muramic acid 6-phosphate) which makes cell wall distinct from other Gram-stain-negative bacteria (Overmann, 2008b). The green color of the culture is because of the presence of bacteriochlorophyll *c* and *d*, carotenoid (β- and γ-carotene) and their derivatives (Takaichi, 2006). The cells of *Choroflexus aurantiacus* acclimatize itself to the low light by increasing the number of chlorosomes by synthesizing more bchl *a* and *c*. The strains of *Heliothrix oregonensis* contain only bchl α and oxo- γ-carotene which impart orange to red color to the cell. Members of the phylum *Chloroflexi* harbors type II reaction center (quinone/pheophytin -type) in the cytoplasmic membrane. The reaction center proteins are having L and M-subunit, whereas H-subunit is absent in GNSB (Glaeser and Overmann, 1999). These bacteria depend on electron donors having more negative redox potential than water. They prefer photoorganoheterotrophy and can use various simple organic substrate (pyruvate, glycerol, glutamate, acetate, and other simple carbon sources). Growth of

Chloroflexus spp. can be enhanced by yeast extract and casamino acids in the media. GNSB metabolize organic carbon by tricarboxylic acid cycle and glyoxylate cycle in anaerobic conditions (Overmann, 2008b). Strains of *Chloroflexus aurantiacus* can grow photoautotrophically by sulfide dependent manner or on hydrogen. The members of the GNSB harbor a unique carbon dioxide fixation pathway known as hydroxypropionate cycle. *Chloroflexus aurantiacus* is the first microorganism reported to harbor hydroxypropionate cycle in which CO_2 is fixed by carboxylation of acetyl-CoA to glyoxylate forming hydroxypropionyl-CoA as intermediate (Hügler and Fuchs, 2005). Similarly, these bacteria can assimilate other organic substrate and can use it as electron donors, which is unique property among the other phototrophs.

Chloroflexus aurantiacus, Chloroflexus aggregans, Heliothrix oregonensis and *Roseiflexus castenholzii* are thermophilic in nature (van Der Meer et al., 2010). *Chloroflexus aurantiacus* can grow at an optimum temperature of 52 °C to 60 °C (ability to grow at 74 °C) and slightly alkaline pH (7.6 to 8.4). *Chloroflexus aurantiacus* is reported to have a gliding velocity of 0.01 to 0.04μm s^{-1} and are resistant to UV-C radiation, which is lethal to many microorganisms (Overmann, 2008b). GNSB can position themselves along the gradient of light and sulfide concentration by gliding motility and tactic response. Microbial mats containing GNSB with long filaments reduce the erosion and makes the sediment matrix a stable structure. The unique properties of *Chloroflexus aurantiacus* help to selectively enrich in the hot spring environment and geothermal environment (brownish-orange layer; van Der Meer et al., 2010). GSB grow beneath the cyanobacterial layer and utilize exudates of cyanobacteria and its anaerobic decomposed products. Metagenomics studies have shown the presence of GNSB in a lower layer of microbial mats from hot spring, which are not cultivated yet indicate the tremendous uncultivable diversity of GNSB (Overmann, 2008b). *Chlorofelxus*-like bacteria can grow chemotrophically in the presence of oxygen where predominant carotenoids are echinenone and myxobactone. The synthesis of bacteriochlorophyll and chlorosomes is decreased, whereas β-carotene and γ-carotene (it's derivative) are more in phototrophic growth (Overmann, 2008b; Takaichi, 2006). The metabolic flexibility of these bacteria gives a selective advantage to colonize in benthic environments, where there is a varying concentration of oxygen and sulfide. Chloroflexi-like microorganisms are reported from the marine microbial mats, hypersaline habitat (up to 16% w/v NaCl concentration) and associated with cyanobacteria and purple sulfur bacteria (Imhoff, 1995). The metagenomics study of the hypersaline microbial mat of Guerrero Negro has shown 40% occurrence of GNSB majority of which belongs to novel phylotypes (Harris et al., 2013). In the microbial mat, GNSB are responsible for respiration till the depth of 1 mm, whereas the role of heterotrophic bacteria is negligible. *Roseiflexus castenholzii* are reported from neutral to alkaline hot springs like Japanese hot spring and Yellow Stone National Park (van Der Meer et al., 2010). Members of the genus *Herpetosiphon* are reported from the benthic sediments, decaying organic matter, soil environments, the mouth of a river, water bodies close to cities, sewage treatment plant, (steppe and semi-desert habitat) and freshwater (Overmann, 2008a). They are responsible for the flocculation in the sewage treatment plant. The members of class *Anaerolineae* reported from rice paddy soil, thermophilic digester and methanogenic granules where they maintain granular structure by forming a web-like structure on the outer layer of sludge granules (Yamada and Sekiguchi, 2018).

GNSB are phylogenetically distinct from other anoxygenic phototrophic bacteria (Imhoff, 1995). The phylogenetic relationship has not shown the deeper branching compared to other bacterial divisions, which indicate that their ancestors are more ancient than other phototrophs (Overmann, 2008b). The light harvesting antenna of *Chloroflexus aurantiacus* is phylogenetically related to that of α-*Proteobacteria* members. It is hypothesized that the reaction center of GNSB is evolved after the divergence of cyanobacteria, whereas the chlorosomes proteins of GSB and GNSB have a common evolutionary origin (Blankenship et al., 1995). Some hypothesis proposes that these are the chimeric microorganisms formed by the fusion of two and not a descent of early evolution. The composition of lipids (verrucosan-2β-ol and verrucosan) and carotenoids are unique in these bacteria (Takaichi, 2006).

Photosynthesis in anoxygenic phototrophic bacteria

Bacterial photosynthesis is the conversion of radiant energy into chemical energy by utilizing carbon dioxide and reductant without evolving the oxygen except in cyanobacteria. The photosynthetic reaction is divided into two types; light reaction and dark reaction. In light reaction, the radiant energy is trapped and converted into the chemical energy in the form of ATP and NADPH, whereas in dark reaction, energy molecules are synthesized from carbon dioxide using ATP and NADPH. Photosynthetic bacteria have photosynthetic reaction center and photosynthetic pigments like bacteriochlorophylls, carotenoids, bacteriorhodopsin, and phylobilins which trap the light energy. All the anoxygenic phototrophic bacteria harbor a bacteriochlorophyll as a photosynthetic pigment which is different from the chlorophyll pigments found in Cyanobacteria, algae, and plants. In 1932, C.B. van Neil first time discovered the Bacteriochlorophylls (Gest and Blankenship, 2004). Chlorophyll and bacteriochlorophyll have the distinct magnesium porphyrin with tetrapyrrole ring and phytol tail. The difference lies in substitution around the ring, presence of double bond, and length of phytol ring. There is eight type of bacteriochlorophyll which is based on the presence or absence of double bond at C_3 and C_4 and can absorb at a longer wavelength. The distribution of different bacteriochlorophyll in anoxygenic phototrophic bacteria is shown in Table 2. Carotenoids are the accessory pigments as well as play a key role in photoprotection in phototrophic bacteria. Carotenoids are lipophilic isoprenoids with hydroxyl or aldehyde group and embedded in a membrane having a red-brown, green, yellow color, and absorb light in the blue region (Takaichi, 2006). Major carotenoid of anoxygenic phototrophic bacteria is listed in Table 3. Some halophiles have a carotenoid like proteins (bacteriorhodopsin) which carry out the light mediated ATP synthesis and cyanobacteria have a water soluble protein called phycobilins as chromophores on the outer surface of the thylakoid membrane.

Different APB harbor diverse antenna structures and pigments (Madigan and Jung, 2008) for performing the process of photosynthesis. Reaction center is surrounded by antenna complexed, which forms photosynthetic apparatus. It is encoded by a cluster of genes arranged in an organized manner known as Photosynthetic gene cluster (PGS). PGS consists of five types of genes involved in bacteriochlorophyll biosynthesis (*bch*), light harvesting complexes (*puf*), proteins of reaction centre assembly (*puh*), carotenoid biosynthesis (*crt*) and regulatory genes (Brinkmann et al., 2018). The specific arrangement of these genes is crucial for the regulation of gene expression and adaptation to the changing environments. PGS helps in lateral transfer of these pathways among different phototrophs and after the divergence of APB from other bacteria, complex recombination events had occurred in PGS region (Frigaard 2016). In *Rhodobacter sphaeroides* PGS region span 40.7 kb consisting of thirty-eight open reading frames (ORFs; Fig 2). PGS helps to tightly regulate anoxygenic photosynthesis, especially at the low oxygen concentration. The quantity and intensity of light also an important parameter which regulate the expression of photosynthetic genes in PGS (Brinkmann et al., 2018). The low intensity of light (<1000 lux) upregulate the expression of light harvesting apparatus. PGS has two proteins *ppsR* and *ppaA,* which makes organisms sensitive to light intensity, oxygen tension and helps to avoid photooxidative damage to the cell. There are low transcript levels of *puf* gens at high oxygen concentration, which helps an organism to acclimatize faster to changing conditions (Brinkmann et al., 2018).

Table 2: Different lineages of APB and their pigments and photochemical reaction center

Phylum/class of APB	Physiological group	Type of reaction center	Light harvesting center		
			Location of photosynthesis	Bchl	carotenoids
α and β proteobacteria	PNSB	II	ICM	a/b	+
	AAPB	II	-	a	-
γ-Proteobacteia	PSB	II	ICM	a/b	+
	AAPB	II	-	a	-
Chlorobi	GSB	I	chlorosomes	a; c/d/e	+
chloroflexi	GNSB	I	chlorosomes	a; c/d/c_s	+
Acidobacteria	-	I	chlorosomes	a/b	-
Firmicutes	Heliobacteria	I	-	g	+

AAPB: Aerobic anoxygenic phototrophic bacteria; ICM: intracytoplasmic membrane; Bchl: bacteriochlorophyll; PNSB: GSB: Green sulfur bacteria; GNSB: Green non-sulfur bacteria; Purple non-sulfur bacteria; PSB: Purple sulfur bacteria

Fig. 2: Arrangement of photosynthetic gene cluster in *Rhodobacter spheroides*. Green color: bacterialchlorophyll biosynthesis genes; red color: *puf* and regulators genes; pink color: *puh* genes; orange color: carotenoid (*crt*) genes; yellow color: *lhaA* gene; blue color: *hem* and *cyc* gene; grey color: hypothetical protein blank color: uncertain or unrelated genes (Data not yet published).

Table 3: Major biosynthetic carotenoid groups produced by microorganisms

Carotenoid series	Carotenoids
Spirilloxanthin series	Lycopene, spirilloxanthin, rhodopin
Alternate spirilloxanthin series	Hydroxyneurosporene, spheroidene, spheroidenone (spirilloxanthin)
Rhodopinal series	Lycopene, lycopenol, lycopenal, rhodopin, rhodopinol rhodopinal
Okenone series	Okenone
Isorenieratene series	Isorenieratene, β-carotene
Chlorobactene	Chlorobactene, γ-carotene

Carbon assimilation is the process of carbon dioxide fixation and converting it into organic compounds. For a long time in history, the Calvin-Benson-Bassham (CBB) pathway was the only autotrophic pathway known to fix the CO_2. Whereas with the progress of science, the new carbon fixation pathway was discovered and elucidated to be operated in different organisms. Apart from CBB pathway, there are five more CO_2 fixation pathway; reductive tricarboxylic acid (rTCA) cycle, hydroxypropionate pathway, 3-hydroxybutyrate/4-hydroxybutyrate pathway, reductive acetyl CoA pathway, dicarboxylate/4-hydroxybutyrate pathway (Hügler and Sievert, 2011). In anoxygenic phototrophic bacteria, CBB pathway, rTCA cycle and hydroxypropionate pathway are reported. CBB pathway which is also known as reductive pentose phosphate pathway, occurred in α-, β-, γ-proteobacteria, cyanobacteria, algae and plants. RubisCO is the key enzyme in the pathway which carboxylate the ribulose-1,5-bisphosphate resulting in the formation of phosphoglycerate. In addition to the CBB pathway, the genome of *Rhodovulum viride* JA756 have the genes of C_4 pathway involved in anaplerotic CO_2 assimilation (Khandavalli et al., 2018). Another CO_2 fixation pathway is rTCA pathway, which was first reported in the green sulfur bacterium, *Chlorobium limicola* DSM 245 and restricted to anaerobic and microaerophilic bacteria (Hügler and Sievert, 2011). rTCA cycle needs eight enzymes to convert CO_2 to acetyl Co-A. Five enzymes in TCA cycle can perform reversible reactions whereas three enzymes ATP-citrate lyase, 2-oxoglutarate synthase, and fumarate reductase carry out an irreversible reaction, which is vital in rTCA cycle (Hügler and Sievert, 2011). The symbiont of tubeworm *Riftia* (uncultured gammaproteobacteria) reported to have both CBB pathway and rTCA cycle and operate simultaneously (Thiel et al., 2012). Green-non-sulfur bacteria, *Chloroflexus aurantiacus* have 3-hydroxypropionate pathway, in which acetyl-CoA carboxylase and propionyl-CoA carboxylase carry out two carboxylation reactions (Hügler and Fuchs, 2005). This pathway needs 13 enzymes which convert bicarbonate to pyruvate. Genome of *Rhodovulum viride* JA756 harbor key genes of all the three pathway (CBB pathway, rTCA cycle and hydroxypropionate pathway), whereas experimental proof of the operation of this pathway is lacking (Khandavalli et al., 2018). Another CO_2 fixation pathway is reductive pentose phosphate pathway, which is also known as Wood-Ljungdahl (WL) pathway present in anaerobic acetogenic bacteria, ammonia oxidizing bacteria, autotrophic sulfate reducing bacteria and few species of phylum Spirochaeta. 3-hydroxypropinate/4-hydroxybutyrate pathway and dicarboxylate/4-hydroxybutyrate pathway are reported from the domain *Archaea* (Hügler and Sievert, 2011).

Applications of APB

APB are metabolically and physiologically versatile organisms with potential to produce diverse group of bioactive molecules and novel metabolites of various biotechnological applications (Table 4). APB has application in wastewater treatment, carbon sequestration, biofuel production, bioaccumulation, bioremediation of pollutants, as food products, cosmetic, nutraceutical, and pharmaceutical industries (Sasikala and Ramana, 1995, Frigaard 2016). APB can produce a wide variety of isoprenoid (quinone, carotenoids, and hopanoids) and carotenoids (lycopene and okenone derivatives) molecules of diverse commercial interest (Takaichi, 2006). Mutant of *Rhodospirillum rubrum* produces a high amount of antioxidant carotenoid, lycopene having commercial value as a colorant, pharmaceutical products to treat cancer and heart disease (Frigaard 2016). A modified green sulfur bacterium, *Chlorobaculum tepidum* can produce lycopene and zeta-carotenoids in high amount. A natural carotenoid biosynthetic mutant of *Rhodobacter viridis* JA737 accumulate neurosporene as major carotenoid, which has commercial value as antioxidant and UV protectant (Ramaprasad et al., 2013). APB have the genetic potential to produce other bioactive terpenoids molecules like quinone and hopanoid (Merugu et al., 2012; Tushar et al., 2014, 2015). As APB are anaerobic organisms, there is no need of carotenoids to protect cells from the detrimental effect of light. Therefore, APB can be genetically engineered to channelize the isoprene molecules to other valuable terpenoids.

APB are capable of producing environment friendly hydrogen biofuel. All the phototrophic microorganisms (oxygenic and anoxygenic) have capability to produce molecular hydrogen, but the mechanisms of production are different. There are two key enzymes which have capability to produce hydrogen are hydrogenase and nitrogenase. In most of the PNSB the photoproduction of hydrogen is carried out by nitrogenase enzyme in the nitrogen limiting conditions. In the absence of nitrogen, nitrogenase-dependent production of hydrogen increases in *Rhodospirillum rubrum* (Kern et al., 1994). In green sulfur bacteria, inorganic sulfur is the source of electron donor for photosynthesis and hydrogen is produced by light and nitrogenase-dependent manner. A syntrophic co-culture of *Chlorobium vibrioforme* and *Desulfuromonas acetoxidans* (acetate oxidizing and sulfur reducing bacterium) produces hydrogen photobiologically (Frigaard 2016). PNSB is also a model host organism for high expression of membrane proteins. Expression of heterologous membrane protein in *Escherichia coli* results in lower titer and form inclusion bodies, which makes it hard to purify these proteins. PNSB have intracytoplasmic membrane vesicles where synthesized foreign proteins accumulate, which makes it easy for the purification of this protein after cell lysis. The functionally intact proteins like Occludin (*Occ*) a human tight junction protein, aquaporin g (*hAQpg*), cellulase synthase and cytochrome c_y were overexpressed in *Rhodobacter sphaeroides*. Similarly, PNSB is widely used to study the function and overexpression of challenging membrane proteins (Frigaard 2016).

Table 4: Biotechnological application of anoxygenic phototrophic bacteria

Sr.No.	Product	Application	Reference
1.	Single- cell protein	Protein source	Sasaki et al., 2002
2.	Metal ion uptake	Metal recovery	Talaiekhozani and Rezania, 2017
3.	Vitamin B_{12}	Vitamin	Merugu et al., 2012
4.	Hydrogen	Fuel	Sasikala et al., 1992
5.	NH_3	Fertilizer	Sasikala and Ramana, 1995
6.	Carotenoid	Natural dye	Noparatnaraporn et al., 2001
7.	Biodegradation of organic and inorganic compounds	Wastewater treatment	Blasco and Castillo, 1993
8.	Hopanoids	Therapeutics	Nagumo *et al.*, 1991
9.	Poly β-hydroxy alkanoaotes	Biodegradable plastic	Brandl *et al.* 1991
11.	Hydrogen	Fuel	Sasikala et al., 1992

APB offers an environment friendly approach to remove the toxic, corrosive, and malodorous sulfide compounds from the wastewater and gas streams (Hurse et al., 2008). APB partially oxidize hydrogen sulfide to elemental sulfur and not to sulfate which has various benefits as; elemental sulfur can be easily removed by sedimentation and has commercial value as fertilizer and feedstock, which generate less corrosive H^+ and operate at low light (Frigaard 2016). Green sulfur bacteria have an advantage over other purple bacteria as it converts more sulfide per light unit, tolerates high sulfide concentration and uptake is not decrease in the presence of organic carbon. GSB, *Chlorobium limicola* DSM 257 was used in pilot-scale study, which can remove 100% sulfide, and 92-95% of elemental sulfur was recovered (Gregersen et al., 2011). Biogas plant also releases hydrogen sulfide in small amounts which can be removed by chemical techniques and chemotrophic sulfide oxidizing bacteria. Photobiological removal using anoxygenic phototrophic bacteria can be eco-friendly and cost-effective alternative as they can operate continuously and at high sulfide concentration (Gregersen et al., 2011).

Purple non-sulfur bacterium *Rhodobacter sphaeroides* can accumulate heavy metals (lead, nickel, cadmium, cobalt, chromium, mercury, and zinc; Talaiekhozani and Rezania, 2017). These bacteria produce a high amount of extracellular polymeric substances on the surface having negative charges which absorb different metal ions. Immobilized cells of the anoxygenic phototrophic bacterium *Rhodobacter sphaeroides* SSI are also able to accumulate radioactive isotopes of strontium (Sr) and cesium (Cs; Sasaki et al., 2013). These phototrophs have the metabolic capability to degrade recalcitrant dyes and pesticides which are xenobiotic. Azo dyes are the synthetic colorant and one of the major pollutant in wastewater streams. Many PNSB have genetic ability to reduce the azo bonds and decolorize them by an enzyme azoreductase. *Rhodopseudomonas palustris* efficiently decolorize and degrades the azo dyes under illuminated anaerobic conditions. Various strains of *Rhodobacter capsulatus, Rhodobacter blasticus, Rhodovulum adriaticum, Rhodovulum strictum,* and *Rhodopseudomonas palustris* can decolorize most of the dyes tested in anaerobic conditions (Frigaard 2016). Similarly, green sulfur bacteria are capable of removing 60 to 90% of the azo dyes from the polluted water streams. Members of the genus *Rhodospirillum* and *Rhodopseudomonas* have the potential to degrade pollutant, halogenated aromatic compounds like 3-chlorobenzoate (Mutharasaiah et al., 2010).

Preservation of APB

APB are difficult to preserve by the conventional methods because they are sensitive to oxygen, sulphide and light. Few purple non-sulphur bacteria can grow in microaerophilic conditions by chemoautotrophically and chemoheterotrophically: therefore can be stored by conventional methods (Malik, 1992). Initially, reducing agents like sulphide and cysteine was used to preserve phototrophic cells but they have lethal effect on cells (Malik, 1990). Porphyrin is photochemically active reagent absorb the light and protect the cells against photooxidation during lyophilization (Malik, 1984). Prolonged exposure of porphyrins generate singlet oxygen, which is toxic to cell. Widely used cryoprotectant like glycerol, and skim milk are not effective for preservation of APB. Dimethyl sulfoxide (DMSO; 5%) in water is mostly used for the preservation of APB. DMSO increases the viability as it has capability to diffuse and interact with phospholipids which maintaining membrane permeability during freezing (Esteves-Ferreira et al., 2013). Activated charcoal creates reduced environment and protect cells from photooxidation and therefore, useful in freeze drying of APB (Malik, 1990).

Conclusion

Versatile nature of APB is demonstrated by diverse physiology, morphology, several types of bacteriochlorophylls, different carotenoids, and phylogenetic relationship. APB are reported from diverse ecological niches such as freshwater lakes, lagoons, estuaries, ponds, sediments of ditches, moist soils, intertidal zones, hypersaline environments, alkaline environment, acid water, hot springs, Arctic and Antarctica. APB are non-homogenously distributed in phylum *Proteobacteria* (purple bacteria in includes class α-, β-, γ-*Proteobacteria* with interspersing chemotrophs), *Chlorobi* (green sulfur bacteria), *Chloroflexi* (green non-sulfur bacteria), *Firmicutes* (Heliobacteria) and *Acidobacteria* (*Chloracidobacterium thermophilum*). APB have a potential for the application in removal of H_2S from wastewater and gas streams, hydrogen production, a host for the production of membrane proteins, source of carotenoids and terpenoids, full applications in agriculture, production of bioactive compounds with antimicrobial, antioxidant, anticancer activity and degradation of xenobiotic, toxic metals and radioisotopes. Though APB have the potential for various biotechnological applications yet not exploited as widely as other chemotrophic bacteria. There is a lot of scopes to explore the diversity, distribution, ecological role, photosynthetic pathway, and biotechnological potential of anoxygenic phototrophic bacteria. So, we hope that the knowledge of photosynthesis and diverse group of APB will be useful for students, teachers and researcher's to find new insights into the anoxygenic bacteria for future exploration.

Acknowledgement

TL and PK thanks Department of Biotechnology (DBT), Government of India for financial support to the project "Establishment of Centre of Excellence for National Centre for Microbial Resoruce".

References

Biebl H, Pfennig N. 1981. Isolation of Members of the Family *Rhodospirillaceae*. In: Starr M.P. Stolp H., Trüper H.G., Balows A., Schlegel H.G. (eds) *The Prokaryotes*. Springer, Berlin Heidelberg. 267–73.

Blankenship RE, Madigan MT, Michael T, Bauer CE. 1995. Anoxygenic photosynthetic bacteria. Kluwer Academic Publishers.

Blasco R. Castillo F. 1993. Characterization of a nitrophenol reductase from the phototrophic bacterium *Rhodobacter capsulatus* E1F1. *Applied and Environmental Microbiology* 59: 1774–1778.

Brandl H, Gross RA, Lenz RW, Lloyd R, Fuller C. 1991. The accumulation of poly (3-hydroxyalkanoates) in *Rhodobacter sphaeroides*. *Arch. Microbiol*. 155: 337-340.

Brinkmann H, Göker M, Koblí•ek M, Wagner-Döbler I, Petersen J. 2018. Horizontal operon transfer, plasmids, and the evolution of photosynthesis in *Rhodobacteraceae*. *The ISME Journal* 12: 1994–2010.

Bryantseva I, Gorlenko VM, Kompantseva EI, Imhoff JF, Suling J, Mityushina L. 1999. *Thiorhodospira sibirica* gen. nov., sp. nov., a new alkaliphilic purple sulfur bacterium from a Siberian soda lake. *Int. J. Syst. Bacteriol.* 49: 697-703.

Castenholz RW, Pierson B.K. 1995. Ecology of thermophilic anoxygenic phototrophs. In: Blankenship R.E., Madigan M.T., Bauer C.E. (eds) Anoxygenic photosynthetic bacteria. advances in photosynthesis and respiration, vol 2. Springer, Dordrecht.

Esteves-Ferreira AA, Corrêa DM, Carneiro APS, Rosa RM, Loterio R, Araújo WL. 2013. Comparative evaluation of different preservation methods for cyanobacterial strains. *J. Appl. Phycol.* 25: 919–929

Evans MC, Buchanan BB, Arnon DI. 1966. A new ferredoxin-dependent carbon reduction cycle in a photosynthetic bacterium. *Proc. Natl. Acad. Sci.* 55: 928–934.

Frigaard NU. 2016. Biotechnology of anoxygenic phototrophic bacteria. *Adv Biochem Eng Biotechnol.* 156: 139–54.

Gandham S, Tushar L, Chintalapati S, Chintalapati VR. 2018. *Rhodobacter alkalitolerans* sp. nov., isolated from an alkaline brown pond. *Arch. Microbiol.* 200:1487-1492.

Gest H, Blankenship RE. 2004. Time line of discoveries: Anoxygenic bacterial photosynthesis *Photosynthesis Research*. 80: 59-70.

Glaeser J, Overmann J. 1999. Selective enrichment and characterization of *Roseospirillum parvum*, gen. nov. and sp. nov., a new purple non-sulfur bacterium with unusual light absorption properties. *Arch. Microbiol.* 171: 405-416.

Gregersen LH, Bryant DA, Frigaard NU. 2011. Mechanisms and evolution of oxidative sulfur metabolism in green sulfur bacteria. *Front. Microbiol.* 24:116.

Hanada S, Pierson BK. 2006. The Family *Chloroflexaceae*. In: Dworkin M., Falkow S., Rosenberg E., Schleifer KH., Stackebrandt E. (eds) *The Prokaryotes*. Springer, New York, NY.

Hansen TA, van Gemerden H. 1972. Sulfide utilization by purple non-sulfur bacteria. *Arch. Mikrobiol.* 86: 49-56.

Harada N, Nishiyama M, Otsuka S, Matsumoto S. 2005. Effects of inoculation of phototrophic purple bacteria on grain yield of rice and nitrogenase activity of paddy soil in a pot experiment. *Soil Sci. Plant Nutr.* 51: 361-367.

Harris JK, Caporaso GJ, Walker JJ, et al. 2013. Phylogenetic stratigraphy in the Guerrero Negro hypersaline microbial mat. *The ISME J.* 7: 50-60.

Hiraishi A, Okamura K. 2017. *Rhodopseudomonas telluris* sp. nov., a phototrophic alphaproteobacterium isolated from paddy soil. *Int. J. Syst. Evol. Microbiol.* 67: 3369-3374.

Hiraishi A, Ueda Y. 2009. Intrageneric structure of the genus *Rhodobacter*: transfer of *Rhodobacter sulfidophilus* and related marine species to the genus *Rhodovulum* gen. nov. *Int. J. Syst. Bacteriol.* 44: 15-23

Hügler M, Fuchs G. 2005. Assaying for the 3-hydroxypropionate cycle of carbon fixation. *Methods Enzymol.* 397: 212-221.

Hügler M, Sievert SM. 2011. Beyond the Calvin cycle: Autotrophic carbon fixation in the ocean. *Ann. Rev. Mar. Sci.* 3: 261-289.

Hurse T.J., Kappler U., Keller J. 2008. Using anoxygenic photosynthetic bacteria for the removal of sulfide from wastewater. In: Hell R, Dahl C, Knaff D, Leustek T. (eds) Sulfur metabolism in phototrophic organisms. *Advances in Photosynthesis and Respiration*, vol 27. Springer, Dordrecht.

Imhoff J.F. 1995. Taxonomy and physiology of phototrophic purple bacteria and green sulfur bacteria. In: Blankenship RE, Madigan MT, Bauer CE. (eds) Anoxygenic photosynthetic bacteria. *Advances in Photosynthesis and Respiration*, vol 2. Springer, Dordrecht

Imhoff JF. 2003. Phylogenetic taxonomy of the family *Chlorobiaceae* on the basis of 16S rRNA and fmo (Fenna-Matthews-Olson protein) gene sequences. *Int. J. Syst. Evol. Microbiol.* 53: 941-951.

Imhoff J.F. 1992. The family *Ectothiorhodospiraceae.* In: Balows A, Trüper HG, Dworkin M, Harder W, Schleifer KH. (eds) *The Prokaryotes*. Springer, New York, NY.

Imhoff JF. 2014a. Biology of Green Sulfur Bacteria. In: eLS. John Wiley & Sons, Ltd: Chichester, 1-12.

Imhoff J.F. 2014b. The family *Chlorobiaceae.* In: Rosenberg E, DeLong EF, Lory S, Stackebrandt E, Thompson F. (eds) *The Prokaryotes*. Springer, Berlin, Heidelberg.

Kern M, Klipp W, Klemme JH. 1994. Increased nitrogenase-dependent H_2 photoproduction by *hup* mutants of *Rhodospirillum rubrum. Appl. Environ. Microbiol.* 60: 1768-1774

Khandavalli LVNS, Tushar L, Abdullah M, Guruprasad L, Chintalapati S, Chintalapati VR. 2018. Insights into the carbonic anhydrases and autotrophic carbon dioxide fixation pathways of high CO_2 tolerant *Rhodovulum viride* JA756. *Microbiol. Res.* 215: 130–40.

Madigan MT. 1986. *Chromatium tepidum* sp. nov., a thermophilic photosynthetic bacterium of the family *Chromatiaceae. Int. J. Syst. Bacteriol.* 36 222–227.

Madigan MT, Jung DO. 2009. An overview of purple bacteria: Systematics, physiology, and habitats. In: Hunter CN, Daldal F, Thurnauer MC, Beatty JT. (eds) The Purple Phototrophic Bacteria. *Advances in Photosynthesis and Respiration*, vol 28. Springer, Dordrecht.

Malik KA. 1992. A new method for preservation of microorganisms by liquid-drying under anaerobic conditions. *J. Microbiol. Methods.* 14: 239-245.

Malik KA. 1990. Use of activated charcoal for the preservation of anaerobic phototrophic and other sensitive bacteria by freeze-drying. *J. Microbiol. Methods.* 12: 117-124.

Martínez-Alonso M, van Bleijswijk J, Gaju N, Muyzer G. 2005. Diversity of anoxygenic phototrophic sulfur bacteria in the microbial mats of the Ebro Delta: A combined morphological and molecular approach. *FEMS Microbiol. Ecol.* 52: 339–350.

Merugu R, Rudra MPP, Girisham S, Reddy SM. 2012. Biotechnological applications of purple non sulphur phototrophic bacteria: A minireview. *Int. J. Appl. Biol. Pharm. Technol.* 3, 376–84.

Mutharasaiah K, Govindareddy V, Karigar C. 2010. Photobiodegradation of halogenated aromatic pollutants. *Adv. Biosci. Biotechnol.* 01, 238–40.

Nagumo A, Takanashi K, Hojo H, Suzuki Y, 1991. Cytotoxicity of bacteriohopane-32-ol against mouse leukemia L1210 and P388 cells *in vitro. Toxicol. Lett.* 58: 309-313.

Noparatnaraporn N, Takeno K, Sasaki K. 2001. Hydrogen and poly- (hydroxy) alkanoate production from organic acids by photosynthetic bacteria. *Biohydrogen II An Approach to Environmentally Acceptable Technology*. 2001: 33-40

Okamura K, Hisada T, Kanbe T, Hiraishi A. 2009. *Rhodovastum atsumiense* gen. nov., sp. nov., a phototrophic alphaproteobacterium isolated from paddy soil. *J. Gen. Appl. Microbiol.* 55: 43-50.

Overmann J. 2008a. Ecology of phototrophic sulfur bacteria. In: Hell R, Dahl C, Knaff D, Leustek T. (eds) Sulfur metabolism in phototrophic organisms. *Advances in Photosynthesis and Respiration*, vol 27. Springer, Dordrecht.

Overmann J. 2008b. Green non-sulfur bacteria. *In: eLS. John Wiley & Sons, Ltd: Chichester* 1–10.

Petri R, Imhoff JF. 2001. Genetic analysis of sea-ice bacterial communities of the Western Baltic Sea using an improved double gradient method. *Polar Biol.* 24: 252–257.

Pfennig N. 1977. Phototrophic green and purple bacteria: a comparative, systematic survey. *Annu. Rev. Microbiol.* 31, 275–90.

Pfennig N, Truper HG. 1971. Higher taxa of the phototrophic bacteria. *Int. J. Syst. Bacteriol.* 21: 17-18.

Pfennig N, Trüper H.G. 1981. Isolation of members of the families *Chromatiaceae* and *Chlorobiaceae.* In: Starr MP, Stolp H, Trüper HG, Balows A, Schlegel HG. (eds) *The Prokaryotes*. Springer, Berlin, Heidelberg 279-289.

Ramana CV, Sasikala C, Arunasri K, et al. 2006. *Rubrivivax benzoatilyticus* sp. nov., an aromatic hydrocarbon-degradiing purple betaproteobacterium. *Int. J. Syst. Evol. Microbiol.* 56: 2157-2164

Ramana VV, Raj PS, Tushar L, Sasikala C, Ramana ChV. 2013. *Rhodomicrobium udaipurense* sp. nov., a psychrotolerant, phototrophic alphaproteobacterium isolated from a freshwater stream. *Int J Syst Evol Microbiol.* 63: 2684–2689.

Ramaprasad EV, Tushar L, Dave B, Sasikala C, Ramana CV. 2016. *Rhodovulum algae* sp. nov., isolated from an algal mat. *Int J Syst Evol Microbiol.* 66: 3367-3371.

Ramaprasad EVV, Sasikala C, Ramana CV. 2013. Neurosporene is the major carotenoid accumulated by *Rhodobacter viridis* JA737. *Biotechnol Lett.* 35: 1093–1097.

Sasaki K, Watanabe M, Tanaka T, Tanaka T. 2002. Biosynthesis, biotechnological production and applications of 5-aminolevulinic acid. *Appl Microbiol Biotechnol.* 58: 23-29.

Sasaki K, Morikawa H, Kisibe T, et al. 2013. Simultaneous removal of cesium and strontium using a photosynthetic bacterium, *Rhodobacter sphaeroides* SSI immobilized on porous ceramic made from waste glass. *Adv. Biosci. Biotechnol.* 4: 6-13.

Sasikala K, Ramana CV. Raghuveer RP. 1992. Photoproduction of hydrogen from the waste water of a distillery by *Rhodobacter sphaeroides* O.U. 001. *International Journal of Hydrogen Energy* 17: 23-27

Sasikala C, Ramana CV. 1995. Biotechnological potentials of anoxygenic phototrophic bacteria. I. production of single-cell protein, vitamins, ubiquinones, hormones, and enzymes and use in waste treatment. *Adv. Appl. Microbiol.* 41, 173–226.

Sattley WM, Madigan MT. 2014. The family *Heliobacteriaceae*. In: Rosenberg E, DeLong EF, Lory S, Stackebrandt E, Thompson F. (eds) *The Prokaryotes*. Springer, Berlin, Heidelberg.

Srinivas A, Vinay Kumar B, Divya Sree B, Tushar L, Sasikala C, Ramana ChV. 2014. *Rhodovulum salis* sp. nov. and *Rhodovulum viride* sp. nov., phototrophic Alphaproteobacteria isolated from marine habitats. *Int J Syst Evol Microbiol.* 64, 957–62.

Sucharita K, Shiva Kumar E, Sasikala C, Panda BB, Takaichi S, Ramana CV. 2010. *Marichromatium fluminis* sp. nov., a slightly alkaliphilic, phototrophic gammaproteobacterium isolated from river sediment. *Int. J. Syst. Evol. Microbiol.* 60: 1103-1107.

Sutcliffe IC. 2010. A phylum level perspective on bacterial cell envelope architecture. *Trends Microbiol.* 18: 464-470.

Takaichi S. (1999) Carotenoids and carotenogenesis in anoxygenic photosynthetic bacteria. In: Frank HA, Young AJ, Britton G, Cogdell RJ. (eds) The photochemistry of carotenoids. *Advances in Photosynthesis and Respiration,* Springer, Dordrecht 8:39-69.

Talaiekhozani A, Rezania S. 2017. Application of photosynthetic bacteria for removal of heavy metals, macro-pollutants and dye from wastewater: A review. *J. Water Process Eng.* 19: 312-321.

Tank M, Bryant DA. 2015. *Chloracidobacterium thermophilum* gen. nov., sp. nov.: An anoxygenic microaerophilic chlorophotoheterotrophic acidobacterium. *Int. J. Syst. Evol. Microbiol.* 65: 1426-1430.

Thiel V, Hügler M, Blümel M, et al. 2012. Widespread occurrence of two carbon fixation pathways in tubeworm endosymbionts: Lessons from hydrothermal vent associated tubeworms from the mediterranean sea. *Front. Microbiol.* 3: 423.

Trüper HG, Pfennig N. 1981. Characterization and identification of the anoxygenic phototrophic bacteria. In: Starr MP, Stolp H, Trüper HG, Balows A, Schlegel HG. (eds) *The Prokaryotes*. Springer, Berlin, Heidelberg 299-312

Tushar LD, Srinivas A, Sasikala C, Ramana CV. 2015. Hopanoid inventory of *Rhodoplanes* spp. *Arch Microbiol.* 197: 861–7.

Tushar LD, Sasikala C, Ramana CV. 2014. Draft genome sequence of *Rhodomicrobium udaipurense* JA643^{T} with special reference to hopanoid biosynthesis. *DNA Res.* 21: 639-647.

van Der Meer MT, Klatt CG, et al. 2010. Cultivation and genomic, nutritional, and lipid biomarker characterization of *Roseiflexus* strains closely related to predominant in situ populations inhabiting yellowstone hot spring microbial mats. *J. Bacteriol.* 192: 3033-3042.

van Niel CB. 1944. The culture, general physiology, morphology, and classification of the non-sulfur purple and brown bacteria. *Microbiol. Mol. Biol. Rev.* 8: 1–118

van Niel CB. 1941. The Bacterial photosyntheses and their importance for the general problem of photosynthesis. In: Nord FF. Werkman CH. *Advances in Enzymology and Related Areas of Molecular Biology*. 1

Vinay Kumar B, Ramprasad EVV, Sasikala C, Ramana CV. 2013. *Rhodopseudomonas pentothenatexigens* sp. nov. and *Rhodopseudomonas thermotolerans* sp. nov., isolated from paddy soils. *Int. J. Syst. Evol. Microbiol.* 63: 200-207.

Yamada T, Sekiguchi Y. 2018. *Anaerolineaceae. Bergey's Manual of Systematics of Archaea and Bacteria*. John Wiley & Sons, Inc., in association with Bergey's Manual Trust.

4

Anaerobic Fungi (phylum *Neocallimastigomycota*) in Ruminant and Non-ruminant Animals: Advances in Taxonomy, Ecology, Cultivation and Biotechnological Potential

Rohit Sharma, Dilip R. Ranade*

National Centre for Microbial Resources (NCMR), National Centre for Cell Science (NCCS), NCCS Complex, Ganeshkhind, Pune-411 021, Maharashtra, India

Abstract

The anaerobic fungi are one of the early organisms in evolution which are commonly found in the digestive tract of ruminating and non-ruminating animals, (including herbivores) deep sea and earth, dumping sites, etc. In animals, these are considered the primary colonizers of the plant biomass ingested by them. Still their taxonomy was unclear for a long period of time. However, with the help molecular techniques, they are now classified under a separate phylum Neocallimastigomycota. *There are 38 species known so far belonging to 11 genera. These have been identified based on both morphological data (like thallus, rhizoids, flagella per zoospore, etc.) and phylogenetic analysis. There is a need to explore the undiscovered anaerobic fungi and look for their potential in industrial applications and solutions to the present environmental problems. It can be achieved by improving techniques of sampling, isolation, maintenance and long term preservation. Animals like herbivores do not have the ability to produce enzymes for degrading the plant biomass and anaerobic fungi help them by producing a range of extracellular enzymes. This ability of anaerobic fungi are of use to humans in bioenergy (biogas production) and biomass degradation (by producing cellulases). Recent advances in metagenomics*

**Corresponding Author: rohit@nccs.res.in*

coupled with new methods of cultivation are helping immensely to know the uncultivated anaerobic fungal populations from the animal gut. The present book chapter discusses the past and present of anaerobic fungi, lacunaes and need for more work on this group of fungi, It focuses on the progress in sampling, isolation and preservation techniques, simultaneously discussing the advances in the application these fungi in biotechnology.

Keywords: *Neocallimastigomycota*, anaerobic fungi, herbivores, animals identification, preservation.

Introduction

Fungi are one of the oldest known living organisms which not only survive but live comfortably in all kinds of habitats. With due course of time and research, we know that some of the important biological activity on earth takes place beneath, both underneath the water as well as soil. Such habitats have been commonly called as deep biosphere. These deep biosphere habitats may be anaerobic or anoxic and range from few metres to several kilometres below the earth or ocean surface (Lin et al. 2006, Drake and Ivarsson 2017). In anaerobic habitats, free or combined oxygen is not available, and in anoxic habitats oxygen is available in combined forms only. In such anoxic reactions, inorganic electron acceptors (nitrate and sulphate) are used in replacement to oxygen. The information about fungi living in such habitats is poor as compared to the bacteria and archaea because in early times it was believed that fungi cannot survive in the absence of oxygen. However, later it has been repeatedly found that they live comfortably in habitats like below soil, rocks cavities and deep oceanic waters. Some examples of deep biosphere anaerobic habitats include marine sediments, deep-sea hydrothermal vents, sub-seafloor igneous rocks, terrestrial sedimentary and igneous rocks, and in the deep continental crust (Parkes et al. 2005, Connell et al. 2009, Le Calvez et al. 2009; Nagano and Nagahama, 2012, Orsi et al. 2013, Sohlberg et al. 2015, Drake et al. 2017a, Schrenk et al. 2009, Pedersen et al. 2008, Wu et al. 2016, Fredrickson et al. 1995, Ivarsson et al. 2012, 2016a).

It is believed that early life was aquatic and has slowly shifted to terrestrial regions and so has various groups of organisms including fungi. The two groups of fungi *viz.*, zygomycetous and zoosporic fungi are considered to be early colonizers or primitive fungi (Berbee et al. 2017). The zygomycetous fungi were non-flagellated, filamentous fungi marking the transition from aquatic to terrestrial which were mainly associated with plants. The other group i.e. zoosporic fungi (flagellated) which are now recognized in three phyla *Chytridiomycota, Blastocladiomycota* and *Neocallimastigomycota* were either saprotrophs or parasite with plant or animals. Although the fossil records of fungi are less as compared to plants, it is thought that these three groups of zoosporic fungi have separated at an early stage (several million years ago) from plants. In the course of time, members of phylum *Neocallimastigomycota* have made several modifications to adapt to anaerobic life *viz.*, absence of mitochondria, cytochromes and oxidative phosphorylation pathway biochemistry (Yarlett et al. 1986, Youssef et al. 2013). However, even today these fungi accomplish many useful functions in various ecosystems with the help of special organelles called hydrogenosomes, i.e. oxidation of glucose to cellular energy in the absence of oxygen. For example, the mammalian animals including herbivores do not produce the cellulolytic and hemi-cellulolytic enzymes by themselves and other microbes (bacteria, methanogenic archaea and protozoa) including anaerobic fungi help in the degradation of complex materials.

The phylum *Neocallimastigomycota* is thought to be derived from aerobic ancestors and inhabit vertebrate digestive tracts and group along with sister group *Chytridiomycota* (James et al. 2006a,b, Hibbett et al. 2007, Spatafora et al. 2016, Strullu-Derrien et al. 2017). The members of the anaerobic gut fungi were originally discovered in sheep but thereafter, they have been reported from rumen, hindgut, and faeces of ruminant and non-ruminant herbivorous mammals and reptilian animals (Orpin 1975, Youssef et al. 2013) (Fig. 1 A-B). Recently, anaerobic fungi have also been reported from new niches like outside of terrestrial animal gut and inside of sea animal gut (Ivarsson et al. 2016a, Edwards et al. 2017, Picard 2017). Earlier these were considered as protozoans because of the popular belief that fungi are obligate aerobes (Liebetanz, 1910). However, later it was showed that their cell wall is composed of chitin which confirmed their placement in fungal kingdom (Orpin 1977a). Due to evolution over a long period, along with occupying the unique niche of digestive tract, these fungi have also developed unique adaptive features like production of enzymes, structures, metabolite production, etc. These fungi are known to produce lot of enzymes essential for plant biomass decomposition including cellulases, xylanases, mannases, esterases, glucosidases and glucanases (Fliegerova et al. 2015). These features help them to survive in the absence of oxygen or at low oxygen and also perform necessary metabolic functions. They might account for 5-20% of the total microbial biomass in the rumen (Rezaeian et al. 2004). Relatively recently, with the advent of new technologies of cultivation and genomic studies, many of these fungi have been sequenced partially and even complete genomes have been sequenced *viz., Orpinomyces* sp. strain C1A (Youssef et al. 2013). A application point of view, these fungi have a lot of biotechnological potential in biofuel processing and biogas production as they have the capacity to colonize and break down the complex ligno-cellulosic plant materials using various enzymes.

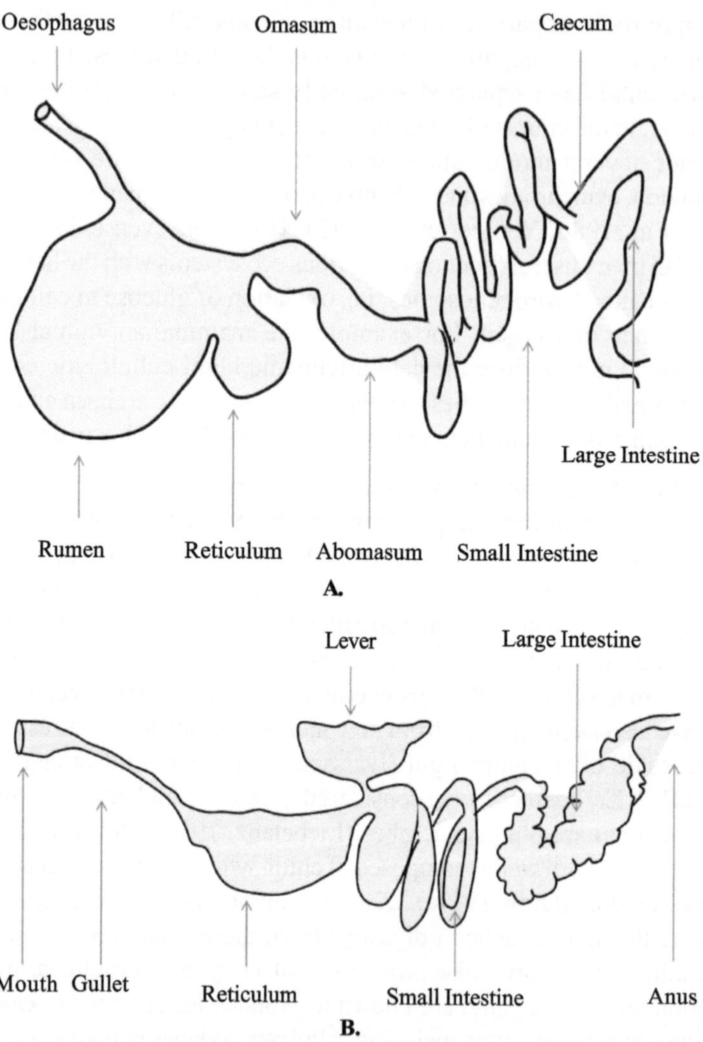

Fig. 1 A-B: Diagrammatic representation of the digestive tract of a ruminant and non-ruminant animals.

The group of fungi are important to study not only because of their evolutionary history but also because of their unique traits like anaerobiosis, presence of hydrogenosomes instead of mitochondria, multiple flagella and extremely low genomic GC content. Even though there have been extensive research that is going on phylum *Neocallimastigomycota,* there are many questions that need to be addressed regarding morphology, taxonomy, diversity, genomics, ecology and the biotechnological importance of gut anaerobic fungi. The present chapter aims to highlight basic techniques used in the isolation and preservation of ruminant and non-ruminant anaerobic fungi, advances in their taxonomy (traditional and present species concepts), their potential in biotechnology and also recent genomic studies in this group of fungi.

Initial Confusion of Anaerobic Fungi as Protozoan

Systematics of any organism covers the classification and identification including its evolutionary history. In this portion of the chapter we discuss the previous incorrect classification of the members of the anaerobic fungi due to lack of information on their structure, physiology and ecology. Earlier these anaerobic fungi were considered as protozoa and thought to be related to zooflagellate parasites. It was first observed by Liebetanz (1910) and later named by Braune (1913) as *Callimastix* (*Callimastix frontalis*) as it was similar to a parasite with the name Cyclops. Later, Vavra and Joyon (1966) formed a new genus *Neocallimastix* and accommodated similar flagella bearing species from the rumen. However, it was Orpin who did extensive studies on their life cycle and chemistry (Orpin 1975, 1976, 1977a, 1977b). Both the genera, i.e. *Callimastix* and *Neocallimastix* were chytrids and Heath et al. (1983) described the species *Neocallimastix frontalis* according to the Botanical Code of Nomenclature. They assigned the genus in the family *Neocallimastigaceae* in order *Spizellomycetales*. Later on researchers studied their ultrastructural, physiological, rRNA gene sequences and ecological differences of anaerobic fungi and they were of the opinion that this group of fungi should be separated from *Spizellomycetales* (Li and Heath 1992, Munn et al. 1988). Later, the family *Neocallimastigaceae* was included in order *Neocallimastigales* in phylum *Chytridiomyota* which consisted of five orders (Li et al. 1993, Heath et al. 1983; Barr et al. 1989). *Neocallimastigales* consisted of all obligately anaerobic fungal species and facultative anaerobes were classified in order *Blastocladiales* (Emerson and Natvig, 1981). The members of phylum *Chytridiomycota* were considered as phylogenetically related to true fungi based on the biochemical characters and 18S rRNA gene sequencing but forms separate clade (Fliegerova et al. 2004, James et al. 2006a; Powell and Letcher, 2012). The members of phylum *Neocallimastigomycota* (present recognized phylum) have many morphological similarities with *Chytridiomycota* but the former differs in their anaerobic physiology and flagellar apparatus. Moreover, the analysis of G + C ratios of the anaerobic fungal DNA and the sequencing of rRNA gene had showed that the rumen fungi are true fungi and not protists (Doré and Stahl 1991, Li and Heath, 1992), and form a monophyletic group with high similarity between species (Orpin and Joblin 1997).

Taxonomy of Anaerobic Fungi and Present Species Concept

The advances in the area of taxonomy and systematics have been influenced by new methods and techniques which help in better resolution of species (Sharma et al. 2015). The taxonomy of anaerobic fungi has been immensely influenced by modern biochemical and molecular methods. The multi-gene dataset phylogenies of DNA sequences of ribosomal RNA regions (18S rRNA gene 28S rRNA gene ITS region) along with protein-coding genes (EF1á, RNA polymerase II subunits RPB1and RPB2) used by Hibbett et al. (2007) led to the separate identity of anaerobic gut fungi and formation of phylum, *Neocallimastigomycota*. At present, taxonomically anaerobic fungi follow the classification proposed by Hibbett et al. (2007) based on the rRNA region phylogeny and evolutionary relationship (Table 1). Box 1 summarizes the present classification in brief.

Box 1

Phylum	: *Neocallimastigomycota*
Class	: *Neocallimastigomycetes*
Order	: *Neocallimastigales*
Family	: *Neocallimastigaceae*

Phylum: *Neocallimastigomycota* M. J. Powell, phylum nov. (MB 501279)

Class: *Neocallimastigomycetes* M. J. Powell, class. nov.

Order: *Neocallimastigales* J. L. Li, I. B. Heath & L. Packer, (1993).

Phylum Characters: *"Thallus monocentric or polycentric; anaerobic, found in digestive system of larger herbivorous mammals and possibly in other terrestrial and aquatic anaerobic environments; lacks mitochondria but contains hydrogenosomes of mitochondrial origin; zoospores posteriorly unflagellate or polyflagellate, kinetosom9e present but non-functional centriole absent, kinetosome- associated complex composed of a skirt, strut, spur and circumflagellar ring, microtubules extend from spur and radiate around nucleus, forming a posterior fan, flagellar props absent; nuclear envelope remains intact throughout mitosis"* (Hibbett et al. 2007).

At present 11 genera of anaerobic fungi are recognised *viz., Anaeromyces, Caecomyces, Cyllamyces, Neocallimastix, Orpinomyces, Piromyces, Buwchfawromyces, Feramyces, Oontomyces, Pecoramyces* and *Liebetanzomyces* (Callaghan et al. 2015, Dagar et al. 2015a, Hanafy et al. 2017, 2018, Joshi et al. 2018). As per index fungorum, there are 38 validly published and recognised species belonging to these 11 genera (Fig. 2, Table 2). To summarize the taxonomy of phylum *Neocallimastigomycota,* it is based on thallus morphology (rhizoidal or bulbous) and zoospore flagellation (monoflagellate or polyflagellate) and their growth patterns (monocentric or

Table 1: The systematics of the chytridiomycetous based on traditional, modern and present classification schemes (Source: modified from Eckart et al. 2010).

Earlier System	Traditional System	Modern System	Present System
Phycomycetes	domain Eukaryota	domain *Eukaryota*	domain *Eukaryota*
	kingdom Fungi (Linnaeus 1753) Nees 1817	kingdom Fungi (Linnaeus 1753) Nees 1817	-
	phylum *Chytridiomycota* von Arx 1967	phylum *Chytridiomycota* von Arx 1967	phylum *Neocallimastigomycota* (Hibbett et al. 2007)
	class *Chytridiomycetes* (de Bary 1863) Sparrow 1958	class *Chytridiomycetes* (de Bary 1863) Sparrow 1958	class *Neocallimastigomycetes*
	order *Chytridiales* Cohn 1879	order *Chytridiales* Cohn 1879	order *Neocallimastigales*
	order *Spizellomycetales* Barr 1980	order *Spizellomycetales* Barr 1980	family *Neocallimastigaceae*
	order *Blastocladiales* Fitzpatrick 1930	order *Rhizophydiales* Letcher 2006	genera *Anaeromyces*
	order *Monoblepharidales* Sparrow 1942	order *Neocallimastigales* Li et al. 1993	*Caecomyces*
		class *Monoblepharidomycetes* Powell 2007	*Cyllamyces*
		order *Monoblepharidales* Sparrow 1942	*Neocallimastix*
		phylum *Neocallimastigomycota* Powell 2007	*Orpinomyces*
		class *Neocallimastigomycetes* Powell 2007	*Piromyces*
		order *Neocallimastigales* Li et al. 1993	*Buwchfawromyces*
		phylum *Blastocladiomycota* James et al. 2006	*Feramyces*
		class *Blastocladiomycetes* James et al. 2006	*Oontomyces*
		order *Blastocladiales* Fitzpatrick 1930	*Pecoramyces*
			Liebetanzomyces

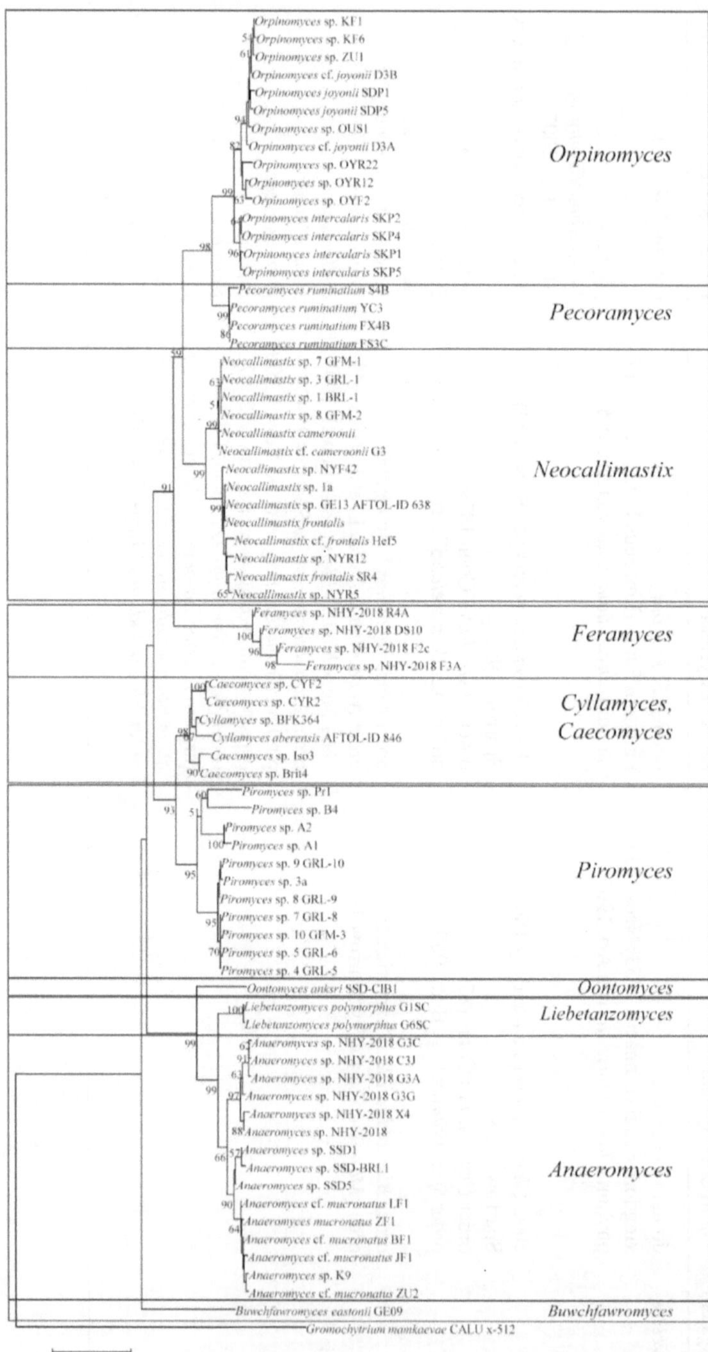

Fig. 2: Phylogenetic tree depicting the present phylogenetic position of all known anaerobic fungi based on large subunit (LSU).

Table 2: Morphological classification of the currently recognised anaerobic fungal genera.

S.No.	Name of Genus	Number of Species	Thallus	Rhizoids	Flagella per zoospore	Any Other Feature
1.	*Anaeromyces*	5	Polycentric	Filamentous	Uniflagellate	-
2.	*Caecomyces*	5	Monocentric	Bulbous	Uniflagellate, sometimes bi- or quadriflagellate	-
3.	*Cyllamyces*	2	Polycentric	Bulbous	Uniflagellate, sometimes bi-or triflagellate	-
4.	*Neocallimastix*	7	Monocentric	Filamentous	Polyflagellate (7–30)	-
5.	*Orpinomyces*	3	Polycentric	Filamentous	Polyflagellate (14–24)	-
6.	*Piromyces*	11	Monocentric	Filamentous	Uniflagellate, sometimes bi- or quadriflagellate	-
7.	*Buwchfawromyces*	1	Monocentric	Filamentous	Uniflagellate	Single terminal sporangium
8.	*Feramyces*	1	Monocentric	Filamentous	Polyflagellate (7–16)	Single terminal sporangium
9.	*Oontomyces*	1	Monocentric	Filamentous	Uniflagellate	Single terminal sporangium
10.	*Pecoramyces*	1	Monocentric	Filamentous	Uniflagellate, sometimes biflagellate	Single terminal sporangium
11.	*Liebetanzomyces*	1	Monocentric	Filamentous	Uniflagellate	Single terminal sporangium

polycentric). Based on growth patterns, the eleven genera within *Neocallimastigomycota* can also be categorized into two groups: monocentric (*Neocallimastix, Piromyces, Caecomyces, Liebetanzomyces, Pecoramyces Oontomyces, Feramyces, Buwchfawromyces*) or polycentric (*Orpinomyces, Anaeromyces, Cyllamyces*). In the monocentric growth patterns the thalli has a single sporangium and in polycentric growth the thalli is complex with multiple sporangia (Ho and Barr 1995, Griffith et al. 2009). Based on zoospore flagellation, the anaerobic fungi can be characterized in to two groups: monoflagellate (*Piromyces, Caecomyces, Anaeromyces, Cyllamyces, Liebetanzomyces, Pecoramyces, Oontomyces, Buwchfawromyces*) or polyflagellate (*Neocallimastix, Orpinomyces, Feramyces*). Based on the morphology, the anaerobic fungi can be characterized as bulbous (*Caecomyces, Cyllamyces*) and rhizoidal (*Anaeromyces, Piromyces, Neocallimastix, Orpinomyces, Liebetanzomyces, Pecoramyces, Oontomyces, Feramyces, Buwchfawromyces*). Table 2 summarizes the characters of all validly published genera and Table 3 lists complete list of validly published species along with their source.

Presently, the taxonomy of phylum *Neocallimastigomycota* is based on polyphasic approach of species concept. To characterize a species and ascertain its taxonomic position one needs to first do microscopic examination for morphological description of the fungus. Along with it phylogenetic analysis is done using internal transcribed spacer (ITS) (~450-500bp) and D1/D2 domains portion of large subunit (LSU) regions (~750bp) of the rRNA of ribosomes. It has been observed that the percent of variation among different species is different in ITS1 and ITS2. Dagar et al. (2015a) observed that in species of *Oontomyces,* ITS1 has 11% variation and ITS 2 has 30% variation in sequences. For morphology, as discussed above the thallus, rhizoids, zoospores, flagella per zoospore and the sporangium are important features needed for the complete description of the novel fungus. Sometimes DAPI (4',6-diamidino-2-phenylindole) staining is done to understand the location and distribution of the nuclei across the thallus (Callaghan et al. 2015). Various researchers have evaluated the substrate utilisation (wheat straw, cellulose, xylan, starch, inulin, raffinose, chitin, alginate, pectin, disaccharide, sucrose, maltose, trehalose, lactose, monosaccharide, xylose, mannose, fructose, arabinose, ribose, glucuronic acid, galactose, peptide, tryptone), enzyme activities (avicelase, CMCase, xylanase, Â-glucosidase) and fermentation product analyses (rice straw, wheat straw, cellulose, xylan, starch, pectin, cellobiose, sucrose, maltose, lactose, glucose, xylose, fructose) for the characterization of various anaerobic fungi (Joshi et al. 2018, Dagar et al. 2018, 2015b, Trinci et al. 1994, 2015b, Edwards et al. 2017).

Table 3: Validly published anaerobic fungal genera and species along with their authority and reference.

S.No.	Name of Genus	Name of Species	Source of Isolations	Reference
1.	*Anaeromyces*	*Anaeromyces contortus* R.A. Hanafy, B. Johnson, M.S. Elshahed & N.H. Youssef (2018)	rumen of *Bos taurus*	Hanafy et al. (2018)
		Anaeromyces elegans (Y.W. Ho) Y.W. Ho (1993)	rumen of *Bos taurus*	Ho et al. (1993)
		A. mucronatus Breton, Bernalier, Dusser, Fonty, B. Gaillard & J. Guillot (1990)	rumen of cow	Breton et al. (1990)
		A. polycephalus (Y.C. Chen, C.Y. Chien & R.S. Hseu) Fliegerová, K. Voigt & P.M. Kirk (2012)	rumen fluid of *Bubalus befullus*	Kirk (2012)
		A. robustus O'Malley, Theodorou & Henske (2016)	digestive tract of ruminants	Li et al. (2016)
2.	*Neocallimastix*	*N. californiae* O'Malley, Theodorou & K.V. Solomon (2016)	digestive tract of ruminants *Capra aegagrushircus*	Li et al. (2016)
		N. cameroonii G.W. Griff., Dollhofer & T.M. Callaghan (2015)	Faeces of Cameroon sheep (*Ovisaries*)	Ariyawansa et al. (2015)
		N. frontalis (RA Braune) Vavra and Joyo´n 1966 ex IB Heath et al. (1983)	-	Heath et al. (1983)
		N. hurleyensis Theodorou & J. Webb (1991)	ovine rumen	Webb and Theodorou (1991)
		N. joyonii Breton, Bernalier, Bonnemoy, Fonty, B. Gaillard & Gouet, (1989)	rumen of sheep	Breton et al. (1989)
		N. patriciarum Orpin & E.A. Munn (1986)	rumen of *Ovis*	Orpin and Munn (1986)
		N. variabilis Y.W. Ho & D.J.S. Barr (1993)	rumen of *Bos indicus*	Ho et al. (1993a)
3.	*Caecomyces*	*C. churrovis* J.K. Henske, S.P. Gilmore & M.A. O'Malley (2018)	faecal pellets of *Ovisaries*	Henske et al. (2017)
		C. communis J.J. Gold, I.B. Heath & Bauchop (1988)	rumen of *Ovis*	Gold et al. (1988)
		C. equi J.J. Gold (1988)	caecum of *Equus*	Gold et al. (1988)
		C. sympodialis Y.C. Chen, S.D. Tsai & C.Y. Chien (2007)	rumen fluid of *Bos indicus*	Chen et al. (2007)

(Contd.)

4.	*Cyllamyces*	*C. aberensis* Ozkose, B.J. Thomas, D.R. Davies, G.W. Griff. & Theodorou 2001	Fresh faeces of cow	Ozkose et al. (2001)
		C. icaris M. Sridhar, Deep.Kumar & Anandan (2014)	rumen of *Bubalus bubalis*	Sridhar et al. (2014)
5.	*Orpinomyces*	*O. bovis* D.J.S. Barr, H. Kudo, Jakober & K.J. Cheng (1989)	tract of Holstein steer	Barr et al. (1989)
		O. intercalaris Y.W. Ho (1994)	rumen contents of *Bos indicus*	Ho et al. (1994)
6.	*Piromyces*	*P. citronii* B. Gaillard, Breton, Dusser & Julliand (1995)	pony and donkey caecum	Gaillard-Martinie et al. (1995)
		P. communis (E. Liebet.) J.J. Gold, I.B. Heath & Bauchop (1988)	stomach of ruminant	Gold et al. (1988)
		P. cryptodigmaticus Fliegerová, K. Voigt & P.M. Kirk (2012)	manure of *Bostaurus*	Kirk (2012)
		P. dumbonicus J.L. Li (1990)	intestine of *Elephantas*	Li (1990)
		Piromyces finnis O'Malley, Haitjema & Gilmore (2016)	digestive tract of horse	Li et al. (2016)
		Piromyces irregularis Fliegerová, K. Voigt & P.M. Kirk (2015)	rumen fluid of slaughtered cow	Ariyawansa et al. (2015)
		P. mae J.L. Li (1990)	intestine of *Equus*	Li (1990)
		P. minutes Y.W. Ho (1993)	rumen contents of *Cervus nippon*	Ho et al. (1993c)
		P. polycephalus Y.C. Chen, C.Y. Chien & R.S. Hseu (2002)	rumen fluid of *Bubalus befullus*	Chen et al. (2002)
		P. rhizinflatus Breton, Dusser, B. Gaillard, Guillot, Millet & Prensier (1991)	dung of ass	Breton et al. (1991)
		P. spiralis Y.W. Ho (1993)	rumen contents of *Capra hircus*	Ho et al. (1993b)
7.	*Buwchfawromyces*	*B. eastonii* T.M. Callaghan & G.W. Griff. 2015	Buffalo	Callaghan et al. (2015a)
8.	*Feramyces*	*F. austinii* R.A. Hanafy, Elshahed & N.H. Youssef 2018	WildBarbary sheep, Fallow deer	Hanafy et al. (2018)
9.	*Oontomyces*	*O. anksri* Dagar, Puniya & G.W. Griff. (2015)	Camel	Dagar et al. (2015)
10.	*Pecoramyces*	*P. ruminantium* Hanafy, N.H. Youssef, G.W. Griff. & Elshahed (2017)	Cattle, Sheep	Hanafy et al. (2017)
11.	*Liebetanzomyces*	*L. polymorphus* Joshi, G.W. Griff. & Dagar (2018)	Goat	Joshi et al. (2018)

The microscopy is done on cultures growing on wheat straw or cellobiose using epifluorescence microscope and images recorded on digital camera through the attached software. Sometimes, the fluorescence microscopy is used to visualize the nuclei which is done and observed by DAPI stain (0.3 mg.ml-1 in 50 mM Tris-HCl, pH 7.2). For visualizing the cell walls and septa Calcofluor white M2R (100 ìM) is used using UV-W filters: 365 nm excitation/420 nm emission (Day et al. 2002, Callaghan et al. 2015). For molecular studies, the mycelial biomass is crushed using pestle mortar or freezing by liquid nitrogen or freeze drying. The DNA is isolated using any of the common methods of fungal DNA isolation like Edwards methods, CTAB method, etc. (Doyle et al. 1987, Griffith and Shaw 1998). For sequencing, ITS and LSU region are amplified and sequenced using various primers. Some of the primers commonly used are mentioned in Table 4 below. The quality of the sequence is checked by Sequence scanner (ABI, USA) and edited using Chromas Pro v1.34 or Geneious (v6.1.6) bioinformatics package (Drummond et al. 2011) or better. Sequence alignment is done by MEGA 7.0.21 or MAFFT (Kumar et al. 2015; Katoh et al. 2002). Phylogenetic analysis can be conducted using TOPALi (v2.5) (Milne et al. 2004) or MEGA (v7.0.21) (Kumar et al. 2016) or PhyML (Guindon et al. 2010) or Mr Bayes (Huelsenbeck and Ronquist 2001) or MrModeltest2 (Nylander 2004). Researchers use Bayesian inference (BI), Neighbour joining (NJ), Maximum Likelihood (ML) and Maximum Parsimony (MP) analyses. Usually most of the researchers use default settings for the phylogenetic analysis.

Diversity

During evolution and course of development of earth, the herbivorous animals developed the symbiosis with microbes which provides enzymes required to digest complex plant structural polysaccharides. In herbivores, there are two types of digestive system, in one the chamber with microbes (called rumen) precedes the stomach and in another the microbial digestion occurs in caecum or large intestine. The research on rumen had started long back in early 1900's, partially because of the importance of domesticated ruminant animals in farming. The ruminant animals of agricultural importance which have been studied are mainly cattle, sheep, camel and goats. However, focus has been more on cattle and sheep as compared to others. According to Van Soest (1994), diet and not the geographical region or species of animal is the determinant factor for microbial diversity inside the rumen. The development of rumen starts after the birth. Initially, the young animal takes milk from the mother and it continues till few months which are digested by stomach. After few months it starts taking some vegetation along with milk. Hence, the microbiota of rumen starts developing with the help of contamination from surrounding vegetation, mother's milk, saliva and even faeces and urine. It develops further with the increase in vegetative

Table 4: Primers and primer sequence commonly used for the sequencing of members of phylum *Neocallimastigomycota.*

S.No.	Region	Name of Primer	Sequence	Reference
1.	LSU	NL1	GCATATCAATAAGCGGAGGAAAAG	Dagar et al. 2011, Fliegerová et al. 2006
		NL4	GGTCCGTGTTTCAAGACGG	
2.	ITS	GM1	TGTACACACCGCCCGTC	Edwards et al. 2008
		MN106	CGTTGTAAAACACTCAWAACC	
		ITS1	TCCGTAGGTGAACCTGCG G	White et al. 1990, Fliegerová et al. 2004, Dagar et al. 2011
		ITS4	TCCTCCGCTTATTGATATGC	

matter of diverse nature ingested by the animal. But in dairies, this may not be true as the young ones and mother are separated at an earlier stage after birth and there is no prolonged contact between them. However, there are animals where no evidence of anaerobic fungi has been found in the gut *viz.*, roe deer (*Capreolus*), muntjac deer (*Muntiacus muntjac*), red brocket deer (*Mazama amaericana*), hippopotamus (*Hippopotamus amphibius*), pigmy hippopotamus (*Choeropsis liberiensis*), giant panda (*Ailuropoda melanoleuca*), domestic pig (*Sus scrofa*), rabbit (*Oryctolagus cuniculatus*), eurpean hare (*Lagus europaeus*), golden hamster (*Mesocricetus* aureus), Mongolian gerbil (*Meriones unguiculatus*), mouse (*Mus domesticus*), rat (*Rattus norvegicus*), coypu (*Myocaster coypus*), koala (*Phascolarctos cinereus*), common brush-tail possum (*Trichosurus vulpecula*), minke whale (*Balaenoptera acutorostrata*), etc. (Orpin 1994). Table 5 summarizes list of anaerobic fungi reported from various ruminating and non-ruminating animals.

The research on anaerobic fungi started way back in early 1990's when Braune (1913) and Hsuing (1930) described *Callimastix frontalis* and *C. equi* as protozoans isolated from horse along with other flagellated protozoans. However in 1950, Weissenberg (1950) suggested that *C. cyclopsis* may be a zoosporic fungus which was confirmed by Vavra and Joyon (1966) by discovering thallus of fungus and new genus *Neocallimastix* was established. The research on these fungi was taken to next level by Orpin (1974, 1975) who showed that these zoosporic were chytrid like fungi and described three anaerobic fungi *viz., Neocallimastix frontalis, Sphaeromonas communis* and *Piromonas communis*. Since the early breakthrough, many workers have been focussing on these group of fungi especially their diversity, taxonomy, physiology and ecology. Metagenomic studies are adding new insight to their physiology and ecology (Wang et al. 2017, Steenbakkers et al. 2001, Orpin and Joblin 1997, Liggenstoffer et al. 2010, Kittelmann et al. 2012, Hanafy et al. 2017, Gruninger et al. 2014, Griffith et al. 2009, 2010, Fliegerova et al. 2010, Drake et al. 2017b). Ivarsson et al. (2016b) had discussed the symbiotic association of anaerobic fungi and bacterial and archaeal methanogens as a good and continuous source of biologically produced hydrogen. Dakar et al. (2015) had discussed about the role of anaerobic fungi in wheat straw degradation and effects of plant feed additives on rumen. Trinci et al. (1994) had given an excellent account of anaerobic fungal taxonomy, physiology and ecology and emphasized their importance in future biotechnology. Relatively recently, there has been growing interest in studying the ecology of anaerobic fungi. It has helped to understand the digestive processes of animal and growth, their food requirements and feed efficiency, fermentation processes of cellulosic material and methane production. With increase in research on human and animal gut microbiome using uncultured

Table 5. Diversity of anaerobic fungi amongst various ruminant and non-ruminant herbivores.

S. No.	Name of Animal	Kind of genera
Ruminant		
1.	Yak (*Bos grunniens*)	*Neocallimastixfrontalis, Orpinomycesjoyonii, Caecomyces*sp.
2.	Camel (*Camelus dromedarius*)	*Oontomycesanksri*
3.	Buffalo (*Bubalus bubalis*)	*Buwchfawromyceseastonii, Piromycesrhizinflatus*
4.	Sheep (*Ovis aries*)	*Callamastixfrontalis, Caecomycescommunis, Neocallimastixfrontalis, N. patriarcum, Piromyces* sp.
5.	Cow (*Bos taurus*)	*Anaeromycesmucronatus, Caecomycessympodialis, Cyllamycesaberensis*
6.	Antelope (*Hippotragus equines*)	-
7.	Goat (*Capra aegagrus hircus*)	*Liebetanzomycespolymorphus*
8.	Fallow deer (*Dama dama*)	*Feramycesaustinii*
9.	Wild Barbary sheep (*Ammotragus lervia*)	*Feramycesaustinii*
10.	Buffalo (*Bubalus bubalis*)	*Buwchfawromyceseastonii*
11.	Arabian oryx (*Oryx leucoryx*)	*Neocallimastix* sp.
12.	Gaur (*Bos gaurus*)	*Neocallimastix* sp.
13.	Grey kangaroo (*Macropus giganticus*)	*Neocallimastix* sp.
14.	Impala (*Aepyceros melampus*)	*Neocallimastix* sp.
15.	Llama (*Lama glama, L. pacos, L. guanicoe*)	*Neocallimastix* sp.
16.	Musk ox (*Ovibos moschatus*)	-
17.	Red deer (*Cervus elaphus*)	-
18.	Redneck wallaby (*M. rufogriseus*)	Unidentified sporangia and vegetative growth observed
19.	Reindeer (*Rangifer tarandus*)	*Neocallimastix* sp.
Non-ruminant		
1.	Elephant (*Elephas maximus, Loxodonta africana*)	*Piromyces* sp.
2.	Horse (*Equus caballus*)	*Caecomyces equi*
3.	Zebra (*Equus burchelli*)	Unidentified sporangia and vegetative growth observed
4.	Donkey (*Equus caballus*)	*Piromyces* sp., *Caecomyces* sp.
5.	Rhinoceros (*Diceros bicornis, Rhinoceros unicornis*)	*Piromyces* sp.
6.	Guinea pig (*Cavia aperea*)	*Sphaeromonas* sp.
7.	Mara (*Diplochotis patagonum*)	*Piromyces* sp.

techniques, it is now known that the kind of feed consumed by the animal is responsible for community structure of animal gut (Liggenstoffer et al. 2010, Fliegerova et al. 2015). Even the anaerobic fungal diversity and richness depends on the kind of feed animal takes (Denman et al. 2008, Kumar et al. 2012, 2013). Generally, two kinds of fungi are found in the gut/ intestine of animals *viz.,* first group consists of facultative anaerobic fungi and aerobic fungi which colonizes initially just as plant biomass enters the gut and second group consists of obligatory anaerobic fungi which are saprotrophic on the dead plant biomass.

In India, anaerobic mycology is in infant stage. There are very few studies focussing on the culturing of anaerobic fungi from various ruminating and non-ruminating animals like camel, Indian yak, Himalayan sheep, goat and indigenous cattle varieties. However, studies on diversity, taxonomy and ecology holds lot of potential in Indian scenario. Because, so far three monotypic genera have been descried from India *viz., Oontomyces anksri* gen. nov., sp. nov., *Liebetanzomyces polymorphus* gen. et sp. nov., *Buwchfawromyces eastonii* gen. nov., sp. nov. isolated from camel, goat and buffalo respectively (Callaghan et al. 2015, Dagar et al. 2015a, Joshi et al. 2018). A long term coordinated program among all the researchers working in the country will help to document the anaerobic diversity of ruminating and non-ruminating animals of India. A categorization between domestic and wild animals, geographical categorization and even at the breed level will help in covering more fungal diversity. There are certain breeds and types of animals which are restricted to a particular area. It will be a good idea to include veterinary clinics, hospitals and biosphere reserve forest office/ veterinary clinics for the same. It is a long gap between first observations in 1910 to 2020, i.e. more than 100 years. We have made more progress in all other fields in much shorter time than in anaerobic fungi.

Cultivation and Preservation

The preliminary techniques developed for the cultivation of anaerobic microbe of gut were first described by Hungate (Hungate 1966, Hungate and Macy, 1973). Later on in due course of time these methods were modified which supplemented the earlier methods of culturing anaerobic fungi (Balch and Wolfe 1976, Bryant 1972, Miller and Wolin 1974, Theodorou et al. 1990, 1995). For culturing anaerobic fungi, serum tubes or bottles sealed with butyl rubber stoppers and aluminium crimp seal or Hungate-type tube with screw cap and butyl rubber septum or Balch Tubes with butyl rubber stopper and aluminium crimp seal to hold stopper in place (Fig. 3 A-D).

a. Sampling and isolation techniques

Fresh faecal samples and/ or solid and liquid contents are obtained from ruminating and non-ruminating animals. Samples are collected either from rumen

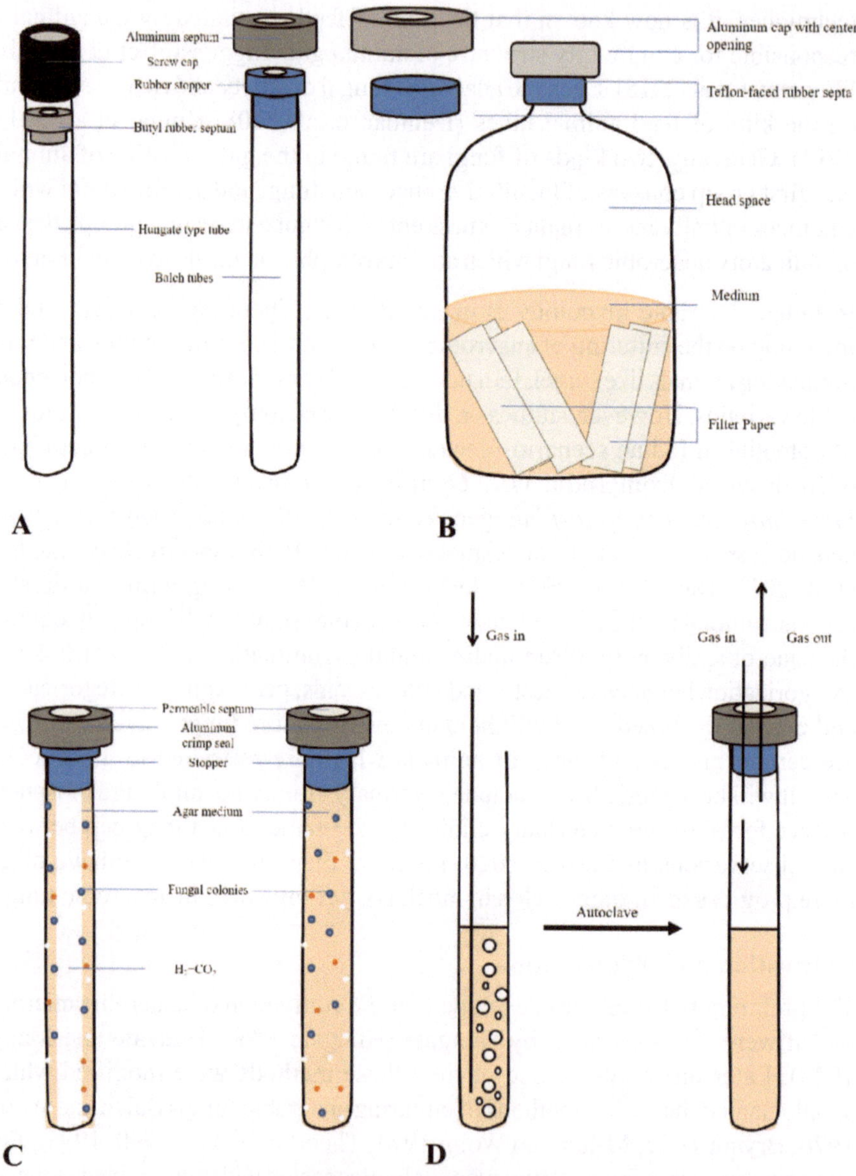

Fig. 3. A-B: Basic requirements of isolation of anaerobic fungi, Hungate-type tube with screw cap and butyl rubber septum and Balch tube with butyl rubber stopper and aluminium crimp seal. **C.** Roll tube is used to isolate anaerobic fungi. It is similar to Petri dish of aerobic fungi wherein cultures are grown on inner wall of tubes coated with thin layer of medium under anaerobic conditions. After incubation, colonies grown on the walls and can be picked up for purification. **D.** For anaerobic culture enrichment, the medium in Hungate tubes are bubbled with an oxygen-free gas before autoclaving and flushing the headspace with a gas like CO_2 to remove excess O_2. Lastly, carbonate, phosphate, trace elements, vitamins are added.

fistulae or using stomach pipe or from butchery or fresh animal faecal sample. Many times, it is easier to collect fresh intestinal ingredients from the butcheries. The samples are brought to the laboratory in a pre-warmed and O_2-free (gassed with N_2 and/ or CO_2) sampling bag or container or tubes (Dagar et al. 2015a, Callaghan et al. 2015) Many researchers fill the sampling tube with the faecal material or intestinal fluid to the top which displaces air keeping the sample near to anoxic conditions. Samples should be transferred to anaerobic enrichment medium as soon as possible, whether the sampling site is local (within 10-15min) or outstation (within 24 h) (Hanafy et al. 2017, 2018). As is known in all studies, sampling, sample storage conditions and their transportation are an essential portion of isolation. Most of the damage to the sample occurs during the above processes, if they are not maintained as per the need of the target microbes. Similarly for anaerobic fungi, maintenance of anaerobic conditions is utmost essential.

For isolations, the faeces or intestinal fluid/ solid are serially diluted and used as inoculums in Hungate tubes with anaerobic medium supplemented with switch grass or filter paper or straw as carbon source (1% v/v) and antibiotic like chloramphenicol. These are than incubated at desired temperature and duration (39±0.1°C for 72-96 h or more) for enrichment (Hooker et al. 2018). Matt growth of fungal cultures are obtained in 3-4 weeks which are then subsequently repeatedly sub-cultured 2-3 times on new enrichment medium to get unclouded fungal cultures. Pure cultures of anaerobic fungi are obtained by inoculating liquid fungal culture in roll tubes prepared by adding agar (2% w/v), glucose (0.45% w/v), and chloramphenicol (25 ìg/ml) to anaerobic medium under 100% CO_2 headspace (Lowe et al. 1985). Some researchers use others antibiotics like benzylpenicillin, streptomycin, kanamycin, penicillin and streptomycin (Joshi et al. 2018; Hanafy et al. 2018). These are again incubated for 39 ± 0.1°C for 3-4 weeks. The incubated roll tubes are observed regularly for the development of the fungal colonies. Since it is easy to observe colonies in roll tubes, morphologically different colonies are picked and purified in fresh medium under anaerobic culture in serum bottles. Usually this process of purification is done 2-3 times for each culture. It is always a good idea to conduct initial subcultures on media made of sampling source material which will help to provide unknown but essential micronutrients during their transition from natural environment to artificial media. Usually, the number of cultures reduces a lot during initial sub culturing as many fungi are not able to sustain the shock of change in habitat and nutrition for long. The protocol for isolation of anaerobic fungi is mentioned in Fig. 4.

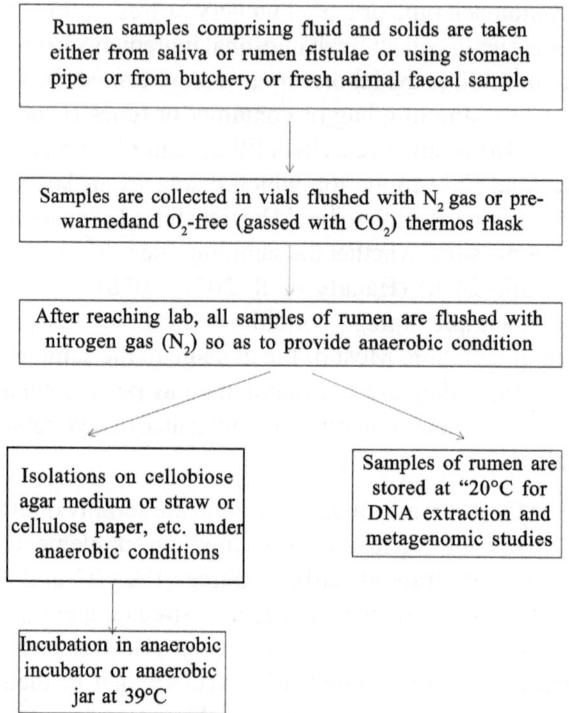

Fig. 4. Isolation of anaerobic fungi from animal digestive system from ruminating and non-ruminating organisms.

b. Media and growth conditions

The anaerobic cultures are grown without agitation in the complete absence of O_2. However, the toxicity of O_2 varies from micro-aerophilic to anoxic to complete anaerobic fungi. In comparison to the aerobic fungi, liquid culture is more preferred for the isolation and maintenance of anaerobic fungi. Even today, anaerobic fungi are more difficult to cultivate and maintained than aerobic fungi. The cultivation requires more careful observation and handling of cultures, monitoring the working of gases and anaerobic conditions during isolations and incubation. Usually well grown colony is considered as inoculums for revival in comparison to zoospores for better revival and preventing initial time of slow growth. Apart from the liquid medium, roll tubes with agar medium is used for the isolation and especially for purification (Fig. 4 C). It is easy to purify the culture from roll tube and is equivalent to Petri dish of aerobic fungal culturing. The media presently used for the isolation are either non-defined rumen material or complex chemically defined media (Davies et al., 1993; Marvin-Sikkema et al., 1992). The chemical reducing agents are added to the culture media either prior to or after the autoclaving. Common agents are sodium sulphide and

L-cysteine hydrochloride. It helps in lowering the redox level supporting the anaerobic growth of fungi. A redox indicator dye like Resazurin is used to ensure proper reduction. Enriching the sample with the rumen material or faeces or the food substrate of the animal and incubating them for a week may help in culturing of more anaerobic fungi. Box 2 lists some common meda used in the isolation of anaerobic fungi.

Media is one of the most important factor affecting the culturing of anaerobic fungi. This is due to lack of new media in recent times. The list of media in Box 2 clearly shows that less media have been developed in recent times. Physico-chemical analysis of the rumen material and/ or use of rumen material either completely or in dilute form will provide better chance of culturing these fungi. It will help to know other minor requirements of these group of fungi. However, it shall vary from animal to animal and rumen to rumen. The other features to be studied critically are O_2 tolerance, pH tolerance, micro-nutrients, vitamins, redox agents, C and N sources. Since, the anaerobic fungi colonize the plant biomass within few minutes of ingestion, use of different kinds of plant biomass in media will also help in culture of anaerobic fungi. Moreover, the gut of ruminating and non-ruminating animals have fungi ranging from micro-aerophilic fungi to anaerobic fungi, it will be good to enrich as well as incubate under different conditions, i.e. micro-aerophilic, anoxic and anaerobic. In general, the pH of rumen in animals varies from 5.8-6.8 which is usually maintained by the bicarbonates and phosphates from the saliva by regularly entering in the rumen. The temperature is usually between 38-40°C and the composition of gases is approximately 62% CO_2, 30% CH_4, 7% N_2, 0.6% O_2, 0.3% other gases and redox potential ranges from -250 to -450 mV (Trinci et al. 1994, Orpin 1975, Hobson 1971).

Box 2

Medium (gl^{-1})

Basal solution A - 830 ml
Glucose solution (37.5 g l^{-1}) - 100 ml
Na_2CO_3 solution (80 g l^{-1}) - 50 ml
Vitamin solution - 10 ml;
Reducing agent solution - 10 ml

Medium M1 modified (g l^{-1})

Solution A - 150 ml Solution B - 150 ml
Centrifuged rumen fluid - 200 ml
Bactocasitone - 10 g
Yeast extract - 2.5 g Na

Rumen-Fluid-Cellobiose (RFC) medium (gl^{-1})

Minerals solution I - 150 ml
Mineral solution II - 150 ml
Wolin's metal solution - 1 ml
Cellobiose - 3.75 g
Sodium bicarbonate - 6 g
Clarified sterile rumen fluid - 150 ml

Cellobiose Agar Medium (g l^{-1})

Yeast extract - 3 g
Tryptone - 5 g
150 ml of each
Solution 1 (0.3% K_2HPO_4),

(Contd.)

HCO_3 - 6 g
L-cysteine.HC1 - 1 g
Fructose - 2 g
Xylose - 2 g
Cellobiose - 2 g
Resazurin solution (0-1 %, w/v) - 1 ml

Basal medium (g I^{-1})
Tryptone - 20.0
Yeast extract - 1.0
Maltose - 0.2
L-cysteine - 0.1
L-ascorbic acid - 0.02
Fresh centrifuged rumen fluid - 100 ml

Medium B(g I^{-1})
Basal solution B - 810 ml
KH2P04 (68 g I^{-1}) 10 ml
Yeast extract solution (50 gl^{-1}) 10 ml
Glucose solution (37.5 g I^{-1}) - 100 ml
Na2CO3 solution (80 g I^{-1}) - 50 ml
Vitamin solution - 10 ml
Reducing agent solution - 10 ml

Orpin's Medium C
Salts solution I
Salts solution II
Cysteine hydrochloride
Yeast extract Bacto-casitone
Ammonium carbonate
Resazurin solution

Solution 2 [0.3% KH_2PO_4, 0.6% $(NH_4)_2SO_4$, 0.6% NaCl, 0.06% $MgSO_4.7H_2O$ and 0.06% $CaCl_2.2H_2O$]
Clarified rumen fluid,
Resazurin (0.1%) - 1 ml
Hemin (0.05%) - 1 ml
$NaHCO_3$ - 6 g
L-cysteine-HCl - 1 g
Cellobiose - 5 g

Medium M2 (g I^{-1})
Cellobiose - 5 g
Tryptone - 5 g
Cysteine-HCl - 1 g
Hemin - 2 ml (stock of 0.05% hemin mixed in 1:1 ethanol and 0.05 M NaOH)

Fungal Culture Medium (g I^{-1})
Solution A - 150 ml
Solution B - 150 ml
Cellobiose - 5 g
NaHCO3 - 6 g
L-cysteine. HCl - 1 g
Trace elements solution - 10 ml
Haemin solution - 10 ml
Resazurin solution (0.1 %, w/v) - 1 ml
Demineralized water - 950 ml

Solution I and II
Solution I - 150 ml
Solution II - 150 ml
Resazurin (0.1%, wt/vol) - 1 ml
Yeast extract - 0.52 g
Trypticase - 2.0 g
Volatile fatty acid mixture - 3.1 ml
Hemin (0.05%, wt/vol) - 2.0 ml
Na_2CO_3 (8%, wt/vol) - 70 ml

c. Long term preservation

Maintenance of anaerobic culture by sub-culturing is not only time consuming but also result in genetically unstable strains. Moreover, to have standard cultures for research, to avoid repeated sub-culturing, to have easy transport of cultures between laboratories and develop culture collection of anaerobic fungi, long term preservation is necessary. Since anaerobic fungi are difficult to isolate it is even more difficult to maintain them for long period. Repeated sub-culturing is required to keep the culture viable on roll tube. Moreover, most of the cultures die during initial sub-culturing. The anaerobic fungi can be maintained by growing them on plant material at 39°C without sub-culturing for up to 3 months (Joblin 1981). Long term preservation is done by cryopreservation at -80°C or -196°C in deep freezers and LN_2 respectively using 5% dimethylsulfoxide (DMSO) or 10% (vol/vol) glycerol as a cryoprotectant (Phillips and Gordon 1988). The survival rate of anaerobic fungi varies a lot under different long term strategies. It ranges from 40% to 80% when the fungi are preserved in glycerol and ethylene glycol (EG) at -80°C respectively (Yarlett et al. 1986, Sakurada et al. 1995). In a relatively recent study, Nagpal et al. (2012) demonstrated that glycerol is a better cryoprotectant than both DMSO and EG with 100% survival rate in a 3-month study. Other cryoprotectant that had been tested for the preservation of anaerobic fungi are dimethyl sulfoxide and propylene glycol. A flowchart of protocol for preservation of anaerobic fungi is given in Fig. 5.

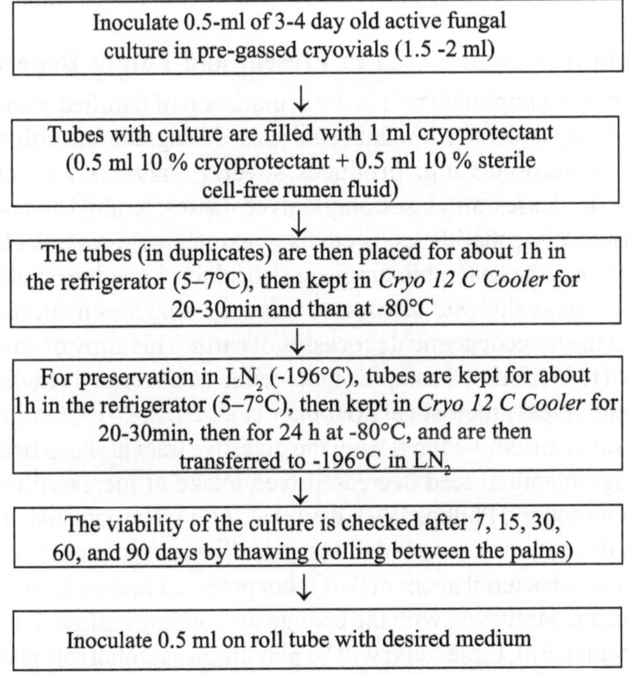

Fig. 5: Methods of long term preservation of anaerobic fungi

The progress in the research of any group of microbes has been due to improvement in their cultivation practices. Development of new media, new practices in sampling and methods of isolation are important factors. Newer methods of isolation by *in vitro* baiting techniques in different habitats may help in the capturing more diversity *viz.,* soil, water, etc. But as far as ruminating non-ruminating animals are concerned, proper sampling collection and sample transport are key to obtaining more anaerobic fungi in culture. This is because, maximum fungi die in the initial exposure to O_2. Present techniques used are either developed by Hungate or modifications of the older techniques. Hence, new and simple techniques of isolation and maintenance need attention for better culturing of anaerobic fungi. Improvement in technology for achieving anaerobic conditions is also a drawback in the progress of the field. Small and low cost anoxic chambers should be developed and promoted by companies so that more researchers take up the group of fungi for their research interests. Recurring cost becomes another factor as the regular availability of gas is required for the maintenance of anaerobic fungi. The slow growth rate of these fungi is another factor for the lack of interest by many researchers. However, some groups of fungi are micro-aerophilic and anoxic and researchers can focus on such groups which do exactly die at low concentrations of O_2 but tolerate it. However, the authors hope that in coming times with the increase of importance of these fungi in applications, it will improve in coming years.

6. Potential of Anaerobic Fungi in Present and Future Biotechnology

Rumen fungi play an important role in the degradation of lignified plant materials in gut of ruminating and non-ruminating animals. To degrade and utilize the plant material, the anaerobic fungi produces several enzymes *viz.,* cellulases, hemicellulases, proteases, amylases, amyloglycosidases, feruloyl and p-coumaryl esterases, various disaccharidases, pectinases, exonucleases or avicelases (Paul et al., 2004). Hence, the anaerobic fungi are critical for the carbon cycling as they utilize several lingo-cellulosic biomasses using the enzymes mentioned above, contributing in the biogeochemical processes of earth. This army of enzymes also help the animal to break down complex lgno-cellulose material (including various feed stocks) and release their sugars (Hooker et al. 2018, 2019, Gruninger et al. 2014). Removal of anaerobic fungi from the digestive tract in sheep or addition of anaerobic fungi in animal feed decreased feed intake or increased weight gain respectively (Gordon and Phillips 1993, Paul et al. 2011). The potential of anaerobic fungi to break down complex animal forage as well as produce natural compounds that can act as anti-bacterial agent makes them potential probiotic candidates for animal husbandry. Moreover, with the change in cropping pattern due to climate and environmental drift, these fungi will be playing an essential role and evolve as per the livestock feed availability to the cattle's.

Apart from this anaerobic fungi will be playing an important role in enhancing the biogas production. As the cellulosic material inside the biogas plants affect the biogas efficiency, the enzymes such as amylase, cellulose, hemicellulose and protease are produced extracellularly by fungi (and anaerobic fungi in this case) (Dollhofer et al. 2015) . Yýldýrým et al. (2017) reported 60% increase in methane production in biogas plants by bioaugmenting of anaerobic fungi. Prochazka et al. (2012) and Kazda et al. (2014) also reported such observations. Akyol et al. (2019) observed that *Orpinomyces* sp. increases the methane production by 15-30% when bioaugmented in the biogas digester. The exploitation of anaerobic fungi is still limited and most of the findings are limited to the laboratory. However, there is a need to have specific research in to the biogas system, in terms of screening of substrate, screening of anaerobic fungi and effect of inocula size on bioaugmentation process and effect of consortia. Due to difficulty in cultivation, such studies become tedious. Anaerobic fungi isolated from rumen such *Anaeromyces*, *Neocallimastix*, *Orpinomyces* and *Piromyces* enhanced the fermentation and degradation of biomass when microalgae was used as substrate (Aydin et al. 2017). The anaerobic fungi isolated from rumen help the process of methane production by physically penetrating the substrate and extracellular enzyme secretion, a prominent property of these organisms. The released enzymes of anaerobic fungi which degrade cell wall are free enzymes and cellulosomes (extracellular multi-enzyme complexes), both acting together as efficient fibre degraders in industries. There is a need to conduct more research on isolating the rumen fungi and testing its potential for biomass degradation and enzyme production. However, there is a need for complete characterization of enzymes, study various anaerobic fungi and methanogens as co-culture for enhanced biogas production. Researchers have observed enhanced biogas production by using co-cultures isolated using enrichment technique (Cheng et al. 2009, 2018). Recent omics techniques are boon to this field as the cultivation and other experiments are difficult to conduct on them. Hence, studying the genome and annotating the genes responsible for enzyme is helping to gain much inside to the physiology and biochemistry of these organisms in much shorter period. The genome and meta-genome data is also helping to study the metabolism and mechanism of enzyme production and substrate degradation. Thus helping to understand the regulation of biotechnologically important genes. The applications of anaerobic fungi from rumen in future are immense and range from second-generation bioethanol production to improved animal health (Hooker et al. 2019, Ranganathan et al. 2017).

Conclusion

Even though the applications of anaerobic fungi are many (Table 6), it is at infant stage and not many groups are working on it. The studies on anaerobic fungi suffered due to many reasons; firstly their taxonomy was not clear and was not recognized under any particular group of organisms, secondly availability of surgically modified or recently deceased animals was not common and thirdly the requirement of growth of anaerobic fungi requires specialized facility of O_2 free environment. However, quite lately the organism got their correct position in *Neocallimastigomycota* and also isolations got a success with fresh faeces and intestinal fluids.

Table 6. List of anaerobic fungi and their uses in industries.

S. No.	Species/ Genus	Applications
1	*Orpinomyces* sp.	Biomass degrader in biofuel production
2	*Neocallimastix frontalis*	Improve biogas production from lignocellulosic substrates.
3	*Anaeromyces* *Neocallimastix* *Orpinomyces* *Piromyces*	Enhanced methane production from microalgae biomass
4	*N. frontalis*	Cellulase production
5	*Piromyces N. patriciarum*	Highly stable and high capacity degradation of microcrystalline cellulose
6	*S. cervesia*	Ethanol production,The production of enzymes and recombinant proteins and the development of drug screening assays
7	*Pecoramyces ruminantium*	Biofuel production
8	*Lentisphaerae, Clostridium, Methanolinea*	Methane production

The difficulty faced by biogas industries in cleaning the cellulosic mass accumulated inside the chamber is unsolved till today. The industries shut their plants and clean the cellulosic material accumulated which affects the gas production capacity of biogas chamber. Cleaning of 40-50 thousand litres biogas plant is a huge task and possesses several difficulties. As per the discussion of the authors with several industrial experts, the need for anaerobic fungi which can degrade the cellulosic mass inside the biogas plants is the need of time. In view of less work in India on anaerobic fungi, it will be a good idea to develop common facilities across India wherein the researchers can book the facility for a particular period on charge basis which will help the present and future researchers working on anaerobic fungi. Moreover, all the culture collections should focus on this area for research on cultivation and preservation of anaerobic fungi. This will help in the development of facility for long term preservation of anaerobic fungi. The funding agency should promote this area of expertise as it will benefit the biogas industries and governments clean energy drive for coming years and betterment of earth climate.

Acknowledgements

The authors thank Department of Biotechnology, New Delhi, for financial support for the establishment of National Centre for Microbial Resource (NCMR), Pune, wide grant letter no. BT/Coord.II/01/03/2016 dated 6th April 2017. RS thank Dr. Om Prakash for his inputs during discussions on coffee table and sharing his experience on working with anaerobic bacteria during his post-doc and NCCS-NCMR. The authors thank Mr. Mahesh S. Sonawane for his help in formatting of media table.

References

Akyol C, Ince O, Bozan M, Ozbayram EG, Ince B. 2019. Fungal bioaugmentation of anaerobic digesters fed with lignocellulosic biomass: what to expect from anaerobic fungus *Orpinomyces* sp. *Bioresource Technology*. doi: https://doi.org/10.1016/j.biortech.2019.01.024

Aydin S, Yýldýrým E, Ince O, Inceb B. 2017. Rumen anaerobic fungi create new opportunities for enhanced methane production from microalgae biomass. *Algal Research* 23:150–160.

Balch WE, Wolfe RS. 1976. New approach to the cultivation of methanogenic bacteria: 2-mercaptoethanesulfonic acid (HS-CoM)-dependent growth of *Methanobacterium ruminantium* in a pressurized atmosphere. *Applied and Environmental Microbiology*, 32(6): 781–791.

Barr DJ, Kudo H, Jakober KD, Cheng KJ. 1989. Morphology and development of rumen fungi: *Neocallimastix* sp., *Piromyces communis*, and *Orpinomyces bovis* gen. nov., sp. nov. *Canadian Journal of Botany*. 67(9):2815–2824.

Berbee ML, James TY, Strullu-Derrien C. 2017. Early diverging fungi: diversity and impact at the dawn of terrestrial life. *Annual Review of Microbiology*. 8:71.

Braune RA. 1913. Untersuchungen über die im Wiederkäuermagen vorkommenden Protozoen. *Archiv für Protistenkunde*. 32:111–170.

Bryant MP. 1972. Commentary on the Hungate technique for culture of anaerobic bacteria. *American Journal of Clinical Nutrition* 25, 1324-1328.

Breton A, Bernalier A, Bonnemoy F, Fonty G, Gaillard B, Gouet P. 1989. Morphological and metabolic characterization of a new species of strictly anaerobic rumen fungus: Neocallimastix joyonii. FEMS Microbiol. *Letts*. 58: 309-314.

Breton A, Bernalier A, Dusser M, Fonty G, Gaillard-Martinie B, Guitlot J. 1990. Anaeromyces mucronatus nov. gen., nov. sp. A new strictly anaerobic rumen fungus with polycentric thallus. *FEMS Microbiology Letters*. 70:177–182.

Breton A, Dusser M, Gaillard-Martine B, Guillot J, Millet L, Prensier G. 1991. Piromyces rhizinflata nov. sp., a strictly anaerobic fungus from faeces of the Saharian ass: a morphological, metabolic and ultrastructural study. FEMS Microbiology Letters. 82:1-8.

Callaghan TM, Podmirseg SM, Hohlweck D, Edwards JE, Puniya AK, Dagar SS, Griffith GW. 2015. Buwchfawromyces eastonii gen. nov., sp. nov.: a new anaerobic fungus (Neocallimastigomycota) isolated from buffalo faeces. *MycoKeys*. 9: 11–28.

Chen Y-C, Hseu R-S, Chein C-Y. 2002. Piromyces polycephalus (Neocallimastigaceae), a new rumen fungus. Nova Hedwigia. 75: 409–414.

Chen YC, Tsai SD, Cheng HL, Chien CY, Hu CY, Cheng TY. 2007. Caecomyces sympodialis sp. nov., a new rumen fungus isolated from Bos indicus. Mycologia. 99:125–130.

Cheng Y, Shi Q, Sun R, Liang D, Li Y, Li Y, Jin W, Zhu W. 2018. The biotechnological potential of anaerobic fungi on fibre degradation and methane production. *World Journal of Microbiology and Biotechnology*. 34:155.

Cheng Y, Zhu W. 2009. Diversity analysis of anaerobic fungi in the co-cultures with or without methanogens by amplified ribosomal intergenic spacer analysis. *Acta Microbiol Sin* 49(4):0504–0511

Connell L, Barrett A, Templeton A, Staudigel H. 2009. Fungal diversity associated with an active deep sea volcano: Vailulu'u Seamount, Samoa. *Geomicrobiology Journal*. 26(8):597-605.

Dagar SS, Kumar S, Griffith GW, Edwards JE, Callaghan TM, Singh R, Nagpal AK, Puniya AK. 2015a. A new anaerobic fungus (*Oontomyces anksri* gen. nov., sp. nov.) from the digestive tract of the Indian camel (*Camelus dromedarius*). *Fungal biology*. 119(8):731–737.

Dagar SS, Kumar S, Mudgil P, Singh R, Puniya AK. 2011. D1/D2 domain of large-subunit ribosomal DNA for differentiation of *Orpinomyces* spp. *Appl. Environ. Microbiol.* 77(18):6722–6725.

Dagar SS, Singh N, Goel N, Kumar S, Puniya AK. 2015. Role of anaerobic fungi in wheat straw degradation and effects of plant feed additives on rumen fermentation parameters *in vitro*. *Benef Microbes*. 6(3):353-360.

Dagar SS, Kumar S, Mudgil P, Puniya AK. 2018. Comparative evaluation of lignocellulolytic activities of filamentous cultures of monocentric and polycentric anaerobic fungi. *Anaerobe*. 50:76–79.

Davies DR, Theodorou MK, Lawrence MI, Trinci AP. 1993. Distribution of anaerobic fungi in the digestive tract of cattle and their survival in faeces. *J Gen Microbiol* 139(Pt 6):1395–1400.

Day AC, Strasik M, McCrary KE, Johnson PE, Gabrys JW, Schindler JR, Hawkins RA, Carlson DL, Higgins MD, Hull JR. 2002. Design and testing of the HTS bearing for a 10 kWh flywheel system. *Superconductor Science and Technology*. 15(5):838.

Denman S, Nicholson M, Brookman J, Theodorou M, McSweeney C. 2008. Detection and monitoring of anaerobic rumen fungi using an ARISA method. *Lett Appl Microbio*. l47: 492–499.

Dollhofer V, Podmirseg SM, Callaghan TM, Griffith GW, Fliegerová K. 2015. Anaerobic fungi and their potential for biogas production. In: Biogas Science and Technology. Springer, Cham. pp. 41-61.

Dore J, Stahl DA. 1991. Phylogeny of anaerobic rumen *Chytridiomycetes* inferred from small subunit ribosomal RNA sequence comparisons. *Canadian Journal of Botany*. 69(9):1964-1971.

Doyle JJ, Doyle JL, Doyle JA, Doyle FJ. 1987. A rapid DNA isolation procedure for small amounts of fresh leaf tissue.

Drake H, Ivarsson M. 2017. The role of anaerobic fungi in fundamental biogeochemical cycles in the deep biosphere. *Fungal Biology Reviews*. 32(1):20–25.

Drake H, Ivarsson M, Bengtson S, Heim C, Siljeström S, Whitehouse MJ, Broman C, Belivanova V, Åström, ME. 2017a. Anaerobic consortia of fungi and sulfate reducing bacteria in deep granite fractures. *Nat Commun*. 8:55.

Drake H, Heim C, Roberts NMW, Zack T, Tillberg M, Broman C, Ivarsson M, Whitehouse MJ, Åström ME. 2017b. Isotopic evidence for microbial production and consumption of methane in the upper continental crust throughout the Phanerozoic eon. *Earth Planet Sci Lett*. 470:108e118.

Drummond AJ, Ashton B, Buxton S, Cheung M, Cooper A, Duran C, Field M, Heled J, Kearse M, Markowitz S, Moir R. 2011. Geneious v5. 4 created by Biomatters. Website http://www. geneious. com.

Eckart M, Fliegerová K, Hoffmann K, Voigt K. 2010. Molecular Identi?cation of Anaerobic Rumen Fungi. In: Molecular Identi?cation of Fungi (eds. Y. Gherbawy and K. Voigt). Springer-Verlag Berlin Heidelberg , pp 297-313.

Edwards J, Huws S, Kingston-Smith A, Jimenez H, Skøt K, Griffith GW, McEwan NR, Theodorou MK. 2008. Dynamics of initial colonisation of nonconserved perennial ryegrass by anaerobic fungi in the bovine rumen. *FEMS Microbiology Ecology*. 66(3):537–546.

Edwards JE, Forster RJ, Callaghan TM, Dollhofer V, Dagar SS, Cheng Y, Chang J, Kittelmann S, Fliegerova K, Puniya AK, Henske JK. 2017. PCR and omics based techniques to study the diversity, ecology and biology of anaerobic fungi: insights, challenges and opportunities. *Frontiers in Microbiology*. 8:1657.

Emerson R, Natvig DO. 1981. Adaptation of fungi to stagnant waters. Mycology series.

Fliegerova K, Hodrova B, Voigt K. 2004. Classical and molecular approaches as a powerful tool for the characterization of rumen polycentric fungi. *Folia Microbiologica*. 49(2):157.

Fliegerova K, Kaerger K, Kirk P, Voigt K. 2015. Rumen Fungi. In: Rumen Microbiology: From Evolution to Revolution (eds. A.K. Puniya et al.). Springer India.

Fliegerova K, Mrazek J, Hoffmann K, Zabranska J, Voigt K. 2010. Diversity of anaerobic fungi within cow manure determined by ITS analysis. *Folia Microbiologica* 55: 319–325.

Fredrickson JK, McKinley JP, Nierzwicki Bauer SA, White DC, Ringelberg DB, Rawson SA, Li SM, Brockman FJ, Bjornstad BN. 1995. Microbial community structure and biogeochemistry of Miocene subsurface sediments: implications for long term microbial survival. *Molecular Ecology*. 4(5):619-26.

Gold JJ, Heath IB, Bauchop T. 1988. Ultrastructural description of a new chytrid genus of caecum an- aerobe, Caecomyces equi gen. nov. sp. nov., assigned to the Neocallimasticaceae. *BioSystems*. 21:403-415.

Gordon GLR, Phillips MW. 1993. Removal of anaerobic fungi from the rumen of sheep by chemical treatment and the effect on feed consumption and in vivo ?bre digestion. *Lett Appl Microbiol*. 17:220–223.

Gaillard-Martinie B, Breton A, Dusser M, Julliand V. 1995. Piromyces citronii sp. nov., a strictly anaerobic fungus from the equine caecum: a morphological, metabolic, and ultrastructural study. *FEMS Microbiology Letters*. 130(2-3):321–326.

Griffith GW, Baker S, Fliegerova K, Liggenstoffer A, Van Der Giezen M, Voigt K, Beakes G. 2010. Anaerobic fungi: *Neocallimastigomycota*. *IMA Fungus*. 1(2):181.

Griffith GW, Ozkose E, Theodorou MK, Davies DR. 2009. Diversity of anaerobic fungal populations in cattle revealed by selective enrichment culture using different carbon sources. *Fungal Ecology*. 2(2):87-97.

Griffith GW, Shaw DS. 1998. Polymorphisms in *Phytophthora infestans*: four mitochondrial haplotypes are detected after PCR amplification of DNA from pure cultures or from host lesions. *Appl. Environ. Microbiol*. 64(10):4007-4014.

Gruninger RJ, Puniya AK, Callaghan TM, Edwards JE, Youssef N, Dagar SS, Fliegerova K, Griffith GW, Forster R, Tsang A, McAllister T. 2014. Anaerobic fungi (phylum *Neocallimastigomycota*): advances in understanding their taxonomy, life cycle, ecology, role and biotechnological potential. *FEMS Microbiology Ecology*. 90(1):1-7.

Guindon S, Dufayard JF, Lefort V, Anisimova M, Hordijk W, Gascuel O. 2010. New algorithms and methods to estimate maximum-likelihood phylogenies: assessing the performance of PhyML 3.0. *Systematic Biology*. 59(3):307-21.

Hanafy RA, Elshahed MS, Liggenstoffer AS, Griffith GW, Youssef NH. 2017. *Pecoramyces ruminantium*, gen. nov., sp. nov., an anaerobic gut fungus from the faeces of cattle and sheep. *Mycologia*. 109(2):231-243.

Hanafy RA, Elshahed MS, Youssef NH. 2018. *Feramyces austinii*, gen. nov., sp. nov., an anaerobic gut fungus from rumen and fecal samples of wild Barbary sheep and fallow deer. *Mycologia*. 110(3):513-525.

Heath IB, Bauchop T, Skipp RA. 1983. Assignment of the rumen anaerobe *Neocallimastix frontalis* to the *Spizellomycetales* (*Chytridiomycetes*) on the basis of its polyflagellate zoospore ultrastructure. *Canadian Journal of Botany*. 61(1):295-307.

Henske JK, Gilmore SP, Knop D. et al. 2017. Transcriptomic characterization of Caecomyces churrovis: a novel, non-rhizoid-forming lignocellulolytic anaerobic fungus. Biotechnol Biofuels. 10:305. https://doi.org/10.1186/s13068-017-0997-4

Hibbett DS, Binder M, Bischoff JF, Blackwell M, Cannon PF, Eriksson OE, Huhndorf S, James T, Kirk PM, Lücking R, Lumbsch HT. 2007. A higher-level phylogenetic classification of the Fungi. *Mycological Research*. 111(5):509-547.

Ho YW, Barr DJ. 1995. Classification of anaerobic gut fungi from herbivores with emphasis on rumen fungi from Malaysia. *Mycologia*. 87(5):655-677.

Ho YW, Barr DJS, Abdullah N, Jalaludin S, Kudo H. 1993a. Neocallimastix variabilis, a new species of anaerobic rumen fungus from cattle. *Mycotaxon*. 46:241–258.

Ho YW, Barr DJS, Abdullah N, Jalaludin S, Kudo H. 1993b. Piromyces spiralis, a new species of anaerobic fungus from the rumen of goat. *Mycotaxon*. 48:59–68.

Ho YW, Barr DJS, Abdullah N, Jalaludin S, Kudo H. 1993c. A new species of Piromyces from the rumen of deer in Malaysia. *Mycotaxon*. 47:285–293.

Ho YW, Barr DJS, Abdullah N. 1993d. Anaeromyces, an earlier name for Ruminomyces. *Mycotaxon*. 47:283–284.

Ho YW, Abdullah N, Jalaludin S. 1994. Orpinomyces intercalaris, a new species of polycentric anaerobic rumen fungus from cattle. *Mycotaxon*. 50:139–150.

Ho YW, Abdullah N, Jalaludin S. 2000. The diversity and taxonomy of anaerobic gut fungi. *Fungal Diversity*. 4:37–51.

Hobson, P. N. (1971). Rumen micro-organisms. *Progress in Industrial Microbiology* 9: 42-77.

Hooker CA, Hillman ET, Overton JC, Ortiz-Velez A, Schacht M, Hunnicutt A, Mosier NS, Solomon KV. 2018. Hydrolysis of untreated lignocellulosic feedstock is independent of S-lignin composition in newly classified anaerobic fungal isolate, Piromyces sp. UH3-1 . *Biotechnol Biofuels*. 11:293.

Hooker C, Lee KZ, Solomon KV. 2019. Leveraging anaerobic fungi for biotechnology. *Current Opinion in Biotechnology* 59:103-110

Hsiung TS. 1930. A monograph on the protozoa of the large intestine of the horst. Iowa State *College Journal of Science*. 4:359-423.

Huelsenbeck JP, Ronquist F. 2001. MRBAYES: Bayesian inference of phylogenetic trees. *Bioinformatics*. 17(8):754-755.

Hungate RE, Macy J. 1973. A roll tube method for cultivation of strict anaerobes. *Bull Ecol Res Comm*. 17:123–125.

Hungate, R. E. 1966. The Rumen and Its Microbes. Academic Press, New York.

Ivarsson M, Bengtson S, Belivanova V, Stampanoni M, Marone F, Tehler A. 2012. Fossilized fungi in sub seafloor Eocene basalts. *Geology*. 40(2):163-166.

Ivarsson M, Bengtson S, Neubeck A. 2016. The igneous oceanic crust–Earth's largest fungal habitat? *Fungal Ecology*. 20:249-255.

Ivarsson M, Schnürer A, Bengtson S, Neubeck A. 2016b. Anaerobic fungi: a potential source of biological H2 in the oceanic crust. *Frontiers in Microbiology*. 7:674.

James TY, Kauff F, Schoch CL, Matheny PB, Hofstetter V, Cox CJ, Celio G, Gueidan C, Fraker E, Miadlikowska J, Lumbsch HT. 2006a. Reconstructing the early evolution of fungi using a six-gene phylogeny. *Nature*. 443(7113):818.

James TY, Letcher PM, Longcore JE, Mozley-Standridge SE, Porter D, Powell MJ, Griffith GW, Vilgalys R. 2006b. A molecular phylogeny of the flagellated fungi (*Chytridiomycota*) and description of a new phylum (*Blastocladiomycota*). *Mycologia*. 98(6):860-871.

Joblin KN. 1981. Isolation, enumeration and maintenance of rumen anaerobic fungi in roll-tubes. *Appl Environ Microbiol*. 42:1119–1122.

Joshi A, Lanjekar VB, Dhakephalkar PK, Callaghan TM, Griffith GW, Dagar SS. 2018. *Liebetanzomyces polymorphus* gen. et sp. nov., a new anaerobic fungus (*Neocallimastigomycota*) isolated from the rumen of a goat. *MycoKeys*. (40):89.

Katoh K, Misawa K, Kuma KI, Miyata T. 2002. MAFFT: a novel method for rapid multiple sequence alignment based on fast Fourier transform. Nucleic Acids Research. 30(14):3059-3066.

Kazda M, Langer S, Bengelsdorf FR. 2014. Fungi open new possibilities for anaerobic fermentation of organic residues. *Energy. Sustain. Soc.* 4:6. doi:10.1186/2192-536 0567-4-6.

Kirk PM. 2012. Anaeromyces polycephalus (Chen YC, Chien CY, Hseu RS) Flieg., K. Voigt & P.M. Kirk. IF550012. *Index Fungorum*. 1: 1.

Kittelmann S, Naylor GE, Koolaard JP, Janssen PH. 2012. A proposed taxonomy of anaerobic fungi (Class *Neocallimastigomycetes*) suitable for large-scale sequence-based community structure analysis. *PLoS One*. 7(5):e36866.

Kumar S, Dagar SS, Puniya AK 2012. Isolation and characterization of methanogens from rumen of Murrah buffalo. *Ann. Microbiol.* 62:345–350. 10.1007/s13213-011-0268-8

Kumar S, Dagar SS, Sirohi SK, Upadhyay RC, Puniya AK. 2013. Microbial profiles, in vitro gas production and dry matter digestibility based on various ratios of roughage to concentrate. *Ann. Microbiol.* 63 541–545. 10.1007/s13213-012-0501-0

Kumar S, Stecher G, Tamura K. 2016. MEGA7: molecular evolutionary genetics analysis version 7.0 for bigger datasets. *Molecular Biology and Evolution*. 33(7):1870-1874.

Le Calvez T, Burgaud G, Mahé S, Barbier G, Vandenkoornhuyse P. 2009. Fungal diversity in deep-sea hydrothermal ecosystems. Appl. Environ. *Microbiol.* 175(20):6415-21.

Li J, Heath B, Bauchop T. 1990. Piromyces mae and Piromyces dumbonica, two new species of uniflagellate anaerobic chytridiomycetes fungi from the hindgut of the horse and elephant. Canad J Bot. 68:1021–1033.

Li J, Heath IB, Packer L. 1993. The phylogenetic relationships of the anaerobic chytridiomycetous gut fungi (*Neocallimasticaceae*) and the *Chytridiomycota*. II. Cladistic analysis of structural data and description of *Neocallimasticales* ord. nov. *Canadian Journal of Botany*. 71(3):393-407.

Li J, Heath IB. 1992. The phylogenetic relationships of the anaerobic chytridiomycetous gut fungi (*Neocallimasticaceae*) and the *Chytridiomycota*. I. Cladistic analysis of rRNA sequences. *Canadian Journal of Botany*. 70(9):1738-1746.

Li GJ, Hyde KD, Zhao RL, Hongsanan S, Abdel-Aziz FA, Abdel-Wahab MA, Alvarado P, et al. 2016. Fungal diversity notes 253-366: taxonomic and phylogenetic contributions to fungal taxa. *Fungal Diversity*. 78(1):1–237.

Liebetanz E. 1910. Die parasitischen Protozoen des Wiederkauermagens. *Archiv für Protistenkunde*. 19:19-80.

Liebetanz, E. (1910). Die Parasitischen Protozoen des wiederkauermagens. *Archiv fur Protistenkunde* 19: 19.

Liggenstoffer A, Youssef N, Couger M. et al. 2010. Phylogenetic diversity and community structure of anaerobic gut fungi (phylum *Neocallimastigomycota*) in ruminant and non-ruminant herbivores. *The ISME J* 4:1225–1235.

Lin LH, Wang PL, Rumble D, Lippmann-Pipke J, Boice E, Pratt LM, Lollar BS, Brodie EL, Hazen TC, Andersen GL, DeSantis TZ. 2006. Long-term sustainability of a high-energy, low-diversity crustal biome. *Science*. 314(5798):479-482.

Lowe SL, Theodorou MK, Trincv APJ, Hespell RB. 1989. Growth of Anaerobic Rumen Fungi on Defined and Semi-defined Media Lacking Rumen Fluid. *Journal of General Microbiology*. 131:2225–2229.

Marvin-Sikkema FD, Lahpor GA, Kraak MN, Gottschal JC, Prins RA. 1992. Characterization of an anaerobic fungus from llama faeces. *J Gen Microbiol.* 138(10):2235–2241.

Miller TL, Wolin MJ. 1974. A serum bottle modification of the Hungate technique for cultivating obligate anaerobes. *Applied Microbiology* 27(5): 985–987.

Milne I, Wright F, Rowe G, Marshall DF, Husmeier D, McGuire G. 2004. TOPALi: software for automatic identification of recombinant sequences within DNA multiple alignments. *Bioinformatics*. 20(11):1806-1807.

Munn EA, Orpin CG, Greenwood CA. 1988. The ultrastructure and possible relationships of four obligate anaerobic chytridiomycete fungi from the rumen of sheep. *Biosystems*. 22(1):67-81.

Nagano Y, Nagahama T. 2012. Fungal diversity in deep-sea extreme environments. *Fungal Ecology*. 5(4):463-471.

Nagpal R, Puniya AK, Sehgal JP, Singh K. 2012. Survival of anaerobic fungus Caecomyces sp. in various preservation methods: a comparative study. *Mycoscience*. 53:427–432.

Nylander JA. 2004. MrModeltest v2. Program distributed by the author. Uppsala University, Sweden.

Orpin CG, Joblin KN. 1997. The rumen anaerobic fungi. In: The rumen microbial ecosystem. Springer, Dordrecht. pp. 140-195.

Orpin CG. 1974. The rumen flagellate *Callimastix frontalis*: does sequestration occur? *Microbiology*. 84(2):395-398.

Orpin CG. 1975. Studies on the rumen flagellate *Neocallimastix frontalis*. *Microbiology*. 91(2):249-262.

Orpin CG. 1976. Studies on the rumen flagellate *Sphaeromonas communis*. *Microbiology*. 94(2):270-280.

Orpin CG. 1977a. The occurrence of chitin in the cell walls of the rumen organisms *Neocallimastix frontalis*, *Piromonas communis* and *Sphaeromonas communis*. *Microbiology*. 99(1):215-218.

Orpin CG. 1977b. The rumen flagellate *Piromonas communis*: its life-history and invasion of plant material in the rumen. *Microbiology*. 99(1):107-117.

Orpin CG. 1994. Anaerobic Fungi: Taxonomy, Biology, and Distribution in Nature. In: Anaerobic Fungi: Biology: Ecology, and Function (eds. Mountfort DO, Orpin CG). CRC Press. pp 304.

Orpin CG, Munn EA. 1986. Neocallimastix patriciarum sp.nov., a new member of the Neocallimasticaceae inhabiting the rumen of sheep. *Transactions of the British Mycological Society*. 86(1):178–181.

Orsi W, Biddle JF, Edgcomb V. 2013. Deep sequencing of sub seafloor eukaryotic rRNA reveals active fungi across marine subsurface provinces. *PLoS One*. 8(2):e56335.

Parkes RJ, Webster G, Cragg BA, Weightman AJ, Newberry CJ, Ferdelman TG, Kallmeyer J, Jørgensen BB, Aiello IW, Fry JC. 2005. Deep sub-seafloor prokaryotes stimulated at interfaces over geological time. *Nature*. 436(7049):390.

Pedersen K, Arlinger J, Eriksson S, Hallbeck A, Hallbeck L, Johansson J. 2008. Numbers, biomass and cultivable diversity of microbial populations relate to depth and borehole-specific conditions in groundwater from depths of 4–450 m in Olkiluoto, Finland. *The ISME Journal*. 2(7):760.

Picard KT. 2017. Coastal marine habitats harbor novel early-diverging fungal diversity. *Fungal Ecology*. 25:1-3.

Powell MJ, Letcher PM. 2012. From zoospores to molecules: the evolution and systematics of *Chytridiomycota*. Systematics and evolution of fungi. CRC, Boca Raton, FL. 10:29-54.

Procházka J, Mrázek J, Štrosová L, Fliegerová K, Zábranská J, Dohányos M. 2012. Enhanced biogas yield from energy crops with rumen anaerobic fungi. *Eng. Life Sci.* 12:343–351.

Ranganathan A, Smith OP Youssef NH, Struchtemeyer CG, Atiyeh HK and Elshahed MS (2017) Utilizing anaerobic fungi for two-stage sugar extraction and biofuel production from lignocellulosic biomass. Front. Microbiol. 8:635. doi: 10.3389/fmicb.2017.00635

Rezaeian M, Beakes GW, Parker DS. 2004. Distribution and estimation of anaerobic zoosporic fungi along the digestive tracts of sheep. *Mycological Research*. 108(10):1227-33.

Sakurada M, Tsuzuki Y, Morgavi DP, Tomita Y, Onodera R. 1995. Simple method for cryopreservation of an anaerobic rumen fungus using ethylene glycol and rumenuid. *FEMS Microbiol Lett*. 127:171–174.

Schrenk MO, Huber JA, Edwards KJ. Microbial provinces in the sub seafloor. 2010. *Annual Review of Marine Science*. 2:279-304.

Sharma R, Polkade AV, Shouche YS. 2015. "Species Concept" in microbial taxonomy and systematics. *Current Science*. 108(10):1804–1814.

Sohlberg E, Bomberg M, Miettinen H, Nyyssönen M, Salavirta H, Vikman M, Itävaara M. 2015. Revealing the unexplored fungal communities in deep groundwater of crystalline bedrock fracture zones in Olkiluoto, Finland. *Frontiers in Microbiology*. 6:573.

Spatafora JW, Chang Y, Benny GL, Lazarus K, Smith ME, Berbee ML, Bonito G, Corradi N, Grigoriev I, Gryganskyi A, James TY. 2016. A phylum-level phylogenetic classification of zygomycete fungi based on genome-scale data. *Mycologia*. 108(5):1028-46.

Sridhar M, Kumar D, Anandan S. 2014. Cyllamyces icaris sp. nov., a new anaerobic gut fungus with nodular sporangiophores isolated from Indian water buffalo (Bubalus bubalis). *Int J Curr Res Aca Rev*. 2(1):07–24.

Steenbakkers PJ, Li XL, Ximenes EA, Arts JG, Chen H, Ljungdahl LG, Op Den Camp HJ. 2001. Noncatalytic docking domains of cellulosomes of anaerobic fungi. *J Bacteriol*. 183(18):5325-5333.

Strullu-Derrien C, Spencer ART, Goral T, Dee J, Honegger R, Kenrick P, Longcore JE, Berbee ML. 2017. New insights into the evolutionary history of Fungi from a 407 Ma Blastocladiomycota fossil showing a complex hyphal thallus. *Phil. Trans. R. Soc*. B 373: 20160502.

Theodorou MK, Davies DR, Nielsen BB, Lawrence MIG, Trinci APJ. 1995. Determination of growth of anaerobic fungi on soluble and cellulosic substrates using a pressure transducer. *Microbiology*-UK. 141(3):671–678.

Trinci AP, Davies DR, Gull K, Lawrence MI, Nielsen BB, Rickers A, Theodorou MK. 1994. Anaerobic fungi in herbivorous animals. *Mycological Research*. 98(2):129-152.

Vavra J, Joyon L. 1966. Etude sur la morphologie le cycle evolutif et la position systematique de *Callimastix cyclopis* Weissenberg. *In: Journal of Protozoology*. (p. 41). 810 E 10th ST, Lawrence, KS 66044: Soc Protozoologists.

Wang X, Liu X, Groenewald JZ. 2017. Phylogeny of anaerobic fungi (phylum Neocallimastigomycota), with contributions from yak in China. *Antonie van Leeuwenhoek*. 110:87–103.

Webb J, Theodorou MK. 1991. Neocallimastix hurleyensis sp.nov., an anaerobic fungus from the ovine rumen. *Canadian Journal of Botany*. 69(6):1220–1224.

Weissenberg R. 1950. The development and affinities of *Callimastix cyclopis* Weissenberg, a parasitic microorganism from the serum of Cyclops. *Proc. Amer. Soc. Protozool*. 1:4-5.

White TJ, Bruns T, Lee S, Taylor JW. 1990. Amplication and direct sequencing of fungal ribosomal RNA genes for phylogenetics. In: Innis MA, Gelfand DH, Sninsky JJ, White TJ (eds), PCR Protocols: A Guide to Methods and Applications. Academic Press, Inc., New York, pp 315–322.

Wu X, Holmfeldt K, Hubalek V, Lundin D, Åström M, Bertilsson S, Dopson M. 2016. Microbial metagenomes from three aquifers in the Fennoscandian shield terrestrial deep biosphere reveal metabolic partitioning among populations. *The ISME Journal*. 10(5):1192.

Yarlett NC, Yarlett N, Orpin CG, Lloyd D. 1986. Cryopreservation of the anaerobic rumen fungus *Neocallimastix patriciarum*. *Letters in Applied Microbiology*. 3(1):1-3.

Yýldýrým E, Ince O, Aydin S, Ince B. 2017. Improvement of biogas potential of anaerobic digesters using rumen fungi. *Renewable Energy*, doi: 10.1016/j.renene.2017.03.021

Youssef NH, Couger MB, Struchtemeyer CG, Liggenstoffer AS, Prade RA, Najar FZ, Atiyeh HK, Wilkins MR, Elshahed MS. 2013. The genome of the anaerobic fungus *Orpinomyces* sp. strain C1A reveals the unique evolutionary history of a remarkable plant biomass degrader. *Appl. Environ. Microbiol*. 79(15):4620-4634.

Van Soest P.J. 2018. Nutritional ecology of the ruminant. Cornell University Press.

5

Methanogens: The Unique, Unicellular Powerhouses of Future

Manasi P. Tukdeo[1], *Kasturi Deore*[2], *Prashant K. Dhakephalkar*[2] *Vikram B. Lanjekar*[2*]

[1]*Department of Microbiology, PES Modern College of Arts, Science and Commerce, Ganeshkhind, Pune, Maharashtra, India*
[2]*Bioenergy Group, MACS-Agharkar Research Institute, Pune, Maharashtra India*

Abstract

Microorganisms that produce methane as their metabolic end product are known as methanogens. Methanogens are prokaryotic, unicellular, obligate anaerobes having fastidious growth requirements making them difficult to isolate and preserve. They catalyse the terminal step in anaerobic breakdown of organic substances. Even though methanogenic archaea or methanogens show very limited metabolic diversity, they are one of the most phylogenetically and ecologically diverse group of Archaea. Methanogenic community varies greatly with respect to temperature. Methanogens have been reported from permanently frozen habitats like permafrost to hyperthermophilic environments such as hydrothermal vents. However, the metabolic and phylogenetic diversity of methanogens in these habitats vary greatly. Methanogens are the biocatalysts, having the ability to provide a solution to help solve energy crisis by producing methane as storable energy carrier. Methane being a flammable gas, can act as a clean and renewable source of energy. The biogenic methane gas thus produced may serve as an efficient alternative for fossil fuels in the future.

**Corresponding Author: vblanjekar@aripune.org*

Introduction

Methanogenesis was first discovered in 1776 by the Italian physicist Alessandro Volta who described 'combustible air' formed in the sediments of streams, bogs and lakes full of rotting organic material. Omelianski subsequently defined the microbial involvement in the generation of the methane gas.Methanogens are a group of microbes that belong to the phylum Euryarchaeota of domain Archaea. These microbes inhabit only anoxic environments where they produce methane by utilizing various substrates. For growth, methanogens require strict anaerobic conditions and low redox potential (less than -200 mV). Phylogenetically, methanogens are divided into five families, comprising more than 33 genera and 160 species, each with its unique characteristics (Boone et al. 2001). Methanogens are found in various environments like wetlands, the digestive tracts of animals, hot springs, anaerobic digesters, oil wells, hydrothermal vents, hot desert soil, lignite reservoirs, etc. These microbes can survive at various temperatures and over broad temperature range. The substrate utilization profiles of methanogens are also quite diverse, and include acetate, formate, alcohols, amines, H_2:CO_2 (80:20), etc. Being a renewable and clean energy source, methane is currently utilized as an alternative to petrol in many European countries though its use in India is not as extensive. Methane is also recognized as one of the most potent greenhouse gas therefore, has environmental implications (Dhakephalkar et al. 2019). As methanogens are present at the end of the anaerobic food chain, the information on type and abundance of methanogens in a particular environment is also indicative of the type of other group of microbes. Hence, the study of community structure of methanogens in such habitats is extremely relevant and important. However, due to obligate anaerobic nature and slow growth rates, the cultivation of methanogens is extremely tedious and challenging. In this direction next generation sequencing techniques have helped providing deeper insights into ecology and functions of such habitats. However, the culture-independent studies are not sufficient alone and must complement the cultivation-dependent approaches. The culturing-based studies will detail the cultivable diversity of methanogens, their substrate requirement, growth rate, the rate of methanogenesis and physicochemical characteristics. Similarly, the culture-independent tools also inform about the population that is not yet cultured. The process of methanogenesis is the final and rate limiting step in the process of anaerobic digestion. If the rate of methanogenesis is enhanced, the process of anaerobic digestion can be improved. Therefore, augmentation of such methanogens is expected to have positive implications in improving the performance of anaerobic digesters towards production of biomethane.

Cultivation techniques for methanogens

Techniques for cultivation of methanogens vary in detail among laboratories; traditionally anaerobic techniques introduced by Hungate further modified by Bryant for preparation of pre-reduced media are followed (Hungate 1950, 1969; Bryant 1972). Further, development of procedures where culturing of methanogens under pressurized atmosphere were followed, where the chances for contamination or loss of reducing potential were essentially eliminated (Balch and Wolfe 1976; Holdeman et al. 1977; Balch et al. 1979). For additional details on the cultivation of methanogens, the methodologies by Sowers and Noll (1995) may be referred. Methanogens are known to be present in low density in their habitats, hence the samples need to be initially enriched in the similar media followed by isolation techniques.

We have summarized the same in Figure 1 and simplified all the modified techniques as stated below:

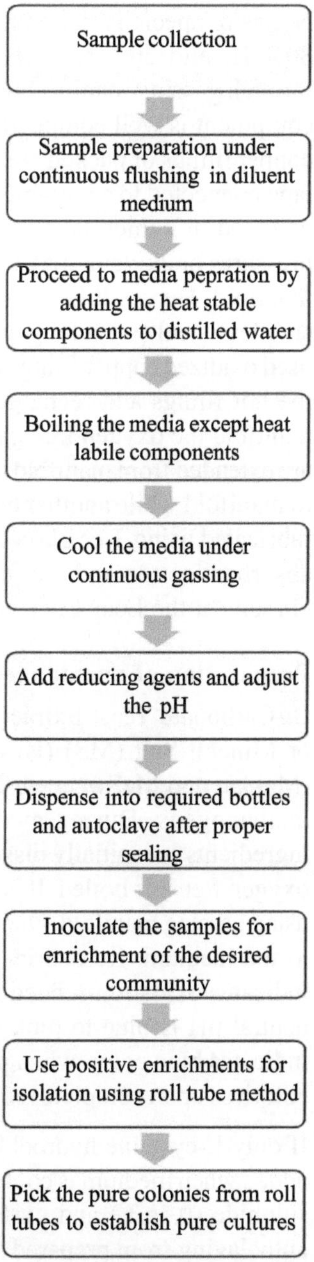

Sample collection

Sample preparation under continuous flushing in diluent medium

Proceed to media pepration by adding the heat stable components to distilled water

Boiling the media except heat labile components

Cool the media under continuous gassing

Add reducing agents and adjust the pH

Dispense into required bottles and autoclave after proper sealing

Inoculate the samples for enrichment of the desired community

Use positive enrichments for isolation using roll tube method

Pick the pure colonies from roll tubes to establish pure cultures

Fig. 1: Media preparation for cultivation & isolation of methanogens

Collection of sample

Sample selection and collection is the most important factor in the cultivation of methanogens. The sample such as deep sediments (river, sea, etc), fecal matter of animals, digestor effluent, etc. are collected aseptically in a sterile, nitrogen filled container/ serum vial. The sample is filled to the brim of the container so as to minimize exposure to oxygen. When the sample is brought to the laboratory, oxygen-free nitrogen is again flushed and the sample is used immediately. In the transit the sample is stored in ice-box.

Usage of oxygen scrubbers

Methanogens are known to be sensitive to oxygen and require a reduced environment for cultivation. Hence, to cultivate the methanogens the mechanism for scrubbing traces of oxygen from commercially available compressed gases is followed. In order to generate oxygen free environment;

N_2 gas or specially calibrated mixtures of 80% N_2 and 20% CO_2 (N_2:CO_2) or 80% H_2 and 20% CO_2 (H_2:CO_2) gases are used. These gases are passed through gassing manifold which acts as an inexpensive oxygen scrubber. This equipment is well equipped with heating glass column that contains reduced-copper filings or packed copper turnings which are heated with electric heating tape connected to a variable transformer. Forinsulation, the cylinder should be wrapped in a thick layer of glass wool. Gas pressures maximum upto 2 atm pressures or 2kg/cm^2 may be used through rotameters equipped to this gassing manifold. Hence, it is essential that a sturdy gas-tight system with copper or stainless-steel (ss) tubing and connecters is to be fabricated/constructed. The used oxidized copper filings are reduced by passing H_2 or H_2:CO_2 slowly through the hot filings and recharged again for media preparation. Through gassing manifold the oxygen free gas is distributed through 5 to 6 outlet probes which are extended from manifold using silicon tubings (one end of tubing is connected to manifold while another end is connected to gassing jet). The gassing jets are fabricated using 3-cc glass Leur-Lock syringe and filled with glass wool /non-absorbent cotton and are further connected with 16-gauge (G) SS needle 12 cm long at the Leur-Lock end.

Preparation of anoxic medium

Bi-Carbonate Yeast Extract Trypticase (BCYT) (Touzel and Albagnac 1983) or Mineral Salt (MS) (Hi et al. 1999), or Bi-carbonate Yeast Extract (BY) (Makkar and McSweeney, 2005) are commonly used media for cultivation of methanogens. During media preparation all the required non-heat labile ingredients are initially dissolved in distilled water. In order make the medium oxygen free it is boiled. It is then cooled under passage of oxygen free nitrogen using gassing manifold. One millilitre of resazurin stock solution (1g/L) is added to the medium. Resazurin- one of the most sensitive redox indicator which indicates oxidation-reduction level of the medium, the indicator end point at neutral pH is blue to pink to colorless. The medium with dissolved oxygen indicates blue color, when partially reduced turns pink color and finally turns colorless upon complete reduction.

If only L-cysteine hydrochloride (1g/L) is to be used as reducing agent, it is added when medium is cooled to around 80°C. Alternatively, mixture of sodium sulphide (0.5g/L) and cysteine hydrochloride (0.5g/L) is to be added after autoclaving from prepared stock solution (mixture of sodium sulphide and L-cysteine hydrochloride generates more redox potential required for cultivation of methanogens as compare to only cysteine hydrochloride). Gassing is continued with nitrogen till the temperature drops to 55°C if the medium is to be used for thermophilic or 40°C for mesophilic or 25°C for psychrophilic methanogens. Gas phase of the medium is change at this stage if it is different from nitrogen.

Gases such as specially calibrated mixtures of 80% N_2 and 20% CO_2 (N_2:CO_2) for acetoclastic and methylotrophic methanogens or 80% H_2 and 20% CO2 (H_2:CO_2) for hydrogentrophic methanogens are used. Temperature of the medium is maintained using serological water bath. Specific salt solution of desired concentration is added in the medium from the prepared stocks. pH of the medium is adjusted after few minutes to the desired level by 1N HCl/1N NaOH when gas phase is N_2; with 1N HCl/sodium bicarbonate (10% w/v) when gas phase is either N_2:CO_2or H_2:CO_2.

The containers or serum vials, in which the medium is to be dispensed, are pregasssed with specific gas phase and then desired volume of the medium is transferred under continuous gassing using same specific gas. The gassing in the vials is continued for a few more minutes to avoid post-transfer oxygen contamination. The vials are then closed securely using butyl rubber stoppers and aluminium seals. They are then autoclaved at 121°C for 20 minutes.

Remaining sterile heat-labile media compounds such as vitamins, volatile compounds (methanogenic substrates-formic acid, or acetic acid, or methanol, etc). or trimethylamine, reducing agent (mix of sodium sulfide and cysteine hydrochloride) are separately added after autoclaving and cooling and sterility checking of the medium. Within a few minutes, the medium gets reduced as indicated by decolorization of resazurin. Medium is now ready to be inoculate either the sample for enrichment or subculture the isolate.

For routine transfer of the methanogenic enrichments or isolates from one vial to another vial under anoxic conditions disposable sterile hypodermic syringes of desired volumes (1, 2, 5, 10, 20ml, etc.) are used. Hypodermic needles of desired pore sizes are used, especially for methanogens like *Methanosarcina* sp. which are clumpy in nature needles of 18 to 20 gauze size are advised.

Vacuum Vortex Method

Vacuum Vortex method is recently described method by Wolfe and Metcalf (2010). This is a quick method in which anaerobic solutions of nonvolatile compounds, anaerobic medium for methanogens are prepared in Hungate tubes or serum vials specifically in small volumes. In this method boiling step for oxygen removal is avoided and involves the use of the equipment with the addition of a common laboratory vortex mixer. Only 10 mL of a solution in a Hungate tube or serum vial also can be rendered anaerobic only in 90 s by three alternate cycles of strong vortexing under vacuum (negative pressure) followed by addition of gas N_2, (or with N_2:CO_2 / H_2:CO_2 for culture media) through 0.2 μm filter. The gassing manifold even with single outlet can be used with the precaution that after each vacuum-vortexing period a positive gas of desired pressure is passed in order to clear remaining liquid from the needle prior the next vacuum–vortex cycle. Accidently, if any liquid enters the 0.2 μm filter, it may seal the filter, preventing gas from passing through the filter. If this happens, replace the filter. This is the most effective method for small volumes of medium for cultivation of methanogens.

Cultivation of methanogens on agar medium

Once the methanogens are enriched upto the desired density and methane gas is detected upto more than 50% (v/v) in the enrichment medium vials, they are isolated using solid agar medium (media with similar composition, only 2 to 2.5% (w/v) agar-agar is additionally added). Hungate's roll tube method (Hungate 1950) is the most suitable method used for isolation of methanogens on agar medium. Alternatively roll bottle method is also popularly used which provides a better surface area for methanogens to get cultivate. Slope bottle method is also used where the agar medium slants are prepared on which the methanogens are isolated using streaking method. An improved Agar plate method described by Herman et al.(1986) can also be used for isolation of methanogens. Later Metcalf et al. (1998) also described a method in which an anaerobic intrachamber incubator is used for growth of *Methanosarcina* spp. on methanol-containing solid media.

Use of an Anaerobic Chamber/ Anaerobic glove box

The isolated colonies of the methanogens are picked using glass sticks with pointed bent tips and transferred from solid agar to liquid broth (respective medium) under anoxic conditions using a sophisticated equipment called anaerobic chamber or anaerobic glove box. Anaerobic glove box is the equipment whose anaerobic conditions are maintained continuously using specially calibrated gas (N_2:H_2:CO_2) (85:10:5). It is well equipped with various filters (palladium-for oxygen scrubbing; charcoal-for carbon absorption; desiccant-for moisture absorption).

Temperature based diversity of methanogens

Methanogenesis is reported over a broad temperature range. Methanogens are isolated from low-temperature ecosystems to hyperthermophilic vents. However at temperatures below 15°C, rate of methanogenesis reduces significantly. Mesophilic organisms are commonly found in freshwater sediments, lakes, marine sediments, intestines and rumen of animals, paddy fields, man-made biogas reactors and anaerobic digesters etc (Liu and Whitman 2008). In thermophilic anoxic habitats, methane is produced either by thermogenic activities or by methanogens as biogenic methane. Since, thermophilic methanogens have higher metabolic activity and higher rate of methanogenesis, therefore, studies on thermophilic methanogens are crucial for both i.e. production of clean energy as well as to limit greenhouse gas emission.Temperature based diversity of methanogens in various habitats is summarised in Table 1 is further discussed below in detail.

Table1: Diversity of methanogens based on temperature

S.N.	Name	Habitat	Substrate (s)	Optimum temperature (°C)	Optimum pH	Optimum salt	Growth factor requirements/ stimulants	Reference
PSYCHROTOLERANT METHANOGENS								
1	*Methanogenium frigidum* Ace 2 (T)	Anoxic waters of Ace Lake, Antartica	Hydrogen- carbon dioxide, formate (slow growth)	15	7.5-7.9	nd	acetate, yeast extract, peptone	Franzmann et al. 1997
2	*Methanococcoides-burtonii* DSM 6242 (T)	Hypolimnion of Ace Lake, Antarctica	methanol	23.4				Franzmann et al. 1992
3	*Methanococcoides alaskense* AK-5 (T)	Anoxic sediment, Skan Bay, Alaska	Trimethylamine	23.6	6·3–7·5	0·3 - 0·4 M	Yeast extract, peptone	Singh et al. 2005
4	*Methanoculleus marisnigri* JR-1 (T)	Marine sediment	Hydrogen- carbon dioxide, formate, carbon dioxide- 2 propanol, carbon dioxide-2 butanol	20-25	6.6	1 g%	ND	Romesser et al. 1979; Maestrojuán et al. 1990
5	*Methanogenium cariaci* JR1 (T)	Sediment from Cariaco Trench	Hydrogen- carbon dioxide, formate	20-25	6.8-7.3	2.69 g%	Yeast extract and acetate	Romesser et al. 1979; Maestrojuán et al. 1990
MESOPHILIC METHANOGENS								
6	*Methanogenium marinum* AK-1 (T)	Permanently cold marine sediments	Hydrogen- carbon dioxide, formate (slow growth)	25	6-6.6			Chong et al. 2002
7	*Methanospirillum-psychrodurum* X-18 (T)	Wetland soil	Hydrogen- carbon dioxide	25	7	nd	yeast extract	Zhou et al. 2014

(Contd.)

8	*Methanosarcina baltica* GS1-A (T)	Baltic Sea sediment	methanol, acetate, methylamine	25	6.5-7.5	0·3–0·4 M		von Klein et al. 2002
9	*Methanosarcina lacustris* ZS (T)	Anoxic lake sediments	Monomethylamine, dimethylamine, hydrogen-carbon dioxide	25	7	nd		Simankova et al. 2002
10	*Methanolobust indarius* Tindari 3 (T)	Marine black sediment	Methanol, methylamines	25	6.5	0.47 M		König and Stetter 1983
11	*Methanobacterium veterum* MK4 (T)	Ancient Siberian permafrost	Hydrogen- carbon dioxide, methanol-hydrogen, methylamine-hydrogen	28	7.0-7.2	0.05 M	acetate	Krivushin et al. 2010
12	*Methanosarcina soligelidi* SMA-21 (T)	Siberian permafrost-affected soil.	Acetate, hydrogen-carbondioxide	28	7.8	0.02 M		Wagner et al. 2013
13	*Methanobrevibacter curvatus* RFM-2 (T)	Hindgut content of termite *Reticulitermes flavipes*	Hydrogen- carbon dioxide	30	7.1-7.2	nd	rumen fluid	Leadbetter and Breznak 1996
14	*Methanobrevibacter filiformis* RFM-3 (T)	Hindgut content of termite *Reticulitermes flavipes*	Hydrogen- carbon dioxide	30	7.0-7.2	nd	Yeast extract	Leadbetter et al. 1998
15	*Methanobacterium lacus* 17A1 (T)	Profundal sediment of a freshwater meromictic lake	Hydrogen- carbon dioxide, methanol-hydrogen	30	6.5	0.1 M	yeast extract, vitamins, acetate, rumen fluid	Borrel et al. 2012
16	*Methanocorpusculum sinense* China Z (T)	Distillery waste water treatment plant	Hydrogen- carbon dioxide, formate	30	7		Rumen fluid	Zellner et al. 1987

(Contd.)

17	*Methanospirillum lacunae* Ki8-1 (T)	Puddly soil	Hydrogen- carbon dioxide, formate	30	7.2	0	acetate, yeast extract	Iino et al. 2010
18	*Methanococcoides vulcanii* SLH33 (T)	Mud volcano	trimethylamine, dimethylamine, monomethylamine, methanol, betaine, N, N-dimethylethanol amine and choline	30	7	0.5 M	Vitamins, yeast extract, peptone, tryptone and Casamino acids	L'Haridon et al. 2014
19	*Methanolobus profundi* MobM (T)	Deep subsurface sediments in a natural gas field	Methanol, methylamines	30	6.5	0.35 M	Magnesium	Mochimaru et al. 2009
20	*Methanohalophilus-halophilus* INMI Z-7982 (T)	Marine cyanobacterial mat	Methylamine, methanol	30		1.2–1.5 M		Zhilina 1983, Wilharm et al. 1991
21	*Methanospirillumstamsii Pt1 (T)*	anaerobic expanded granular sludge bed bioreactor operated at low temperature	Hydrogen- carbon dioxide, formate (poor growth)	20-30	7-7.5	0	nd	Parshina et al. 2014
22	*Methanoculleus chikugoensis* MG62 (T)	Paddy field soil	Hydrogen- carbon dioxide, formate, carbon dioxide- 2 propanol, carbon dioxide-2 butanol, cyclopentanol-carbon dioxide	25-30			Trypticase, peptone, casitone, acetate	Dianou et al. 2001
23	*Methanosphaerula palustris* E1-9c (T)	Minerotrophic fen peatland	Hydrogen- carbon dioxide, formate	28-30	5.5	<0.2%	Vitamin, CoM, acetate	Cadillo-Quiroz et al. 2009

(Contd.)

24	*Methanoplanusendosymbiosus* MC1 (T)	Marine ciliate	Hydrogen- carbon dioxide, formate	32	6.8-7.3	1.5 g%		van Bruggen et al. 1986
25	*Methanobacterium movilense* MC-20 (T)	Subsurface lake in Romania	Hydrogen- carbon dioxide, 2-propanol, 2-butanol, formate	33	7.4	0.08 M	None	Schirmack et al. 2014
26	*Methanosarcinas pelaei* MC-15 (T)	Sulfurous subsurface lake	Hydrogen- carbon dioxide, methanol, monomethylamine, dimethylamine, trimethylamine, acetate	33	6.6	0.05 M	None	Ganzert et al. 2014
27	*Methanobrevibacter acididurans* ATM (T)	Acidogenic digestor	Hydrogen- carbon dioxide	35	6	nd	acetate, rumen fluid, amino acids	Savant et al. 2002
28	*Methanobacterium espanolae* GP9 (T)	Sludge of a bleachcraft mill	Hydrogen- carbon dioxide (2-propanol, 2-butanol)	35	5.6-6.2	nd	Cells were grown in vitamin free medium containing acetate	Patel et al. 1990
29	*Methanobacterium petrolearium* Mic5c12 (T)	Sludge deposited in crude oil storage tank	Hydrogen- carbon dioxide	35	6.5	0-0.68 M	yeast extract, acetate	Mori and Harayama 2011
30	*Methanomicrobium antiquum* MobH (T)	Deep sedimentary natural gas bearing aquifers	Hydrogen- carbon dioxide, formate	35	7.0-7.5	2%	tungsten, yeast extract and acetate	Mochimaru et al. 2016
31	*Methanocalculu spumilus* MHT-1 (T)	Leachate of a sea-based site for solid waste disposal	Hydrogen- carbon dioxide, formate	35	6.5-7.5	1%	Acetate	Mori et al. 2000

(Contd.)

32	*Methanocalculus alkaliphilus* AMF-2 (T)	Soda lake, Russia	Hydrogen- carbon dioxide, formate, acetate	35	9.5	0.3-0.6 M	carbonates	Sorokin et al. 2015
33	*Methanosarcina subterranea* HC-2 (T)	Deep subsurface diatomaceous shale formation	Monomethylamine, dimethylamine, dimethylsulfide	35	6.6-6.8	0.1-0.2 M		Shimizu et al. 2015
34	*Methanohalophiluslevihalophilus* GTA13 (T)	Natural gas bearing confines aquifers	dimethylamine and trimethylamine	35	7.0–7.5	0.35–0.40 M	Vitamins, sodium, magnesium	Katayama et al. 2014
35	*Methanohalophilus mahii* SLP (T)	Saline lake sediment	Methylamine, methanol	35	7.5	2 M	Sodium, magnesium, iron, potassium	Paterek and Smith 1988
36	*Methanocalculusnatronophilus* Z-7105 (T)	Soda lake, Russia	Hydrogen- carbon dioxide, formate	35	9.0-9.5	1.4-1.9 M	carbonates	Zhilina et al. 2013
37	*Methanoregula formicica* SMSP (T)	Mesophilic UASB reactor	Hydrogen- carbon dioxide, formate	30-33	7.4	nd	Yeast extract, acetate, CoM	Yashiro et al. 2011
38	*Methanogenium organophilum* CV (T)	Marine mud	Hydrogen- carbon dioxide, formate, carbon dioxide- 2 propanol, carbon dioxide-2 butanol, carbon dioxide-1 propanol, carbon dioxide-ethanol	30-35	6.4-7-3	2 g %	biotin, 4 amino benzoate, vitamin B12, tungstate, Magnesium chloride	Widdel et al. 1988
39	*Methanosarcinasemesiae* MD1 (T)	Mangrove sediment	dimethylsulfide, methanethiol, methanol and methylamines	30-35	6.5-7.5			Lyimo et al. 2000
40	*Methanococcoides methylutens* TMA-10 (T)	Submarine canyon sediment, California	methanol, N, N-dimethylethanol amine	30-35				Sowers and Ferry 1983

(Contd.)

41	*Methanobrevibacter gottschalkii* HO (T)	Horse feces	Hydrogen- carbon dioxide	37	7	nd	acetate or yeast extract or trypticase	Miller and Lin 2002
42	*Methanobrevibacter thaueri* CW (T)	Cow feces	Hydrogen- carbon dioxide	37	7	nd	acetate or yeast extract or trypticase	Miller and Lin 2002
43	*Methanobrevibacterwo linii* SH (T)	Sheep feces	Hydrogen- carbon dioxide	37	7	nd	acetate or yeast extract or trypticase	Miller and Lin 2002
44	*Methanobrevibacter woesei* GS (T)	Goose feces	Hydrogen- carbon dioxide, formate (poor growth)	37	7	nd	acetate or yeast extract or trypticase	Miller and Lin 2002
45	*Methanobrevibacter cuticularis* RFM-1 (T)	Hindgut content of termite *Reticulitermes flavipes*	Hydrogen- carbon dioxide, formate (poor growth)	37	7.7	nd	none	Leadbetter and Breznak 1996
46	*Methanobacterium beijingense* 8–2 (T)	Anaerobic digestor	Hydrogen- carbon dioxide, formate	37	7.2		Yeast extract	Ma et al. 2005
47	*Methanobacterium alcaliphilum*WeN4 (T)	Alkaline lake	Hydrogen- carbon dioxide	37	8.1 - 9.1	nd	trypticase or yeast extract	Worakit et al. 1986
48	*Methanobacterium arcticum* M2 (T)	Holocene Arctic Permafrost	Hydrogen- carbon dioxide, formate	37	6.8-7.2	0.1 M	None	Shcherbakova et al. 2011
49	*Methanoculleus hydrogenitrophicus* HC (T)	Wetland soil	Hydrogen- carbon dioxide	37	6.6	0.2 M	yeast extract	Tian et al. 2010
50	*Methanoculleus taiwanensis* CYW4 (T)	Deep-sea sediment	Hydrogen- carbon dioxide, formate	37	8.1	0.08 M	acetate	Weng et al. 2015
51	*Methanoculleus sediminis* S3Fa (T)	Submarine mud volcano	Hydrogen- carbon dioxide, formate	37	7.1	0.17 M	acetate	Chen et al. 2015
52	*Methanoculleus olentangyi* RC/ER (T)	River sediment	Hydrogen- carbon dioxide, formate	37		0.17 M	Acetate	Corder et al., 1983;

(Contd.)

								Maestrojuán et al. 1990
53	*Methanofollisaquaemaris* N2F9704 (T)	Aquaculture fish pond	Hydrogen- carbon dioxide, formate	37	6.5	0.50%	Tungsten	Lai and Chen 2001
54	*Methanofollis ethanolicus* HASU (T)	Lotus field	Ethanol, 1-propanol, 1-butanol, hydrogen-carbon dioxide and formate	37	7	0	yeast extract	Imachi et al. 2009
55	*Methanolaciniapetro-learia* SEBR (T)	African off shore oil field	Hydrogen- carbon dioxide, formate, carbon dioxide- 2 propanol	37	7	1-3 g%	Acetate, yeast extract	Ollivier et al. 1997; Göker et al. 2014
56	*Methanocorpusculu-mlabreanum* Z (T)	Surface sediments of La Brea Tar Pits, LA, USA	Hydrogen- carbon dioxide, formate	37	7	d"1.5 %	Trypticase, peptone, yeast extract, acetate	Zhao et al. 1989
57	*Methanocorpusculum-bavaricum* SZSXXZ (T)	Anaerobic waste water treatment pond at a sugar factory	Hydrogen- carbon dioxide, 2 propanol, 2 butanol, formate	37	7		Rumen fluid	Zellner et al. 1987
58	*Methanocorpusculum parvum* XII (T)	Waste water	Hydrogen- carbon dioxide, formate, carbon dioxide- 2 propanol, carbon dioxide-2 butanol	37	6.8-7.5	0-4.7 g%	Tungsten	Zellner et al. 1987
59	*Methanocalculus chung hsing ensis* K1F9705b (T)	Estuary and a marine fishpond	Hydrogen- carbon dioxide, formate	37	7.2	1%	Acetate	Lai et al. 2004
60	*Methanocalculustaiw-anense* P2F9704a (T)	Estuary in Taiwan	Hydrogen- carbon dioxide, formate	37	6.7	0.50%	acetate	Lai et al. 2002
61	*Methanosarcina horonobensis* HB-1 (T)	Deep subsurface Miocene	Acetate, methanol, dimethylamine,	37	7-7.25	0.1 M	Bacto yeast extract,	Shimizu et al. 2011

(Contd.)

		formation	trimethylamine, dimethylsulfide				Bactocasitone (BD), vitamins and trace minerals	
62	*Methanolobus bombayensis* B-1 (T)	Arabian sea sediments	methylamines, methanol, and dimethyl sulfide	37	7.2	0.5 M	divalent cations (magnesium and calcium), yeast extract and peptone	Kadam et al. 1994
63	*Methanolobuschelungpuianus* St545Mb (T)	Fault gouge sample	methanol, methylamines	37	7	0-0.08 M	Magnesium	Wu and Lai 2011
64	*Methanolobus vulcani* PL-M (T)	Sea sediment	methylamines and methanol	37	7.2	0.5 M	biotin and yeast extract	Stetter 1989; Kadam and Boone 1995
65	*Methanolobus taylorii* GS-16 (T)	Estuarine sediments	Methylamine, methanol, dimethylsulfide, methanethiol	37	8	0.5 M	biotin	Oremland and Boone 1994
66	*Methanosalsumnatronophilum* AME-2 (T)	Hypersaline soda lake in Russia	methanol, methylamines and dimethylsulfide	37	9.5	1.5 M	Yeast extract	Sorokin et al. 2015
67	*Methanolinea mesophila* TNR (T)	Rice field	Hydrogen- carbon dioxide, formate	37	7	0 g/l	None	Sakai et al. 2012
68	*Methanoregulaboonei* 6A8 (T)	Acidic peat bog	Hydrogen- carbon dioxide	37	5	nd	Yeast extract and acetate	Bräuer et al. 2006
69	*Methanomassiliicoccus fluminyensis* B10 (T)	Human feces	Hydrogen-Methanol, monomethylamine, dimethylamine and trimethylamine	37	7.6	1%	yeast extract, methylamine, one or more B vitamins, tungstate, and selenite	Dridi et al. 2012

(Contd.)

70	*Methanomethylovorans-hollandica*DMS-1 (T)	Eutrophic lake	Methanol, dimethylsulfide, methanethiol	37	6.5-7	0-40 mM	ferrous chloride	Lomans et al. 2004
71	*Methanomethylovoran-suponensis* EK1 (T)	Wetland sediment	Methanol, mono-, di- and trimethylamine, dimethyl sulfide and methanethiol	37	6-6.5	0	Cobalt chloride, ferrous chloride	Cha et al. 2013
72	*Methanothrix soehngenii* Opfikon (T)	mesophilic sewage digester	Acetate	37	7.4-7.8			Husar et al. 1982
73	*Methanobacterium arbophilicum* DH1(T)	Wetwood of living trees	Hydrogen- carbon dioxide	30-37	7.5-8.0	nd	rumen fluid, vitamins	Zeikus and Henning 1975
74	*Methanospirillum hungatei* JF-1 (T)	Sewage sludge	Hydrogen- carbon dioxide, formate	30-37	6.6-7.4	0	yeast extract	Ferry et al. 1974
75	*Methanobrevibacterar boriphilus* DH1 (T)	Decaying cottonwood tissue	Hydrogen- carbon dioxide, formate (not by all strains)	30–37	7.5–8	nd	acetate, vit B	Zeikus and Henning 1975;
76	*Methanolobus psychrotolerans* YSF-03 (T)	Saline meromictic Lake Shira in Siberia	Methanol, methylamines	30–37		7-7.4	0.17 M	Chen et al. 2018
77	*Methanobacterium paludis* SWAN1 (T)	Peatlands	Hydrogen- carbon dioxide	32–37	5.4–5.7	nd	CoM and Acetate	Cadillo-Quiroz et al. 2014
78	*Methanobacterium palustre* F (T)	Peat bog	Hydrogen- carbon dioxide, 2-propanol, formate (2-butanol)	33-37	7		None	Zellner et al. 1988
79	*Methanosaeta harundinacea* 8Ac (T)	Anaerobic digestor treating beer manufacture wastewater	Acetate	34-37	7.2-7.6	nd	Yeast extract, peptone	Ma et al. 2006

(Contd.)

80	*Methanobrevibacter smithii* PS (T)	Sewage sludge	Hydrogen- carbon dioxide, formate (poor growth)	34-46	5.5-7.0	nd	acetate, vit B	Balch et al. 1979
81	*Methanolobus oregonensis* WAL1 (T)	Saline, alkaline aquifer	methanol, methylamines, dimethylsulfide	35-37	8.6	0.48 M	Biotin, thiamine, yeast extract, peptone	Liu et al. 1990
82	*Methanocella paludicola* SANAE (T)	Rice field	Hydrogen- carbon dioxide, formate	35-37	7	0	yeast extract	Sakai et al. 2008
83	*Methanomicrobium mobile* BRM16	Bovine rumen	Hydrogen- carbon dioxide, formate	38	6.5-7	nd	Yeast extract and acetate	Paynter and Hungate 1968; Jarvis et al. 2000
84	*Methanocalculus halotolerans* SEBR 4845 (T)	French oil well	Hydrogen- carbon dioxide, formate	38	7.6	5 g%	Acetate, yeast extract	Ollivier et al. 1998
85	*Methanobrevibacter oralis* ZR (T)	Human subgingival plaque	Hydrogen- carbon dioxide	35-38	6.9–7.4		Fecal extract	Ferrari et al., 1994
86	*Methanobacterium flexile* GH (T)	Sediments of Gahai lake, China	Hydrogen- carbon dioxide, formate	35-38	7.0–7.5	0–0.1 M	yeast extract	Zhu et al. 2011
87	*Methanobacteriumm-ovens* TS-2 (T)	Sediments of Tuosu lake, China	Hydrogen- carbon dioxide	35-38	7.2–7.5	0–0.3 M	Yeast extract	Zhu et al. 2011
88	*Methano micrococcus blatticola*PAT (T)	Hindgut of the cockroach Periplanetaam-ericana	Methnaol, methylamine with molecular hydrogen	39	7.2-7.7	<100 Mm	Acetate, coenzyme M, yeast extract, tryptic soy broth and vitamins	Sprenger et al. 2000
89	*Methanobacterium oryzae* Fpi (T)	Rice field	Hydrogen- carbon dioxide, formate	40	7	nd	None	Joulian et al. 2000

(Contd.)

90	*Methanobacterium ferruginis* Mic6co5 (T)	Interior of pipe transporting natural gas containing brine	Hydrogen- carbon dioxide	40	6-7.5	0.34 M	Vitamins	Mori and Harayama 2011
91	*Methanobacteriumkan-agiense* 169 (T)	Rice field	Hydrogen- carbon dioxide	40	7.5-8.5	5 g/L	None	Kitamura et al. 2011
92	*Methanobacterium aggregans* E09F.3 (T)	Biogas plant	Hydrogen- carbon dioxide, formate	40	6.7-7	6.8 mM	none	Kern et al. 2015
93	*Methanoculleus palmolei* INSLUZ (T)	Anaerobic digester treating waste water of a palm oil plant	Hydrogen- carbon dioxide, formate, carbon dioxide- 2 propanol, carbon dioxide-2 butanol, cyclopentanol-carbon dioxide	40	6.9–7.5	ND	Acetate	Zellner et al. 1998
94	*Methanofollisformosanus* ML15 (T)	Aquaculture fish pond	Hydrogen- carbon dioxide, formate	40	6.6	3%	Tungsten	Wu et al. 2005
95	*Methanofollisliminatans* GKZPZ (T)	Sludge	Hydrogen- carbon dioxide, formate, carbon dioxide- 2 propanol, carbon dioxide-2 butanol, cyclopentanol-carbon dioxide	40	7	0	Acetate	Zellner et al. 1990, 1999; Zellner and Boone 2001
96	*Methanolaciniapaynteri* G2000 (T)	Marine sediment from mangrove swamp	Hydrogen- carbon dioxide, 2 propanol, 2 butanol	40	7	0.15 M	acetate	Rivard et al. 1983; Zellner et al. 1989
97	*Methanoplanus limicola* M3 (T)	Swamp	Hydrogen- carbon dioxide, formate	40	6.5-7.5	1 g%	acetate	Wildgruber et al. 1982

(Contd.)

98	*Methanosarcina siciliae* DSM 3028 (T)	Submarine canyon sediments	Monomethylamine, dimethylamine, trimethylamine, methanol	40	6.5–6.8	0.4–0.6 M		
99	*Methanosarcina vacuolata* DSM 1232 (T)	Anaerobic digestor	Hydrogen- carbon dioxide, methanol, monomethylamine, dimethylamine, trimethylamine, acetate	40	7.5	0.1 M		Zhilina and Zavarzin 1987
100	*Methanohalophilus portucalensis*FDF-1 (T)	Salt pond	Methylamine, methanol	40	6.5-7.5	0.5 - 2 M	p-aminoben-zoate, biotin	Boone et al. 1993a
101	*Methanolobus siciliae* T4/M (T)	Gas and oil well in the Gulf of Mexico	methanol, trimethylamine, diaminesulfide	40	6.5-6.8	0.4-0.6 M	Magnesium ions, yeast extract	Ni and Boone 1991
102	*Methanoculleus submarinus* Nankai-1 (T)	Ocean sediment that contain methane hydrates	Hydrogen- carbon dioxide, formate	43	6.0-7.5	0.1-0.4	Acetate	Mikucki et al. 2003
103	*Methanobacterium subterraneum* A8p (T)	Deep granitic ground water	Hydrogen- carbon dioxide, formate	20-40	7.8-8.8		None	Kotelnikova et al. 1998
104	*Methanohalobiumevesti-gatum* Z-7303 (T)	Sediments of saline lagoons in Sivash	Methylamine, methanol	30-40	7.0-8.0	2-4 M		Zhilina and Zavarzin 1987
105	*Methanosphaera cuniculi* 1R7 (T)	Rabbit rectum	Hydrogen- carbon dioxide, methanol	35-40	6.8		acetate	Biavati et al. 1988
106	*Methanococcus vannielii* SB (T)	Marine sediments	Hydrogen- carbon dioxide, formate	35-40	7.0-8.0	0.6-2	Selenium	Stadtman and Barker 1951
107	*Methanococcus voltae* PS (T)	Marine sediments	Hydrogen- carbon dioxide, formate	35-40	6.0-7.0	1.0-2.0	Acetate, amino acids, selenium	Balch et al. 1979; Whitman et al. 1982

(Contd.)

108	*Methanococcus maripaludis* JJ (T)	Salt marsh sediments	Hydrogen- carbon dioxide, formate	35-40	6.8-7.2	0.6-2	Acetate, amino acids, selenium	Jones et al. 1983b
109	*Methanoculleus bourgense* MS-2 (T)	Digestor	Hydrogen- carbon dioxide, formate	35-40	6.7	0.17 M	Acetate	Ollivier et al. 1986; Maestrojuan et al. 1990
110	*Methanosarcina acetivorans* C2A (T)	Marine canyon sediments	Acetate, monomethylamine, dimethylamine, trimethylamine, methanol	35-40	6.5-7	0.2 M		Sowers et al. 1984
111	*Methanosaeta concilii* GP6 (T)	Anaerobic sewage digestor	Acetate	35-40	7.1-7.5		Sludge fluid, calcium, magnesium, biotin, thiamine hydrochloride	Patel and Sprott 1990
112	*Methanobrevibacter olleyae* KM1H5-1P (T)	Ovine rumen	Hydrogen- carbon dioxide, formate	36-40	7.5		acetate	Rea et. al. 2007
113	*Methanosphaera stadtmanae* MCB3 (T)	Human feces	Hydrogen- carbon dioxide, methanol	36-40	6.5-6.9		acetate, Thiamine, isoleucine, leucine	Miller and Wolin 1985
114	*Methanobrevibacter millerae* ZA-10 (T)	Bovine rumen	Hydrogen- carbon dioxide, formate	36–42	7.0-8.0		acetate, yeast extract or trypticase	Rea et. al. 2007
115	*Methanobrevibacter ruminantium* M1 (T)	Bovine rumen	Hydrogen- carbon dioxide, formate (poor growth)	37-39	6.0-7.0	nd	acetate, vit B, 2-methylbutyric acid and CoM	Smith and Hungate 1958;
116	*Methanobacterium bryantii* M.o.H. (T)	Anaerobic digestor	Hydrogen- carbon dioxide (2-propanol,	37-39	6.9-7.2	nd	None	Boone 1987

(Contd.)

			2-butanol, cyclopentanol)					
117	*Methanobrevibacter-boviskoreani*JH1 (T)	Rumen of Korean cattle	Hydrogen- carbon dioxide, formate	37-40	6.5-7	nd	yeast extract, CoM, VFAs	Lee et al. 2013
118	*Methanobacterium ulig inosum*P2St (T)	Marshy soil	Hydrogen- carbon dioxide	37-40	6.0-7.5		None	König 1984
119	*Methanofollistationis* Chile 9 (T)	Solfataric field on Mount Tatio, Chile	Hydrogen- carbon dioxide, formate	37-40	7	8-12 g/l	Yeast extract and acetate	Zabel et al. 1984; Zellner et al. 1999
120	*Methanobacteriumcong-olense* C (T)	Anaerobic digestor	Hydrogen- carbon dioxide (2-propanol, 2-butanol, cyclopentanol)	37-42	7.2	nd	none	Cuzin et al. 2001
121	*Methanoculleus horono bensis* T10 (T)	Groundwater sampled from a deep diatomaceous shale formation	Hydrogen- carbon dioxide, formate	37–42	6.7–6.8	0.1–0.2 M	CoM, acetate, yeast extract	Shimizu et al. 2013
122	*Methanobacterium form icicum* MF (T)	Sewage sludge	Hydrogen- carbon dioxide, formate	37-45	7.0-7.5	nd	none	Boone 1987
123	*Methanobacteriumivano-vii*Ivanov (T)	Rock core	Hydrogen- carbon dioxide	37–45	7.0 - 7.4		None	Belyaev et al. 1983; Jain et al. 1987
124	*Methanosarcinamazei* S-6 (T)	Anaerobic digestor	Hydrogen-carbondioxide, monomethylamine, acetate	40-42	6.0-7.0	0.1-0.3 M		Barker 1936; Mah and Kuhn 1984
THERMOPHILIC AND HYPERTHERMOPHILIC METHANOGENS								
125	*Methanobacteriumaar-husense* H2-LR (T)	Marine sediment	Hydrogen- carbon dioxide	45	7.5-8		None	Shlimon et al. 2004

(Contd.)

126	*Methanoculleusoldenburgensis* CB-1 (T)	River sediment	Hydrogen- carbon dioxide, formate	45	8	0	Acetate	Blotevogel et al. 1991
127	*Methanosarcina flavescens* E03.2 (T)	Full-scale anaerobic digester	Hydrogen- carbon dioxide, Acetate, methanol, monomethylamine, dimethylamine, trimethylamine	45	7	6.8 mM	Yeast extract, vitamins	Kern et al. 2016
128	*Methanosarcina barkeri* MS1 (T)	Anaerobic sewage digestor	Hydrogen- carbon dioxide, methanol, monomethylamine, dimethylamine, trimethylamine, acetate	45	7	< 0.2 M		Maestrojuán and Boone, 1991
129	*Methanosalsumzhilinae* WeN5 (T)	Alkaline, saline lake sediment, Egypt	methanol, methylamines, dimethylsulfide	45	9.2	0.7 M	Yeast extract, trypticase, peptone, rumen fluid	Mathrani et al. 1988; Boone and Baker 2002
130	*Methanocellaarv oryzae* MRE50 (T)	Rice field	Hydrogen- carbon dioxide, formate	45	7	0-2 g/l	None	Sakai et al. 2010
131	*Methanococcusaeolicus* Nankai-3 (T)	Marine sediments	Hydrogen- carbon dioxide, formate	46	7	1.0-2.0	Selenium	Kendall et al. 2006
132	*Methanosarcinathermophila* TM-1 (T)	Thermophilic anaerobic digestor	Acetate, monomethylamine, dimethylamine, trimethylamine, methanol	50	6	<0.05 M		Zinder et al. 1985
133	*Methanolinea tarda* NOBI-1 (T)	Methanogenic sludge obtained from muncipal sewage treatment plant	Hydrogen- carbon dioxide, formate	50	7	0	Yeast extract and acetate	Imachi et al. 2008

(Contd.)

134	*Methanomethylovorans-thermophila*L2FAW (T)	Thermophilic UASB reactor	Methanol, methylamines	50	6.5	<100 mM	Cobalt	Jiang et al. 2005
135	*Methanolobus zinderi* SD1 (T)	Deep subsurface coal seam	Methanol, monomethylamine, dimethylamine or trimethylamine	40-50	7.0-8.0	0.2-0.6 M	Magnesium	Doerfert et al. 2009
136	*Methanothermobacter-thermoplexus* IDZ (T)	Anaerobic digestor	Hydrogen- carbon dioxide, formate	55	7.9–8.2		CoM	Kotelnikova et al. 1993
137	*Methanoculleus thermophilus* CR-1 (T)	Soil sediment	Hydrogen- carbon dioxide, formate	55	7	0.25 M	Trypticase, vitamin solution	Rivard and Smith 1982; Maestrojuan et al. 1990
138	*Methanocellacon radii* HZ254 (T)	Rice field	Hydrogen- carbon dioxide	55	6.8	0-1 g/l	yeast extract	Lü and Lu 2012
139	*Methanoculleus receptaculi* ZC-2 (T)	Chinese oil wells (Shengli)	Hydrogen- carbon dioxide, formate	50–55	7.5–7.8	0.2M	Acetate	Cheng et al. 2008
140	*Methanothermo bacter thermophilus* M (T)	Sludge of methane tank	Hydrogen- carbon dioxide	57	7.5		CoM	Laurinavichyus et al. 1988
141	*Methanothermo bacter-defluvii* ADZ (T)	Anaerobic digestor	Hydrogen- carbon dioxide, formate	60	7		CoM	Kotelnikova et al. 1993
142	*Methanosaeta thermophila* PT (T)	Anaerobic digestor	Acetate	55-60	6.7	nd	Sludge fluid	Kamagata and Mikami 1991; Kamagata et al. 1992; Boone and Kamagata 1998
143	*Methanothermobacter marburgenesis* Marburg (T)	Sewage sludge	Hydrogen- carbon dioxide	65	6.8–7.4			Wasserfallen et al. 2000

(Contd.)

144	*Methanothermo bacter crinale* Tm2 (T)	Oil sands and oil production water in the Chinese oil field (Shengli)	Hydrogen- carbon dioxide	65	6.9	0.025 g/L	Acetate	Cheng et al. 2011
145	*Methanobacterium thermaggregans* DSM 3266 (T)	Mud taken from a cattle pasture	Hydrogen- carbon dioxide	65	7-7.5	0	yeast extract	Blotevogel and Fischer 1985; Wasserfallen et al. 2000
146	*Methermicoccus shengliensis* ZC-1	Oil-production water from Chinese oil field	Methanol, methylamine and trimethylamine	65	6.0-6.5	0.3-0.5 M	Magnesium chloride, HS-CoM	Cheng et al. 2007
147	*Methanothermobacter wolfeii*VKM-B 1829 (T)	Sewage sludge and river sediment	Hydrogen- carbon dioxide	55-65	7.0-7.5			Winter et al. 1984
148	*Methanothermococcus thermolithotrophicus* SN1 (T)	Coastal geothermally heated sea sediments	Hydrogen- carbon dioxide, formate	60-65	5.1 - 7.5	2.0 - 4.0		Huber et al. 1982
149	*Methanothermococcus-okinawensis* IH1 (T)	Deep-sea hydrothermal vent	Hydrogen- carbon dioxide, formate	60-65	6.0 - 7.0	2.5 - 5.0 (Sea salts)	Selenium	Takai et al. 2002
150	*Methanothermobacter-thermautotrophicus* ΔH (T)	Sewage sludge	Hydrogen- carbon dioxide	65-70	7.2–7.6			Zeikus and Wolfe 1972
151	*Methanotorrisformicicus* Mc-S-70 (T)	Deep-sea black smoker chimney	Hydrogen- carbon dioxide, formate	75	6.7	2.4	None	Takai et al. 2004
152	*Methanocaldococcus-villosus*KIN24-T80 (T)	Submarine hydrothermal system	Hydrogen- carbon dioxide	80	6.5	2.50%	Selenate, tungstate and yeast extract	Bellack et al. 2011
153	*Methanocaldococcus-vulcanius* M7 (T)	Deep-sea hydrothermal vent	Hydrogen- carbon dioxide	80	6.5	2.5	Selenium, yeast extract	Jeanthon et al. 1999; Whitman 2001

(Contd.)

154 *Methanocaldococcus-bathoardescens* JH 146 (T)	Volcanically active deep-sea hydrothermal vent	Hydrogen- carbon dioxide	82	7	2.90%	None	Stewart et al. 2015
155 *Methanocaldococcus jannaschii* JAL-1 (T)	Deep-sea hydrothermal vent	Hydrogen- carbon dioxide	85	6	3	Selenium	Jones et al. 1983a; Whitman 2001
156 *Methanocaldococcus fernus* ME (T)	Deep-sea hydrothermal vent	Hydrogen- carbon dioxide	85	6.5	2	Selenium, yeast extract	Jeanthon et al. 1998; Whitman 2001
157 *Methanocaldococcus fervens*AG86 (T)	Deep-sea hydrothermal vent	Hydrogen- carbon dioxide	85	6.5	3	Selenium, yeast extract	Jeanthon et al. 1999; Zhao et. al. 1988; Whitman 2001
158 *Methanocaldococcus indicus* SL 43 (T)	Deep-sea hydrothermal vent	Hydrogen- carbon dioxide	85	6.5	3	Selenium, yeast extract	L'Haridon et al. 2003
159 *Methanothermus social bilis* Kf1-F1 (T)	Icelandic hot spring	Hydrogen- carbon dioxide, formate	88	6.5	nd	None	Lauerer et al. 1986
160 *Methanotorris igneus* Kol 5 (T)	Shallow marine hydrothermal vent	Hydrogen- carbon dioxide	88	5.7	1.8	None	Burggraf et al. 1990
161 *Methanothermus fervidus* V24S (T)	Icelandic hot spring	Hydrogen- carbon dioxide, formate	80-85	6.5	nd	None	Stetter et al. 1981
162 *Methanopyrus kandleri* AV19 (T)	Hydrothermally heated deep-sea sediments and from a shallow marine hydrothermal system	Hydrogen- carbon dioxide	98	6.5	2 g%	None	Kurr et al. 1991

Psychrophilic methanogens

An organism growing optimally at 15°C or lower with cardinal temperatures of growth between 0°C or lower and below 20°C is called a psychrophilic organism (Madigan et al. 2012). Franzmann et al. (1997) first time reported psychrophilic methanogen viz., *Methanogenium frigidum* from Ace Lake, Antarctica. Many different methanogens have been reported from different natural as well as man-made habitats. *Methanobacterium arcticum, M. veterum* and *Methanosarcina mazei* have been reported from Siberian permafrost (Rivkina et al. 2007, Krivushin et al. 2010 and Shcherbakova et al. 2011). Methanogenesis was reported from permafrost of Bunger Hills Oasis, Antarctica and the dominant genus was *Methanosarcina* sp. (Vishnivetskaya et al. 2018). The metabolic versatility allows members of genus *Methanosarcina* exhibit diverse metabolic activities thus allowing them to inhabit varied environments including Siberian and Antarctic permafrost (Liu and Whitman 2008). Recently, Car et al. (2018) reported predominance of another acetoclastic genus viz., *Methanosaeta* from organic-rich Antarctic marine sediments.

Psychrotolerant methanogens

Organisms that grow optimally between 20–40°C but can grow at 0°C are called psychrotolerant organisms (Madigan et al. 2012). Psychrotolerant methanogens growing optimally at 18–35°C have been isolated from the boreal as well as polar region. The first psychrotolerant methanogen viz., *Methanococcoides burtonii* was isolated from Ace Lake, Antarctica (Franzmann et al. 1992). Methanogenic pathways in a few cold habitats have been studied earlier. In most cold environments, 67% of total methane production is contributed by acetoclastic methanogenesis. Kotsyurbenko et al. (2004) showed that Methanosarcinaceae, Methanomicrobiaceae and Rice cluster II were dominant in an acidic Siberian peat bog. Methylotrophic methanogenesis is a predominant pathway in Zoige wetland at Tibetan Plateau (Jiang et al. 2010). In northern acidic peatlands, hydrogenotrophic methanogenesis is dominant in methane production (Kotsyurbenko et al. 2007, Yavitt et al. 2012). Studies on different types of borealwetlands with acidic peat bogs in Finland have revealed distinct methanogenic pathways and different methanogenic communities (Galand et al. 2005). Three different methanogens viz., *Methanococcoides alaskense*, *Methanogenium boonei* and *M.marinum* have been isolated from methane hydrates of Skan Bay, Alaska (Chong et al. 2002, von Klein et al. 2002, Singh et al. 2005, Kendall et al. 2007). Another psychrotolerant hydrogenotrophic methanogenic strain, *Methanospirillum psychrodurum* X-18, has been isolated from the Madoi wetland at Qinghai-Tibetan plateau (Zhou et al. 2014). Zhang et al. (2008) reported a novel uncultured methanogen cluster, Zoige cluster I (ZC-I) affiliated to Methanosarcinales from Zoige wetland, using culture-independent approach.

Mesophilic methanogens

A mesophile is an organism that grows optimally between 20 and 45 °C (Prescott et al. 2005). Methanogens have been reported from many mesophilic habitats detailed of which are included below.

The rumen microbiota consists of an assemblage of anaerobic rumen fungi, anaerobic bacteria as well as some protozoa that not only affect the health but also the productivity of animals. Establishment of rumen microbiota takes place within a short span after the birth of an animal. Methanogens are an integral part of rumen microbiota and show a lot of diversity. Initiation of establishment of the methanogenic population is initiated within 30 h in the rumen of lamb. And the population density reaches upto 10^9 organisms/g rumen fluid within 1–3 weeks of age (Morvan et al.1994, Skillman et al. 2004). Interestingly, methanogens begin to colonize even before the consumption of forage material is started (Skillman et al. 2004).

The methanogenic diversity in ruminant animals does not change with the geographical locations and the host (Sundset et al. 2009a, b). Methanogens reported from adult ruminants areusually dominated by the genus *Methanobrevibacter* (Janssen and Kirs 2008). Based on the *mcr*A gene sequence-based analysis, Sirohi et al. (2013) reported dominance of methanogens belonging to the orders Methanobacteriales, Methanosarcinales and Methanomicrobiales in ruminants. Nicholson et al. (2007) reported equivalent levels of uncultured methanogens and *Methanobrevibacter ruminantium* like methanogens from the rumen of sheep. Studies carried out on the rumen of Murrah buffaloes (*Bubalus bubalis* in North India showed presence of methanogens belonging to the orders Methanomicrobiales, Methanobacteriales and Methanomassiliicoccales (Kumar et al. 2018). The study further revealed *Methanomicrobium* and *Methanobrevibacter* to be the dominant genera. Similar results were also obtained by Chaudhary and Sirohi (2009). *Methanobrevibacter* was the most dominant methanogen not only in the rumen of dairy cows (Kumar et al. 2015) but also in the bovine rumens (Miller and Lin 2002, Rea et al. 2007). Apart from *Methanobrevibacter*, methanogens belonging to other genera viz. *Methanomicrobium, Methanosarcina*, and *Methanobacterium*, are also reported from bovine rumen (Jarvis et al. 2000). *Methanobrevibacter* was also found to be a dominant methanogen in the rumen of Norwegian reindeer (Sundset et al. 2009a, b). Recent studies on cultivation of methanogens from the rumen, faeces and gut of buffalo, blue bull, goat and sheep showed presence of three hydrogenotrophic methanogenic genera, namely, *Methanocorpusculum*, *Methanobrevibacter,* and *Methanobacterium* (Joshi et al. 2018).

Dominance of *Methanobrevibacter* in the gut cannot only be seen in ruminants but also in humans and other non-ruminants. *Methanobrevibacter cuticularis* RFM-1, *Methanobrevibacter curates* RFM-2 and *Methanobrevibacter filiformis* RFM-3 have been isolated from the hindgut of termites *Reticulitermessperatus* (Shinzato et al. 1999, Guerrero 2001, Miller and Lin 2002). *Methanobrevibacter* was also isolated from the faeces of rats (Maczulak et al. 1989). Culture based reports on methanogens from faeces of Indian star tortoise and termite gut also confirmed dominance of *Methanobrevibacter* in non-ruminants (Joshi et al. 2018). *Methanomicrococcus* was isolated from the hindgut of cockroach (Sprenger et al. 2007) while *Methanosphaera cuniculi*1R7 was reported from the rectum of rabbit (Guerrero, 2001).

Presence of methanogenic archaea in the human gastrointestinal tract was described more than 30 years ago. The commonly reported methanogens from human gut include members of the order Methanobacteriales, *Methanobrevibacter smithies* and *Methanosphaera stadtmanae* (Miller et al. 1982, Druid et al. 2012, Gabi et al. 2014). Recently, Chaudhary et al. (2018) reported that the human micro biota harbours methanogenic species represented by *Methanobrevibacterin orals*, *Methanobrevibacter smithies*, *Methanomassiliicoccus luminyensis*, *Methanosphaera stadtmanae*, Candidatus *Methanomassiliicoccus intestinalis* and Candidatus *Methanomethylophilusalvus*. *Methanobrevibacter smithies* strain KB11 was reported from a Korean faecal sample and was found to be a major colonizer of the human gut (Kim and Jing 2018). Methanogens are also observed to associated with the oral cavity (Kulik et al. 2001). The study stated that the genetic diversity of methanogens from human oral cavity is quite low and is dominated by *M. oralis*. Methanogenicarchaea are also a part of human skin microbiome and are proposed to play a role in ammonia turnover. The dominant methanogens from skin include members of genera *Methanobrevibacter*, *Methanosphaera*, *Methanosarcina* (Proust et al. 2013).

Marine sediments produce 0.7–14 Tg methane year making them the largest deposits of methane on Earth (Valentine 2011). Generally, a major part of methane is produced biogenically by methanogens. Sedimentary methanogens include representatives of the order Methanosarcinales, Methanomicrobiales, Methanococcales, Methanopyrales, Methanobacteriales and Methanomassiliicoccales (Katayama et al. 2016). Desiccation of natural as well as man-made coastal environments has resulted into a wide variety of hypersaline habitats. These include transient salt pans, permanently hypersaline sabkhas, inland salt lakes and springs, gas and oil deposits and many industrial waste waters. Salt deposits containing brines vary from a cubic micrometer in volume to many cubic meters have been found below the Earth's crust.

Dissolution of ancient evaporates has resulted in formation of large lakes on the floor of the Red Sea, Mediterranean Sea and Gulf of Mexico. NOMAD instrument detected presence of methane as the major gas detected in the Martian atmosphere (Vandal et al. 2015). Recent studies have also shown that the sabkhas and Martian environments are similar (McKay et al. 2016). Methanogenesis is an important microbial process in saline, hypersaline deep zones as well as soda lakes where there is depletion of sulphate, surplus hydrogen production and availability of non-competitive carbon sources (Wilms et al. 2007, Buckley et al. 2008). Methylotrophic methanogenesis is almost always the key methanogenic process in saline and hypersaline environment. A few marine strains of the genus *Methanococcoides* carry out methanogenesis from choline and glycine betaine as substrates without the help of amyotrophic partner (L'Haridon et al. 2014, Watkins et al. 2014). In depth studies on one of the *Methanococcoides* strain showed partial demethylation of glycine beanie to N,N dimethylglycine, probably because excess energy is gained from the first demethylation step than the subsequent steps and the product may also be used as a compatible solute (Watkins et al. 2014). Methanogenesis directly from glycine betaine as a substrate has not yet been demonstrated from hypersaline environments. Studies on methanogens isolated from saline environments show that methylotrophic methanogens contribute the highest (about 30%) followed byhydrogenotrophic methanogens (around 12%) while acetoclastic methanogens contributed the least (only 4%). This highlights the involvement of methylotrophic substrates to methanogenesis at different salinities (Oren 2011). Another hybrid pathway of methanogenesis has been proposed by Barrel et al. (2014). It involves methylation of single carbon (C1) compounds which serve as electron acceptors, while H^+ serves as an electron donor. It was demonstrated by Sorokin et al. (2017) that methanogens follow a novel methyl reduction pathway when sediment slurries from hypersaline soda lakes were incubated in presence of methanol or trim ethylamine + formate or H^+, under high temperature, pH and salt-saturating conditions. Based on its ribosomal protein phylogeny, the novel methanogens are classified to form a novel class 'Methanonatronarchaea' and are closely related to the class *Halobacterium* (Sorokin et al. 2017). It has also been suggested that in salt-saturated conditions both the classical methylo- and hydrogenotrophic methanogenic pathways would be displaced by this pathway (McGenity and Sorokin 2018). Alkaliphilic methanogens have been reported from Lonar lake in India (Thakkar and Ranade, 2002; Surakasi et al. 2007).

Geological evidence suggests that majority of the Earth'spetroleum reservoirs has been degraded by microorganisms over the years, leading to the methane formation (Jones et al. 2008). Many laboratory studies on hydrocarbon metabolism undermethanogenic conditions have been published (Toth and Gieg 2018). Methanogens reported from oil reservoirs across the globe include hydrogenotrophic, acetoclastic as well as methylotrophic methanogens commonly belonging to the orders viz., Methanosarcinales, Methanobacteriales, Methanococcales and Methanomicrobiales. Apart from these, members of the order Methanocellales, Methanomassiliicoccales have also been reported. On account of technical as well as practical difficulties to sampling, majority of theculture-dependent and culture-independent diversity studies are carried out using either formation water orproduced water obtained from well heads of oil wells. The commonly reported mesophilic methanogenic genera from oil reservoirs include *Methanobacterium, Methanosarcina, Methanoplanus, Methanosaeta* and *Methanolobus* (Varjani et al. 2018).

Marine methane hydrates are cages of the crystalline lattice of water molecules in which methane molecules are trapped. Methane hydrates have been projected to contain approximately 4,000 times more methane than today's atmospheric content. As majority of the methane trapped in the known hydratesis of biogenic origin, they may serve as a potential new source of natural gas.The commonly reported orders of methanogens from methane hydrates include *Methanobacteriales* and *Methanosarcinales* (Marchesi et al. 2001). Dabir et al. (2014) recently reported draft genome sequence of a novel species of methanogenic genus *Methanoculleus*. This isolate was obtained from deep sub-seafloor sediment in the Krishna Godavari Basin off the eastern coast of India. This methanogen is thought to be mainly responsible for the production of methane that forms submarine methane hydrate deposits.

Rice fields are the major anthropogenic sources of methane emission.Methane in rice fields is derived from both hydrogenotrophic as well as aceticlastic methanogenesis. Dominant methanogens reported from rice fields belong to the families Methanomicrobiaceae, Methanobacteriaceae, Methanosarcinaceae, Methanosaetaceae, and RC-I (Chin et al. 2004, Lu et al. 2005). The methanogenic community structure in rice fields changes significantly with time as well as location. Methane production from H_2:CO_2 predominates immediately after flooding, while aceticlastic methanogenesis increases later, reaching a maximum at about 70–80 days after flooding (Roy et al. 1997, Kruger et al. 2005). The methanogenic community on the roots of rice is different from those in the bulk soils. RC-I is usually the dominant organism on rice roots, and about 60–80% of methane is mainly derived from H_2:CO_2 (Chin et al. 2004, Lu et al. 2005). In contrast, aceticlastic methanogenesis is predominant in the bulk soil and around 50–83% methane is derived from cleavage of acetate (Joulian et al. 1998, Lu et al. 2005). The commonly reported methanogens from the roots of rice plants belong to the orders Methanomicrobiales, Methanocellales and Methanobacteriales (Lehmann-Richter et al. 1999). Phylogenetic analysis of the clones obtained from rhizospheric rice soils showed presence of diverse group of methanogens closely related to Methanomicrobiaceae, Methanosarcinaceae, Methanosaetaceae and Methanocellales (Singh and Dubey 2012, Singh et al.2012). Members from the family Methanosarcinaceae and orders Methanomicrobiales, Methanocellales, Methanobacteriales and Methanosaetaceae have been reported from non- rhizospheric rice field soils (Ramakrishnan et al. 2001, Watanabe et al. 2010, Liu et al. 2012). The first report on isolation and characterization of hydrogenotrophic and aceticlastic methanogens from rice field was by Kawakawa et al. (1993, 1995). They characterized *Methanobrevibacterin arboriphilus* strain SA and *Methanosarcina mazei* strain TMA isolated from rice fields. Cultures of *Methanogenium, Methanosarcina*, *Methanobacterium* and *Methanosaeta* were isolated from Japanese rice fields by Kudo et al. (1997) while Großkopf et al. (1998) reported prevalence of *Methanosaeta, Methanosarcina* and *Methanobacterium* species in Italian rice field soils. Using 16S renal-based technique, Joulian et al. (1998) investigated microbial cultures from 13 rice field soils and reported presence of *Methanobacterium formicicum, Methanoculleus marisnigri. Methanobacterium bryantii, Methanosarcina mazei* and *Methanosarcina barkeri*. Many novel methanogens such as *Methanobacterium oryzae, Methanoculleus chikugonsis* and *Methanobacterium kanagiense* was also reported from paddy field (Joulian et al. 2000, Dianou et al. 2001, Kitamura et al. 2011).Molecular approaches suggest that uncultivated lineages of archaea could be ubiquitous in various mesophilic terrestrial environments such as agroecosystems (Schüttle et al. 2008). Großkopf et al. (1998) reported 22 novel methanogenic lineages (RC I through V) from rice roots and anoxic soils using culture independent approach. Surprisingly, none of the lineages were found to be closely related to any of the cultured archaea reported. Lineages RC I and II showed closest affiliation between Methanosarcinales and Methanomicrobiales while lineage RC III showed distinct relationship with *Thermoplasma acidophilum*. Further, RC IV branched deep within the crenarchaeotal assemblage while RC V has been distantly related to Euryarchaeota. The study played pivotal role in showing that many novel lineages of methanogens inhabit the rice field, and opened new avenues for further isolation and characterisation of novel methanogens from these lineages. The RC-I lineage, comprising solely of 16S rRNA gene sequence, is suggested to play the crucial role in methane production in rice fields (Conrad et al. 2006). Fastidious growth conditions and slow growth have been the main constrains in isolation of RC-I in pure culture. The uncultivated organisms of this group are not only ecologically but also of biochemical importance. The successful enrichment of RC-I methanogens was carried out by Sakai et al. (2007). They carried out co-culture with syntrophic, propionate oxidizing and H_2 producing syntroph *Syntrophobacter fumaroxidans* and reported

Methanocellapaludicola as the first novel isolate of lineage RC-I. They further added that the methanogen belongs to the novel archaealorder Methanocellales (Sakai et al. 2008). Another novel species, *Methanocellaar oryzae*, belonging to RC I was reported from Italian rice fields by Sakai et al. (2010).

Till recently, all the known methanogenic archaea were assigned to the phylum Euryarchaeota, and it was well established that archaeal methane production originated within the Euryarchaeota. However, with the discovery of new phylum Bathyarchaeota, this theory has been challenged. Evans et al. (2015) assembled genomes of two distinct lineages of Bathyarchaeota (BA1 and BA2) originating from coal-bed methane wells in Australia. They observed that both the genomes had the complete set of genes for methanogenesis as present in the phylogenetically distinct Euryarchaeota. Butthere is a conspicuous phylogenetic divide betweenmcrA sequences of Bathyarchaeota and Euryarchaeota. It is because of this divergence that bathyarchaeotal *mcr*A remainedun amplifiable by the oligonucleotides used for euryarchaeotal *mcr*A. Further analysis indicated BA1 and BA2 to be methylotrophs with additional 12 genes encoding methyltransferase on top of gene encoding enzymes arranged by the euryarchaeotal methanogens. BA1 could use lactate, while BA2 peptides, monosaccharides, and pyruvate for energy generation in contrary to euryarchaeotal methanogens. Thus, BA1 and BA2 seem to have the wider substrate range than euryarchaeotal methanogens, but yet, fail to catabolise H_2/CO_2 and acetate like euryarchaeotal methanogens. Evans et al. (2015) reported similar *mcr*A sequences in BA1 and BA2 growing in a wide range of environments indicating the more widespread presence of Bathyarchaeota, and other Archaea divisions that could have the potential for methane production. This opens up a completely new vista for exploring new methanogenic archaea from the anoxic sediments.

Thermophilic methanogens

Thermophilic ecological niches are hot springs, oil wells, thermophilic digestors, /treatment plants, etc. Approximately, 334 hot springs are present in India, and most of these are not yet studied for thermophilic methanogens. Oilwells are another interesting site, where microbial communities are involved in biodegradation of oil, and hence, the presence of methanogens is anticipated. Hydrogenotrophic methanogen are mostly reported from thermophilic environment on the other hand aceticlastic methanogens are reported mostly from mesophilic environment. This is well evident from the recent findings from Grimalt-Alemany et al. (2019) where *Methanothermobacter* sp. was the dominant methanogen in the thermophilic enrichment responsible for conversion of syngas to methane and aceticlastic methanogenesis (mesophilic) contributed to maximum methane production. *Methanosphaera* sp. was found to be dominant in co-digestion of solid waste with palm oil meal effluent under thermophilic condition (Seengenyoung et al. 2019). Sakai et al. (2019) reported *Methanofervidicoccus abyssi* gen. nov., sp. nov., a hydrogenotrophic methanogen from a hydrothermal vent chimney in the Mid-Cayman Spreading Center, the Caribbean Sea. *Methanothermobacter thermautotrophicus* which was isolated from anaerobic sewage digester and grows optimally between 65 to 70°C is the most commonly reported methanogen from these environments. The diversity of thermophilic methanogenic archaea from oil wells has also been explored in some oil fields around the world. *Methanoplanus petrolearius* (Ollivieret al. 1997) and *Methanocalculus halotolerans* (Ollivier et al. 1998) were isolated from oil producing well from Africa and France. Cheng et al. (2007, 2008) have reported taxonomically different and novel methanogenic archaea like *Methermicoccuss hengliensis, Methanoculleus receptaculi* and *Methanothermobacter crinale* isolated from Shengli oil fields in China. *Methanothermobacter tenebrarum*, a novel species of thermophilic methanogen was isolated from oil fields in Japan (Nakamura et al. 2013). *Methanothermobacter thermautotrophicum* was isolated from thermophilic oil reservoirs in India (Nerlekar 2014). *Methanobacterium wolfei* (optimum temperature 55 to 65°C) from sludge and river sediment (Winter et al. 1984). Over the past few years, oil wells around the world are explored for the presence of microbes along with methanogens including the oil fields of Japan, France, and Africa. The physical parameters of the environment in oil well are such that they support the growth of thermophilic methanogens. Likewise, the physical parameters in hot springs like high temperature and absence of dissolved oxygen in water support the growth of thermophilic methanogens. The thermophilic methanogenesis in a hot spring algal, thebacterial mat was studied which showed optimum methanogenesis at 45ºC (Ward, 1978). Since then a few hot springs around the world are investigated for the existence of microbial diversity. The exploration of these hot springs has led to the discovery of thermophilic methanogens like *Methanothermus fervidus* (Stetter et al. 1981) from Icelandic hot spring and *Methanothermobacter marburgensis* from a hot spring in China (Ding et al. 2010). There are very few studies carried on themophilic methanogens from India. Joshi et al. (2018) recently reported isolation of *Methanothermobacterm arburgensis* strain HS31 an thermophilic hydrogenotrophic methanogen from Rajwadi hotsprings. Several reports are also available regarding isolation of thermophilic anaerobic cellulolytic bacteria from environments like compost (Sizova et al. 2011), digesters (Leschine 1995) etc., and isolation of thermophilic aerobic cellulolytic and other bacteria from hot springs. But only limited studies have been conducted for isolation of thermophilic methanogens from hot springs and other elevated temperature locations. Thermophilic methanogens are more efficient than mesophilic and psychrophilic methanogens as they have less doubling time.

Several studies have been conducted worldwide to explore the diversity of thermophilic methanogens from petroleum and oil reservoirs, using various molecular techniques. These studies have revealed the dominance of several genera of methanogens including the thermophilic group. The technique of clone library construction and sequencing reveals the dominance of *Methanobacterium* and *Methanococcus* in oil fields of California (Orphan et al.2000), *Methanococcus* and *Methanolobus* in North Sea oil fields (Dahle et al. 2008), and *Methanothermobacter* in oil fields of Japan (Yamane et al. 2011). Using quantitative PCR, Chapelle et al. (2002) revealed the dominance of methanogenic archaea in hot springs. A pyrosequencing-based study has revealed the presence of genera *Methanothermobacter, Methanosaeta, Methanomassiliicoccus* and *Methanocella*in hot springs of Yellowstone national park (Leon et al. 2013) and *Methanospirillum, Methanoregula, Methanosaeta* and *Methanomethylovorans* in Armenian hot springs (Hedlund et al. 2013). In one PCR DGGE based study, genera *Methanothermobacter, Methanothrix, Methanomethylovorans,* and *Methanothermobacter* were found dominant in terrestrial hot spring of Kamchatka (Merkel et al. 2016). Recently, Plenge et al. (2016) have identified putative novel methanogenic orders with diverse metabolic capabilities, including Methanobacteriales, Methanomicrobiales, and Methanosarcinales from El tatio geyser field, Chile using pyrosequencing.

Hyperthermophilic methanogens

The methanogens which thrives in extremely hot environments from 60°C (140°F) upwards are known as hyperthermophilic methanogens. Some methanogens also can grow at temperatures higher than 100°C at large depths in sea where water does not boil because of high pressure (Stetter et al. 2006).

Over the years many hyperthermophilic methanogens have been isolated from various anaerobic habitats. *Methanopyruskandleri* is a hyperthermophile, discovered on the wall of a black smoker from the Gulf of California at a depth of 2000 m, at temperatures of 84–110°C (Kurr et al. 1991). It was also discovered in black smoker fluid of the Kairei hydrothermal field; and found to survive and reproduce at 122°C (Takai et al. 2008). It utilizes H_2:CO_2 primarily for methane production. It is the only well-established species under the family Methanopyraceae; genus *Methanopyrus* (Huber and Stetter 2001). Other hyperthermophilic methanogens which is extensively studied is *Methanocaldococcus jannaschii* in the class Methanococci (optimum temperature- 85ºC) (formerly *Methanococcus jannaschii*) isolated from white smoker chimney on east pacific rise (Jones et al. 1989). It was the first archaeon whose complete genome sequenced (Bult et al. 1996). The sequencing identified many genes unique to the archaea. Many of the synthesis pathways for methanogenic cofactors were worked out biochemically in this organism as were several other archaealspecific metabolic pathways (Grochowski et al. 2006).The genome includes many hydrogenases, such as a 5,10-methenyl tetrahydro methanopterin hydrogenase, a ferredoxin hydrogenase (eha), and a coenzyme F420 hydrogenase. Proteomic studies showed that *M. jannaschii* contains a large number of inteins: 19 were discovered by one study (Wallin and Heijne 1998). Many novel metabolic pathways have been worked out in *M. jannaschii*, including the pathways for synthesis of many methanogenic cofactors, riboflavin, and novel amino acid synthesis pathways. Many information processing pathways have also been studied in this organism, such as an archaealspecific DNA polymerase family. Information about singlepass transmembrane proteins from *M. jannaschii* was compiled in Membranome database (Andrade et al. 1997). *Methanothermus sociabilis* (optimum temperature -88°C) from Icelandic continental geothermal areas (Lauerer et al. 1986). Similarly, *Methanocaldococcus bathoardescens* sp. nov., a hyperthermophilic methanogen is reported from a volcanically active deepsea hydrothermal vent (Stewart et al. 2015).

Preservation of methanogens

Preservation of pure cultures is important for their longterm storage and maintenance. Since a long time, various methods like, serial passage, refrigeration, cryopreservation and lyophilisation or liquid drying are routinely used for bacteria and few archaea (Garrity and Holt 2001). Of these listed techniques, serial passage and refrigeration at 4°C (liquid broth or slants) fall under the short-term storage whereas, cryopreservation and lyophilisation fall under the long term methods which can preserve the cultures for up to 30 years. Unlike the prokaryotes, preservation of methanogens and overall archaea is not very well explored. There is a lack of the universal methods and protocols which shall help in both long term and short-term preservation of these microbes. The reasons for this are, lack of knowledge, tedious processes, strict anaerobic nature, very wide temperature range and diverse groups of organisms with separate needs. Though some methods have been tried for the preservation of methanogens, their revival rate remains very low. Methanogenic archaea are slow growers and many take 15 to 30 days to grow, exceptions being the thermophilic methanogens. Serial passage for their maintenance every 30 to 45 days is a very widely used routine used preservation method. Many of them can also be refrigerated at four degree Celsius in liquid broth or agar slants for up to 3 months. Though the short term storage is possible for methanogens, there is a need for a robust method of their long term storage for the feasibility of storage at culture collections. A few trials of preservation using various methods and cryoprotectants have been carried out by some researchers, which reveal the following results.

The long-term storage includes freezing (at -80°C and -196°C in liquid nitrogen) and lyophilisation. For their effective revival after a long time, it is a prerequisite that all the methanogenic cultures must be in the growing log phase. For long term storage, methanogens have generally been preserved by freezing. Eleven species of methanogens were preserved at -18°C by Winter in 1983. Further, Koga and Ohga(1992) preserved fifteen species of methanogens by freezing in liquid nitrogen. For freezing the cultures for a long term, a cryoprotectant must be added to the culture which helps the cell withstand the stress of freezing and thawing. Cryoprotectants are osmoprotectants which are low molecular weight, hydrophilic, nontoxic molecules that assist a cell under osmotic stress to stabilize its concentration of internal solutes (Cleland et al. 2004). The most commonly used cryoprotectants for freezing are, glycerol, ethylene glycol and dimethyl sulfoxide (DMSO) but a recent study has shown the use of glycine betaine and sucrose/bovine serum albumin as cryoprotectants gave better revival rates in case of *Methanothermobacter thermautotrophicus*, a hydrogenotrophicthermophilic methanogen (Cleland et al. 2004).

Lyophilisation or liquid drying is another method widely used for long term preservation of cultures. Lyophilisation is preferred over other methods as it is suitable for postal and parcel transportation (Sakane et al. 1992). Another advantage of this method over the other methods is the stability of the cultures in powder form at room temperature. The tiny size of a lyophilised vial and the powder form also reduces the cost of transport. The most commonly used protectants for this process include skimmed milk, trehalose, dextran and combination of sucrose with bovine serum albumin (Simione and Brown 1991, Hubalek 2003). The successful revival of methanogens, *Methanobrevi-bacter arboriphilus, Methanosarcinamazei, Methanoculleusc -hikugoensis* and *Methanothermobacter thermautotrophicus* after lyophilisation is reported by Iino and Suzuki in 2006. They have proved the increased efficiency of revival if the strict anaerobic conditions are maintained throughout the procedure (Iino and Suzuki 2006).

Table 2: Various methods used for preservation of methanogens

Sr.No.	Cryoprotectant and concentration	Technique	Methanogen	Reference
1.	0.1 M potassium phosphate buffer (pH 7.0), 3% sodium glutamate, 1.5% adonitol, and 0.05% L-cysteine monohydrate	Liquid drying	*Methanobrevibacterar boriphilus* *Methanosarcina mazei* *Methanothermobacter thermautotrophicus* *Methanoculleus chiku goensis*	Iino and Suzuki 2006
2.	Glycine betaine (12%)Sucrose/BSA(20/10%) Dextran/Trehalose(10/5%)	Freeze drying/ lyophilisation	*Methanothermobacter thermautotrophicus*	Cleland et al. 2004
3.	Glycerol (20%)Glycine betaine (12% for few)	Liquid nitrogen	*Methanothermobacter thermautotrophicus*	
4.	Trehalose (10% & more effective) and glycerol	Freeze drying	Consotrium	Castro et al. 2002
5.	Glucose (10%) gave best results	Freeze drying	Anaerobic biomass	Colleran et al. 1992
6.	DMSO (5%)	-800.C	Methanogens	Makkar, Joblin
7.	Glycerol (25%)	-800.C	*Methanoplanus limicola* *Methanoplanus endosymbiosis* *Methanolaciniapaynteri* *Methanosarcinabarkeri* *Methanogeniumbourgense* *Methanofollistationis* *Methanocorpusculum labreanum*	Archaea, Whitman et al

To conclude, methanogenic archaea, being strict anaerobes, their sensitivity to oxygen makes the process of long-term storage very tedious and time consuming with low successful revival of the cultures. The studies conducted so far include the mesophilic as well as thermophilic methanogens utilizing various substrates like methanol, acetate and H_2:CO_2. This shows that the methods of preservation don't change for different groups of methanogens and they can be preserved by any of the above-mentioned method. An experimental trial is always advised before implementing any method because of the paucity of information regarding the preservation studies of methanogens. All specific methods for preservation of methanogens are listed in Table 2.

Implications

Anaerobic digestion of biomass to energy product can be fissible under psychrophilic, mesophilic and thermophilic conditions. Compare to mesophilic, thermophilic anaerobic digestion enhances both methanation yield and methane content of biogas (Gao et al. 2018,Tsapekos et al. 2019; Arras et al. 2019). Similar observation was recently reported by Grimalt-Alemany et al. (2019) where thermophilic culture based mixed consortia compare to mesophilic contributed for maximum biomethanation from syngas. Differences in hydrolysis and acidogenesis were examined in batch mode using anaerobic sludge at mesophilic, thermophilic and hyperthermophilic condition with all three temperature mixed cultures as inoculum. Production of volatile fatty acids (VFAs) -acetic acid at mesophilic, butyric acid at thermophilic as compare to lactic acid production at hyperthermophilic condition revealed that the substrate degradation efficiency is not affected by different operating temperatures but biotransformation pathways do vary with the differences in temperature (Arras et al. 2019).

References

Andrade M, et al. 1997. Sequence analysis of the *Methanococcus jannaschii* genome and the prediction of protein function. *Computer Applications in the Biosciences* 13(4): 481-483.

Arras W., et al., 2019. Mesophilic, thermophilic and hyperthermophilic acidogenic fermentation of food waste in batch: Effect of inoculum source. *Waste Management* 87:279-287.

Asakawa S, Akagawa-Matsushita M, Hiroyuki M, Yosuke K, Koichi H. 1995. Characterization of *Methanosarcina mazeii* TMA isolated from a paddy field soil. *Current Microbiology* 31:34–38.

Asakawa S, Hiroyuki M, Akagawa-Masttushita M, Yosuke K, Koichi H. 1993. Characterization of *Methanobrevibacterarboriphilicus* SA isolated from a paddy field soil and DNA DNA hybridization among *M. arboriphilicus* strains. *International Journal of Systematic and Evolutionary Biology* 43:683–686.

Balch WE, Fox GE, Magrum LJ, Woese CR, Wolfe RS. 1979. Methanogens: Reevaluation of a unique biological group. *Microbiological Reviews* 43:260–296.

Balch WE, Wolfe RS. 1976. New approach to the cultivation of methanogenic bacteria: 2-Mercaptoethanesulfonic acid (HS-CoM)-dependent growth of *Methanobacterium ruminantium* in a pressurized atmosphere. *Applied and Environmental Microbiology* 32:781–791.

Barker HA. 1936. Studies upon the methane-producing bacteria. *Archives of Microbiology* 7:420-438.

Bellack A, Huber H, Rachel R, Wanner G, Wirth R. 2011. *Methanocaldococcus villosus* sp. nov., a heavily flagellated archaeon that adheres to surfaces and forms cell-cell contacts. *International Journal of Systematic and Evolutionary Biology* 61:1239–1245.

Belyaev SS, Wolkin R, Kenealy WR, Deniro MJ, Epstein S, Zeikus JG. 1983. Methanogenic bacteria from the Bondyuzhskoe oil field: general characterization and analysis of stable-carbon isotopic fractionation. *Applied and Environmental Microbiology* 45:691–697.

Biavati B, Vasta M, Ferry JG. 1988. Isolation and characterization of *Methanosphaera cuniculi* sp. nov. Applied and Environmental Microbiology 54:768–771.

Blotevogel KH, Fischer U. 1985. Isolation and characterization of a new thermophilic and autotrophic methane producing bacterium: *Methanobacterium thermo aggregans* spec. nov.*Archives of Microbiology* 142: 21

Blotevogel KH, Gahl-Janßen R, Jannsen S, Fischer U, Pilz F, Auling G, Macario AJL, Tindall BJ. 1991. Isolation and characterization of a novel mesophilic, fresh-water methanogen from river sediment *Methanoculleus oldenburgensis* sp. nov.*Archives of Microbiology* 157:54–59.

Boone DR, Baker CC: Genus VI. *Methanosalsum* gen. nov.*In*: D.R. Boone, R.W. Castenholz and G.M. Garrity (editors): Bergey's Manual of Systematic Bacteriology, second edition, vol. 1 (The *Archaea* and the deeply branching and phototrophic *Bacteria*), Springer-Verlag, New York, 2001, pp. 287-289.

Boone DR, Castenholz RW Garrity GM. 2001. Bergey's Manual of Systematic Bacteriology, second edition, vol. 1 (The *Archaea* and the deeply branching and phototrophic *Bacteria*), Springer-Verlag, New York

Boone DR, Kamagata Y. 1998. Rejection of the species *Methanothrix soehngenii*^{VP} and the genus *Methanothrix*^{VP} as *nominaconfusa*, transfer of *Methanothrix thermophila*^{VP} to the genus *Methanosaeta*^{VP} as *Methanosaeta thermophile* comb. Nov. Request for an Opinion. *International Journal of Systematic and Evolutionary Biology* 48:1079-1080

Boone DR, Mathrani IM, Liu YT, Menaia J, Mah RA, Boone JE. 1993a. Isolation and characterization of *Methanohalophilus portucalensis* sp. nov. and DNA reassociation study of the genus *Methanohalophilus*. *International Journal of Systematic and Evolutionary Biology*43:431–437.

Boone DR, Whitman WB, Rouvieìre P. 1993. Diversity and taxonomy of methanogens. In *Methanogenesis*, pp. 35–80. Edited by J. G. Ferry. New York: Chapman & Hall.

Boone DR. 1987. Request for an opinion: replacement of the type strain of *Methanobacterium formicicum* and reinstatement of *Methanobacterium bryantii* sp. nov. nom. rev. (ex Balch and Wolfe, 1981) with M.o.H. (DSM 863) as the type strain. *International Journal of Systematic and Evolutionary Biology* 37:172–173.

Borrel G, Joblin K, Guedon A, Colombet J, Tardy V, Leyhours AC, Fonty G. 2012. *Methanobacteriumlacus* sp. nov., isolated from the profundal sediment of a freshwater meromictic lake. *International Journal of Systematic and Evolutionary Biology* 62: 1625-1629

Borrel G, Parisot N, Harris HM, Peyretaillade E, Gaci N, Tottey W, Bardot O, Raymann K, Gribaldo S, Peyret P. 2014. Comparative genomics highlights the unique biology of Methanomassiliicoccales, a Thermoplasmatales related seventh order of methanogenic archaea that encodes pyrrolysine. *BMC Genome* 15(1):679

Brauer SL, Cadillo-Quiroz H, Yashiro E, Yavitt JB, Zinder SH. 2006. Isolation of a novel acidiphilic methanogen from an acidic peat bog. *Nature* 442:192–194.

Bryant MP. 1972. Commentary on the Hungate technique for culture of anaerobic bacteria. *American Journal of Clinical Nutrition* 25:1856–1859.

Buckley DH, Baumgartner LK, Visscher PT. 2008. Vertical distribution of methane metabolism in microbial mats of the Great Sippewissett Salt Marsh. *Environmental Microbiology* 10(4):967–977

Bult CJ, et al. 1996. Complete genome sequence of the methanogenic archaeon, *Methanococcus jannaschii*. *Science*, 273(5278):1058-1073.

Burggraf S, Fricke H, Neuner A, Kristjansson J, Rouvier P, Mandelco L, Woese CR, Stetter KO. 1990. *Methanococcus igneus* sp. nov., a novel hyperthermophilic methanogen from a shallow submarine hydrothermal system. Systematics and Applied Microbiology 13:263–269.

Cadillo-Quiroz H, BräuerSL, Goodson N, YavittJB, Zinder SH.2014. *Methanobacterium paludis* sp. nov. and a novel strain of *Methanobacterium lacus* isolated from northern peatlands *International Journal of Systematic and Evolutionary Biology*64:1473–1480

Cadillo-Quiroz H, Yavitt JB, Zinder SH 2009 *Methanosphaerula palustris* gen. nov., sp. nov., a hydrogenotrophic methanogen isolated from a minerotrophic fen peatland*International Journal of Systematic and Evolutionary Biology* 59: 928-935

Carr SA, Schubotz F, Dunbar RB, et al. 2018. Acetoclastic*Methanosaeta* are dominant methanogens in organic-rich Antarctic marine sediments. *ISME Journal*. 12(2):330–342. doi:10.1038/ismej.2017.150.

Castro H, Queirolo M, Quevedo M and Mux′i L. 2002. Preservation methods for the storage of anaerobic sludges (2002) *H.Biotechnology Letters* 24: 329–333, 2002

Cha IT, Min UG, Kim SJ, Yim KJ, Roh SW, Rhee SK. 2013. *Methanomethylovoransuponensis* sp. nov., a methylotrophic methanogen isolated from wetland sediment. *Antonie Van Leeuwenhoek* 104:1005-1012.

Chapelle FH, O'Neill K, Bradley PM, Methé BA, Ciufo SA, Knobel LL, Lovley DR. 2002. A hydrogen-based subsurface microbial community dominated by methanogens. *Nature*. 415(6869):312-5.

Chaudhary PP, Conway PL, Schlundt J. 2018. Methanogens in humans: potentially beneficial or harmful for health. *Applied Microbiology and Biotechnology*. 102(7):3095–3104.

Chen SC, Chen MF, Lai MC, Weng CY, Wu SY, Lin S, Yang TF, Chen PC. 2015. *Methanoculleus sediminis* sp. nov., a methanogen from sediments near a submarine mud volcano. *International Journal of Systematic and Evolutionary Biology*65:2141–2147

Chen SC, Huang HH, Lai MC, Weng CY, Chiu HH, Tang SL, Rogozin DY, Degermendzhy AG. 2018. *Methanolobuspsychrotolerans* sp. nov., a psychrotolerantmethanoarchaeon isolated from a saline meromictic lake in Siberia. *International Journal of Systematic and Evolutionary Biology*68:1378-1383

Cheng L, Dai L, Li X, Zhang H, Lu Y. 2011. Isolation and characterization of *Methanothermobacter crinale* sp. nov., a novel hydrogenotrophic methanogen from the Shengli oil field. *Applied and Environmental Microbiology* 77:5212-5219

Cheng L, Qiu TL, Li X, Wang WD, Deng Y, Yin XB, Zhang H. 2008 Isolation and characterization of *Methanoculleus receptaculi* sp. nov. from Shengli oil-feld, China. *FEMS Microbiol Lett* 285:65–71

Cheng L, Qiu TL, Yin XB, Wu XL, Hu GQ, Deng Y, Zhang H. 2007. *Methermicoccus shengliensis* gen. nov., sp. nov., a thermophilic, methylotrophic methanogen isolated from oil-production water, and proposal of Methermicoccaceae fam. nov. *International Journal of Systematic and Evolutionary Microbiology* 57(12):2964-2969.

Chin KJ, Lueders T, Friedrich MW, Klose M, Conrad R. 2004. Archaeal community structure and pathway of methane formation on rice roots. *Microbial Ecology* 47:59–67.

Chong SC, Liu YT, Cummins M, Valentine DL, Boone DR. 2002. *Methanogenium mari numsp* nov., a H_2-using methanogen from Skan Bay, Alaska, and kinetics of H_2 utilization. *Antonnie van Leeuwenhoek* 81:263–270

Cleland D, Krader P, McCree C, Tang J, Emerson D. 2004. Glycine betaine as a cryoprotectant for prokaryotes. *Journal of Microbiological Methods* 58(1):31-38.

Colleran E, Concannon F, Golden T, Geoghegan F, Crumlish B, Killilea E, Henry M and Coates J. 1992. Use of methanogenic activity tests to characterize anaerobic sludges, screen for anaerobic biodegradability and determine toxicity thresholds against individual anaerobic trophic groups and species. *Water Science Technology.* 25: 31–40

Conrad R, Erkel C, Liesack W. 2006. Rice Cluster I methanogens, an important group of Archaea producing greenhouse gas in soil. *Current Opinion in Biotechnology* 17:262–267.

Corder RE, Hook LA, Larkin JM, Frea JI. 1983. Isolation and characterization of two new methane-producing cocci: *Methanogeniumolentangi*, sp. nov., and *Methanococcusdeltae*, sp. nov.*Archives of Microbiology* 134:28–32.

Cuzin N, Ouattara AS, Labat M, Garcia JL. 2001. *Methanobacterium congolense* sp. nov., from a methanogenic fermentation of cassava peel. *International Journal of Systematic and Evolutionary Biology* 51:489–493.

Dabir A, Honkalas V, Arora P, Pore S, Ranade D, Dhakephalkar PK. 2014. Draft genome sequence of *Methanoculleus*sp. MH98A, a novel methanogen isolated from sub-seafloor methane hydrate deposits in Krishna Godavari basin. *Marine Genomics* 18:139–140

Dahle H, Garshol F, Madsen M, Birkeland NK. 2008. Microbial community structure analysis of produced water from a high-temperature North Sea oil-field. *Antonnie van Leeuwenhoek.*93:37–49.

Dhakephalkar PK, Prakash O, Lanjekar VB, Tukdeo MP, Ranade DR. 2019. Methanogens for Human Welfare: More Boon Than Bane. In: Satyanarayana T., Das S., Johri B. (eds) Microbial Diversity in Ecosystem Sustainability and Biotechnological Applications. Springer, Singapore

Dianou D, Miyaki T, Asakawa S, Morii H, Nagaoka K, Oyaizu H, Matsumoto S. 2001. *Methanoculleuschikugoensis* sp. nov., a novel methanogenic archaeon isolated from paddy field soil in Japan, and DNA-DNA hybridization among *Methanoculleus* species *International Journal of Systematic and Evolutionary Microbiology*51:1663–1669.

Ding X, Yang WJ, Min H, Peng XT, Zhou HY, Lu ZM. 2010. Isolation and characterization of a new strain of *Methanothermobacter marburgensis* DX01 from hot springs in China. *Anaerobe*16(1):54-59.

DoerfertSN, Reichlen M, IyerP, Wang M, Ferry JG.2009. *Methanolobuszinderi* sp. nov., a methylotrophic methanogen isolated from a deep subsurface coal seam. *International Journal of Systematic and Evolutionary Biology*59:1064–1069.

Dridi B, Fardeau ML, Ollivier B, Raoult D, Drancourt M. 2012. *Methanomassiliicoccus-luminyensis*gen. nov., sp. nov., a methanogenic archaeon isolated from human faeces. *International Journal of Systematic and Evolutionary Microbiology* 62(8):1902–1907

Evans PN, Parks DH, Chadwik GL, Robbins SJ, Orphan VJ, Golding SD, Tyson GW. 2015. Methane metabolism in the archaeal phylum Bathyarchaeota revealed by genome-centric metagenomics. *Science* 350:434–438.

Ferrari A, Brusa T, Rutili A, Canzi E, Biavati B. 1994. Isolation and characterization of *Methanobrevibacteroralis* sp. nov.*Current Microbiology* 29:7-12.

Ferry JG, Smith PH, Wolfe RS. 1974. *Methanospirillum*, a new genus of methanogenic bacteria, and characterization of *Methanospirillum hungatii* sp. nov. *International Journal of Systematic and Evolutionary Biology*24:465–469.

Franzmann PD, Liu YT, Balkwill DL, Aldrich HC, Conway de Macario E, Boone DR. (1997). *Methanogeniumfrigidum* sp. nov., a psychrophilic, H_2-using methanogen from Ace Lake, Antarctica. *International Journal of Systematic and Evolutionary Microbiology* 47:1068–1072

Franzmann PD, Springer N, Ludwig W, Conway DeMacario E, Rhode M. 1992. A methanogenic Archaeon from Ace Lake, Antarctica: *Methanococcoides burtonii* sp. nov.*Systematic and Applied Microbiology* 15:573-581.

Gaci N, Borrel G, Tottey W, O'Toole PW, Brugère JF. 2014. Archaea and the human gut: new beginning of an old story. *World Journal of Gastroenterology* 20(43):16062

Galand PE, Fritze H, Conrad R, Yrjala K. 2005. Pathways for methanogenesis and diversity of methanogenic archaea in three boreal peatland ecosystems. *Applied and Environmental Microbiology* 71:2195–2198

Ganzert L, Schirmack J, Alawi M, Mangelsdorf K, Hillebrand-Voiculescu A, Wagner D. 2014. *Methanosarcinaspelaei* sp. nov., a methanogenic archaeon isolated from a floating biofilm of a subsurface sulphurous lake. *International Journal of Systematic and Evolutionary Biology*64:3478–3484.

Gao, Y., et al., 2018. Digestion performance and microbial metabolic mechanism in thermophilic and mesophilic anaerobic digesters exposed to elevated loadings of organic fraction of municipal solid waste. *Energies* 11(4):952.

Garrity GM, Holt JG. 2001. Phylum AII Euryarchaeotaphy. nov., In Boone, D.R. &Castenholz, R.W. (eds.), Bergey¼ s Manual of Systematic Bacteriology, second edition. vol. 1, p. 211-355, Springer-Verlag, New York.

Göker M, Lu M, Fiebig A, Nolan M, Lapidus A, Tice H, et al. 2014. Genome sequence of the mud-dwelling archaeon *Methanoplanus limicola* type strain (DSM 2279^T), reclassification of *Methanoplanuspetrolearius* as *Methanolaciniapetrolearia* and emended descriptions of the genera *Methanoplanus* and *Methanolacinia*. *Standards in Genomic Sciences* 9(3):1076–1088.

Grimalt-Alemany A, £ê¿yk M, Kennes-Veiga DM, Skiadas IV, Gavala HN 2019. Enrichment of mesophilic and thermophilic mixed microbial consortia for syngas biomethanation: the role of kinetic and thermodynamic competition. *Waste and Biomass Valorization*. https://doi.org/10.1007/s12649-019-00595-z.

Grochowski LL, Xu H, White RH. 2006. *Methanocaldococcus jannaschii* uses a modified mevalonate pathway for biosynthesis of isopentenyl diphosphate. *Journal of Bacteriology* 188(9):3192-3198.

Großkopf, R, Subner S, Liesack W. 1998. Novel euryarchaeotal lineages detected on rice roots and in the anoxic bulk soil of flooded rice microcosms. *Applied and Environmental Microbiology* 64:4983–4989.

Guerrero R (2001) Bergey's Manuals and the classification of prokaryotes. *International Microbiology* 4(2):103–109.

Hedlund BP, Dodsworth JA, Cole JK, Panosyan HH. 2013. An integrated study reveals diverse methanogens, Thaumarchaeota, and yet-uncultivated archaeal lineages in Armenian hot springs. *Antonie Van Leeuwenhoek*. 104(1):71-82.

Herman M, Noll KM, Wolfe RS. 1986. Improved agar bottle plate for isolation of methanogens or other anaerobes in a defined gas atmosphere. *Applied and Environmental Microbiology* 51:1124–1126.

Hi M, Well O, Ni S, Boone DR. 1999. Isolation and characterization of a dimethylsulfide-degrading sulfide-degrading methanogen, *Methanolobussiciliae* HI350, from an oil well, characterization of *M. siciliae* T4/MT, and Emendation of *M. siciliae*. *International Journal of Systematic and Evolutionary Biology* 41(3):410–416.

Holdeman LV, Cato EP, Moore WEC. 1977. Anaerobe Laboratory Manual. 4th edn. V. P. I. Anaerobe Laboratory, Blacksburg, VA.

Hubalek Z. 2003. Protectants used in the cryopreservation of microorganisms. *Cryobiology* 46: 205– 229.

Huber H, Stetter K. Family I. Methanopyraceae fam. nov. 2001. In Boone, D.R. &Castenholz, R.W. (eds.), Bergey¼ s Manual of Systematic Bacteriology, second edition. vol. 1, p. 211-355, Springer-Verlag, New York.

Huber H, Thomm M, Konig H, Thies G, Stetter KO. 1982. *Methanococcus thermo lithotrophicus*, a novel thermophilic lithotrophic methanogen. *Archives of Microbiology* 132:47-50

Hungate RE. 1950. The anerobic mesophilic cellulolytic bacteria. *Bacteriological Review* 14: 1–49.

Hungate RE. 1969. A roll-tube method for cultivation of strict anaerobes. *Methods in Microbiology* 3B:117–132.

Husar BA, Wuharmann K, Zehnder AJB. 1982. *Methanothrixsoehngenii* gen. nov. sp. nov., a new acetotrophic non-hydrogen-oxidizing methane bacterium *Microbiology*. 132:1-9

Iino T, Mori K, Suzuki K. 2010. *Methanospirillum lacunae* sp. nov., a methane-producing archaeon isolated from a puddly soil, and emended descriptions of the genus *Methanospirillum* and *Methanospirillumhungatei. International Journal of Systematic and Evolutionary Biology*60:2563–2566.

Iino T, Suzuki KI. 2006. Improvement of the L-drying procedure to keep anaerobic conditions for long-term preservation of methanogens in a culture collection. *Microbiology and Culture Collection* 22:99-104.

Imachi H, Sakai S, Nagai H, Yamaguchi T, Takai K. 2009. *Methanofollis ethanolicus* sp. nov., an ethanol-utilizing methanogen isolated from a lotus field. *International Journal of Systematic and Evolutionary Biology* 59: 800-805

Imachi H, Sakai S, Sekiguchi Y, Hanada S, Kamagata Y, Ohashi A, Harada H. 2008. *Methanolineatarda* gen. nov., sp. nov., a methane producing archaeon isolated from a methanogenic digeser sludge. *International Journal of Systematic and Evolutionary Biology* 58:294-301

Jain MK, Thompson TE, Conway De Macario E, Zeikus JG. 1987. Speciation of *Methanobacterium* strain Ivanov as *Methanobacteriumivanovii*, sp. nov." *Systematic and Applied Microbiology* 9:77-82.

Janssen PH, Kirs M (2008) Structure of the archaeal community of the rumen. *Applied and Environmental Microbiology* 74(12):3619–3625

Jarvis GN, Strömpl C, Burgess DM, Skillman LC, Moore ER, Joblin KN. 2000 Isolation and identification of ruminal methanogens from grazing cattle. *Current Microbiology* 40(5):327–332

Jeanthon C, L'Haridon S, Reysenbach AL, Corre E, Vernet M, Messner P, Sleytr UB, Prieur D. 1999. *Methanococcusvulcanius* sp. nov., a novel hyperthermophilic methanogen isolated from East Pacific Rise, and identification of *Methanococcus* sp. DSM 4213T as *Methanococcusfervens* sp. nov. *International Journal of Systematic and Evolutionary Biology* 49:583–589.

Jeanthon C, L'Haridon S, Reysenbach AL, Vernet M, Messner P, Sleytr UB, Prieur D. 1998. *Methanococcus infernus* sp. nov., a novel hyperthermophilic lithotrophic methanogen isolated from a deep-sea hydrothermal vent. *International Journal of Systematic and Evolutionary Biology* 48:913–919

Jiang B, Parshina SN, van Doesburg W et al. 2005. *Methano methyl ovorans thermophila* sp nov., a thermophilic, methylotrophic methanogen from an anaerobic reactor fed with methanol. *International Journal of Systematic and Evolutionary Biology* 55:2465–2470

Jiang N, Wang Y, Dong X. 2010. Methanol as the primary methanogenic and acetogenic precursor in the cold Zoige wetland at Tibetan Plateau. *Microbial Ecology* 60:206–213

Jones D, Head I, Gray N, Adams J, Rowan A, Aitken C, Bennett B, Huang H, Brown A, Bowler B. 2008. Crude-oil biodegradation via methanogenesis in subsurface petroleum reservoirs. *Nature* 451(7175):176

Jones W, Stugard C, Jannasch H. 1989. Comparison of thermophilic methanogens from submarine hydrothermal vents. *Archives of Microbiology* 151(4):314-318.

Jones WJ, Leigh JA, Mayer F, Woese CR, Wolfe RS. 1983a. *Methanococcus jannaschii* sp. nov., an extremely thermophilic methanogen from a submarine hydrothermal vent. *Archives of Microbiology* 136:254–261.

Jones WJ, Paynter MJB, Gupta R. 1983b. Characterization of *Methanococcus maripaludis* sp. nov., a new methanogen isolated from salt marsh sediment. *Archives of Microbiology* 135: 91–97.

Joshi A, Lanjekar V, Dhakephalkar PK, Dagar SS. 2018. Cultivation of multiple genera of hydrogenotrophic methanogens from different environmental niches. *Anaerobe* 50:64–68

Joulian C, Ollivier B, Patel BKC, Roger PA. 1998. Phenotypic and phylogenetic characterization of dominant culturable methanogens isolated from rice field soils. *FEMS Microbiology Ecology* 25:135–145.

Joulian C, Patel BKC, Ollivier B, Garcia JL, Roger PA. 2000. *Methanobacteriumoryzae* sp. nov., a novel methanogenic rod isolated from a Philippines ricefield. *International Journal of Systematic and Evolutionary Biology* 50: 525–528.

Kadam PC, Boone DR. 1995. Physiological characterization and emended description of *Methanolobusvulcani. International Journal of Systematic and Evolutionary Biology* 45:400–402.

Kadam PC, Ranade DR, Mandelco L, Boone DR. 1994. Isolation and characterization of *Methanolobusbombayensis* sp. nov., a methylotrophic methanogen that requires high concentrations of divalent cations. *International Journal of Systematic and Evolutionary Biology*44:603–607

Kamagata Y, Kawasaki H, Oyaizu H, Nakamura K, Mikami E, Endo G, Koga Y, Yamasato K. 1992. Characterization of three thermophilic strains of *Methanothrix* ("*Methanosaeta*") *thermophile* sp. nov. and rejection of *Methanothrix* ("*Methanosaeta*") *thermoacetophila. International Journal of Systematic and Evolutionary Biology* 42:463-468

Kamagata Y, Mikami E. 1991. Isolation and characterization of a novel thermophilic *Methanosaeta* strain. *International Journal of Systematic and Evolutionary Biology* 41:191–196

Katayama T, Yoshioka H, Mochimaru H, Meng X-Y, Muramoto Y, Usami J, Ikeda H, Kamagata Y, Sakata S. 2014. *Methanohalophiluslevihalophilus* sp. nov., a slightly halophilic, methylotrophic methanogen isolated from natural gas-bearing deep aquifers, and emended description of the genus *Methanohalophilus. International Journal of Systematic and Evolutionary Biology* 64:(6)2089–2093.

Katayama T, Yoshioka H, Takahashi HA, Amo M, Fujii T, Sakata S. 2016. Changes in microbial communities associated with gas hydrates in subseafloor sediments from the Nankai Trough. *FEMS Microbiology Ecology* 92(8):fiw093

Kendall MM, Liu Y, Sieprawska-Lupa M, Stetter KO, Whitman WB, Boone DR. 2006. *Methanococcusaeolicus* sp. nov., a mesophilic, methanogenic archaeon from shallow and deep marine sediments. *International Journal of Systematic and Evolutionary Biology* 56:1525–1529.

Kendall MM, Wardlaw GD, Tang CF et al 2007. Diversity of Archaea in marine sediments from Skan Bay, Alaska, including cultivated methanogens, and description of *Methanogeniumboonei* sp. nov.v*Applied and Environmental Microbiology* 73:407–414

Kern T, Fischer MA, Deppenmeier U, Schmitz RA, Rother M. 2016. *Methanosarcina flavescens* sp. nov., a methanogenic archaeon isolated from a full-scale anaerobic digester *International Journal of Systematic and Evolutionary Biology* 66:1533-1538

Kern T, Linge M, Rother M.2015. *Methanobacteriumaggregans* sp. nov., a hydrogenotrophic methanogenic archaeon isolated from an anaerobic digester *International Journal of Systematic and Evolutionary Biology* 65:1975–1980

Kim BC, Jeong H. 2018. Complete genome sequence of *Methanobrevibactersmithii*Strain KB11, isolated from a Korean Fecal Sample. *Genome Announc*ements 6(7):e00038–e00018

Kitamura K, Fujita T, Akada S, Tonouchi A. 2011. *Methanobacterium kanagience* sp. nov., a hydrogenotrophic methanogen, isolated from rice-field soil. *International Journal of Systematic and Evolutionary Microbiology* 61:1246–1252.

Koga Y, Ohga M. 1992. Maintenance of methanogenic bacteria in liquid nitrogen. *HakkokogakuKaishi* 70: 303-309 (in Japanese).

König H, Stetter K O. 1982. Isolation and characterization of *Methanolobustindarius*, sp. nov., a coccoid methanogen growing only on methanol and methylamines. *ZentralblBakteriolParasitenkdInfektionskrHygAbt 1 Orig*C3, 478–490.

König H. 1984. Isolation and characterization of *Methanobacteri umuliginosum* sp. nov. from a marshy soil. *Canadian Journal of Microbiology* 30:1477–1481.

Kotelnikova S, Macario AJL, Pedersen K. 1998. *Methanobacterium subterraneum* sp. nov., a new alkaliphilic, eurythermic and halotolerant methanogen isolated from deep granitic groundwater. *International Journal of Systematic and Evolutionary Biology* 48:357–367.

Kotelnikova SV, Obraztsova AY, Gongadze GM, Laurinavichius KS. 1993. *Methanobacteriumthermoflexum* sp. nov. and *Methanobacterium defluvii* sp. nov.: Thermophilic rod-shaped methanogens isolated from anaerobic digestor sludge. *Systematic and Applied Microbiology* 16:427–435.

Kotsyurbenko OR, Chin KJ, Glagolev MV et al. 2004. Acetoclastic and hydrogenotrophic methane production and methanogenic populations in an acidic West-Siberian peat bog. *EnvironmentalMicrobiology* 6:1159–1173

Kotsyurbenko OR, Friedrich MW, Simankova MV Nozhevnikova AN, Golyshin PN, Timmis KN, Conrad R. 2007. Shift from acetoclastic to H_2- dependent methanogenesis in a west Siberian peat bog at low pH values and isolation of an acidophilic *Methanobacterium* strain. *Applied and Environmental Microbiology* 73:2344–2348

Krivushin KV, Shcherbakova VA, Petrovskaya LE, Rivkina EM. 2010. *Methanobacterium veterum* sp. nov., from ancient Siberian permafrost. *International Journal of Systematic and Evolutionary Microbiology* 60(2):455–459

Krüger M, Frenzel P, Kemnitz D, Conrad R. 2005. Activity, structure and dynamics of the methanogenic archaeal community in a flooded Italian rice field. *FEMS Microbiology Ecology* 51:323–331.

Kudo Y, Shibata S, Miyaki T, Aono T, Oyaizu H. 1997. Peculiar archaea found in Japanese paddy soils. *Bioscience Biotechnology and Biochemistry* 61:917–920.

Kulik EM, Sandmeier H, Hinni K, Meyer J. 2001. Identification of archaeal rDNA from subgingival dental plaque by PCR amplification and sequence analysis. *FEMS Microbiology Letters* 196(2):129–133

Kumar S, Dagar SS, Agrawal R, Puniya AK. 2018. Comparative diversity analysis of ruminal methanogens in Murrah buffaloes (*Bubalusbubalis*) in four states of North India. *Anaerobe* 52:52–63

Kumar S, Indugu N, Vecchiarelli B, Pitta DW. 2015 Associative patterns among anaerobic fungi, methanogenic archaea, and bacterial communities in response to changes in diet and age in the rumen of dairy cows. *Frontiers in Microbiology* 6:78

Kurr, M., Huber R, König H, Jannasch HW, Fricke H, Trincone A, Kristjansson JK, Stetter KO. 1991. *Methanopyrus kandleri*, gen. and sp. nov. represents a novel group of hyperthermophilic methanogens, growing at 110°C. *Archives of Microbiology* 156(4):239-247.

L'Haridon S, Chalopin M, Colombo D, Toffin L. 2014. *Methanococcoides vulcani* sp. nov., a marine methylotrophic methanogen that uses betaine, choline and N, N dimethylethanolamine for methanogenesis, isolated from a mud volcano, and emended description of the genus *Methanococcoides. International Journal of Systematic and Evolutionary Microbiology*64(6):1978–1983

L'Haridon S, Reysenbach AL, Banta A, Messner P, Schumann P, Stackebrandt E, Jeanthon C. 2003. *Methanocaldococcusindicus* sp. nov., a novel hyperthermophilicmethanogen isolated from the Central Indian Ridge. *International Journal of Systematic and Evolutionary Biology* 53:1931–1935.

Lai MC, Chen SC, Shu CM, Chiou MS, Wang CC, Chuang MJ, Hong TY, Liu CC, Lai LJ, Hua JJ. 2002. *Methanocalculustaiwanensis* sp. nov., isolated from an estuarine environment. *International Journal of Systematic and Evolutionary Biology* 52:1799-1806

Lai MC, Chen SC. 2001. *Methanofollisaquaemaris* sp. nov., a methanogen isolated from an aquaculture ûsh pond. *International Journal of Systematic and Evolutionary Biology* 51:1873–1880.

Lai MC, Lin CC, Yu PH, Huang YF, Chen SC. 2004. *Methano calculus chung hsing ensis* sp. nov., isolated from an estuary and a marine fishpond in Taiwan. *International Journal of Systematic and Evolutionary Biology* 54:183-189

Lauerer, G. Kristjansson JK, Langworthy TA, Koenig H, Stetter KO. 1986. *Methanothermus sociabilis* sp. nov., a second species within the Methanothermaceae growing at 97 °C. *Systematic and Applied Microbiology* 8(1-2):100-105.

Laurinavichyus KS, Kotelnikova SV, Obraztsova AY. 1988. New species of thermophilic methane producing bacteria *Methanobacteriumthermophilum. Mikrobiologiya* 57:1035–1041.

Leadbetter JR, Breznak JA. 1996. Physiological ecology of *Methanobrevibacter cuticularis* sp. nov. and *Methanobrevibactercurvatus* sp. nov., isolated from the hindgut of the termite *Reticulitermes flavipes*. *Applied and Environmental Microbiology* 62:3620–3631.

Leadbetter JR, Crosby LD, Breznak JA. 1998. *Methanobrevibacterfiliformis* sp. nov., a filamentous methanogen from termite hindguts. *Archives of Microbiology* 169: 287–292.

Lee JH, Kumar S, Lee GH, Chang DH, Rhee MS, Yoon MH, Kim BC. 2013. *Methanobrevibacterboviskoreani* sp. nov., isolated from the rumen of Korean native cattle. *International Journal of Systematic and Evolutionary Biology* 63:4196–4201

Lehmann-Richter S, Grosskopf R, Liesack W, Frenzel P, Conrad R. 1999. Methanogenic archaea and CO_2-dependent methanogenesis on washed rice roots. *Environmental Microbiology* 1:159–166.

León JB, Sullivan CM, Sehgal AR. 2013. The prevalence of phosphorus-containing food additives in top-selling foods in grocery stores. *Journal of Renal Nutrition.* 23(4):265-270.

Leschine SB. 1995. Cellulose degradation in anaerobic environments. *Annual Review In Microbiology.* 49: 399-426.

Liu GC, Tokida T, Matsunami T, Nakmura H, Okada M, Sameshima R, Hasegawa T, Sugyama S. 2012. Microbial community composition controls the effects of climate change on methane emission from rice paddies. *Environmental Microbiology* 4:648–654.

Liu Y, Boone DR, Choy C. 1990. *Methanohalophilus oregonense* sp. nov., a methylotrophic methanogen from an alkaline, saline aquifer. *International Journal of Systematic and Evolutionary Biology* 40:111–116.

Liu Y, Whitman WB. 2008. Metabolic, phylogenetic, and ecological diversity of the methanogenic archaea. *Annals of the New York Academy of Sciences* 1125:171–189

Lomans BP, Maas R, Luderer R, Op den Camp HJM, Pol A, van der Drift C, Vogels GD. 1999. Isolation and characterization of *Methanomethylovorans hollandica* gen. nov., sp. nov., isolated from freshwater sediment, a methylotrophic methanogen able to grow on dimethyl sulfide and methanethiol. *Applied and Environmental Microbiology* 65:3641–3650

Lu Y, Lueders T, Friedrich MW, Conrad R. 2005. Detecting active methanogenic populations on rice roots using stable isotope probing. *Environmental Microbiology* 7:326–336.

L Z, Lu Y. 2012. *Methanocellaconradii* sp. nov., a thermophilic, obligate hydrogenotrophic methanogen, isolated from Chinese rice field soil. *PLoS One* 7(4): e35279.

Lyimo TJ, Pol A, Op den Camp HJM, Harhangi HR, Vogels GD. 2000. *Methanosarcina semesiae* sp. nov., a dimethylsulfide-utilizing methanogen from mangrove sediment. *International Journal of Systematic and Evolutionary Biology* 50:171–178.

Ma K, Liu X, Dong X. 2005. *Methanobacteriumbeijingense* sp. nov., a novel methanogen isolated from anaerobic digesters. *International Journal of Systematic and Evolutionary Biology* 55: 325–329.

Ma K, Liu X, Dong X. 2006. *Methanosaetaharundinacea* sp. nov., a novel acetate-scavenging methanogen isolated from a UASB reactor. *International Journal of Systematic and Evolutionary Biology* 56:127–131.

Maczulak AE, Wolin M, Miller TL. 1989. Increase in colonic methanogens and total anaerobes in aging rats. *Applied and Environmental Microbiology* 55(10):2468–2473

Madigan, M., Martinko, J., and Stahl, D., Clark D. P. (2012) *Brock biology of Microorganisms*. 13th ed. Boston: Pearson. pp.134.

Maestrojuan GM, Boone DR, Xun L, Mah RA, Zhang L. 1990. Transfer of *Methanogeniumbourgense, Methanogeniummarisnigri, Methanogeniumolentangyi*, and *Methanogeniumthermophilicum* to the genus *Methanoculleus* gen. nov., emendation of *Methanoculleus*. *International Journal of Systematic and Evolutionary Biology* 40: 117–122.

Maestrojuán GM, Boone DR. 1991. Characterization of *Methanosarcinabarkeri* MS ^{T} and 227, *Methanosarcinamazei* S-6 ^{T}, and *Methanosarcina vacuolata* Z-761 ^{T}. *International Journal of Systematic and Evolutionary Biology*41:267–274.

Mah RA, Kuhn DA. 1984. Transfer of the Type Species of the Genus Methanococcus to the Genus Methanosarcina, Naming it Methanosarcinamazei (Barker 1936) comb. nov. et emend. and Conservation of the Genus Methanococcus (Approved Lists 1980) with Methanococcusvannielii (Approved Lists 1980) as the Type Species Request for an Opinion *International Journal of Systematic and Evolutionary Biology* 34:260-265

Makkar PS and McSweeney CS. 2005. Methods in gut microbial ecology for ruminants. In Springer. 47-53. https://doi.org/https://doi.org/10.1007/1-4020-3791-0

Marchesi JR, Weightman AJ, Cragg BA, Parkes RJ, Fry JC. 2001. Methanogen and bacterial diversity and distribution in deep gas hydrate sediments from the Cascadia margin as revealed by 16S rRNA molecular analysis. *FEMS Microbiology Ecology* 34:221-228.

Mathrani IM, Boone DR, Mah RA, Fox GE, and P. P. Lau. 1988. *Methano halophilus zhilin ue*sp. nov., an alkaliphilic, methylotrophic methanogen. *International Journal of Systematic and Evolutionary Biology* 38:139-142.

McGenity TJ, Sorokin DY. 2018. Methanogens and methanogenesis in hypersaline environments. *Biogenesis of hydrocarbons*. Cham, Cham Springer 1–27

McKay CP, Rask JC, Detweiler AM, Bebout BM, Everroad RC, Lee JZ, Chanton JP, Mayer MH, Caraballo AA, Kapili B. 2016. An unusual inverted saline microbial mat community in an Interdune Sabkha in the Rub'al Khali (the Empty Quarter), United Arab Emirates. *PLoS One* 11(3):e0150342.

Merkel AY, Podosokorskaya O, Sokolova T, Osmolovskaya E A B. 2016. Diversity of methanogenic archaea from the 2012 terrestrial hot spring (Valley of Geysers, Kamchatka). *Microbiology* 85(3):342-349.

Metcalf WW, Zhang JK, Wolfe RS. 1998. An anerobic, intrachamber incubator for growth of *Methanosarcina* spp. on methanol-containing solid media. *Applied and Environmental Microbiology* 64:768–770.

Mikucki JA, Liu Y, Delwiche M, Colwell FS, Boone DR. 2003. Isolation of a methanogen from deep marine sediments that contain methane hydrates, and description of *Methanoculleussubmarinus* sp. nov. *Applied and Environmental Microbiology* 69: 3311–3316.

Miller TL, Lin C. 2002. Description of *Methanobrevibactergottschalkii*sp. nov., *Methanobrevibacterthaueri*sp. nov., *Methanobrevibacterwoesei*sp. nov. and *Methanobrevibacter wolinii* sp. nov.*International Journal of Systematic and Evolutionary Microbiology* 52(3):819–822

Miller TL, Wolin M, de Macario EC, Macario A. 1982. Isolation of *Methanobrevibacter smithii* from human feces. *Applied and Environmental Microbiology* 43(1):227–232

Miller TL, Wolin MJ. 1985. *Methanosphaerastadtmanae* gen. nov., sp. nov.: a species that forms methane by reducing methanol with hydrogen. *Archives of Microbiology* 141:116–122.

Mochimaru H, Tamaki H, Hanada S, Imachi H, Nakamura K, Sakata S, Kamagata Y. 2009. *Methanolobus profundi* sp. nov., a methylotrophic methanogen isolated from deep subsurface sediments in a natural gas field. *International Journal of Systematic and Evolutionary Biology* 59:714–718.

Mochimaru H, Tamaki H, Katayama T, Imachi H, Sakata S, Kamagata Y. 2016. *Methanomicrobium antiquum* sp. nov., a hydrogenotrophic methanogen isolated from deep sedimentary aquifers in a natural gas field. *International Journal of Systematic and Evolutionary Biology* 6 6:4873-4877

Mori K, HarayamaS. 2011. *Methanobacteriumpetrolearium* sp. nov. and *Methanobacterium ferruginis* sp. nov., mesophilic methanogens isolated from salty environments. *International Journal of Systematic and Evolutionary Biology* 61:138–143.

Mori K, Yamamoto H, Kamagata Y, Hatsu M, Takamizawa K. 2000. *Methanocalculuspumilus* sp. nov., a heavy-metal-tolerant methanogen isolated from a waste-disposal site. *International Journal of Systematic and Evolutionary Biology* 50:1723-1729

Morvan B, Dore J, Rieu-Lesme F, Foucat L, Fonty G, Gouet P. 1994. Establishment of hydrogen-utilizing bacteria in the rumen of the newborn lamb. *FEMS Microbiology Letters* 117(3): 249–256

Nakamura K. Takahashi A, Mori C, Tamaki H, Mochimaru H, Nakamura K, Takamizawa K, Kamagata Y. 2013. *Methanothermobacter tenebrarum* sp. nov., a hydrogenotrophic, thermophilic methanogen isolated from gas-associated formation water of a natural gas field. *International Journal of Systematic and Evolutionary Microbiology* 63(2):715-722.

Nerlekar MR. 2014. Diversity of methanogens from oil reservoirs in India (Doctoral Dissertation). Savitribai Phule Pune University

Ni SS, Boone DR. 1991. Isolation and characterization of a dimethyl sulfide-degrading methanogen, *Methanolobussiciliae* HI350, from an oil well, characterization of *M. siciliae* T4/M [T], and emendation of *M. siciliae*. *International Journal of Systematic and Evolutionary Biology*41:410–416.

Nicholson MJ, Evans PN, Joblin KN. 2007 Analysis of methanogen diversity in the rumen using temporal temperature gradient gel electrophoresis: identification of uncultured methanogens. *Microbial Ecology* 54(1):141–150

Ollivier B, Fardeau ML, Cayol JL, Magot M, Patel BK, Prensier G, Garcia JL. 1998. *Methanocalculus halotolerans* gen. nov., sp. nov., isolated from an oil-producing well. *International Journal of Systematic and Evolutionary Microbiology* 48(3):821-828.

Ollivier B, Cayol JL, Patel BKC, Magot M, Fardeau ML, Garcia JL. 1997. *Methanoplanus petrolearius* sp. nov., a novel methanogenic bacterium from an oil-producing well. *FEMS Microbiology Letters*147(1):51-56.

Ollivier BM, Mah RA, Garcia JL, Boone DR. 1986. Isolation and characterization of *Methanogeniumbourgense* sp. nov. *International Journal of Systematic and Evolutionary Biology* 36:297–301.

Oremland RS, Boone DR. 1994. *Methanolobustaylorii* sp. nov., a new methylotrophic, estuarine methanogen. *International Journal of Systematic and Evolutionary Biology* 44:573–575.

Oren A. 2011. Thermodynamic limits to microbial life at high salt concentrations. *Environmental Microbiology* 13(8):1908–1923

Orphan V, Taylor L, Hafenbradl D, Delong E. 2000. Culture-dependent and culture-independent characterization of microbial assemblages associated with high-temperature petroleum reservoirs. *Applied and Environmental Microbiology*. 66:700–711.

Parshina SN, Ermakova AV, Bomberg M, Detkova EN. 2014. *Methanospirillumstamsii* sp. nov., a psychrotolerant, hydrogenotrophic, methanogenic archaeon isolated from an anaerobic expanded granular sludge bed bioreactor operated at low temperature, *International Journal of Systematic and Evolutionary Biology.* 64:180–186.

Patel GB, Sprott GD, Fein JE. 1990. Isolation and characterization of *Methanobacteriumespanolae* sp. nov., a mesophilic, moderately acidiphilic methanogen. *International Journal of Systematic and Evolutionary Biology* 40:12–18.

Patel GB, Sprott GD. 1990. Methanosaetaconcilii gen. nov. sp. nov. ("Methanothrixconcilii") and *Methanosaetathermoacetophila* nom. rev., comb. nov. *International Journal of Systematic and Evolutionary Biology* 40:79-82

Paterek JR, Smith PH. 1988. *Methanohalophilusmahii* gen. nov., sp. nov., a methylotrophic halophilic methanogen. *International Journal of Systematic and Evolutionary Biology* 38:122–123.

Paynter MJB, Hungate RE. 1968. Characterization of *Methanobacteriummobilis*, sp. n., isolated from the bovine rumen. *Journal of Bacteriology* 95:1943–1951.

Plenge M, Summers E, Engel AS, Omelon C, Bennett PC. 2016. Thermophilic archaeal diversity and methanogenesis from El Tatio Geyser Field, Chile. *Geomicrobiology*. DOI: 10.1080/01490451.2016.1168496.

Prescott LM, Harley JP, Klein DA 2002 *Microbiology*. 5th edition. The McGraw-Hill companies New York. pp 122

Probst AJ, Auerbach AK, Moissl-Eichinger C. 2013. Archaea on human skin. *PloS one* 8:e65388

Ramakrishnan B, Lueders T, Dunfield PF, Conrad R, Friedrich MW. 2001. Archaeal community structures in rice soils from different geographical regions before and after initiation of methane production. *FEMS Microbiology Ecology* 37:175–186.

Rea S, Bowman JP, Popovski S, Pimm C, Wright A-DG. 2007. *Methanobrevibacter millerae* sp. nov. and *Methanobrevibacterolleyae*sp. nov., methanogens from the ovine and bovine rumen that can utilize formate for growth. *International Journal of Systematic and Evolutionary Microbiology* 57(3):450–456

Rivard CJ, Henseon JM, Thomas MV, Smith PH. 1983. Isolation and characterization of *Methanobacteriumpaynteri* sp. nov., a mesophilic methanogen isolated from marine sediments. *Applied and Environmental Microbiology* 46:484-490

Rivard CJ, Smith PH 1982. Isolation and characterization of a thermophilic marine methanogenic bacterium *Methanogeniumthermophilicum* sp. nov. *International Journal of Systematic and Evolutionary Biology* 32:430–436.

Rivkina E, Shcherbakova V, Laurinavichius K, Petrovskaya L, Krivushin K, Kraev G, Pecheritsina S, Gilichinsky D. 2007. Biogeochemistry of methane and methanogenic archaea in permafrost. *FEMS Microbiology Ecology* 61:1–15

Romesser JA, Wolfe RS, Mayer F, Spiess E, Walther-Mauruschat A. 1979. *Methanogenium,* a new genus of marine methanogenic bacteria, and characterization of *Methanogeniumcariaci* sp. nov. and *Methanogenium marisnigri* sp. nov.*Archives of Microbiology* 12:147-153.

Roy R, Kluber HD, Conrad R. 1997. Early initiation of methane production in anoxic rice soil despite the presence of oxidants. *FEMS Microbiology Ecology* 24:311–320.

Sakai S, Conrad R, Liesack W, Imachi H. 2010. *Methanocellaarvoryzae*spnov., a hydrogenotrophic methanogen isolated from rice field soil. *International Journal of Systematic and Evolutionary Microbiology* 60:2918–2923.

Sakai S, Ehara M, Tseng IC, Yamaguchi T, Bräuer SL, Quiroz HC, Zinder SH, Imachi H. 2012. *Methanolineamesophila* sp. nov., a hydrogenotrophic methanogen isolated from rice field soil, and proposal of the archaeal family Methanoregulaceae fam. nov. within the order Methanomicrobiales. *International Journal of Systematic and Evolutionary Biology* 62:1389-1395

Sakai S, Imachi H, Hanada S, Ohashi A, Harada H, Kamagata Y. 2008. *Methanocella paludicola* gen. nov., sp. nov., a methane-producing archaeon, the first isolate of the lineage 'Rice Cluster I', and proposal of the new archaeal order Methanocellales ord. nov.*International Journal of Systematic and Evolutionary Microbiology* 58:929–936.

Sakai S, Imachi H, Sekiguchi Y, Ohashi A, Harada H, Kamagata Y. 2007. Isolation of key methanogens for global methane emission from rice paddy fields: a novel isolate affiliated with the clone cluster rice cluster I. *Applied and Environmental Microbiology* 73:4326–4331.

Sakai S, Takaki Y, Miyazaki M, Ogawara M, Yanagawa K, Miyazaki J, Takai K. 2019. *Methanofervidicoccus abyssi* gen. nov., sp. nov., a hydrogenotrophic methanogen, isolated from a hydrothermal vent chimney in the Mid-Cayman Spreading Center, the Caribbean Sea. *International Journal of Systematic and Evolutionary Microbiology* 69(4):1225-1230.

Sakane T, Fukuda I, Itoh T, Yokota A. 1992. Long-term preservation of halophilic archaebacteria and thermoacidophilic archaebacteria by liquid drying. *Journal of Microbiological Methods* 16:281-187.

Savant DV, Shouche YS, Prakash S, Ranade DR. 2002. *Methanobrevibacter acididurans* sp. nov., a novel methanogen from a sour anaerobic digester. *International Journal of Systematic and Evolutionary Biology* 52: 1081–1087.

SchirmackJ, Mangelsdorf K, Ganzert L, Hillebrand-VoiculescuA, Wagner D. 2014. *Methanobacterium movilense* sp. nov., a hydrogenotrophic, secondary-alcohol-utilizing methanogen from the anoxic sediment of a subsurface lake. *International Journal of Systematic and Evolutionary Biology* 64:522–527.

Schüttle UME, Abdo Z, Bent SJ, Shyu C, Williams CJ, Pierson JD, Forney LJ. 2008. Advances in the use of terminal restriction fragment length polymorphism (T-RFLP) analysis of 16S rRNA genes to characterize microbial communities. *Applied Microbiology and Biotechnology* 80:365–380.

Seengenyoung J, Mamimin C. Prasertsan P,O-ThongS. 2018. Enhancement of biohythane production from solid waste by co-digestion with palm oil mill effluent in two-stage thermophilic fermentation. *International Journal of Hydrogen Energy*. DOI: 10.1016/j.ijhydene.2018.08.021

Shcherbakova V, Rivkina E, Pecheritsyna S, Laurinavichius K, Suzina N, Gilichinsky D. 2011. *Methanobacteriumarcticum*sp. nov., a methanogenic archaeon from Holocene Arctic permafrost. *International Journal of Systematic and Evolutionary Microbiology* 61(1):144–147

Shimizu S, Ueno A, Naganuma T, Kaneko K. 2015. *Methanosarcinasubterranea* sp. nov., a methanogenic archaeon isolated from a deep subsurface diatomaceous shale formation*International Journal of Systematic and Evolutionary Biology* 65:1167-1171

Shimizu S, Uneo A, Tamamura S, Naganuma T, Kaneko K. 2013. *Methanoculleus horonobensis* sp. nov., a methanogenic archaeon isolated from a deep diatomaceous shale formation. *International Journal of Systematic and Evolutionary Biology* 63:4320-4323

Shimizu S, Upadhye R, Ishijima Y, Naganuma T. 2011. *Methanosarcina horonobensis* sp. nov., a methanogenic archaeon isolated from a deep subsurface Miocene formation. *International Journal of Systematic and Evolutionary Biology* 61:2503–2507.

Shinzato N, Matsumoto T, Yamaoka I, Oshima T, Yamagishi A. 1999. Phylogenetic diversity of symbiotic methanogens living in the hindgut of the lower termite *Reticulitermessperatus* analyzed by PCR and in situ hybridization. *Applied and Environmental Microbiology* 65(2):837–840

Shlimon AG, Friedrich MW, Niemann H, Ramsing NB, Finster K. 2004. *Methanobacterium Aarhu Sense* sp. nov., a novel methanogen isolated from a marine sediment (Aarhus Bay, Denmark). *International Journal of Systematic and Evolutionary Biology* 54:759–763.

Simankova MV, Parshina SN, Tourova TP, Kolganova TV, Zehnder AJB, Nozhevnikova AN. 2001. *Methanosarcinalacustris* sp. nov., a new psychrotolerant methanogenic archaeon from anoxic lake sediments. *Systematic and Applied Microbiology* 24:362–367.

Simione FP, Brown EM. (Eds.), 1991. ATCC Preservation Methods: Freezing and Freeze-Drying, American Type Culture Collection, Rockville, MD, pp. 7– 13

Singh A, Dubey SK. 2012. Temporal variations in methanogenic community structure and methane production potential of tropical rice ecosystems. *Soil Biology and Biochemistry* 48:162–166.

Singh A, Singh RS, Upadhyay SN, Joshi CG, Tripathi AK, Dubey SK. 2012. Community structure of methanogenic archaea and methane production associated with compost treated tropical rice-field soil. *FEMS Microbiology Ecology* 82:118–134.

Singh N, Kendall MM, Liu YT, Boone DR. 2005. Isolation and characterization of methylotrophic methanogens from anoxic marine sediments in Skan Bay, Alaska: description of *Methanococcoides alaskense* sp. nov., and emended description of *Methanosarcina baltica. International Journal of Systematic and Evolutionary Microbiology* 55:2531–2538

Sirohi SK, Chaudhary PP, Singh N, Singh D, Puniya AK. 2013. The 16S rRNA and mcrA gene based comparative diversity of methanogens in cattle fed on high fibre based diet. *Gene* 523(2):161–166.

Sizova MV, Izquierdo JA, Panikov NS, Lynd LR. 2011. Cellulose- and Xylan-Degrading Thermophilic Anaerobic Bacteria from Biocompost. *Applied & Environmental Microbiology*. 77(7): 2282–2291.

Skillman LC, Evans PN, Naylor GE, Morvan B, Jarvis GN, Joblin KN. 2004. 16S ribosomal DNA-directed PCR primers for ruminal methanogens and identification of methanogens colonising young lambs. *Anaerobe* 10(5):277–285

Smith PH, Hungate RE. 1958. Isolation and characterization of *Methanobacterium ruminantium* N. SP. *Journal of Bacteriology* 75:713–718.

Sorokin DY, Abbas B, Merkel AY, Rijpstra WIC, Damsté JSS, Sukhacheva MV, van Loosdrecht MCM. 2015. *Methanosalsumnatronophilum* sp. nov., and *Methanocalculusalkaliphilus* sp. nov., haloalkaliphilic methanogens from hypersaline soda lakes. *International Journal of Systematic and Evolutionary Biology.* 65:3739-3745.

Sorokin DY, Makarova KS, Abbas B, Ferrer M, Golyshin PN, Galinski EA, Ciordia S, Mena MC, Merkel AY, Wolf YI. 2017. Discovery of extremely halophilic, methyl-reducing euryarchaea provides insights into the evolutionary origin of methanogenesis. *Nature Microbiology* 2(8):17081

Sowers K R, Ferry JG 1983. Isolation and characterization of a methylotrophic marine methanogen, *Methanococcoidesmethylutens*gen. nov., sp. nov. *Applied and Environmental Microbiology* 45:684-690.

Sowers KR, Baron SF, Ferry JG. 1984. *Methanosarcinaacetivorans* sp. nov., an acetotrophic methane-producing bacterium isolated from marine sediments. *Applied and Environmental Microbiology* 47:971–978.

Sowers KR, Noll KM. 1995. Techniques for anaerobic growth, methanogenic archeae. In Archaea, a Laboratory Manual (F. T. Robb, A. R. Place, K. R. Soweres, H. J. Schreier, S. DarSarma, and E. M. Fleischmann, eds.), pp. 17–47. Cold Spring Harbor Laboratory Press, NY.

Sprenger WW, Hackstein JH, Keltjens JT. 2007. The competitive success of *Methanomicrococcus blatticola,* a dominant methylotrophic methanogen in the cockroach hindgut, is supported by high substrate affinities and favorable thermodynamics. *FEMS Microbiology Ecology* 60(2):266–275

Sprenger WW, van Belzen MC, Rosenberg J, Hackstein JHP, Keltjens JT. 2000. *Methanomicrococcusblatticola* gen. nov., sp. nov., a methanol- and methylamine-reducing methanogen from the hindgut of the cockroach *Periplanetaamericana*. *International Journal of Systematic and Evolutionary Biology*50:1989–1999.

Stadtman TC, Barker HA. 1951. Studies on the methane fermentation X.: a new formate-decomposing bacterium, *Methanococcus vannielii*. *Journal of Bacteriology* 62:269–280.

Stetter KO, Thomm M, Winter J, Wildgruber G, Huber H, Zillig W. 1981. *Methanothermus fervidus*, sp. nov., a novel extremely thermophilic methanogen isolated from an Icelandic hot spring. *Zentralblatt für Bakteriologie Mikrobiologie und Hygiene: I. Abt. Originale C: Allgemeine, angewandte und ökologische Mikrobiologie*, 2(2):166-178.

Stetter KO. 1989. Genus 11. MethanolobusKonig and Stetter 1983, 439'' (effective publication: Konig and Stetter 1982, 488), p. 2205-2207. In J. T. Staley, M. P. Bryant, N. Pfennig, and J. G. Holt (ed.), Bergey's manual of systematic bacteriology, vol. 3. The Williams & Wilkins Co., Baltimore

Stetter KO. 2006. History of discovery of the first hyperthermophiles. *Extremophiles* 10(5):357-362.

Stewart LC, Jung JH, Kim YT, Kwon SW, Park CS, Holden JF. 2015. *Methanocaldococcus bathoardescens* sp. nov., a hyperthermophilic methanogen isolated from a volcanically active deep-sea hydrothermal vent. *International Journal of Systematic and Evolutionary Microbiology*, 65(4):1280-1283.

Sundset MA, Edwards JE, Cheng YF, Senosiain RS, Fraile MN, Northwood KS, Præsteng KE, Glad T, Mathiesen SD, Wright A-DG. 2009a. Molecular diversity of the rumen microbiome of Norwegian reindeer on natural summer pasture. *Microbial Ecology* 57(2):335–348

Sundset MA, Edwards JE, Cheng YF, Senosiain RS, Fraile MN, Northwood KS, Præsteng KE, Glad T, Mathiesen SD, Wright A-DG. 2009b. Rumen microbial diversity in Svalbard reindeer, with particular emphasis on methanogenic archaea. *FEMS Microbiology Ecology* 70(3):553–562

Surakasi VP, Wani AA, Shouche YS, et al. 2007. Phylogenetic analysis of methanogenic enrichment cultures obtained from Lonar Lake in India: isolation of *Methanocalculus* sp. and *Methanoculleus* sp. *MicrobEcol*54:697-704

Takai K, Inoue A, Horikoshi K. 2002. *Methanothermococcusokinawensis* sp. nov., a thermophilic, methane producing archaeon isolated from a Western Pacific deep-sea hydrothermal vent system. *International Journal of Systematic and Evolutionary Biology* 52:1089–1095.

Takai K, Nakamura K, Toki T,Tsunogai U, Miyazaki M, Miyazaki J, Hirayama H, Nakagawa S,Nunoura T,Horikoshi K. 2008. Cell proliferation at 122°C and isotopically heavy CH_4 production by a hyperthermophilic methanogen under high-pressure cultivation. *Proceedings of the National Academy of Sciences* 105(31):10949-10954.

Takai K, Nealson KH, Horikoshi K. 2004. *Methanotorrisformicicus* sp. nov., a novel extremely thermophilic, methane-producing archaeon isolated from a black smoker chimney in the Central Indian Ridge. *International Journal of Systematic and Evolutionary Biology* 54:1095–1100.

Thakkar C, Ranade DR. 2002. Alkalophilic *Methanosarcina* isolated from Lonar Lake. *Current Science* 82(4):455-458.

Tian J, Wang Y, Dong X. 2010. *Methanoculleus hydrogenitrophicus* sp. nov., a methanogenic archaeon isolated from wetland soil. *International Journal of Systematic and Evolutionary Biology* 60:2165–2169.

Toth CR, Gieg LM. 2018. Time course-dependent methanogenic crude oil biodegradation: dynamics of fumarate addition metabolites, biodegradative genes, and microbial community composition. *Frontiers in Microbiology* 8:2610

Touzel JP, Albagnac G. 1983. Isolation and characterization of *Methanococcusmazei* strain MC3. *FEMS Microbiology Letters* 16:241-245

Tsapekos, P., et al., 2019. Co-digestion of municipal waste biopulp with marine macroalgae focusing on sodium inhibition. *Energy Conversion and Management* 180:931-937.

Valentine DL. 2011. Emerging topics in marine methane biogeochemistry. *Annual Review of Marine Science* 3:147–171

van Bruggen JJA, Zwart KB, Hermans JGF, van Hove EM, Stumm CK, Vogels GD. 1986. Isolation and characterization of *Methanoplanusendosymbiosus* sp. nov., an endosymbiont of the marine sapropelic ciliate *Metopuscontortus* Quennerstedt. *Archives of Microbiology* 144: 367-374

Vandaele AC, Neefs E, Drummond R, Thomas IR, Daerden F, Lopez-Moreno J-J, Rodriguez J, Patel MR, Bellucci G, Allen M. 2015. Science objectives and performances of NOMAD, a spectrometer suite for the ExoMars TGO mission. *Planet Space Sci*ence 119:233–249

Varjani S, Shrivastava VK, Sindhu R, Takur IS, Gnansounou E. 2018. Culture based approaches, dependent and independent, for microbial community fractions in petroleum reservoirs. *Indian Journal of Experimental Biology* 56:444-450

Vishnivetskaya TA, Buongiorno J, Bird J, Krivushin K, Spirina EV, Oshurkova V, Shcherbakova VA, Wilson G, Lloyd KG, Rivkina EM. 2018. Methanogens in the Antarctic Dry Valley Permafrost. *FEMS Microbiology Ecology* 94(8):fiy109

von Klein D, Arab H, Volker H, Thomm M. 2002. *Methanosarcinabaltica*, spnov., a novel methanogen isolated from the Gotland Deep of the Baltic Sea. *Extremophiles* 6:103–110

Wagner D, Schirmack J, Ganzert L, Morozova D, Mangelsdorf K. 2013. *Methanosarcinasoligelidi* sp. nov., a desiccation- and freeze-thaw-resistant methanogenic archaeon from a Siberian permafrost-affected soil. *International Journal of Systematic and Evolutionary Biology* 63:2986–2991.

Wallin E, Heijne GV. 1998. Genome wide analysis of integral membrane proteins from eubacterial, archaean, and eukaryotic organisms. *Protein Science*7(4):1029-1038.

Ward DM. 1978. Thermophilic methanogenesis in a hot-spring algal-bacterial mat (71 to 30 degrees C). *Applied and Environmental Microbiology* 35(6):1019-1026.

Wasserfallen A, Nolling J, Pfister P, Reeve J, Conway de Macario E (2000) Phylogenetic analysis of 18 thermophilic *Methanobacterium* isolates supports the proposals to create a new genus, *Methanothermobacter* gen. nov., and to reclassify several isolates in three species, *Methanothermobacterthermautotrophicus* comb. nov., *Methanothermobacter wolfeii* comb. nov., and *Methanothermobacter marburgensis* sp. nov.*International Journal of Systematic and Evolutionary Biology* 50: 43–53.

Watanabe T, Wang G, Taki K, Ohashi Y, Kimura M, Asakawa S. 2010. Vertical change in bacterial and archaeal communities with soil depth in Japanese paddy fields. *Soil Science and Plant Nutrition* 56:705–715.

Watkins AJ, Roussel EG, Parkes RJ, Sass H. 2014. Glycine betaine as a direct substrate for methanogens (*Methanococcoides* spp.). *Applied and Environmental Microbiology* 80(1): 289–293

Weng CY, Chen SC, Lai MC, Wu SY, Lin S, Yang TF, Chen PC. 2015. *Methanoculleus taiwanensis* sp. nov., a methanogen isolated from deep marine sediment at the deformation front area near Taiwan*International Journal of Systematic and Evolutionary Biology* 65:1044-1049

Whitman WB, Ankwanda E, Wolfe RS. 1982. Nutrition and carbon metabolism of *Methanococcusvoltae*. *Journal of Bacteriology* 149:852–863.

Whitman, W.B. "Genus I. Methanocaldococcus gen. nov." In: Bergey's Manual of Systematic Bacteriology, 2nd ed., vol. 1 (D.R. Boone and R.W. Castenholz, eds.), Springer-Verlag, New York (2001). pp. 243-245.

Whitman WB, Bowen TL and Boone DR. 2006. The Prokaryotes, Archaea. Bacteria: *Firmicutes, Actinomycetes - The Methanogenic Bacteria*. 3:165-207.

Widdel F, Rouvihre PE, Wolfe RS. 1988. Classification of secondary alcohol-utilizing methanogens including a new thermophilic isolate. *Archives of Microbiology* 150:477-481.

Wildgruber G, Thomm M, König H, Ober K, Ricchiuto T, Stetter KO. 1982. *Methanoplanuslimicola*, a plate shaped methanogen representing a novel family, Methanoplanaceae. *Archives of Microbiology* 132:31-36.

Wilharm T, Zhilina TN, Hummel P. 1991. DNA-DNA Hybridization of methylotrophic halophilic methanogenic bacteria and transfer of *Methanococcushalophilus*[vp] to the genus *Methanohalophilus* as *Methanohalophilushalophilus* comb. nov.*International Journal of Systematic and Evolutionary Biology* 41:558–562.

Wilms R, Sass H, Köpke B, Cypionka H, Engelen B. 2007. Methane and sulfate profiles within the subsurface of a tidal flat are reflected by the distribution of sulfate-reducing bacteria and methanogenic archaea. *FEMS Microbiology Ecology* 59(3):611–621

Winter J, Lerp C, Zabel H.-P, Wildenauer FX, Ko$nig H, Schindler F. 1984. *Methanobacteriumwolfei* sp. nov., a new tungsten-requiring, thermophilic, autotrophic methanogen. *Systematic and Applied Microbiology* 5:457–466.

Winter J. 1983. Maintenance of stock cultures of methanogens in the laboratory. *Systematic and Applied Microbiology* 4:558-563

Winter J. Lerp C, Zabel H-P, Wildenauer FX, Koenig H, Schindler F. 1984. *Methanobacterium wolfei*, sp. nov., a new tungsten-requiring, thermophilic, autotrophic methanogen. *Systematic and Applied Microbiology* 5(4):457-466.

Wolfe RS, Metcalf WW. 2010. A vacuum-vortex method for preparation of anoxic solutions or liquid culture media in small amounts for cultivation of methanogens or other strict anaerobes. *Anaerobe* 16:216–219.

Worakit S, Boone DR, Mah RA, Abdel-Samie M-E, El-Halwagi MM. 1986. *Methanobacteriumalcaliphilum* sp. nov., an H_2-utilizing methanogen that grows at high pH values. *International Journal of Systematic and Evolutionary Biology* 36:380–382.

Wu SY, Chen SC, Lai MC. 2005. *Methanofollisformosanus* sp. nov., isolated from a fish pond. *International Journal of Systematic and Evolutionary Biology* 55:837–842.

Wu SY, Lai MC. 2011. Methanogenic Archaea isolated from Taiwan's Chelungpu Fault. *Applied and Environmental Microbiology* 77: 830-838.

Yamane K, Hattori Y, Ohtagaki H, Fujiwara K, Fujiwara K. 2011. Microbial diversity with dominance of 16S rRNA gene sequences with high GC contents at 74 and 98°C subsurface crude oil deposits in Japan. *FEMS Microbiol Ecology.* 76(2):220-35.

Yashiro Y, Sakai S, Ehara M, Miyazaki M, Yamaguchi T, Imachi H. 2011. *Methanoregulaformicica* sp. nov., a methane producing archaeon isolated from methanogeni sludge. *International Journal of Systematic and Evolutionary Biology* 61:53-59.

Yavitt JB, Yashiro E, Cadillo-Quiroz H, Zinder SH. 2012. Methanogen diversity and community composition in peatlands of the central to northern Appalachian Mountain region, North America. *Biogeochemistry* 109:117–131.

Zabel HP, Konig H, Winter J. 1984. Isolation and characterization of a new coccoid methanogen, *Methanogeniumtatii*, spec. nov. from a solfataric field on Mount Tatio. *Archives of Microbiology* 137:308–315.

Zeikus JG, Henning DL. 1975. *Methanobacteriumarbophilicum* sp. nov. An obligate anaerobe isolated from wetwood of living trees. *Antonie Van Leeuwenhoek* 41:543–552.

Zeikus JG, Wolfe RS. 1972. *Methanobacteriumthermoautotrophicus*sp. n., an anaerobic, autotrophic, extreme thermophile. *Journal of Bacteriology*109:707-713.

Zellner G, Alten C, Stackebrandt E, Conway de Macario E, Winter J. 1987. Isolation and characterization of *Methanocorpusculum parvum* gen. nov., spec. nov., a new tungsten requiring, coccoid methanogen. *Archives of Microbiology* 147:13-20.

Zellner G, Bleicher K, Braun E, Kneifel H, Tindall BJ, Macario EC, Winter J. 1988. Characterization of a new mesophilic, secondary alcohol-utilizing methanogen, *Methanobacteriumpalustre* sp. nov. from a peat bog. *Archives of Microbiology* 151:1–9.

Zellner G, Boone DR, Keswani J, Whitman WB, Woese CR, Hagelstein A, Tindall BJ, Stackebrandt E. 1999. Reclassification of *Methanogenium tationis* and *Methanogenium liminatans* as *Methanofollistationis* gen. nov., comb. nov. and *Methanofollisliminatans* comb. nov. and description of a new strain of *Methanofollisliminatans*. *International Journal of Systematic and Evolutionary Biology*49:247–255.

Zellner G, Boone DR. 2001. Genus III. *Methanofollis*. In *Bergey's Manual of Systematic Bacteriology*, 2nd edn, vol. 1, pp. 253–255. Edited by D. R. Boone & R. W. Castenholz. New York: Springer.

Zellner G, Messner P, Kneifel H, Tindall BJ, Winter J, Stackebrandt E. 1989. *Methanolacinia* gen nov., incorporating *Methanomicrobiumpaynteri* as *Methanolaciniapaynteri* comb. nov.*Journal of General and Applied Microbiology*35:185-202.

Zellner G, Messner P, Winter J, Stackebrandt E. 1998. *Methanoculleuspalmolei* sp. nov., an irregularly coccoid methanogen from an anaerobic digester treating wastewater of a palm oil plant in North-Sumatra, Indonesia. *International Journal of Systematic and Evolutionary Biology* 48: 1111–1117.

Zellner G, Sleytr UB, Messner P, Kneifel H, Winter J. 1990. *Methanogeniumliminatans* sp. nov., a new coccoid, mesophilic methanogen able to oxidize secondary alcohols. *Archives of Microbiology* 153:287–293.

Zhang GS, Tian JQ, Jiang N et al 2008. Methanogen community in Zoige wetland of Tibetan plateau and phenotypic characterization of a dominant uncultured methanogen cluster ZC-I. *Environmental Microbiology* 10:1850–1860.

Zhao H, Wood AG, Widdel F, Bryant MP. 1988. An extremely thermophilic *Methanococcus* from a deep-sea hydrothermal vent and its plasmid. *Archives of Microbiology* 150:178–183.

Zhao L, Liu X, Dong X. 2014. *Methanospirillumpsychrodurum* sp. nov., isolated from wetland soil *International Journal of Systematic and Evolutionary Biology* 64: 638-641.

Zhao Y, Boone DR, Mah RA, Boone JE, Xun L.1989.Isolation and characterization of *Methanocorpusculum labreanum* sp. nov. from the LaBrea tar pit. *International Journal of Systematic and Evolutionary Biology*39:10–13

Zhilina TN, Zavarzin GA. 1987. *Methanohalobium evestigatus* n. gen., n. sp. The extremely halophilic methanogenic Archaebacterium. DoklAkadNauk SSSR 293, 464–468 (in Russian).

Zhilina TN, Zavarzin GA. 1987. *Methanosarcinavacuolata* sp. nov., a vacuolated methanosarcina. *International Journal of Systematic and Evolutionary Biology* 37:281–283.

Zhilina TN, Zavarzina DG, Kevbrin VV, Kolganova TV. 2013. *Methanocalculusnatronophilus* sp. nov., a new alkaliphilic hydogenotrophic methanogenic archaeon from soda lake and proposal of the new family *Methanocalculaceae Microbiology (English translation of Mikrobiologiia)* 82:698–706.

Zhilina TN. 1983. A new obligate halophilic methane-producing bacterium. *Mikrobiologiya.* 52:375-382. (In Russian.).

Zhou L, Liu X, Dong X. 2014. *Methanospirillum psychrodurum* sp. nov., isolated from wetland soil. *International Journal of Systematic and Evolutionary Microbiology* 64:638–641

Zhu J, Liu X, Dong X. 2011. *Methanobacteriummovens* sp. nov. and *Methanobacterium flexile* sp. nov., isolated from lake sediment. *International Journal of Systematic and Evolutionary Biology* 61:2974–2978.

Zinder SH, Sowers KR, Ferry JG. 1985. *Methanosarcina thermophila* sp. nov., a thermophilic, acetotrophic, methane-producing bacterium. *International Journal of Systematic and Evolutionary Biology* 35:522–523.

6

Anaerobic Oxidation of Methane A Brief Overview of Anaerobic Methanotrophy

Monali C. Rahalkar[1,2], Kumal Khatri[1,2], Pranitha S. Pandit[1,2] Jyoti A. Mohite[1,2], Rahul A. Bahulikar[3]

[1]C2, Bioenergy group, MACS Agharkar Research Institute, G.G. Agarkar Road Pune-411004, Maharashtra, India

[2]Savitribai Phule Pune University, Ganeshkhind Road, Pune 411007 Maharashtra, India

[3]BAIF Development Research Foundation, Central Research Station Urulikanchan, Pune-412202, Maharashtra, India

Abstract

Methane is the second most important green-house gas and has a major role in global warming. Aerobic oxidation of methane is known for more than 100 years. Anaerobic methane oxidation (AOM) was recently discovered before ~25 years, first in marine sediments. The first discovered AOM was found to be mediated by anaerobic methanotrophic archaea (ANME) in association with sulphate reducing bacteria (SRB) and termed as sulphate dependent anaerobic methane oxidation (S-DAMO). Three main groups of ANME are present which perform methane oxidation by reverse methanogenesis pathway. Later on, methane oxidation by using nitrate or nitrite as the electron acceptor was discovered from a freshwater sediment. N-DAMO or nitrate/ nitrite dependent AMO can be carried out by an archaeal or a bacterial member, respectively. Members of Ca. *Methylomirabilis oxyfera are bacteria, and use nitrite as the electron acceptor and methane monooxygenase to carry out methane oxidation. Whereas, members of* Ca. *Methanoperedens nitroreducens are archaea, and use reverse methanogenesis coupled to nitrate reduction. Metals like FeIII or Mn IV can also be used as electron acceptors for AOM, a process*

**Corresponding Author: monalirahalkar@aripune.org*

recently discovered (M-DAMO). In this case, one archaeal partner and one bacterial partner could be involved, although more research is going in this area. S-DAMO is the main methane oxidation process in the marine sediments and accounts for the major methane oxidation in the marine sediments. In fresh water sediments, aerobic methane oxidation and processes like N-DAMO seem to play an important role.

Introduction

Methane is the second most important greenhouse gas (GHG) after CO_2 and contributes to 30% of the net anthropogenic radiative warming force Conrad (2009a). Methane is estimated to have global warming potential of 28-36 over 100 years (https://www.epa.gov/ghgemissions/understanding-global-warmingpotentials). The concentration of methane has increased from 0.71 ppm from the pre-industrial times to 1.86 ppm till date. The global methane budget is in the range of 500-600 Tg of methane per year (Conrad 2009a). The net balance of the methane emitted is the balance of the sources and sinks. Major natural sources of methane are wetlands, termites, oceans and gas hydrates, which contribute to around 40% of the budget. Remaining sources are anthropogenic: ruminants, rice fields, landfills, sewage treatment plants and fossil fuels.

The major sink for methane is the photo chemical oxidation of methane by the hydroxyl radicals present in the troposphere (~450 Tg/yr^{-1}) and by microbial methane oxidation in soils (~10 Tg/yr^{-1}) and diffusion in the stratosphere. The actual production of methane could be larger than the net emissions. Microbial methane oxidation, both aerobic and anaerobic accounts for mitigating the produced methane and keeping the emissions lower before it escapes to the atmosphere.

Discovery of anaerobic oxidation of methane (AOM)

Methane can be oxidized aerobically by methane by methane oxidizing bacteria or methanotrophs (Hanson and Hanson 1996). Methanotrophs are aerobic or microaerophilic bacteria which can use methane as the sole source of C and energy. Oxygen is required for the first step in methane oxidation in this case. Aerobic methanotrophs were first described and the first methanotrophic isolate was obtained by Soehngen in 1906 (Hanson and Hanson 1996). Anaerobic oxidation of methane was predicted and thought to be thermodynamically possible in 1990s. The first evidence of anaerobic oxidation of methane (AOM) was given (Boetius, 2000) where fluorescence hybridization (FISH) was used and it was found out that aggregates of archaea and sulphate reducing bacteria were responsible for AOM. Later on using FISH coupled to secondary ion mass spectrometry it was clear that the archaea were using methane as the C source.

Later, these consortia were found to consist of ANME-2 cluster (archaeal partner) and sulphate reducing bacteria (*Desulfosarcina*/ *Desulfococcus*). Since the methane oxidation was coupled to sulphate reduction, the process is later described as S-DAMO, i.e. sulphate dependent anaerobic oxidation of methane.

Organisms coupling AOM to sulphate reduction (S-DAMO)

Sulphate dependent anaerobic oxidation of methane (s-DAMO) is carried out by ANME (anaerobic methanotrophic) archaea in syntrophic cooperation with sulphate reducing bacteria. In marine habitats, S-DAMO or anaerobic methane oxidation by ANME is the major process. The metabolic process of AOM is proposed to be a reverse methanogenesis coupled to the reduction of sulphate involving syntrophically associated ANME and sulphate-reducing bacteria (Hoehler et al. 1994); (Nauhaus et al. 2002); (Nauhaus et al. 2005); (Li et al. 2019). Resultant consortia oxidize methane and sulphate equally, and bicarbonate and sulfide are produced as a byproduct. Sulfide supports chemosynthetic communities which derive energy from sulfide oxidation in cold seeps.

$$CH_4 + SO_4^{2-} = HCO_3^- + HS^- + H_2O \ (\Delta G^{0'} = -34 \text{ kJ mol}^{-1} CH_4)$$

The presence of syntrophic partner depends on the type of environmental habitat, physio-chemical conditions of the environment and also on the sulphate-methane transition zone (SMTZ). SMTZ is a widespread high microbial diversity zone of continental margin in which AOM process is often observed where pore water sulphate and methane are both depleted to near-zero concentrations (Chevalier et al. 2013); (Egger et al. 2018); (Treude et al. 2005); (Niewöhner et al. 1998); (Yoshinaga et al. 2014); (Harrison et al. 2009a). This SMTZ is the main methane barrier which prevents the CH_4 release from the deep seafloor. It is a kind of niche interface of overlying sulphate-reducing bacterial population and the underlying methanogen population (D'Hondt et al. 2009); (Harrison et al. 2009b); (Knittel and Boetius 2009); (Reeburgh 2007). Biochemical and molecular signature of marine sediments shows that methane is probably the primary electron donor in SMTZ. It is estimated that an unknown substrate is produced during the AOM process, which is utilized by sulphate-reducing bacteria (SRB). Other possible interspecies electron carriers include hydrogen, acetate, methanol, and formate (Thomsen et al. 2001).

ANME (Anaerobic methanotrophic/methane oxidizing archaea)

Three groups have been identified within ANME. ANME I (subclusters a and b), ANME-2 (subclusters a, b and c) and ANME-3. ANME-1 shows affiliation to *Methanomicrobiales* and *Methanosarcinales*. ANME-2 methanotrophs are related to *Methanosarcinales*, whereas ANME-3 methanotrophs are most related to *Methanococcoides* (Timmers et al. 2017).

SRB associated with ANME: The thermophilic ANME-1 archaea show syntrophy with deep-branching Hot Seep-1 cluster of sulphate-reducing bacteria, while the psychrophilic ANME-3 archaea associated with *Desulfobulbus* bacterial cluster (Holler et al. 2011); (Wegener et al. 2015); (Niemann et al. 2006). ANME-1 and -2 archaea are associated with Seep-SRB1a or Seep-SRB2 cluster of *Desulfosarcina/ Desulfococcus* sulphate-reducing bacteria (Orphan et al. 2001); (Michaelis et al. 2002); (Knittel et al. 2003); (Kleindienst et al. 2012); (Milucka et al. 2012).

Table 1: Characters of ANME

Features	ANME-1	ANME-2	ANME-3
Closest affiliation	Methanomicrobiales & Methanosarcinales	Methanosarcinales	*Methanococcoides* sp.
Partners	SEEP-SRB-1 & SEEP -SRB-2 (Desulfosarcina/ Desulfococcus) group	Desulfobulbus group	
Environment	Marine	Marine & Freshwater	Marine
Morphology of individual cell	Rectangular	Coccoid	Coccoid
Morphology in aggregation	mat-type association	mixed- and shell-type	shell-type
Reverse methanogenesis enzymes	-> only lacks the gene for N5, N10-methylene tetrahydromethanopterin reductase (mer)	All enzymes present	-
Core lipids	glycerol dialkyl glycerol tetraether (GDGT)	tetramethylhexadecane (crocetane)	pentamethylicosenes
Sn-2-OH-archaeol /archaeol ratio	0–0.8	0.1 to 5.5	1.1 to 5.5
Reference	(Knittel et al. 2005); (Schreiber et al. 2010); (Blumenberg et al. 2004); (Niemann and Elvert 2008); (Kurth et al. 2019) ; (Boetius et al. 2000); (Knittel and Boetius 2009); (Niemann et al. 2006); (Losekann et al. 2007); (Mueller et al. 2015)		

New advances in S-DAMO process studies

Sulphate coupled AOM has been believed to be a syntrophic process so far, however the underlying mechanism was not clear. Recently, it has been shown a new dissimilatory sulphate reduction pathway present in ANME archaea could carry out the sulphate reduction resulting on the formation of zerovalent S (0). Therefore, AOM may be carried out only by ANME. The SRB associated with the ANME may be involved in the disproportionation of the S(0) (Milucka et al. 2012).

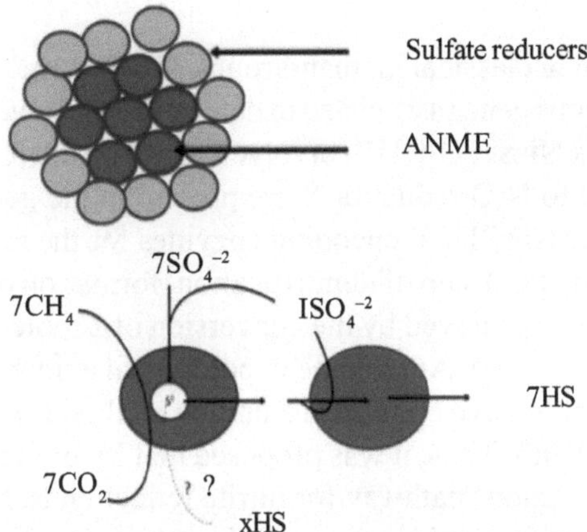

Fig. 1: Schematic representation of a revised model of anaerobic oxidation of methane coupled to sulphate reduction proposed by Milucka et al 2012. ANME-2 oxidize methane with a concomitant reduction of sulphate to zero-valent sulphur (S^0, elemental sulphur) that is partially deposited or bound intracellularly.

Anaerobic oxidation of methane coupled to denitrification (N-DAMO)

Methane and ammonium are the final products of anaerobic degradation of organic matter. Nitrite and nitrate are formed by aerobic ammonia oxidizing bacteria. Anaerobic oxidation of methane coupled to denitrification is also known as N-DAMO, nitrate/ nitrite dependent anaerobic methane oxidation (Jetten, M. 2008). N-DAMO process has been found to occur in many freshwater environments. The process can be carried out by two different groups of organisms: *Ca.* Methanoperedens nitroreducens-like archaea and *Ca.* Methylomirabilis oxyfera (from the NC10 phylum). *Ca.* Methanoperedens nitroreducens can carry the reduction of nitrate coupled to methane oxidation, whereas *Ca.* Methylomirabilis oxyfera can couple nitrite reduction coupled to methane oxidation.

$CH_4 + 4NO_3 \rightarrow CO_2 + 4NO_2^- + 2H_2O$ (Nitrate reduction) ($\Delta G^{0'}$ = - 503 kJ mol^{-1} CH_4)

$3CH_4 + 8NO_2^- + 8H^+ \rightarrow 3CO_2 + 4N_2 + 10H_2O$ (Nitrite reduction) ($\Delta G^{0'}$ = - 928 kJ mol^{-1} CH_4)

In natural and manmade systems, *Ca.* Methanoperedens nitroreducens carries out the methane oxidation using reverse methanogenesis pathway and conversion of nitrate to nitrite. *Ca.* M. oxyfera uses nitrite and methane and converts it to nitric oxide and disproportionates into oxygen and nitrogen gas. This intracellularly generated oxygen is used for the methane oxidation reaction (Ettwig et al. 2010b). This was discovered by genome analysis and by experimental evidence. M. oxyfera is known to possess particulate methane monooxygenase enzyme which is present in classical methanotrophs. The genome, when studied for the presence of various genes associated to denitrification, revealed some interesting facts. The genes NirSJFD/GH/L involved in nitrite to NO reduction and NorZ involved in NO to N_2O reduction were present in the genome and expressed. The gene cluster NosZDFY encoding enzymes for the reduction of N_2O to N_2 was missing. The final step of denitrification, formation of dinitrogen gas, was hypothesized to be achieved by the conversion of 2 molecules of NO to O_2 and N_2 (Ettwig et al. 2010a). After some experimental evidences, it was found that the candidate could have employed a unique method for N_2 and O_2 production (Ettwig et al. 2010a). Thus, it was proposed that M. oxyfera used a novel 'intra aerobic denitrification' pathway for nitrite reduction to N_2 production and the conversion of nitrogen oxide using NO dismutase for O_2 production (Ettwig et al. 2010a).

"*Ca.* Methylomirabilis oxyfera" possesses a unique, atypical polygonal shape with an outer sheath of (glyco) protein surface layer (Wu et al. 2012). The intra-cytoplasmic membrane (ICM) was absent in case of M. oxyfera and thus was an exception to the classical proteobacterial methanotroph (Wu et al. 2012).

'Candidatus Methanoperedens nitroreducens', an archaeon couples the anaerobic oxidation of methane to nitrate reduction. In natural and man-made ecosystems, this organism is often found at oxic-anoxic interfaces. In these areas, nitrate is available along with methane.

The discovery of these dominant n-DAMO bacteria that employs unique pathways for denitrification and anaerobic methane oxidation has opened the doors to new research in exploring organisms capable of surviving under harsh environmental conditions. The phylogenetic and evolutionary basis of the early age existing organisms can thus be deciphered.

Anaerobic oxidation of methane coupled to metal reduction (M-DAMO)

Metal reduction coupled to anaerobic oxidation of methane had also been thought be to be possible. Fe (III) reduction by anaerobic methane oxidizing archaea could be possible by direct electron transfer or by soluble electron shuttles such as humic acids. Also, there is a possibility that iron reducing organisms such as *Geobacter* or *Shewanella* and anaerobic ANME can couple these reactions (Deutzmann 2018).

The predicted reactions for M-DAMO are as follows:

$CH_4 + 8\ Fe\ (OH)_3 + 16\ H+ \rightarrow CO_2 + 8\ Fe^{2+} + 22\ H_2O\ \Delta\ G^{0'}$ = "571.2

$CH_4 + 4\ MnO_2 + 8\ H+ \rightarrow CO_2 + 4\ Mn^{2+} + 6\ H_2O\ \Delta\ G^{0'}$ = "763.2

Table 2: Various AOM processes

Process	Electron acceptors	Phylogenetic groups
S-DAMO (Sulphate-dependent anaerobic methane oxidation	Sulphate	ANME 1, 2, 3 and SRB (sulphate reducing bacteria)
N-DAMO (Nitrate/nitrite-dependent anaerobic methane oxidation)	NitrateNitrite	*Methylomirabilis oxyfera*-like bacteria (NC-10 phylum) *Methanoperedens nitroreducens*-like archaea
M-DAMO (Metal- dependent anaerobic methane oxidation)	Fe (III) Mn (IV)	ANME and Delta-proteobacteria (predicted), Unknown

Micro-aerobic oxidation of methane coupling with denitrification by methanotrophs (aerobic methane oxidizing bacteria)

Methanotrophs oxidize the methane aerobically or under micro-oxic conditions. They dwell in diverse environments where methane is produced (Knief 2015). One such environment are the rice ecosystems where the methane-oxidizing bacteria utilize the methane generated by anaerobic respiration of the methanogens releasing methane to the atmosphere (Conrad 2009b). Methanotrophic taxa form a unique clade in the microbial classification. Its specific requirement for gaseous carbon source from the atmosphere separates them from other proteobacterial species. Methanotrophs with valid published names belong to two classes; *Gammaproteobacteria* (also termed Type I and Type X methanotrophs) and *Alphaproteobacteria* (also termed Type II methanotrophs) (Bowman 2018). Together in both the classes 23 genera and 56 species are validly described (Dedysh and Knief 2018). Gammaproteobacterial methanotrophs belong to the order *Methylococcales*, which constitutes the families *Methylococcaceae*, *Methylothermaceae*, and *Crenotrichaceae* (Orata et al. 2018). Most of the identified isolates belong to the family *Methylococcaceae*, which consists of 16 genera and 39 species (Orata et al. 2018). Recently, many reports have indicated that aerobic methanotrophs have been found in anoxic environments of lakes, fresh water sediments and other habitats. Also, in many of the rice field associated methanotrophs cultured from India (Pandit et al. 2016, Pandit et al. 2016b) show the nitrite or nitrate reduction pathways (Rahalkar et al. 2016, Rahalkar and Pandit 2018, Rahalkar et al. 2019). Many novel methanotrophs are being discovered from the rice field habitats which show denitrification genes, e.g. *Methylocucumis oryzae, Methlomagnum ischizawaii and Methyloterricola oryzae* (Pandit et al. 2018, Pandit and Rahalkar 2018, Frindte et al. 2017).

Methylomonas denitrificans FJG1 is the bacterium which has been demonstrated to have denitrification coupled to methane oxidation (Kits et al.

2015). In lake sediments, aerobic methanotrophs have been repeatedly found in anaerobic zones (Rahalkar 2007, Oswald et al. 2016).

Future research

Oceans cover a large portion of the Earth and have an annual rate of 85-300 Tg methane yr^{-1}, of which more than 90% is consumed by AOM. However, an enormous portion of methane is stored in the form of gas hydrates (~10^7 Tg of C). These methane hydrates are sensitive to increase in temperature and the methane from this source is likely to escape without consumption as the release might be rapid. Thus, it would be important to study the effect of global warming on methane hydrates and on factors that control AOM (Knittel and Boetius 2009). The recent advances in the discovery of zero-valent S produced by ANME questions the role of syntrophy in S-DAMO. Future research in this area would throw more light on the process. In fresh water environments such as lake sediments, agricultural canals, etc., N-DAMO could be an important process and could be a major sink. However, more studies are required in this direction to obtain a clear picture of the impact of AMO on methane emissions in these habitats (Deutzmann 2018). Also, more advances are required in cultivation of the members involved in AOM in the laboratory which would help us in studying various parameters involved in these processes.

Acknowledgements

MCR acknowledges SERB, Department of Science and Technology for providing the funds and the research fellowship for JAM (EMR/2017/002817). PSP acknowledges Department of Science and Technology for providing her women scientist (WOS-A) fellowship and funds (SR/WOS-A/LS-410/2017). KK acknowledges CSIR for the research fellowship.

References

Blumenberg M, Seifert R, Reitner J, Pape T, Michaelis W. 2004. Membrane lipid patterns typify distinct anaerobic methanotrophic consortia. *Proceedings of the National Academy of Sciences* 101: 11111-11116.

Boetius A, Ravenschlag K, Schubert CJ, Rickert D, Widdel F, Gieseke A, Amann R, Jørgensen BB, Witte U, Pfannkuche O. 2000. A marine microbial consortium apparently mediating anaerobic oxidation of methane. *Nature* 407: 623.

Bowman JP. 2018. Methylococcales. *Bergey's Manual of Systematics of Archaea and Bacteria*: 1-4.

Chevalier N, Bouloubassi I, Birgel D, Taphanel MH, Lopez-Garcia P. 2013. Microbial methane turnover at Marmara Sea cold seeps: a combined 16S rRNA and lipid biomarker investigation. *Geobiology* 11: 55-71.

Conrad R. 2009a. The global methane cycle: recent advances in understanding the microbial processes involved. *Environmental Microbiology Reports* 1: 285-92.

Conrad R. 2009b. The global methane cycle: recent advances in understanding the microbial processes involved. *Environmental microbiology reports* 1: 285-292.

D'Hondt S, Spivack AJ, Pockalny R, Ferdelman TG, Fischer JP, Kallmeyer J, Abrams LJ, Smith DC, Graham D, Hasiuk F. 2009. Subseafloor sedimentary life in the South Pacific Gyre. *Proceedings of the National Academy of Sciences* 106: 11651-11656.

Dedysh SN, Knief C 2018. Diversity and phylogeny of described aerobic methanotrophs. *Methane Biocatalysis: Paving the Way to Sustainability.* Springer.

Deutzmann JS 2018. Anaerobic Methane Oxidation in Freshwater Environments. *In:* Boll, M (ed.) *Anaerobic Utilization of Hydrocarbons, Oils, and Lipids, Handbook of Hydrocarbon and Lipid Microbiology,*. Springer International.

Egger M, Riedinger N, Mogollón JM, Jørgensen BB. 2018. Global diffusive fluxes of methane in marine sediments. *Nature Geoscience* 11: 421.

Ettwig KF, Butler MK, Le Paslier D, Pelletier E, Mangenot S, Kuypers MM, Schreiber F, Dutilh BE, Zedelius J, de Beer D, Gloerich J, Wessels HJ, van Alen T, Luesken F, Wu ML, van de Pas-Schoonen KT, Op den Camp HJ, Janssen-Megens EM, Francoijs KJ, Stunnenberg H, Weissenbach J, Jetten MS, Strous M. 2010b. Nitrite-driven anaerobic methane oxidation by oxygenic bacteria. *Nature* 464: 543-8.

Frindte K, Maarastawi SA, Lipski A, Hamacher J, Knief C. 2017. Characterization of the first rice paddy cluster I isolate, Methyloterricola oryzae gen. nov., sp. nov. and amended description of Methylomagnum ishizawai. *International Journal of Systematic and Evolutionary Microbiology* 67: 4507-4514.

Hanson RS, Hanson TE. 1996. Methanotrophic bacteria. *Microbiology and Molecular Biology Reviews* 60: 439-471.

Harrison BK, Zhang H, Berelson W, Orphan VJ. 2009b. Variations in archaeal and bacterial diversity associated with the sulphate-methane transition zone in continental margin sediments (Santa Barbara Basin, California). *Appl. Environ. Microbiol.* 75: 1487-1499.

Hoehler TM, Alperin MJ, Albert DB, Martens CS. 1994. Field and laboratory studies of methane oxidation in an anoxic marine sediment: Evidence for a methanogen sulphate reducer consortium. *Global Biogeochemical Cycles* 8: 451-463.

Holler T, Widdel F, Knittel K, Amann R, Kellermann MY, Hinrichs K-U, Teske A, Boetius A, Wegener G. 2011. Thermophilic anaerobic oxidation of methane by marine microbial consortia. *The ISME journal* 5: 1946.

Kits KD, Klotz MG, Stein LY. 2015. Methane oxidation coupled to nitrate reduction under hypoxia by the Gammaproteobacterium *Methylomonas denitrificans*, sp. nov. type strain FJG1. *Environmental Microbiology*.

Kleindienst S, Ramette A, Amann R, Knittel K. 2012. Distribution and in situ abundance of sulphate-reducing bacteria in diverse marine hydrocarbon seep sediments. *Environmental Microbiology* 14: 2689-710.

Knief C. 2015. Diversity and habitat preferences of cultivated and uncultivated aerobic methanotrophic bacteria evaluated based on pmoA as molecular marker. *Frontiers in Microbiology* 6: 1346.

Knittel K, Boetius A. 2009. Anaerobic oxidation of methane: progress with an unknown process. *Annual Review of Microbiology* 63: 311-34.

Knittel K, Boetius A, Lemke A, Eilers H, Lochte K, Pfannkuche O, Linke P, Amann R. 2003. Activity, distribution, and diversity of sulphate reducers and other bacteria in sediments above gas hydrate (Cascadia Margin, Oregon). *Geomicrobiology Journal* 20: 269-294.

Knittel K, Losekann T, Boetius A, Kort R, Amann R. 2005. Diversity and distribution of methanotrophic archaea at cold seeps. *Applied and Environmental Microbiology* 71: 467-79.

Kurth JM, Smit NT, Berger S, Schouten S, Jetten MS, Welte CU. 2019. Anaerobic methanotrophic archaea of the ANME-2d clade feature lipid composition that differs from other ANME archaea. *FEMS Microbiology Ecology* 95: fiz082.

Li L, Xue S, Xi J. 2019. Anaerobic oxidation of methane coupled to sulphate reduction: Consortium characteristics and application in co-removal of H2S and methane. *Journal of Environmental Sciences* 76: 238-248.

Losekann T, Knittel K, Nadalig T, Fuchs B, Niemann H, Boetius A, Amann R. 2007. Diversity and abundance of aerobic and anaerobic methane oxidizers at the Haakon Mosby Mud Volcano, Barents Sea. *Applied and Environmental Microbiology* 73: 3348-62.

Michaelis W, Seifert R, Nauhaus K, Treude T, Thiel V, Blumenberg M, Knittel K, Gieseke A, Peterknecht K, Pape T. 2002. Microbial reefs in the Black Sea fueled by anaerobic oxidation of methane. *Science* 297: 1013-1015.

Milucka J, Ferdelman TG, Polerecky L, Franzke D, Wegener G, Schmid M, Lieberwirth I, Wagner M, Widdel F, Kuypers MM. 2012. Zero-valent sulphur is a key intermediate in marine methane oxidation. *Nature* 491: 541-6.

Mueller TJ, Grisewood MJ, Nazem-Bokaee H, Gopalakrishnan S, Ferry JG, Wood TK, Maranas CD. 2015. Methane oxidation by anaerobic archaea for conversion to liquid fuels. *Journal of Industrial Microbiology & Biotechnology* 42: 391-401.

Nauhaus K, Boetius A, Krüger M, Widdel F. 2002. In vitro demonstration of anaerobic oxidation of methane coupled to sulphate reduction in sediment from a marine gas hydrate area. *Environmental Microbiology* 4: 296-305.

Nauhaus K, Treude T, Boetius A, Krüger M. 2005. Environmental regulation of the anaerobic oxidation of methane: a comparison of ANME I and ANME II communities. *Environmental Microbiology* 7: 98-106.

Niemann H, Elvert M. 2008. Diagnostic lipid biomarker and stable carbon isotope signatures of microbial communities mediating the anaerobic oxidation of methane with sulphate. *Organic Geochemistry* 39: 1668-1677.

Niemann H, Lösekann T, De Beer D, Elvert M, Nadalig T, Knittel K, Amann R, Sauter EJ, Schlüter M, Klages M. 2006. Novel microbial communities of the Haakon Mosby mud volcano and their role as a methane sink. *Nature* 443: 854.

Niewöhner C, Hensen C, Kasten S, Zabel M, Schulz H. 1998. Deep sulphate reduction completely mediated by anaerobic methane oxidation in sediments of the upwelling area off Namibia. *Geochimica et Cosmochimica Acta* 62: 455-464.

Orata FD, Meier-Kolthoff JP, Sauvageau D, Stein LY. 2018. Phylogenomic analysis of the gammaproteobacterial methanotrophs (Order Methylococcales) calls for the reclassification of members at the genus and species levels. *Frontiers in Microbiology* 9: 3162.

Orphan VJ, Hinrichs KU, Ussler W, 3rd, Paull CK, Taylor LT, Sylva SP, Hayes JM, Delong EF. 2001. Comparative analysis of methane-oxidizing archaea and sulphate-reducing bacteria in anoxic marine sediments. *Applied and Environmental Microbiology* 67: 1922-34.

Oswald K, Milucka J, Brand A, Hach P, Sten Littmann, Wehrli B, Kuypers MMM, Schubert CJ. 2016. Aerobic gammaproteobacterial methanotrophs mitigate methane emissions from oxic and anoxic lake waters. *Limnology and Oceanography*: 1-18.

Pandit PS, Hoppert M, Rahalkar MC. 2018. Description of '*Candidatus* Methylocucumis oryzae', a novel Type I methanotroph with large cells and pale pink colour, isolated from an Indian rice field. *Antonie Van Leeuwenhoek* 111: 2473-2484.

Pandit PS, Rahalkar MC, Dhakephalkar PK, Ranade DR, Pore S, Arora P, Kapse N. 2016. Deciphering community structure of methanotrophs dwelling in rice rhizospheres of an Indian rice field using cultivation and cultivation-independent approaches. *Microbial Ecology* 71: 634-644.

Pandit PS, Ranade DR, Dhakephalkar PK, Rahalkar MC. 2016b. A pmoA-based study reveals dominance of yet uncultured Type I methanotrophs in rhizospheres of an organically fertilized rice field in India. *3 Biotech* 6: 135.

Rahalkar M. 2007. Aerobic methanotrophic bacterial communities from sediments of Lake Constance. PhD, University of Konstanz.

Rahalkar MC, Khatri K, Pandit PS, Dhakephalkar PK. 2019. A putative novel Methylobacter member (KRF1) from the globally important Methylobacter clade 2: cultivation and salient draft genome features. *Antonie Van Leeuwenhoek* 112: 1399-1408.

Rahalkar MC, Pandit P. 2018. Genome-based insights into a putative novel Methylomonas species (strain Kb3), isolated from an Indian rice field. *Gene Reports* 13: 9-13.

Rahalkar MC, Pandit PS, Dhakephalkar PK, Pore S, Arora P, Kapse N. 2016. Genome characteristics of a novel Type I methanotroph 'Sn10-6' isolated from a flooded Indian rice field. *Microbial Ecology* 71: 519-523.

Reeburgh WS. 2007. Oceanic methane biogeochemistry. *Chemical reviews* 107: 486-513.

Schreiber L, Holler T, Knittel K, Meyerdierks A, Amann R. 2010. Identification of the dominant sulphate-reducing bacterial partner of anaerobic methanotrophs of the ANME-2 clade. *Environmental Microbiology* 12: 2327-40.

Thomsen TR, Finster K, Ramsing NB. 2001. Biogeochemical and molecular signatures of anaerobic methane oxidation in a marine sediment. *Applied and Environmental Microbiology* 67: 1646-56.

Timmers PHA, Welte CU, Koehorst JJ, Plugge CM, Jetten MSM, Stams AJM. 2017. Reverse Methanogenesis and Respiration in Methanotrophic Archaea. *Archaea* https://doi.org/10.1155/2017/1654237: 1-21.

Treude T, Niggemann J, Kallmeyer J, Wintersteller P, Schubert CJ, Boetius A, Jørgensen BB. 2005. Anaerobic oxidation of methane and sulphate reduction along the Chilean continental margin. *Geochimica et Cosmochimica Acta* 69: 2767-2779.

Wegener G, Krukenberg V, Riedel D, Tegetmeyer HE, Boetius A. 2015. Intercellular wiring enables electron transfer between methanotrophic archaea and bacteria. *Nature* 526: 587.

Wu ML, van Teeseling MC, Willems MJ, van Donselaar EG, Klingl A, Rachel R, Geerts WJ, Jetten MS, Strous M, van Niftrik L. 2012. Ultrastructure of the denitrifying methanotroph "Candidatus Methylomirabilis oxyfera," a novel polygon-shaped bacterium. *Journal of Bacteriology* 194: 284-291.

Yoshinaga MY, Holler T, Goldhammer T, Wegener G, Pohlman JW, Brunner B, Kuypers MM, Hinrichs K-U, Elvert M. 2014. Carbon isotope equilibration during sulphate-limited anaerobic oxidation of methane. *Nature Geoscience* 7: 190.

7

Microbial Sulfur Compound Respiration

Sheetal Shirodkar, Geetika Sharma and Nancy Garg*

Amity Institute of Biotechnology, Amity University, Uttar Pradesh, India

Abstract

Anaerobic sulfur compound reduction processes make use of sulfur, thiosulfate, trithionate, tetrathionate, and sulfite during the reduction process. Complete reduction of these sulfur compounds leads to production of H_2S. Isolation of sulfur compound reducers requires a suitable electron donor and one of the sulfur compound as a sole electron acceptor during the reduction process carried out in anaerobic conditions. Anaerobic sulfur, thiosulfate and tetrathionate reductases identified to date are molybdopterin binding oxidoreductases that extract out electron during the reduction process. The electron transport chain during the reduction process involves cytochromes and menaquinones. Dissimilatory sulfite reductases include conventional siroheme dissimilatory sulfite reductase and the nonconventional sulfite reductases including ASR sulfite reductase, F_{420} dependent sulfite reductase and the MccA sulfite reductase. Anaerobic sulfur compound reduction process are almost ubiquitous in nature and are important part of the biogeochemical cycle. They occurs in aquatic systems, are found in extreme environments such as hydrothermal vents, acid mine drainage sites and are also found in important human and animal pathogens.

Introduction

Sulfur is one of the essential elements for living systems making about 1% of the total dry weight of the bacterial cells. Sulfur occurs in nature in varied oxidation numbers of -2, 0, +2, +4 and +6 and therefore, sulfur furnishes very good examples of oxidation and reduction in living systems. In biological life,

*Corresponding Author: sshirodkar@amity.edu, geetikasharma49@gmail.com, gnancy987@gmail.com

sulfur is present in proteins (methionine and cysteine or as iron-sulfur clusters), sulfate esters of polysaccharides, vitamins like thiamin and biotin, antioxidant like glutathione, sulfur containing coenzymes like *S*-adenosyl-L-methionine (SAM), coenzyme A, the molybdenum cofactor (MoCo) and in certain hormones. In aminoacid cysteine, sulfur is present in its reduced sulfide form, allowing it to form inter and intra chain disulfide bonds, which plays an important role in maintaining protein structure via proper protein-folding. Sulfur in the cells is also present in oxidized forms in compounds such as glucose sulphate, choline sulphate etc. Sulfur respiration is the sole energy source for some lithotrophic bacteria and archaea. Some strictly anaerobic bacteria use sulfate as the ultimate electron acceptor for respiration. Since sulfate is thermodynamically stable and is available in abundance in the biosphere hence it forms the foundation of the biological sulfur cycle. Sulfate (SO_4^{2-}) is the primary soluble form of sulfur in the soil and is the principle source of sulfur for plants nevertheless most of the living plants and microbes incorporate reduced form of sulfur i.e. sulfide(S^2) into biomolecules. There are two biological pathways of sulfate reduction. The process of bioconversion of sulfate from the environment to its bioactive reduced form at the cellular level by living systems is called sulfur assimilation. The sulfate assimilation pathway is essential for the synthesis of the many sulfur-containing moeities especially thiol group of aminoacid cysteine which acts as the principle starting metabolite for the synthesis of other sulfur-containing organic metabolites. In many bacteria, sulfate assimilation constitutes three reactions: sulfate uptake and synthesis of APS (adenosine 5' phosphosulfate), 3' phosphorylation of APS to produce PAPS, whose sulfuryl group is reduced to sulfide and incorporated into other metabolites. The biological significance of sulfate assimilation is that these plants and microorganisms have the ability to reduce sulfate and incorporate the reduced sulfur into Cys and Met, and these organisms serve as the nutritional source for animals that lack the ability to synthesize these amino acids. As opposed to assimilatory metabolism dissimilatory sulfur compound reduction processes includes sulfur respiration and dissimilatory sulfate reduction. Dissimilation of sulfur compounds is one of the earliest biological strategies used by primitive organisms to obtain energy. Strict anaerobic conditions are required for dissimilatory sulfate reduction and it represents an important process within the biogeochemical sulfur cycle. In the dissimilatory mechanism, sulfate is the terminal electron acceptor producing large quantities of sulphide. By oxidation processes sulfide or elemental sulfur can be reoxidised to form sulfate. The chemical sulphide oxidation does not directly generate sulfate, but leads to production of thiosulfate and elemental sulfur as intermediates.

The assimilatory sulfate reduction process produces negligible amount of H_2S. However death and decay of living systems leads to conversion of organic sulfur compounds into H_2S by mineralization. Interestingly significantly higher amount of H_2S is produced during biological dissimilatory sulfur compound reduction processes as compared to the H_2S released upon death an decay of same amount of biomass. Perhaps the dissimilatory reduction processes in living system provide them advantage in terms of energy production and with ability to survive in diverse and varied environmental conditions. In this book chapter we will be discussing in more details about the anaerobic dissimilatory sulfur compounds reduction and sulfur respiration processes.

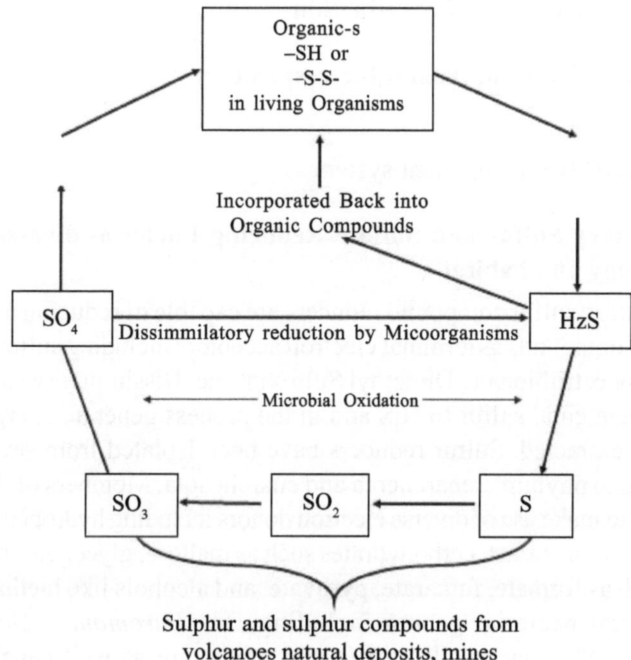

Fig. 1: Sulfur Cycle

Sulfate and Sulfur reduction

- Dissimilatory Sulfur and Sulfate Reducing Bacteria- diversity, morphology and habitat
- Cultivation of Sulfate Reducing Bacteria

Sulfite respiration

- Mechanism of Sulfite respiration
- Siroheme
- The Dsr (Siroheme Dissimilatory Sulfite Reductase) Complex
- ASR Sulfite reductase

- Fsr or F420 dependent sulfite reductase
- MccA/SirA Sulfite reductase

Mechanism of Sulfur and other sulfur compound respiration

- **Dissimilatory Thiosulfate reduction**
- **Dissimilatory Tetrathionite reduction**
- **Dissimilatory Sulfur reduction**

Global Regulators of Anaerobic respiration

- FNR
- ArcAB two component system

Dissimilatory Sulfur and Sulfate Reducing Bacteria- diversity, morphology and habitat

Dissimilatory sulfur compound reducers are capable of reducing a wide variety of sulfur compounds as terminal electron acceptors including sulfur, thiosulfate, tetrathionate, trithionate, Dimethyl Sulfoxide etc. Dissimilatory sulfur reducers convert elemental sulfur to H_2S and in the process generate energy out of the electrons extracted. Sulfur reducers have been isolated from several archaea belonging to phylum crenarcheota and euarcheaota. Members of these phylum are found to make use of diverse electron donors including hydrogen, aminoacids, peptones, yeast extract, carbohydrates such as maltose, glycogen, starch, organic acids such as formate, fumarate, pyruvate, and alcohols like methanol, ethanol etc. Several bacterial genera including *Desulfuromonas, Desulfovibrio, Geobacter, Shewanella, Wollinella* and others are as well capable of using identical sources of electron donors. Some of the archeal and eubacterial sulfur reducers are listed in table 1

Dissimilatory sulfate reducing bacteria make use of sulphate anions in the reduction process and convert it to sulfite and H_2S. The sulfite generated is further reduced augmenting H_2S production in conjugation with extracting energy out of it. Sulfate-reducing bacteria (SRB) are ubiquitously present in soils, and are able to obtain energy by the dissimilatory reduction of sulfate. Based on studies of 16SrRNA gene sequencing the sulfate reducing bacteria can be divided into 5 bacterial and 2 archeal lineages (Muyzer and Stams, 2008). Bacterial sulfate reducing lineages include Deltaproteobacteria with 23 generas including *Desulfovibrio* sps, Clostridia including generas *Desulfotomaculum, Desulfosporomusa Desulfosporosinus,* nitrospirae consisting of thermophilic SRBs (*Thermodesulfo vibrio),* Thermodesulfobacteria (*Thermodesul-*

fobacterium genus) and Thermodesulfobiaceae (*Thermodesulfobium* genus). Archeal sulfate reducing bacteria can be divided into euarcheaota including genus *Archeoglobus* and crenarcheota consisting of genus *Thermocladium* and *Caldirvirga*. Based on ability to utilize the organic substrate sulfate reducers are classified into two groups. The first group is of bacteria are able to completely oxidise the substrate to CO_2 while the second group of organisms can carry our incomplete oxidation of substrate to acetate. Some of the common thiosulfate, tetrathionate and sulfite reducing bacteria and the substrates utilized by them are listed in the table 2, 3 and 4 respectively.

On the basis of cellular characteristics SRBs can be divided into gram positive eubacteria, gram negative eubacteria and archaebacterial. Morphologically these bacteria show varied features including cocci, bacilli, vibrio etc. Classical methods of identification of these bacteria relied on the biochemical characterization of the bacterium (Widdel and Pfennig, 1984; Campbell and Singleton, 1986). Modern methods of bacterial identification makes use of molecular techniques such as use of DNA or RNA probes, 16SrRNA gene sequencing and its BLAST analysis, immune based reactions, FAME (Fatty acid methyl ester) analysis and profiling (Vainshtein et al., 1992), DNA hybridization studies and others.

Sulfate occurs ubiquitously in environment and so are sulfate reducing bacteria such as sea water, sediments and natural water rich in organic matter. They are also present in extreme environments such as hydrothermal vents, acid mine drainage sites, oil fields and deep subsurface. Sulfate reduction occurs when anaerobic conditions become prevalent. In such anaerobic environments these bacteria make use of the organic matter as substrate and completely or partially oxidise it thus contributing towards oxidation of organic matter and hence bioremediation. Some sulfate reducing bacteria are also capable of utilizing hydrocarbons as substrate whereby they may contribute in bioremediation of hydrocarbon contaminated soils. Their presence is also used to detect the possible presence of important hydrocarbons such as petroleum during mining. Sulfate reducing bacteria may also be applied to treat the acid mine drainage.

Some of the deleterious effects of sulfate reduction process includes production of toxic gaseous end product H_2S and a neurotoxin methyl mercury through methylation if inorganic mercury is present in the surroundings. The H_2S produced may also lead to biocorrosion of metal surfaces and concrete. Sulfate reduction process is also found in important human and animal pathogen suggesting that the process may contribute to the pathogenicity of the bacterium.

Table 1: List of dissimilatory sulfur reducing bacteria

Genus	Electron donors used	References
Archea: Crenarchaeota		
Acidianus	H2	Segerer et al., 1986
Pyrobaculum	H2, peptone, extracts of meat and yeast,	Huber et al., 1987
Desulfurococcus	Peptides, starch, pectin, glycogen, yeast extract,casein hydrolysate	Osmolovskaya et al., 1988, Zilling et al., 1982
Archea: Euryarchaeota:		
Thermococcus	Peptides, amino acids, sugars, starch, chitin, Pyruvate	Dirmeier et al., 1998, Neuner et al., 1990
Methanococcus	H2, formate	Stetter and Gaag, 1983
Bacteria		
Aquifex	H2, sulfur, thiosulfate	Huber et al., 1992
Desulfuromonas	Acetate, pyruvate, ethanol	Pfennig and Biebl, 1976
Desulfovibrio	Organic acids, alcohols	Biebl and Pfennig, 1977
Fervidobacterium	Sugars, pyruvate, yeast extract	Huber et al., 1990, Patel et al., 1985
Geobacter	Acetate	Caccavo et al., 1994
Shewanella	Lactate	Moser and Nealson, 1996
Thermotoga	Sugars, peptone, yeast extract, bacterial andarchaeal cell homogenates	Windberger et al., 1989, Huber et al., 1986
Thermosipho	Yeast extract, brain heart infusion, peptone, Tryptone	Huber et al., 1989, Antoine *et al.*, 1997
Wolinella	H2, formate	Macy et al., 1986
S. oneidensis MR-1	Lactate	Burns and DiChristina, 2009
T. neapolitana, T. maritima, F. islandicum	Glucose	Ravot et al., 1995

Table 2: List of Bacterial thiosulfate reducing bacteria

Salmonella typhi	Sodium formate, Glycerol, H_2	Stoffels et al., 2011
S. oneidensis MR-1	Lactate	Burns and DiChristina, *2009*
D. baculatus	Lactate	Rozanova and Nazina, 1976.
D. thermophiles	Lactate	Rozanova and Khudyakova, 1974.
D. africanus	Lactate	Jones, 1971.
Desulfotomaculum ruminis	Lactate	Coleman, 1960.
Thermodesulfobacterium commune	Lactate	Zeikus et al., 1983
Desulfobulbus propionicus	Propionate	Widdel and Pfennig, 1982.
Desulfobacter postgatei	Acetate	Thauer, 1982; Widdel and Pfennig 1981.

(Contd.)

T. Neapolitana, T. maritima, F. islandicum	Glucose	Ravot et al., 1995
Anaerobaculum spps (Archea)	Yeast extract	Liang et al., 2014
Coprothermobacter sps	Glucose	Etchebehere and Muxi, 2000
P. vulgaria	Peptides	Tarr, 1933
Providencia spp.	Peptides	Clarke, 1953

Table 3: List of bacterial tetrathionate reducing bacteria

Salmonella typhi	Ethanolamine, 1,2-propanediol	Marian et al., 2001
Citrobacter spps	Glucose, Galactose, Pyruvate	Kapralek, 1972
C. jejuni	Formate	Liu et al., 2013
S. oneidensis MR-1	Lactate	Burns and DiChristina, 2009

Table 4: List of bacterial sulfite reducing bacteria

S. oneidensis MR-1	Lactate	Burns and DiChristina, 2009
Wolinella succinogenes	Formate	Kern et al., 2011
Salmonella typhimurium	Peptones, Glucose	Huang and Barrett, 1990
Methanocaldococcus jannaschii	H_2	Johnson and Mukhopadhyay, 2005
Thioalkalivibrio nitratireducens	Not known	Trofimov et al., 2010

Cultivation of Sulfate Reducing Bacteria

Sulphur compound reducers can be either facultative or obligatory anaerobic microorganism. For isolation of anaerobic sulfur compound reducers we make use of minimal medium supplemented with an electron donor (lactate /formate) and one of the electron acceptor (Sulfur/ Sulfite/ Thiosulfate/ Tetrathionite) depending on which reducer one wants to isolate. Briefly for anaerobic culture we make use of minimal medium that contains the following in gms/liter: $NaHCO_3$ (2.5), NH_4Cl (1.5), NaH_2PO_4 (0.6), KCl (0.1), yeast extract (0.1), 10 ml vitamin solution and 10 ml mineral solution. The medium is supplemented with 30mM Glycerol / or any other suitable electron donor. Medium is supplemented with one of the following as a sole electron acceptor: Sodium Thiosulfate, Sodium Sulfite, Elemental Sulfur / polysulphide solution/ Sodium tetrathionite. The soluble electron donor can be sterilized by filteration and supplemented to the autoclaved medium. Sterile polysulfide solution is prepared as described below. In general electron acceptor at 10mM concentration can be used. The medium also contains 0.015% $FeSO_4$ which is used for detection of H_2S produced as a byproduct of sulfur compound respiration process. The H_2S produced reacts with $FeSO_4$ to form a black color precipitate indicating reduction of sulfur compounds and subsequent H_2S production. For preparation of solid media for plates we supplement the above medium with 2% agar solution. Following inoculation the medium is incubated in an anaerobic chamber for 2-5 days or until the black color of the medium develops. Alternatively the solid medium in petriplates or liquid medium can also be incubated in an anaerobic jar by making use of commercial anaeropacks. The plates or other contained for cultivation of anaerobes are first transferred into the jars and the anaeropacks pouch opened and activated inside the jar by addition of water. The jar is immediately closed and sealed tightly. The addition of water to the anaeropack allows reaction with in its components generating anaerobic atmosphere in the jars.

Mineral Solution composition (Wolin et al., 1963): Mineral solution contains the following salts in grams per liter of the mineral solution: nitrilotriacetic acid (1.5), $MgS0_4$(3.0), $MnSO_4$ (0.5), NaCl(1.0), $FeS0_4$ (0.1), CaCl (0.1), $CoC1_2$ (0.1), $ZnS0_4$ (0.1), $CuS0_4$ (0.01), $AlK(SO_4)_2$ (0.01), H_3BO_3 (0.01) Na_2Mo0_4 (0.01).

Vitamin solution composition (Wolin et al., 1963): The vitamin stock solution contains the following in mgs per liter of distilled water: biotin (2), folic acid (2), pyridoxine hydrochloride (10), riboflavin (5), thiamine (5), nicotinic acid (5), pantothenic acid (5), vitamin B12 (0.1), p-aminobenzoic acid (5), thioctic acid (5).

Preparation of polysulfide stock solution

Alkaline polysulfide stock solution (2.25M) can be prepared by procedure descried previously (Widdel and Pfennig. 1992). Briefly 7.2 gms of sulfur flowers and 24 gms of $Na_2S.9H_2O$ and mixed and boiled with stirring for atleast 30 minutes to a final volume of 100 ml. This results in reddish solution which is considered sterile and is stored in bottles with minimum headspace. For preparation of sulfur plates the minimal medium is supplemented with this polysulfide solution as electron acceptors. The petriplates are kept for overnight in air for ageing before inoculation.

Mechanism of Sulfite reduction

The sulfite reduction process is poorly understood and there is much dispute regarding its exact mechanism. In general two pathways are proposed for reduction of sulfite to H_2S. According to the trithionite pathway of sulfite reduction there is formation of trithionite and thiosulphate as intermediates. The reduction process is proposed to be carried out by three separate enzymes. The sulfite reductase (converts sulfite to trithionite), the trithionite reductase (converts trithionite to thiosulfate) and the thiosulfate reductase (converts thiosulfate to H_2S). Evidence for the first pathway is majorly based on in vitro studies performed using reduced methyl viologen based enzyme activity assays. During these experiments trithionite and thiosulfate were detected along with H_2S. The enzymes involved in this pathway are the trithionite reductase

(Kim and Akagi. 1985, Lee and Peck. 1971) and thiosulfate reductase (Nakatsukasa and Akagi. 1969; Fitz and Cypionka, 1990). The sulfite reductases identified to date are cytosolic enzymes and it is not understood how they contribute to energy generation. On the contrary the trithionite and thiosulfate reductases are membrane bound and hence may contribute to energy generation.

Thus trithionite pathway appears to be important for energy generation. However, arguments against trithionite pathway suggest chemical formation of these intermediates through interaction between sulfite and the byproduct of its reduction H_2S.

The second proposed pathway of sulfite reduction is the six electron reduction mechanism that converts sulfite to H_2S in a single step. Thiosulfite and trithionite were detected only in small amounts in the reconstitution assays with the desulfoviridin from *D. desulfuricans*. (Steuber et al., 1995). Experiments performed using labelled sulfate with the washed cells of *Desulfovibrio desulfuricans* also did not produce any detectable levels of labeled thiosulfate or trithionate. These experiment support direct conversion of sulfite to H_2S without any formation of thiosulfate or trithionite intermediates. The six-electron reduction pathway appears to be more likely considering that both sulfite and nitrite reductases require siroheme as cofactor and hence might be using similar mechanism for reduction.

Siroheme

Siroheme (or sirohaem) is a heme-like prosthetic group present at the active sites of some enzymes that undertake the six-electron reduction of sulfur and nitrogen. Siroheme is an iron containing isobacteriochlorin where the macrocyclic ligand in siroheme is derived from uroporphyrinogen III. The methylation of uroporphyrinogen III to precorrin-2, its further oxidation to sirohydrochlorin followed by chelation of ferrous iron leads to biosynthesis of sirohaem. The sirohaem containing sulfite and nitrite reductases contain a conserved consensus sequence consisting of two cysteine clusters (Cys-x,-Cys and Cys-x,-Cys) (Campbell and Kinghorn, 1990). These cysteine clusters are important for the 4Fe4S sirohaem cofactor binding (Ostrowski et al., 1989; Crane et al., 1995). Siroheme functional group has been identified in both assimilatory and dissimilatory type sulfite and nitrite reductases (Stroupe and Getzoff, 2009).

$HSO_3^- + 6e^- + 6H^+ \rightarrow HS^- + 3H_2O$ E_0^2 $(HSO_3^-\ HS^-) = -116mV$

Six electron reduction reaction of Sulfite to H_2S

Siroheme sulfite reductase *DSRAB*

The dissimilatory sulfite reductases (dSirs) have been isolated from several sulfate reducing bacteria and have been studies for their molecular, spectroscopic and kinetic properties. The dSirs contain a core consisting of α2β2 tetramer and 2 additional subunits (differing in different organisms) γ and δ, resulting in an α2β2γnδn multimeric complex. The subunits α, β, γ and δ have molecular mass of about 45, 43, 10, and 11 kDa, respectively. The α2β2 core tetramer binds the 2-4 siroheme [4Fe–4S] centers and additional 3-6 [4Fe–4S] clusters. The α2β2 core tetramer is associated with the sulfite reductase activity while the subunit γ with its redox active disulfide bonds is thought to be involved in electron transport to the core and the δ subunit because of its winged helix motif might to be involved in DNA binding and regulation. The dsrA and dsrB code for the α- and β-subunits respectively if the dissimilatory sulphite reductase. These genes are paralogous to each other as they appear to be generated by duplication of the ancient dsr gene. Evidence suggests that Dsr complex is part of the same metabolic pathway as sulphite reductase. Reduction of sulfate to H_2S is an eight electron reduction process which involves formation of intermediates such as sulphite, trithionite etc.

A typical structural property of enzymes of the Sir family is their trilobar structure. The first 2 lobes are composed of mixed β-sheet followed by α-helices while the third lobe is made of 2 attached ferrodoxin like domains. Towards the interacting points of the 3 lobes is present the siroheme attached to the 4Fe-4S cluster via a cysteine thiolate bond (Schiffer et al., 2008). The [4Fe–4S] cluster of the ferredoxins [4Fe–4S] cluster are thought to be terminal redox sites and to be involved in electron transport to the siroheme of dSir (Simon and Kroneck, 2013).

Electron delivery to dSirs is combined to different electron transport systems used to reduce or oxidize the siroheme–Fe/S center. In dSir, the molecular machinery involved in electron transfer has been fully understood, however, both dSir-α and dSirβ subunits show presence of ferrodoxin like domains containing a [4Fe–4S] cluster. The X-ray crystallographic studies suggest that [4Fe–4S] clusters might accept electron from yet unidentified electron donor and serve as immediate electron carrier to the siroheme center of the terminal dSir-α catalytic subunit. The dsrA, B and C (dSir-α, β and γ) are soluble cytosolic proteins and may involve several proteins during electron transport chain of their reduction. Molecular analysis of electron transport chain components of during sulfite reduction by *A. fulgidus* and other sulfate reducers suggests a inner membrane associated reduced quinone pool is generated by utilizing electrons extracted from organic substrates such as lactate through activity of corresponding dehydrogenases (Lactate dehydrogenase). The reduced quinol pool is thought to be oxidized by one or more of the membrane complexes DsrMKJOP, TmcABCD, and HmcABCDEF (Pereira et al., 2011). The complex DsrMKJOP has been isolated from D. desulfuricans (Pires et al., 2006) and from A. fulgidus (designated Hme) (Mander et al., 2002). A model for localization of dissimilatory sulfite reductase subunits and its electron transport chain is as proposed in the figure 2.

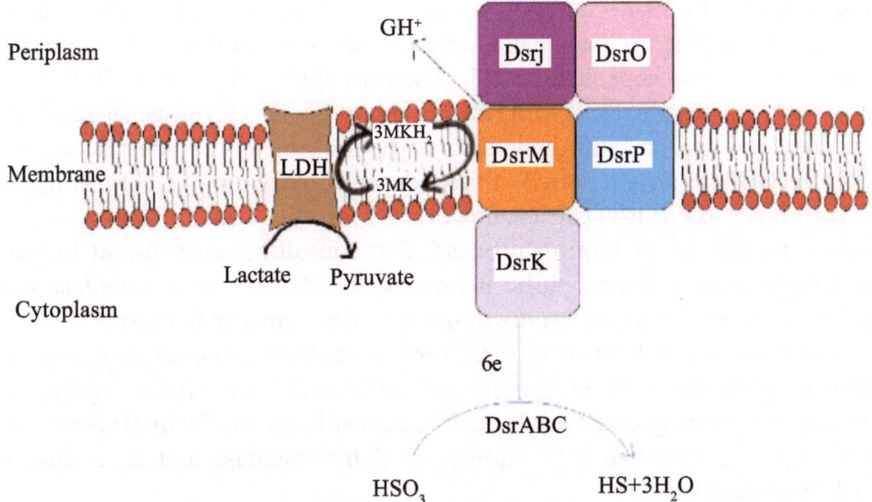

Fig. 2: Siroheme dissimilatory sulfite reductase (dSir) electron transport chain Other Siroheme Reductases

ASR

Anaerobic sulfite reductase (AsrABC) is best studied in *S. typhimurium* in which the functional ASR operon consists of 3 asr genes *asrA, asrB,* and *asrC*. These genes catalyse the dissimilatory reduction of sulfite to hydrogen sulfite. Similar to the siroheme containing enzymes the AsrC subunit contains the conserved cysteine pattern (C- (X)3-C and C- (X)5- C) which are thought to coordinate siroheme binding (Ostrowsk et al., 1989). Both the asrA and the asrC subunits also contained the conserved cysteine residues through which they coordinate with the [4Fe4S]–ferredoxin binding domains indicating their role in electron transfer activities.(Huang and Barrett, 1991).

The earlier reported asr genes encode a 65 kDa siroheme binding protein (Hallenbeck et al., 1989) however the *Salmonella* asr genes are comparatively much smaller in size. Interestingly the *Salmonella* asrC protein which is much smaller in size encodes the typical siroheme binding conserved cysteine patterns suggesting it to be the siroheme containing subunit of asr.

Presence of Cysteine does not appears to effect asr expression, which suggest that asr is principally not a biosynthetic enzyme (Hallenbeck et al., 1989; Huang and Barrett, 1990) however under anaerobic conditions asr genes have been found to perform some assimilatory sulfite reduction in absence of the functional biosynthetic genes (Huang and Barrett, 1990)

Fsr: F420 dependent sulfite reductase

Fsr a novel F420 dependent sulfite reductase was first reported in hyperthermophilic methanogenic archaeon *Methanocaldococcus jannaschii* (Johnson and Mukhopadhyay, 2005). The Fsr polypeptide is about 70 KDa in size with an N terminal half with H2F420 (FqoF/FpoF) dehydrogenase activity and the C-terminal half with dissimilatory-type siroheme sulfite reductase activity (Figure 3). Previously reported sulfite reductases make use of nicotinamide or cytochromes as electron carriers however Fsr uses reduced F420 as source of electrons for the reduction process. Fsr homologs were found in two hydrogenotrophic thermophilic methanogens *Methanopyrus kandleri* and *Methanothermobacter thermautotrophicus.* Although Fsr is present only in four methanogens a small putative sulfite reductase is present in all methanogens. This suggests that methanogenesis and sulfite reduction existed together in primitive methanogens. Probably later during evolution the Fsr might have split into separate enzymes with FqoF/FpoF dehydrogenase and DsrA sulfite reductase activity.

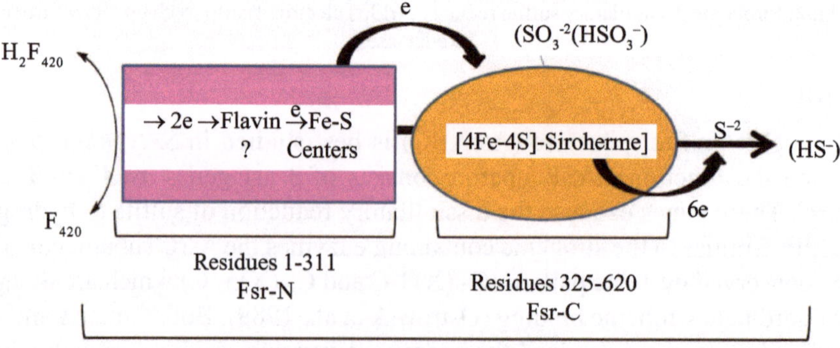

Fig. 3: Fsr: F420 dependent sulfite reductase from *Methanocaldococcus jannaschii*

Other Unconventional Sulfite reductase

MccA Sulfite reductase

Dissimilatory Sulfite reductases identified to date are siroheme protein highly similar to the assimilatory type nitrite and sulfite reductases. Crane et al.,1997;

Hipp et al., 1997; Wagner et al., 1998). However the MccA family of sulfite reductases are different from the typical siroheme sulfite reductases. Till now several members of MccA family proteins have been identified from proteobacterial species. Of these only MccA from *W. succinogenes* (Kern et al., 2011) and SirA from *S. oneidensis* (Shirodkar et al., 2011) have been studies to identify their function in dissimilatory sulfite reduction. The sequence of SirA contains an heme binding motif that requires a specialized haem lyase system for maturation. This haem lyase system also appears to be present with in the same locus encoded by *SirEFG*. Thus in brief SirA sulfite reductase is an octahaem c cytochrome and is predicted to be located in the periplasm. SirC is also predicted to be cytoplasmic and contain two two CxxCxxCxxxC typical of ferrodoxins.

SirCD proteins in the locus show similarity to the NrfCD that function in nitrite reduction in *E.coli* (Hussain et al., 1994) These are probably involved in electron transport to the terminal sulfite reductase SirA. Within the *sirA* locus are also present *sirJKLM* (previously annotated *nosFDYL*). Copper transport genes (Wunsch et al., 2003) were induced in presence of thiosulfate in *Shewanella oneidensis* MR-1 (Beliaev et al., 2005). Interestingly the MccA (homologue of *S. oneidensis* SirA) from *W. succinogenes* also shows presence of *nosDFYL* genes adjacent to it. The presence of copper transport genes in the sulfite reductase operons of *Shewanella* and *Wollinella* species suggests that they require a copper containing component. In conclusion considering the fact that SirA sulfite reductase requires a specialized haem lyase system for its maturation, carries out six electron reduction reaction, but does not share sequence similarity with either sirohaem sulfite reductase or nrfA nitrite reductase suggest that SirA sulfite reductase represents a new class of sulfite reductases.

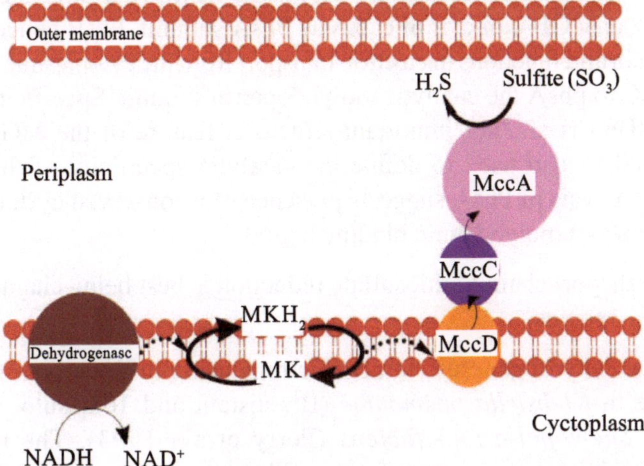

Fig. 4: Proposed model for SirA (MccA) sulfite reductase electron transport chain (Shirodkar *et al.*, 2010)

Thiosulfate reduction

Sulfur compounds such as thiosulfate undergoes transformations from one oxidation state to other either spontaneously or through biological activities occurring in the oxic and anoxic marine environments. In oxic conditions thiosulfate can be oxidized by chemolithotrophs to sulfuric acid, while in anoxic niche the thiosulfate can be reduced to sulfite and H_2S by sulfate reducing bacteria. The thiosulfate reduction process increases with sediment depth suggesting the importance of the process in anoxic environments. Thiosulfate respiration is typically reported to involve a thiosulfate reductase enzyme.

$S_2O_3^{2-}$ + $2H^+$+ $2e^-$ thiosulfate reductase HS^- + HSO_3^-
DsrC + SO_3^{2-} + $2H^+$ + 2e dissimilatory sulfate reductase DsrC – trisulfide + H_2O

Many sulfate reducing bacteria are able to use thiosulfate as an electron acceptor including *Proteus, Citrobacter,* and *Salmonella* species. The thiosulfate reductase from *S. typhimurium* (PhsABC) is similar to other molybdopterin oxidoreductases including the MGD- binding polysulfide reductase (PhsB) from *W. succinogenes*. PhsA sequence analysis suggests its homology to catalytic subunit of several anaerobic reductases. The N terminal region of PhsA contains a sequence (S-R-R-S-F-L-Q) which is similar to the consensus sequence of proteins secreted by the twin arginine transport (TAT) system (Berks, 1996; Hensel et al., 1999). TAT is a specialized system that secretes proteins in their natïve folded form across the inner membrane. Following the twin arginine signal sequence the catalytic subunit of thiosulfate reductase contains 4 cysteine residues which are thought to be important in coordination with the 4 Fe-4S clusters that function in electron transfer from the iron sulfur clusters of phsB subunit to the MGD cofactor of catalytic PhsA subunit. PhsB hence is an iron sulfur protein and functions in electron transport from the cytoplasmic membrane bound phsC to phsA the catalytic molybdopterin subunit. Specificity defining regions (SDR) is another important structural feature of the MGD binding proteins and is also used to define the catalytic specificity of the enzyme. Sequence analysis pf PhsA suggests presence of a conserved cysteine residue to be present as a molybdenum binding ligand.

Electron transport chain of thiosulfate reduction is best being characterized in *Salmonella* species.

Respiratory dehydrogenases extract electrons using different substrates such as formate in *Klebsiella pnemoniae* (Braunstein and Tomasulo, 1976) and lactate in *Shewanella putrifaciens* (Perry et al., 1993). The membrane associated quinones further transfer these electrons to the terminal reductases. Menaquinone mutants in *S. enterica* are unable to reduce thiosulfate (Stoffels et al., 2011) while the ubiquinone mutants reduced wildtype levels of thiosulfate (Clark and Barrett, 1987). These results suggest the role of menaquinone in electron transport to the thiosulfate reductase. Furthermore heme mutants in *S. typhimurium* lose the ability to reduce thiosulfate to H_2S and lack presence of thiosulfate reductase enzyme activity as detected by methyl viologen based activity assay. These results suggest that cytochromes are required for thiosulfate reduction (Clark and Barrett, 1987). Based on these finding a model has been proposed for thiosulfate respiration in *Salmonella* as depicted in figure.

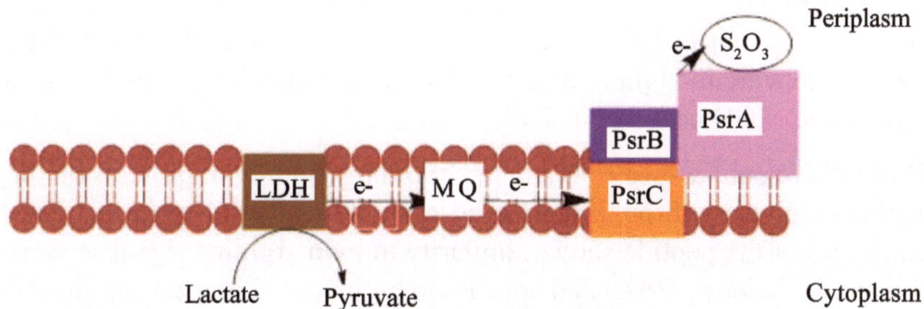

Fig. 5: Model for electron transport during thiosulfate reduction in *Salmonella* (Barrett and Clark, 1987)

Role of Thiosulfate respiration in growth and energetics

Thiosulfate reductases have been detected and studies in several anaerobic and facultative anaerobic bacteria such as *Citrobacter, Proteus, Salmonella* and others. However the reduction process does not seems to support growth in these organisms. The thiosulfate reductases from *D. vulgaris* (Aketagawa et al., 1985, Haschke et al., 1971), D. gigas (Hatchikian, 1975) and *Desulfotomaculum nigrificansall* have been found in the soluble fraction and the enzyme from *D. vulgari* has been found to be localized on the cytosolic side of the inner membrane (Badziong and Thauer, 1980). Thus thiosulfate reduction in these bacteria may not contribute towards PMF or energy generation. Infact in *Salmonella* thiosulfate reduction was found to be dependent on Proton Motive Force (PMF) across the cytoplasmic membrane. Possibly formate hydrogen lyase pathway may shuttle electrons for thiosulfate reduction generating a net PMF making the process energetically viable. In contrast, hyperthermophiles *Thermotoga neopolitana* and *Thermotoga maritime* showed improved growth and yield during thiosulfate reduction (Ravot et al. 1995).

Tetrathionate reduction

Tetrathionate reduction involves 2 electron reduction of tetrathionate to 2 molecules of thiosulfate which is further reduced by thiosulfate reduction pathway. Tetrathionate reductase shows much similarity to the thiosulfate reductase. The TtrABC reductase consists of the following 3 subunts TtrA, TtrB and TtrC.

Similar to the thiosulfate reductase PhsA the TtrA precursor protein at the N terminal end contains a signal sequence (T-R-R-x-x-L-K) which is similar to conserved sequence of the Twin Arginine transfer peptides (S/T-R-R-x-F-L-K) (Berks, 1996). This suggest that the TtrA is localized in the periplasmic side of the membrane. The putative Twin Arginine signal sequence is followed by the cluster of 4 conserved cysteine residues which have been thought to bind 4Fe-4S clusters cluster (Breton et al., 1994; Boyington et al., 1997). These 4Fe-4S clusters are thought to mediate electron transfer from 4Fe-4S clusters in TtrB to the MGD cofactor in TtrA.

TtrB from *Salmonella* species is a member of large family of proteins binding four 4Fe-4S clusters through conserved cysteine as ligand residues. The N terminal end of the peptide shows similarity to twin arginine signal sequence (Keon and Voordouw, 1996) and appears to be involved in electron transport from the TtrC subunit to the Fe-S clusters of the TtrA subunit.

TtrC from salmonella is predicted to be an integral membrane protein with nine transmembrane helices (amino acids 22–40, 58–77, 94–112, 132–156, 161–182, 200–215, 230–253, 267–287 and 309–327) and the N terminal end of the protein is the periplasmic side.

It shows similarity to the polysulphide reductase PsrC from W. succinogenes (Krafft et al., 1992) and nitrite reductase system protein nrfD from E.coli and H influenza. (Hussain et al., 1994; Fleischmann et al., 1995). It appears that similar to the above mentioned PsrC and nrfD subunits TtrC may be involved in electron transport from membrane associated menaquinones to the 16 ferrodoxins of PsrB subunit. Additionally conserved histidine residues are absent in TtrC/PsrC/NrfD family proteins that act as ligands for haem Fe binding. Consistent with the observation the haem deficient mutants in *Salmonella* were capable of tetrathionate respiration suggesting the noninvolvement of cytochromes in the reduction process (Barrett and Clark, 1987).

Role of tetrationite reduction in growth and energetics.

Unlike thiosulfate reduction tetrathionate reduction appears to support growth in *P mirabilis, C freundii* and *S. typhimurium* (Barrett and Clark, 1987). These studies suggest that tetrathionate reduction could be coupled to oxidative phosphorylation. One study by Kapralek showed that per mole of tetrathionite reduced produced 1.3 moles of ATP (Barrett and Clark, 1987). Several compounds were used as electron donors during tetrathionate reduction such as formate, hydrogen and NADH by *P. mirabilis* and *C. freundii,* glycerol by *P. mirabilis* and *S typhi* while lactate by *P mirabilis.* Cytochromes do not appear to be involved in tetrathionate reduction as the heme mutants were not deficient in tetrathionite reduction. Furthermore menaquinones and ubiquinones also are not required for tetrathionite reduction in *S. typhi*

Sulfur respiration

Anaerobic elemental sulfur/ polysulfide reduction is performed by several eubacteria and archaea. Elemental sulfur being insoluble is difficult to be reduced directly, however it is polysulfide in presence of sulfide which is water soluble. Hence polysulfide appears to be an intermediate during sulfur respiration process. Sulfur respiration is well studied in gram negative bacterium *Wolinella succinogenes*. The isolated enzyme psr (Polysulfide reductase) consisted of 3 subunits (PsrA, B and C) (Krafft et al., 1992). Similar to the tetrathionite or the thiosulfate reductases the catalytic subunit PsrA belons to family of molybdopterin dinucleotide containing oxidoreductases (Jankielewicz et al., 1994). The enzyme is present embedded the inner membrane with the catalytic subunit exposed to the periplasmic side. The other 2 subunits are involved in electron transport to the catalytic subunit along with a dehydrogenease (Formate dehydrogenase or hydrogenase) (Schroeder et al., 1988). In some organism multifunctional enzymes have been recognized such as the psr from *Shewanella* is capable of respiring both thiosulfate and sulfur.

$HCO_2^- + S_n^{2-} \quad + H_20 \rightarrow HCO^-_3 + HS^- + S_{n-1}^{2-} + H^+$

$\Delta G°' = -30$ KJ/mol H_2

(Reiner et al., 1999)

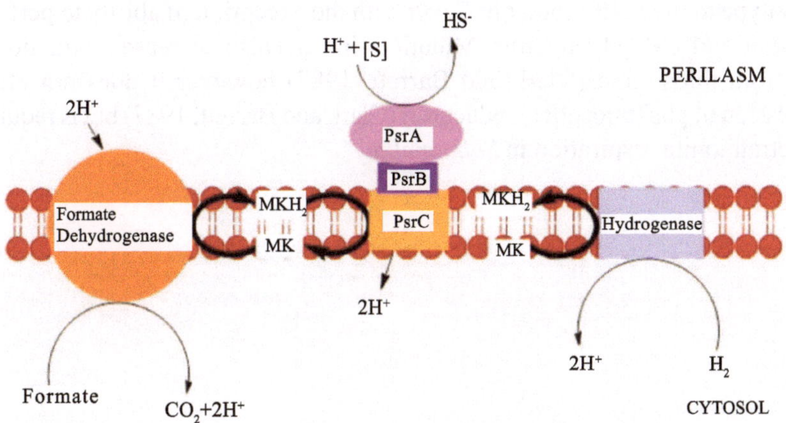

Fig. 6: Electron transport chain during polysulfide reduction *linella succinogens*

Global Regulators of bacterial anaerobic respiration

Transcriptional regulators are sequence specific DNA binding proteins which are known to function in accordance with the environmental clues. These can either activate or repress the transcription of the downstream gene.

FNR: It is a transcriptional factor which control expression of a large number of genes in response to environmental signals and are known as global regulators. One such regulatory protein is FNR (fumarate nitrate reductase regulation protein) which was first identified as a global regulator of anaerobic respiration in *E.coli.* FNR was initially identified in pleiotropic mutants of *E. coli* that cannot use fumarate and nitrate as terminal electron acceptors during anaerobic respiration (Spiro and Guest, 1990). The FNR is a homodimeric protein consists of C terminal helix turn helix DNA binding domain and N terminal sensory domain. The *E.coli* FNR contains 4 essential cysteine residues (C20, C23, C29, and C122) which are important for their *in vivo* function and serve as ligands for binding the 4Fe-4S and 2Fe-2S clusters which are important for oxygen sensing mechanism. Each FNR monomer is associated with 2Fe-2S cluster. On exposure to anaerobic conditions the 2Fe-2S clusters are converted to 4Fe-4S cluster as the monomers are converted to dimers. The dimeric form of FNR is active and it binds the DNA sequences promoting gene expression of anaerobically induced genes.

FNR in *E.coli* can have both positive and negative effect on gene expression of several genes. FNR positively regulates expression of several genes including fumarate reductase, nitrate reductase, nitrite reductase, TMAO reductase, DMSO reductase, formate dehydrogenase, glycerol 3 phosphate dehydrogenase, Fumarase, cytochrome d oxidase, cytochrome a1 and molybdate reductase. A mutation in OxyA (homologue of FNR in *Salmonella typhimurium* shows similar phenotype to the FNR mutant in *E.coli* with the exception of ability to perform fumarate and TMAO reduction. Mutations in *oxrA* also decrease production of H_2S from thiosulfate (Clark and Barrett, 1987) however it does not effect expression of phs (thiosulfate reductase) (Clark and Barrett, 1987) but is required for tetrathionite respiration in *Salmonella*.

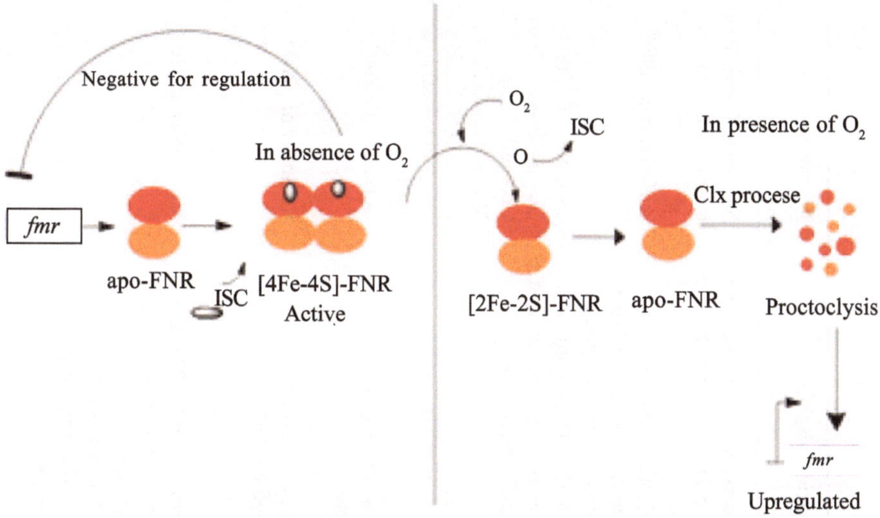

Fig. 7: Modulation of FNR activity in *E.coli*

EtrA homologue of FNR is involved in partially and positively regulating fumarate and nitrate growth in *Shewanella oneidensis* MR-1 although it is able to complement the FNR mutant in *E.coli (*Maier and Myers, 2001*)*. Interestingly *in Shewanella oneidensis* MR-1 cyclic adenosine monophosphate (CRP) acts as activator of anaerobic respiration and is found to regulate anaerobic respiration of several electron acceptors (Saffarini et al., 2003) such as DMSO, TMAO, nitrate, nitrite, fumarate, sulfite, sulfur, thiosulfate etc.

Another global regulatory protein is ArcAB a two component regulatory system. This system controls gene expression upon transition of bacteria from aerobic to anaerobic conditions and is active in micro aerobic conditions. It regulates about 100 operons of TCA cycle and energy metabolism (Gunsalus and Park, 1994, Chang et al., 2002; Georgellis et al., 2001; Luchi and Lin, 1988, Luchi et al., 1989; Luchi et al., 1990; Liu and Wulf, 2004). The ArcB is a sensor kinase that undergoes activation by autophosphorylation at the His292 residue under anaerobic conditions. Oxidized quinones act as negative regulator of this activation process under aerobic conditions. (Georgellis et al., 2001). Activated ArcB subsequently activates its associated transcriptional regulator ArcA by phosphorylating ArcA at Asp54 . The response regulator ArcA then represses expression of aerobic metabolism genes (e.g. citrate synthase and isocitrate lyase) and activates genes required for anaerobic metabolism (e.g. pyruvate formate lyase and hydrogenase) (Oshima et al., 2003; Georgellis et al., 2001; Georgellis et al., 1997; Malpica et al., 2006; Luchi, 1993; Luchi and Lin, 1992; Jeon et al., 2001). Some of the global transcriptional regulators that respond in oxygen deprivation conditions is listed below in table 5

Table 5: List of global transcriptional regulators functional in oxygen deprivation conditions

Regulatory protein	Example Organisms	Biological function	Sensor	Biological response	Reference
FNR	*E.coli, Salmonella typhimurium*	Transcriptional activator or repressor	4Fe-4S clusters	Conversion of 4Fe-4S clusters to 2Fe-2S leading to FNR inactivation	Spiro and Guest, 1990; Crack *et al.*, 2008
CRP	*Shewanella oneidensis MR-1*	Transcriptional activator	cAMP	CRP activity increases by addition of cAMP (encoded by adenylate cyclase gene).	Saffarini *et al.*, 2003
ArcA	*E.coli*	Regulator Repressor and Activator	ArcB (Sensor Kinase)	ArcB (His292) activation by autophosphorylation inihibited by oxidized quinones	Oshima *et al.*, 2002; Georgellis *et al.*, 2001; Georgellis *et al.*, 1997; Malpica *et al.*, 2006; Luchi, 1993; Luchi and Lin, 1992; Jeon *et al.*, 2001.
ResD	*Bacillus*	Transcriptional activator of genes involved in nitrate respiration	ResE (Sensor Kinase)	Unknown mediator of sensing oxygen	Baruah *et al.*, 2004
NreC	*Staphylococcus*	Transcriptional activator of nitrate and nitrite reductase and putative nitrate transporter	NreB sensor kinase	Presence of O2 sensitive 4Fe-4S clusters	Kamps *et al.*, 2004
PpsR	*Rhodobacter*	Transcriptional repressor of photosystem development genes	Thiol groups	In presence of O2 Thiols in PpsR is oxidized resulting in formation of intramolecular disulfide bonds resulting in active conformation of the repressor.	Moskvin *et al.*, 2005

(Contd.)

AppA	*Rhodobacter sphaeroides*	Antirepressor of PpsR	Novel FAD binding domain	Activation via light sensing. Unknown mechanism of O2 sensing	Moskvin *et al.*, 2005
FixJ	*Rhizobia*	Transcriptional activator of nitrogen fixing genes	FixL(Histidine kinase sensor)	Oxygen sensing requires the heme moiety (H) attached to the central, cytoplasmic domain of FixL	Fischer, 1994
NifA	*K pneumoniae*	Transcriptional activator of nitrogen fixing genes	NifL in presence of O2 is inhibitor of NifA	Sensing of cellular oxygen and nitrogen conditions	Fischer, 1994
HlyX	*Actinobacillus pleuropneumoniae*	Transcriptional activator of Hemolytic activity	4Fe-4S clusters	Conversion of 4Fe-4S clusters to 2Fe-2S leading to HlyX inactivation	Spiro and Guest, 1990

Conclusion

Sulfur compounds reduction forms an important part of biogeochemical cycles. Sulfur compound reduction process makes use of several different electron acceptors including sulfur, sulfite, thiosulfate, tetratahionite, trithionite etc. Sulfur compound reducers are ubiquitous in aquatic systems and are also found in extreme environments such as hydrothermal vents, acid mine drainage sites and others. They also form important human and animal pathogens. Dissimilatory sulfur, thiosulfate and tetrathionate reducers make use of the enzymes sulfur reductase, thiosulfate reductase and tetrathionite reductase respectively. All three enzymes are molybdopterin containing oxidoreductases. On the contrary diverse types of sulfite reducers have been identified amongst bacteria. These include the conventional siroheme dissimilatory sulfite reductase and the nonconventional sulfite reductases including ASR sulfite reductase, F_{420} dependent sulfite reductase and the MccA sulfite reductase. The presence of such wide diversity of dissimilatory sulfite reductases is interesting and suggestive of the importance of sulfite reduction during sulfur compound transformations. In fact it is know that that other sulfur compounds such as trithionite and tetrathionate also make sulfite as an intermediate during the reduction process. The sulfite produced may be the result of biological reduction or chemical interactions during the reduction of other sulfur compounds. Thus sulfite reduction seems to be indispensable during the sulfur compound reduction processes. Sulfur compound reduction processes are also regulated in bacteria suggesting their importance in survival of the microbes in unfavorable conditions and niche environments. Perhaps it forms an important process in facultative and obligate anaerobes to sustain in extreme environments and form an important component in sulfur cycling as well as in energy generation.

Acknowledgement

We are thankful to Dr Ashok Chauhan (Founder Amity University) and Dr Atul Chauhan for their support and guidance.

References

Aketagawa J, Kobayashi K, and Ishimoto M. 1985. Purification and properties of thiosulfate reductase from *Desulfovibrio vulgaris*, Miyazaki F. *The Journal of Biochemistry* 97 (4): 1025-1032.

Antoine E, Cilia V, Meunier J, Guezennec J, Lesongeur F, and Barbier G. 1997. *Thermosipho melanesiensis* sp. nov., a new thermophilic anaerobic bacterium belonging to the order Thermotogales, isolated from deep-sea hydrothermal vents in the southwestern Pacific Ocean. *International Journal of Systematic and Evolutionary Microbiology* 47 (4): 1118-1123.

Badzjong W, and Thauer RK. 1980. Vectorial electron transport in *Desulfovibrio vulgaris* (Marburg) growing on hydrogen plus sulfate as sole energy source. *Archives of Microbiology* 125 (1-2): 167-174.

Barrett EL, and Clark MA. 1987. Tetrathionate reduction and production of hydrogen sulfide from thiosulfate. *Microbiological reviews* 51 (2): 192.

Baruah A, Lindsey B, Zhu Y, and Nakano MM. 2004. Mutational analysis of the signal-sensing domain of ResE histidine kinase from *Bacillus subtilis*. *Journal of Bacteriology* 186 (6): 1694-1704.

Beliaev, A. S., D. M. Klingeman, J. A. Klappenbach, L. Wu, M. F. Romine, J. M. Tiedje, K. H. Nealson, J. K. Fredrickson, and J. Zhou. 2005. Global transcriptome analysis of *Shewanella oneidensis* MR-1 exposed to different terminal electron acceptors. *Journal of Bacteriology* 187:7138-7145

Berks BC. 1996. A common export pathway for proteins binding complex redox cofactors? *Molecular Microbiology* 22 (3): 393-404.

Biebl H, and Pfennig N. 1977. Growth of sulfate-reducing bacteria with sulfur as electron acceptor. *Archives of Microbiology* 112 (1): 115-117.

Bonch-Osmolovskaya EA, Slesarev AI, Miroshnichenko ML, Svetlichnaya TP and Alekseev VA. 1988. Characteristics of *Desulfurococcus amylolyticus* n. sp.- a new extremely thermophilic archaebacterium isolated from thermal springs of Kamchatka and Kunashir Island. *Mikrobiologiya* 57:94-101.

Boyington JC, Gladyshev VN, Khangulov SV, Stadtman TC, and Sun PD. 1997. Crystal structure of formate dehydrogenase H: catalysis involving Mo, molybdopterin, selenocysteine, and an Fe4S4 cluster. *Science* 275:1305– 1308.

Braunstein H, and Tomasulo M. 1976. Indole-positive Strains of *Klebsiella pneumoniae* Producing Hydrogen Sulfide in Iron–Agar Slants. *American journal of Clinical Pathology* 65 (5): 702-705.

Breton J, Berks BC, Reilly A, Thomson AJ, Ferguson SJ, and Richardson DJ. 1994. Characterization of the paramagnetic iron-containing redox centres of *Thiosphaera pantotropha* periplasmic nitrate reductase. *FEBS Lett* 345: 76–80.

Burns JL, and DiChristina TJ. 2009. Anaerobic respiration of elemental sulfur and thiosulfate by *Shewanella oneidensis* MR-1 requires psrA, a homolog of the phsA gene of *Salmonella enterica serovar typhimurium* LT2. *Appl. Environ. Microbiol* 75 (16): 5209-5217.

Caccavo F, Lonergan DJ, Lovley DR, Davis M, Stolz JF, and McInerney MJ. (1994). *Geobacter sulfurreducens* sp. nov., a hydrogen-and acetate-oxidizing dissimilatory metal-reducing microorganism. *Appl. Environ. Microbiol* 60 (10): 3752-3759.

Campbell LL, and Singleton RJr. 1986. Genus IV. *Desulfotomaculum,* in: *Bergey's Manual of Systematic Bacteriology* (P.H.A. Sneath, N. S. Mair, M. E. Sharpe and J. G. Holt, eds.) Williams & Wilkins, Baltimore, pp. 1200-1205.

Campbell, W. H. & Kinghorn, 1. R. 1990. Functional domains of assimilatory nitrate reductases and nitrite reductases. *Trends in Biochemical Science* 15,315-319.

Chang DE, Smalley DJ, Conway T.2002. Gene expression profiling of *Escherichia coli* growth transitions: an expanded stringent response model. *Molecular Microbiology* 45:289-306.

Cheng J. Huang And Ericka L. Barrett. 2012. .Identification and Cloning of Genes Involved in Anaerobic Sulfite Reduction by *Salmonella typhimurium*. *Journal Of Bacteriology* 172(7):4100-2.

Cheng J. Huang And Ericka L. Barrett.1991. Sequence Analysis And Expression Of The *Salmonella Typhimurium* Asr Operon Encoding Production Of Hydrogen Sulfide From Sulfite. *Journal of Bacteriology* 173 1544-1553

Clarke PH. 1953. Hydrogen sulphide production by bacteria. *J. Gen. Microbiol,* 8:397-407.

Clark, M. A., and E. L. Barrett. 1987. The phs gene and hydrogen sulfide production by *Salmonella typhimurium*. *Journal of.Bacteriology* 169:2391-2397.

Coleman G. 1960. A sulphate-reducing bacterium from the sheep rumen. *Microbiology* 22 (2): 423-436.

Crack, J. C, A. J. Jervis, A. A. Gaskell, G. F. White, J. Green, A. J. Thomson, and N. E. Le Brun. 2008. Signal perception by FNR: the role of the iron-sulfur cluster. *Biochem Soc Trans* 36:1144-8.

Crane, B. R., Siegel, L. M. & Getzoff, E. D. 1995. Sulfite reductase structure at 1.6 A: evolution and catalysis for reduction of inorganic anions. *Science* 270,5967.

Crane B, Siegel L, and Getzoff E. 1997. Structures of the siroheme- and Fe4S4-containing active centers of sulfite reductase in different states of oxidation: heme activation via reduction-gated exogenous ligand exchange. *Biochemistry* 36: 12101–12119.

Dirmeier R, Keller M, Hafenbradl D, Braun F-J, Rachel R, Burggraf S, and Stetter KO. 1998. *Thermococcus acidaminovorans* sp. nov., a new hyperthermophilic alkalophilic archaeon growing on amino acids. *Extremophiles* 2 (2): 109-114.

Etchebehere C, and Muxí L. 2000. Thiosulfate reduction and alanine production in glucose fermentation by members of the genus *Coprothermobacter*. *Antonie van Leeuwenhoek* 77 (4): 321-327.

Fitz R, and Cypionka H. 1990. Formation of thiosulfate and trithionate during sulfite reduction by washed cells of *Desulfovibrio desulfuricans*. *Archives of Microbiology* 154:400-406

Fischer H-M. 1994. Genetic regulation of nitrogen fixation in rhizobia. *Microbiol. Mol. Biol. Rev.,* 58 (3): 352-386.

Fleischmann RD, Adams MD, White O, Clayton RA, Kirkness EF, Kerlavage AR, et al . 1995. Whole-genome random sequencing and assembly of *Haemophilus influenzae* Rd. *Science* 269: 496–511.

Georgellis D, Kwon O, Lin EC. 2001. Quinones as the redox signal for the Arc two-component system of bacteria. *Science* 292:2314-2316.

Georgellis D, Lynch AS, Lin EC. 1997. In vitro phosphorylation study of the Arc two component signal transduction system of *Escherichia coli*. *Journal of Bacteriology* 179:5429-5435.

Gunsalus RP, Park SJ.1994.Aerobic-anaerobic gene regulation in *Escherichia coli:* control by the ArcAB and Fnr regulons. *Research in Microbiol,* 145:437-450.

Hallenbeck, P. C., M. A. Clark, and E. L. Barrett. 1989.Characterization of anaerobic sulfite reduction by *Salmonella typhimurium* and purification of the anaerobically induced sulfite reductase. *Journal of Bacteriology.* 171:3008-3015.

Haschke RH, and Campbell LL. 1971. Thiosulfate reductase of *Desulfovibrio vulgaris*. *Journal of Bacteriology* 106 (2): 603-607.

Hatchikian E. 1975. Purification and properties of thiosulfate reductase from *Desulfovibrio gigas*. *Archives of Microbiology* 105 (1): 249-256.

Hensel M, Hinsley AP, Nikolaus T, Sawers G, and Berks BC. 1999. The genetic basis of tetrathionate respiration in *Salmonella typhimurium*. *Molecular Microbiology* 32 (2): 275-287.

Hipp W, Pott A, Thum-Schmitz N, Faath I, Dahl C, and Truper H. 1997. Towards the phylogeny of APS reductases and sirohaem sulfite reductases in sulfate-reducing and sulfur-oxidizing bacteria. *Microbiology* 143: 2891–2902.

Huang CJ, and Barrett EL. 1990. Identification and cloning of genes involved in anaerobic sulfite reduction by *Salmonella typhimurium*. *Journal of Bacteriology* 172 (7): 4100-4102.

Huber R, Kristjansson J, and Stetter K. 1987. *Pyrobaculum* gen. nov., a new genus of neutrophilic, rod-shaped archaebacteria from continental solfataras growing optimally at 100 C. *Archives of Microbiology* 149 (2): 95-101.

Huang CJ, and Barrett EL.1991. Sequence analysis and expression of the *Salmonella typhimurium* asr operon encoding production of hydrogen sulfide from sulfite. *J Bacteriol* 173(4):1544-53.

Huber R, Langworthy TA, König H, Thomm M, Woese CR, Sleytr UB, and Stetter KO. 1986. *Thermotoga maritima* sp. nov. represents a new genus of unique extremely thermophilic eubacteria growing up to 90 C. *Archives of Microbiology* 144 (4): 324-333.

Huber R, Wilharm T, Huber D, Trincone A, Burggraf S, König H, Reinhard R, Rockinger I, Fricke H, and Stetter KO. 1992. *Aquifex pyrophilus* gen. nov. sp. nov., represents a novel group of marine hyperthermophilic hydrogen-oxidizing bacteria. *Systematic and Applied Microbiology,* 15(3): 340-351.

Huber R, Woese C, Langworthy TA, Fricke H, and Stetter KO. 1989. *Thermosipho africanus* gen. nov., represents a new genus of thermophilic eubacteria within the "Thermotogales". *Systematic and Applied Microbiology* 12 (1): 32-37.

Huber R, Woese CR, Langworthy TA, Kristjansson JK, and Stetter KO. 1990. *Fervidobacterium islandicum* sp. nov., a new extremely thermophilic eubacterium belonging to the "Thermotogales". *Archives of Microbiology* 154 (2): 105-111.

Hussain, H., J. Grove, L. Griffiths, S. Busby, and J. Cole. 1994. A seven-gene operon essential for formate-dependent nitrite reduction to ammonia by enteric bacteria. *Molecular Microbiology* 12:153-163.

Jankielewicz, A., Schmitz, R.A., Klimmek, O. and Kroë ger, A.1994 Polysulphide reductase and formate dehydrogenase from *Wolinella succinogenes* contain molybdopterin guaninedinucleotide. *Archives of Microbiology* 162, 238-242

Jeon Y, Lee YS, Han JS, Kim JB, Hwang DS.2001. Multimerization of phosphorylated and non-phosphorylated ArcA is necessary for the response regulator function of the Arc two component signal transduction system. *Journal Biological Chemistry* 276:40873-40879.

Johnson EF, and Mukhopadhyay B. 2005. A new type of sulfite reductase, a novel coenzyme F420-dependent enzyme, from the methanarchaeon *Methanocaldococcus jannaschii. Journal of Biological Chemistry* 280 (46): 38776-38786.

Jones H. 1971. A re-examination of *Desulfovibrio africanus*. *Archiv für Mikrobiologie,* 80 (1): 78-86.

Kamps A, Achebach S, Fedtke I, Unden G, and Götz F. 2004. *Staphylococcal* NreB: an O2 sensing histidine protein kinase with an O2 labile iron–sulphur cluster of the FNR type. *Molecular Microbiology* 52 (3): 713-723.

Kapralek F. 1972. The physiological role of tetrathionate respiration in growing *Citrobacter. Microbiology* 71 (1): 133-139.

Keon RG, and Voordouw G. 1996. Identification of the HmcF and topology of the HmcB subunit of the Hmc complex of *Desulfovibrio vulgaris*. *Anaerobe* 2: 231–238.

Kern M, Klotz MG, and Simon J. 2011. The *Wolinella succinogenes* mcc gene cluster encodes an unconventional respiratory sulphite reduction system. *Molecular Microbiology* 82 (6): 1515-1530.

Kim, J.H , and J.M. Akagi. 1985. Characterization of a trithionate reductase system from *Desulfovibrio vulgaris*. *Journal of. Bacteriology* 163:472-475

Krafft, T., Bokranz, M., Klimmek, O., Schroë der, I., Fahrenholz,F., Kojro, E. and Kroë ger, A. 1992.Cloning and nucleotide sequence of the psrA gene of *Wolinella succinogenes* polysulphide reductase. *European Journal of Biochemistry* 206, 503-510.

Liang R, Grizzle RS, Duncan KE, McInerney MJ, and Suflita JM. 2014. Roles of thermophilic thiosulfate-reducing bacteria and methanogenic archaea in the biocorrosion of oil pipelines. *Frontiers in Microbiology* 5: 89.

Lee, J.P. and H.D. Peck, Jr. 1971. Purification of the enzyme reducing bisulfite to trithionate from *Desulfovibrio gigas* and its identification as desulfoviridin. Biochem. Biophys. *Res. Commun.* 45:583-589.

Liu X, De Wulf P.2004. Probing the ArcA-P modulon of *Escherichia coli* by whole genome transcriptional analysis and sequence recognition profiling. *Journal of Biological Chemistry* 279:12588-12597.

Liu Y, Denkmann K, Kosciow K, Dahl C, and Kelly DJ. 2013. Tetrathionate reduction and thiosulphate oxidation in *C. jejuni*. *Molecular Microbiology* 88:173-188.

Luchi S.1993. Phosphorylation/dephosphorylation of the receiver module at the conserved aspartate residue controls transphosphorylation activity of histidine kinase in sensor protein ArcB of *Escherichia coli*. *Journal of Biological Chemistry* 268:23972-23980.

Luchi S, Cameron DC, Lin EC. 1989. A second global regulator gene (*arcB*) mediating repression of enzymes in aerobic pathways of *Escherichia coli. Journal of Bacteriology* 171:868-873.

Luchi S, Lin EC. 1988. *arcA* (*dye*), a global regulatory gene in *Escherichia coli* mediating repression of enzymes in aerobic pathways. *Proc Natl Acad Sci USA* 85:1888-1892.

Luchi S, Lin EC.1992. Mutational analysis of signal transduction by ArcB, a membrane sensor protein responsible for anaerobic repression of operons involved in the central aerobic pathways in *Escherichia coli. Journal of Bacteriology* 1992, 174:3972-3980.

Luchi S, Matsuda Z, Fujiwara T, Lin EC.1990.The *arcB* gene of *Escherichia coli* encodes a sensor-regulator protein for anaerobic repression of the *arc* modulon. *Molecular Microbiology* 4:715-727.

Macy JM, Schröder I, Thauer RK, and Kröger A. 1986. Growth the *Wolinella succinogenes* on H 2 S plus fumarate and on formate plus sulfur as energy sources. *Archives of Microbiology* 144 (2): 147-150.

Maier, T.M and C.R Myers. 2001. Isolation and characterization of a *Shewanella putrefaciens* MR-1 electron transport regulator *etr*A mutant reasseement of the role of EtrA. *Journal of Bacteriology* 183 :4918-26.

Malpica R, Sandoval GR, Rodriguez C, Franco B, Georgellis D. 2006. Signaling by the *arc* two-component system provides a link between the redox state of the quinone pool and gene expression. *Antioxidants and Redox Signaling* 8:781-795.

Mander GJ, Duin EC, Linder D, Stetter KO, & Hedderich R. 2002. Purification and characterization of a membrane-bound enzyme complex from the sulfate-reducing archaeon *Archaeoglobus fulgidus* related to heterodisulfide reductase from methanogenic archaea. *European Journal of Biochemistry* 269:1895–1904.

Moser DP, and Nealson KH. 1996. Growth of the facultative anaerobe *Shewanella putrefaciens* by elemental sulfur reduction. *Appl. Environ. Microbiol.,* 62 (6): 2100-2105.

Moskvin OV, Gomelsky L, and Gomelsky M. 2005. Transcriptome analysis of the *Rhodobacter sphaeroides* PpsR regulon: PpsR as a master regulator of photosystem development. *Journal of Bacteriology* 187 (6): 2148-2156.

Muyzer G. and Stams AJM. 2008. The ecology and biotechnology of sulphate-reducing bacteria. *Nature Reviews Microbiology* 6(6): 441-454.

Nakatsukasa,W. and J.M. AKagi 1969. Thiodulfate reductase isolated from *Desulfotomaculum nigrificans*. *Journal of Bacteriology* 98:429-433.

Neuner A, Jannasch HW, Belkin S, and Stetter KO. 1990. *Thermococcus litoralis* sp. nov.: a new species of extremely thermophilic marine archaebacteria. *Archives of Microbiology* 153 (2): 205-207.

Oshima T, Aiba H, Masuda Y, Kanaya S, Sugiura M, Wanner BL, Mori H, Mizuno T. 2002. Transcriptome analysis of all two-component regulatory system mutants of *Escherichia coli* K-12. *Molecular Microbiology* 46:281-291.

Ostrowski, J., Wu, I.-Y., Rueger, D. C., Miller, B. E., Siegel, L M. & Kredich, N. M. 1989. Characterization of the *cysJZH* regions of *Salmonella typhimurium* and *Escherichia coli* B. *Journal of Biological Chemistry* 264(26): 15726-1 5737.

Patel B, Morgan HW, and Daniel RM. 1985. *Fervidobacterium nodosum* gen. nov. and spec. nov., a new chemoorganotrophic, caldoactive, anaerobic bacterium. *Archives of Microbiology* 141 (1): 63-69.

Pereira IAC, Ramos AR, Grein F, Marques MC, da Silva SM, and Venceslau SS, 2011. A comparative genomic analysis of energy metabolism in sulfate reducing bacteria and archaea. *Frontiers in Microbiology* 2:69.

Pires RH, Venceslau SS, Morais F, Teixeira M, Xavier AV, and Pereira IAC. 2006. Characterization of the *Desulfovibrio desulfuricans* ATCC 27774 DsrMKJOP complex—A membrane bound redox complex involved in the sulfate respiratory pathway. *Biochemistry* 45: 249–262.

Perry K, Kostka J, Luther G, and Nealson K. 1993. Mediation of sulfur speciation by a Black Sea facultative anaerobe. *Science* 259 (5096): 801-803.

Pfennig N, and Biebl H. 1976. *Desulfuromonas acetoxidans* gen. nov. and sp. nov., a new anaerobic sulfur-reducing, acetate-oxidizing bacterium. *Archives of Microbiology* 110 (1): 3-12.

Price-Carter M, Tingey J, Bobik TA, and Roth JR. 2001. The Alternative Electron Acceptor Tetrathionate Supports B12-Dependent Anaerobic Growth of *Salmonella enterica Serovar Typhimurium* on Ethanolamine or 1, 2-Propanediol. *Journal of Bacteriology* 183 (8): 2463-2475.

Ravot G, Ollivier B, Magot M, Patel B, Crolet J, Fardeau M, and Garcia J. 1995. Thiosulfate reduction, an important physiological feature shared by members of the order thermotogales. *Appl. Environ. Microbiol.,* 61 (5): 2053-2055.

Reiner Hedderich A, Oliver Klimmek B, Achim Kroëger B, Reinhard Dirmeier C, Martin Keller C, Karl O. Stetter C .1999,Anaerobic Respiration With Elemental Sulfur And With Disulfides. *FEMS Microbiology Reviews* 22: 353-381

Rozanova E, and Khudyakova A. 1974. New nonspore-forming thermophilic sulfate-reducing organism, *desulfovibrio*-thermophilus nov-sp. *Microbiology* 43 (6): 908-912.

Rozanova E, and Nazina T. 1976. Mesophilic, sulfate-reducing, rod-shaped, non-spore-forming bacterium. *Microbiology,* 45 (5): 711-716.

Saffarini, D. A., R. Schultz, and A. Beliaev. 2003. Involvement of cyclic AMP (cAMP) and cAMP receptor protein in anaerobic respiration of *Shewanella oneidensis. Journal of Bacteriology* 185:3668-71.

Schiffer A, Parey K, Warkentin E, Diederichs K, Huber H, Stetter KO, Kroneck PMH and Ermler U, 2008. Structure of the Dissimilatory Sulfite Reductase from the Hyperthermophilic Archaeon *Archaeoglobus fulgidus*. *Journal of Molecular Biology* 379(5):1063-1074.

Schroeder, I., Kroeger, A. and Macy, J.M. 1988 Isolation of the sulphur reductase and reconstitution of the sulphur respiration of *Wolinella succinogenes*. *Archives of Microbiol.* 149, 572-579.

Segerer A, Neuner A, Kristjansson JK, and Stetter KO. 1986. *Acidianus infernus* gen. nov., sp. nov., and Acidianus brierleyi comb. nov.: facultatively aerobic, extremely acidophilic thermophilic sulfur-metabolizing archaebacteria. *International Journal of Systematic and Evolutionary Microbiology* 36 (4): 559-564.

Simon J and Kroneck PM. 2013. Microbial sulfite respiration. *Adv Microb Physiol,* 62:45-117.

Shirodkar S, Reed S, Romine M, Saffarini D. 2011. The octahaem SirA catalyses dissimilatory sulfite reduction in *Shewanella oneidensis* MR 1. *Environmental Microbiology* 13 (1): 108-115.

Spiro S, and Guest JR. 1990. FNR and its role in oxygen-regulated gene expression in *Escherichia coli. FEMS Microbiology Reviews* 6 (4): 399-428.

Steuber, J., A. F. Arendsen, W. R. Hagen, and P. M. Kroneck. 1995. Molecular properties of the dissimilatory sulfite reductase from *Desulfovibrio desulfuricans* (Essex) and comparison with the enzyme from *Desulfovibrio vulgaris* (Hildenborough). *European Journal of Biochemistry* 233:873-879.

Stette KO, and Gaag G. 1983. Reduction of molecular sulphur by methanogenic bacteria. *Nature* 305 (5932): 309.

Stoffels L, Krehenbrink M, Berks BC, and Unden G. 2011. Thiosulfate reduction in *Salmonella enterica* is driven by the proton motive force. *Journal of Bacteriology* 194 (2): 475-485.

Stroupe M.E., Getzoff E.D. 2009. The Role of Siroheme in Sulfite and Nitrite Reductases. In: Tetrapyrroles. Molecular Biology Intelligence Unit. Springer, New York, NY

Tarr HLA. 1933. The enzymic formation of hydrogen sulphide by certain heterotrophic bacteria. *Biochemical Journal* 27 (6): 1869.

Thauer R. 1982. Dissimilatory sulphate reduction with acetate as electron donor. *Philosophical Transactions of the Royal Society of London. B, Biological Sciences* 298 (1093): 467-471.

Trofimov AA, Polyakov KM, Boyko KM, Tikhonova TV, Safonova TN, Tikhonov AV, Popov AN, and Popov VO. 2010. Structures of complexes of octahaem cytochrome c nitrite reductase from *Thioalkalivibrio nitratireducens* with sulfite and cyanide. *Acta Crystallographica Section D: Biological Crystallography* 66 (10): 1043-1047.

Vainshtein M, Hippe H, Kroppenstedt RM. 1992. Cellular fatty acid composition of *Desulfovibrio* species and its use in classification of sulfate-reducing bacteria. *Systematic and Appl. Microbiol* 15:554-566.

Wagner M, Roger A, Flax J, Brusseau G, and Stahl D. 1998. Phylogeny of dissimilatory sulfite reductases supports an early origin of sulfite respiration. *J Bacteriol.* 180: 2975–2982.

Widdel F, and Pfennig N. 1984. Dissimmilatory sulfate- or sulfur-reducing bacteria, in: *Bergey's Manual of Systematic Bacteriology* (N. R. Krieg and j. G. Holt, eds.) Williams & Wilkins, Baltimore, pp. 663-679.

Widdel F,.Pfennig N.1992. The genus *Desulfuromonas* and other gram-negative sulfur-reducing eubacteria In: *The Prokaryotes* Balows A.Trüper H.G.Dworkin M.Harder W.Schleifer K.H., Eds) Second Edition, 3379-3389

Wolin Ea, Wolin Mj, Wolfe Rs. 1963. Formation Of Methane By Bacterial Extracts. *Journal of Biological Chemistry* 238:2882-6

Wunsch, P., M. Herb, H. Wieland, U. Schiek, and W. Zumft. 2003. Requirements for Cu_A and Cu-S center assembly of nitrous oxide reductase deduced from complete periplasmic enzyme maturation in the nondenitrifier *Pseudomonas putida. Journal of Bacteriology* 185:887-896.

Widdel F, and Pfennig N. 1981. Studies on dissimilatory sulfate-reducing bacteria that decompose fatty acids. *Archives of Microbiology* 129 (5): 395-400.

Widdel F, and Pfennig N. 1982. Studies on dissimilatory sulfate-reducing bacteria that decompose fatty acids II. Incomplete oxidation of propionate by *Desulfobulbus propionicus* gen. nov., sp. nov. *Archives of Microbiology* 131 (4): 360-365.

Widdel F, and Pfennig N. 1992. The genus *Desulfuromonas* and other gram-negative sulfur-reducing eubacteria *The prokaryotes* (pp. 3379-3389): Springer.

Windberger E, Huber R, Trincone A, Fricke H, and Stetter KO. 1989. *Thermotoga thermarum* sp. nov. and *Thermotoga neapolitana* occurring in African continental solfataric springs. *Archives of Microbiology* 151 (6): 506-512.

Wolin E, Wolin MJ, and Wolfe R. 1963. Formation of methane by bacterial extracts. *Journal of Biological Chemistry* 238 (8): 2882-2886.

Zeikus J, Dawson M, Thompson T, Ingvorsen K, and Hatchikia E. 1983. Microbial ecology of volcanic sulphidogenesis: isolation and characterization of *Thermodesulfobacterium commune* gen. nov. and sp. nov. *Microbiology* 129 (4): 1159-1169.

8

Anaerobes: The Essential Commensal Microbiota of Ruminants

*Shailendra Kumar and Sarah Fatima

Centre of Excellence, Department of Microbiology
Dr. Ram Manohar Lohia Avadh University, Faizabad, Uttar Pradesh, India

Abstract

Grazing animals have developed symbiotic relationship with anaerobic microorganisms, which survive in their rumen and help in the oxidation of complex molecules of feed containing lignocellulosic materials into volatile fatty acids. The consortium comprises anaerobic bacteria, fungi and protozoans. The rich microbial diversity of rumen remains constant after its establishment and does not get affected by intruding contaminants along with feed or water. The rumen contains a wide range of microbes such as Fibrobacter spp., Ruminococcus spp., Clostridium spp., Prevotella spp., *Eubacterium* spp., *Methanobacterium* spp., *Treponema* sp. *Lachnospira* sp., *Bacterioides* spp., *fungi: Anaeromyces* spp., *Neocallimastix* spp., *and protozoa: Enoplastron sp., Eudiplodinium* spp., *Ostracodinium* sp. *They often have a variety of enzymes, glucanases, xylanase, glucuronidase, pectinase and amylase, etc. These enzymes inhabit the microbes and even break down complex molecules into simpler forms like; glucose and volatile fatty acids. The present chapter discusses on the role of microorganisms of rumen.*

Introduction

Grazing mammals that acquires food from plant based source and ferments them in a specialized pouch prior to digestion is called ruminants. It consists of four-chambered stomach (rumen). These ruminants are fed on agricultural by-products like, cereal straws, grass, sugarcane bagasse, stovers, tree foliages and oil cakes of cotton, ground nut, Neem and mustard. Ruminants themselves

---

**Corresponding Author:shailendrakumar@rmlau.ac.in*

do not produce the enzymes required for digestion of complex variety of feed ingredients, rather, they have developed a symbiotic association with anaerobic bacteria, fungi and protozoa. As per Woese's classification the rumen microbial ecosystem comprises Bacteria, Archaea (methanogens), and Eucarya (protozoa and fungi).These bacteria alone account for about 95% of rumen microorganisms. Each milligram of rumen content consists of about 10^{10} to 10^{11} bacteria comprising more than 50 genera, nearly 10^5 to 10^6 ciliated protozoa (25 genera), 10^3 to 10^5 anaerobic fungi (5 genera) and 10^7 to 10^9 bacteriophage particles in each millilitre of rumen contents (Kamra 2005). These symbionts establish themselves in the ruminant digestive tract within first few weeks of birth. It consists a complex network of microbiota, where range of microorganisms co-exist. These microorganisms ferment plant carbohydrates such as lignin, cellulose, hemicellulose ingested by the animal. Several enzymes celloxylanase, xylanase, xylodextrinase, glucuronidase, xylosidase, etc. are involved in rumen fermentation (Hespell 1988). Rumen microbes generally stick on the particles of feed and digest plant material by making biofilm (McAllister *et al.*, 1994).

The rumen have semipermeable inner epithelial wall that permits selective transports of small molecules to and from the blood of the animal. For example, volatile fatty acids are transported from the rumen to the blood, while bicarbonate ions are transported in the reverse direction. Neural responses triggered by the specific stimuli activate physiological mechanisms for eructation, initiate contractions to mix and move rumen contents, and activate the mechanism for regurgitation of solids. The grazing animals retch up the regurgitated solids into the oral cavity ("chewing the cud") to allow continued comminution of solid substrates even after they enter the fermentation chamber and fermentation is initiated. The biological features of the rumen provide an excellent mechanism for fermentation of solid substrates, product removal, maintenance of pH, and disposal of fermentation gases (Wolin 1979). Rumen is a biological fermentation unit that functions properly only in a well-defined environmental conditions that are extremely important regulators of the kinds, amounts, and physiological activities of the microorganisms in the ecosystem. Since rumen environment is anaerobic, the gas composition is approximately 65% Co_2, 35% CH_4 and trace amounts of other gases (H_2, N_2, and O_2) have been reported. CH_4 arises strictly as a fermentation product. CO_2 is produced both from fermentation and by neutralization of acids by bicarbonate ions entering the rumen from the saliva and blood. Any small amounts of O_2 entering the system are quickly used up by oxidative metabolism by rumen commensal.

Generally the pH of the rumen is near 6.5. When the ratio of fermentation end products, acetate: propionate, declines in the rumen, CH_4 production also declines (Russell 1998). A decline in the acetate: propionate ratio typically occurs when the diet is switched from high forage to a high grain one, and is the result of increased ruminal propionate production (Russell 1998). Activity of microbes sufficiently produces adverse effects on nutrient recovery by the host animal, although, even in healthy animals, pH may temporarily drop lower during the day depending on ration and feeding pattern. The amount of time spent at sub-optimal pH, rather than the lowest values reached is of practical importance (Alzahal *et al.* 2007). Alternatively, prediction of rumen pH from and rumen temperatures was also analysed Alzahal *et al.* (2008). Neutral plant polymers are fermented predominantly to acid products, and the pH would drop significantly during fermentation if it is not regulated by the strong bicarbonate-carbon dioxide buffering system of the rumen. Salivary bicarbonate is the major neutralizing agent for the acids produced during fermentation. The temperature of the rumen is maintained approximately at 39°C through physiological regulation by the animal (Bryant 1959; Hungate *et al.*, 1964; Hungate 1966; Janis 1976; Bryant *et al.*, 1977). The biochemical pathways playing important role in rumen of grazing animals is schematically shown in Fig. 1. Various important role players rumen microorganisms are described in following sections:

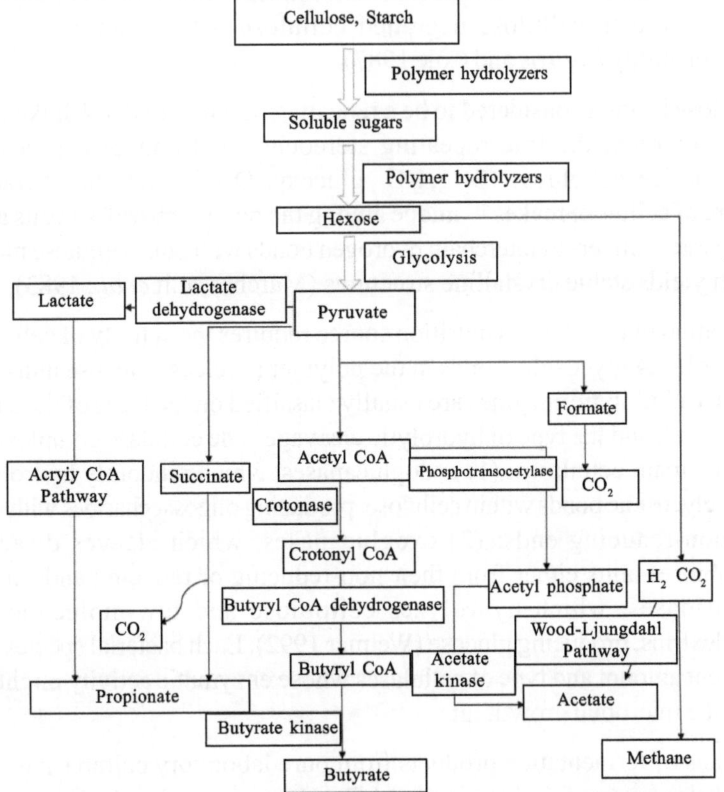

Fig. 1: Schematic flow diagram to represent the biochemical pathways involved in digestion of cellulose and/or starch and production of various products in the rumen of grazing animals.

Cellulolytic Microorganisms

The major composition of plant cell wall is cellulose, it ranges from 20% (in certain grasses) to 90% (in the cotton fibres) of the dry mass of plant tissue. Three bacterial species are predominant cellulolytic organisms in the rumen: *Fibrobacter succinogenes* (Montgomery *et al.*, 1988), *Ruminococcus flavefaciens,* and *R. albus* (Montgomery *et al.*, 1988; Dehority 1991; Hobson and Stewart 1997; Kamra 2005). Eight other cellulolytic species are thought to be of relatively much lesser importance. Less predominant species are *Butyrivibrio fibrisolvens, Clostridium longisporum, C. Lochheadii, C. Chartatabidium, C. Polysaccharolyticum, Eubacterium cellulosolvens, Micromonospora ruminantium.*

The adhesion of rumen cellulolytic bacteria to plant tissues is important for the fibre degradation. Microbial populations associated with feed particles are estimated to be responsible for 88 to 91% of ruminal endoglucanase and xylanase activity. To support the relationship between adhesion ability and subsequent cellulose degradation, demonstrated that isolates of *Ruminococcus albus* strains that adhered to cellulose degraded cellulose better than strains lacking adhesionability (Morris and Cole 1987).

Cellulose is often considered to be a repeating polymer of β-1,4-linked glucose units, although the true repeating stereochemical unit of cellulose is the disaccharide cellobiose (β- 1,4-D- glucosyl-D-glucose).The characteristic feature of cellulose makes it unique among the polysaccharides i.e. its ability to form large numbers of interchain hydrogen bonds with other cellulose molecules, which yields stable crystalline structures (Marchessault *et al.*, 1983).

Utilization of cellulose as nutrition source requires the activity of cellulase that cleaves b-1,4 glycosidic bonds in the polymer to release glucose units (Behera 2017). Cellulolytic enzymes are usually classified on the basis of the substrates hydrolyzed and the type of hydrolytic cleavage. The cellulase complex consists of three main cellulases: (1) endoglucanases, which randomly hydrolyzes the 1,4-β-glycosidic bonds within cellulose, producing oligosaccharides with reducing and non-reducing ends; (2) exoglucanases, which cleaves disaccharide (cellobiose) units either from their non-reducing or reducing ends and (3) β-glucosidases, which hydrolyzes cellobiose and low-molecular-weight cellodextrins, producing glucose (Weimer 1992). Each bacterial species produce different amount and type of cellulases whose enzymatic activity might depend upon the microbial growth rate.

Analysis of fermentation products from pure laboratory cultures has revealed that, *Fibrobacter. succinogenes* and *Ruminococcus. flavefaciens* are among the major producers of both acetate and succinate (a precursor of propionate),

Succinate is an important extracellular intermediate in the overall rumen fermentation (Ellis *et al.*, 1989). It is decarboxylated essentially as fast as it is formed (Blackburn and Hungate 1963). *Selenomonas. ruminantium* can decarboxylate succinate to propionate and carbon dioxide. While the nonsuccinogenic *R. albus* is thought to produce mainly acetate and H_2; the latter product serves as a major energy source for the methanogenic population (Bryant 1970).

Eight species of fungus has also been reported, they include *Anaeromyces mucronatus, Caecomyces communis, Neocallimastix frontalis, N. Joyonii, N. Patriciarum , Orpinomyces bovis, Piromyces communis, Ruminomyces elegans*, they are known to carry out cellulose degradation in rumen (Orpin 1975; Brenton *et al.*, 1989; Barr *et al.*, 1989; Brenton *et al.*, 1990; Ho *et al.*, 1990). Although the amount of enzyme secreted is relatively small, it now appears that the rumen fungi are much more important in rendering the polysaccharides accessible to other microbial agents than they are to the actual fermentation of carbohydrates. Electron microscopic studies of rumen fungi grown on forage materials have revealed that the fungi, like the bacteria, preferentially attach to specific regions of plant particles, particularly physically damaged surfaces (e.g., cut ends) or stomata. The fungi readily move into the plant cell wall lumen and in the middle lamellae between individual plant cell walls. This suggests that the fungi are particularly well adapted to penetrating the cell walls and the extensive hyphal growth that ramifies throughout the interior of plant cell walls following fungal entry is thought to result in physical disruption of the cell wall, which ultimately exposes susceptible biopolymers to further degradation by the entire fibrolytic microbial population. In addition to their ability to physically disrupt plant tissues, the ruminal cellulolytic fungi also produce enzymes that attack resistant covalent linkages. Foremost among these are p- coumaroyl and feruloyl esterases that can degrade p-coumaroyl- and ferulolyl-arabinoxylobiosides (and presumably the homologous substituted xylan polymers that are thought to be limiting factors in plant cell digestion).The primary products of pure-culture carbohydrate fermentation is formate, acetate, ethanol, lactate, hydrogen, and CO_2, - vary somewhat among different species and strains. Succinate is produced only in trace amounts (Borneman *et al.*, 1989).

Twelve species of protozoa, *Diplodinium pentacanthum, Enoploplastron caudatum, Epidinium caudatum, Entodinium caudatum, Eudiplodinium bovis, E. Magii, Ophryoscolex caudatus, Ophryoscolex tricoronatus, Ostracodinium dilobum, O. gracile, Polyplastron multivesiculatum* are likely involved in cellulolytic activity (Dehority 1991 ; Coleman 1985; Bailey and Clarke 1963). This activity seems however to differ across the different protozoan groups: Large *Ophyroscolecidae* such as *Epidinium, Polyplastron* and

Eudiplodinium have greater endoglucanase and xylanase activity. On the other hand, *Entodinium* spp. have only weak activity (Williams and Coleman 1992). Similarly, *Dasytricha* has glucosidase and cellobiosidase activity but negligible fibrolytic activity (Takenaka *et al.*, 2004).

Hemicellulolytic Microorganisms

Hemicellulose constitutes xyloglucans, xylans, mannans and glucomannans, and 1,4-β- and 1,3-β-glucans. Among hemicellulosic substrates, xylan is the most abundant. Xylan structure is more heterogeneous than cellulose and consists of 1,4-linked b-D-xylopyranose backbone that bears many different side chains. The backbone is often partially acetylated and the type of substitution depends on botanical origin. Arabinoxylan is common type of xylan found in grasses and cereals having L-arabinose side chains that can be esterified by ferulic or caumaric acid. Hardwood xylan is glucuronoxylan (GX) with methylated or non-methylated D-glucuronic acid sidechains (Heredia *et al.*, 1995; Zorec *et al.*, 2014). *Prevotella ruminicola* (Marinšek-Logar, 1999), *Eubacterium xylanophilum, Eubacterium uniformis* actively carry out hemicellulolytic activity. Due to xylan's complex structure with diverse nature of sidechains, degradation is also carried out by the synergistic action of many different enzymes. The enzyme system usually consists of endoxylanases and β-xylosidases that cleave the polyxylopyranosyl backbone and many different side-chain-removing enzymes like α-L-arabinofuranosidases, α-glucuronidases, acetylxylan esterases, p-coumaric and ferulic acid esterases. Microorganisms usually possess the whole set of enzymes, often producing multiple forms of certain catalytic property with subtle differences in order to facilitate the solubilization of a variety of xylans (Marinšek - Logar 1994).

Acetogenic Microorganisms

Reductive acetogens are a group of bacteria that can produce acetate from hydrogen and carbon dioxides ($4H_2+2CO_2 \rightarrow CH_3COOH+2H_2O$) using the acetyl-CoA pathway also called Wood-Ljungdahl pathway (Drake *et al.*, 2006). This suggests that acetogens may serve as hydrogen sinks in experiments, which seek to lower methane emissions (van Nevel and Demeyer, 1995; Morvan *et al.*, 1996; Joblin, 1999) in cattle. The synthesis of acetate by these microorganisms is thermodynamically favourable only when the partial pressure of H_2 is very low and is normally achieved in the presence of a hydrogen-utilizing methanogen or sulfate reducer. *Acetitomaculum ruminis*, *Ruminococcus* sp., *Ruminococcus obeum*, *Eubacterium limosum*, *Butyrivibrio schinkii*, *Butyrivibrio hydrogenotrophica*, and *Butyrivibrio wexlerae* are responsible for acetogenesis in rumen. Chaucheyras-Durand *et al.*, (1995) demonstrated the stimulation of hydrogen utilization by acetogens and enhanced acetogenesis in presence of yeast cells in co-cultuer of acetogens and methogens.

Methanogens

Fermentation of feedstuff in the rumen results in methane production. Although, the methane gas can be produced in the lower gastrointestinal tract (GI) of nonruminants. While, 89% of total methane emitted by grazing animals is produced in rumen and CH_4 emitted by ruminants is produced in the rumen and released through mouth and nostrils (Johnson and Ward 1996). In the rumen, CH_4 appears to be almost exclusively produced from the reduction of carbon dioxide by H_2. The Volatile fatty acids (VFA) are generally not used as substrates for methanogenesis due to their lengthy process of conversion in to CO2 and H2, the long process is repressed by rumen turnover (Hobson and Stewart, 1997). Since methanogenesis often employ CO_2 and H_2 produced by carbohydrate fermentation. Methanogens support the other microorganisms involved in substrate fermentation to function optimally by removing the hydrogen. If H_2 is not removed by methanogenesis, it inhibits the metabolism of rumen microorganisms (Sharp *et al.*, 1998). The methanogens have huge capacity to use H_2. Methane is produced by rumen contents at a rate of approximately 5.5 µmol/g / hour (Carroll and Hungate, 1955). Since 4 mol of hydrogen are used to form 1 mol of CH_4, the rate of H_2 utilization is 4 times the rate of production of methane. The hydrogen-production rate is the same as the hydrogen utilization rate. No hydrogen accumulation occurs in the rumen; the partial pressure of hydrogen is approximately 10^{-4} atm (Hungate, 1967). The partial pressure of hydrogen has an important influence on the course of the rumen fermentation and other methanogenic fermentations. Production and loss of methane is usually considered as a loss of energy for the ruminant. Approximately 10 to 15% of the energy of ingested food is lost by eructation of methane (Demeyer *et al.*, 1975). *Methanobacterium formicicum, Methanobacterium bryantii, Methanobrevibacter ruminatium, Methanobrevibacter smithi* are some methanogenic bacteria found in rumen of grazing animals.

Methane mitigation strategies

Methane production through enteric fermentation is of great concern worldwide due to its role as one of the major green house gases responsible for global warming. The feed energy converted to methane is waste as it can't be used by the animal. worldwide, nearly 60% of methane is emitted from the agriculture sector, mostly from livestock development operations; the major source of CH_4 in the environment is grazing animals (Ellis *et al.*, 2007). The methane has been responsible for global warming and several methane abatement strategies should be adopted to bring down methane emission by ruminants to reduce its impact on global warming.

Starch content of diet is used for the formation of propionate, the conversion is mediated by amylolytic bacteria and decreased pH, that decreases methanogenesis (Van Kessel and Russell 1996). Like starch constituents the lipid in feed is considered to decrease methanogenesis due to inhibition of protozoa, bio-hydrogenation of unsaturated fatty acids (UFA) and raised production of propionic acid (Johnson and Johnson 1995). Unsaturated fatty acids may be used as a hydrogen acceptor as an alternative to the reduction of carbon dioxide. Soliva *et al.*, (2004) observed that addition of oil in feed provides dramatic decrease in methane production than individual fatty acids. Methanogens and protozoa in rumen share symbiotic relationship, with this relationship methanogen benefit by receiving hydrogen required for reduction of carbon dioxide to merhane. For this reason, treatments like copper sulphate, acids, triazine, lipids, tanins, ionophores and saponins act on protozoa and decrease their population, which in turn decreases protozoa associated methane production in rumen (Hobson and Stewart 1997). Another resort could be immunization of ruminants with the vaccines against methanogens to inhibit the methanogens. (Wright *et al.*, 2004).

Proteolytic Microorganisms

Bacteria are mainly responsible for dietary protein breakdown, while ciliate protozons break down particulate feed protein of appropriate size and also bacterial protein. The prime mechanism of protein digestion is biphasic, i)dipeptidyl aminopeptidases break dipeptides from larger peptides, the action is followed by activity of dipeptidase. *Prevotella ruminicola* is the only species which shows dipeptidyl aminopeptidase activity. While, dipeptidase has been reported active in several species including *P. ruminicola*. The dipeptidase activity is predominantly high in rumen protozoa (Wallace 1996). The process of deamination is mediated jointly by bacteria and protozoa. The ammonia is formed from protein in the rumen which is subsequently absorbed across the rumen wall and excreted as urea (Leng and Nolan, 1984; Wallace and Cotta, 1988; Wallace, 1994) or the soluble nitrogen is used for microbial biosynthesis. Almost all of the major species of rumen bacteria can use ammonia as the major precursor of the cellular nitrogen compounds (Bryant and Robinson, 1962; Bryant, 1974). In fact, several important species use only ammonia and are incapable of growth with amino acids, presumably because they are impermeable to amino acids (Allison, 1969). Proteins are hydrolyzed to peptides that are small enough to enter bacterial cells, where they are used as nitrogen sources and also catabolized to yield ammonia without accumulation of consequential amounts of extracellular amino acids. The nitrogen catabolized from the amino acid taken up by the bacteria gets incorporated into bacterial protein (Wright, 1967). Further the carbon skeletons of the amino acids are fermented to derivative acids, fatty acids from aliphatic amino acids and aromatic acids from aromatic amino acids. (el-Shazly, 1952; Scott *et al.*, 1964). The products of fermentation of branched chain amino acids are branched-chain fatty acids with one carbon less than the parent amino acids. Several bacterial populations require branched-chain fatty acid as a nutritional requirement which is fulfilled by those bacteria that could synthesize them (Bryant, 1974, Allison, 1969). The *Prevotella ruminicola* has been reported as predominant species in the rumen of most animals (Fulghum and Moore, 1963; Hazlewood and Nugent, 1978; Wallace and Brammall, 1985), *Ruminobacter amylophilus, Clostridium bifermentans* (Wallace and Brammall, 1985). The rumen bacterial population as whole typically possess cell associated enzyme with mainly cysteine protease activity. Rumen ciliate protozoa demonstrate a variety of protease activities, the most important of which are cysteine and aspartic proteases (Forsberg *et al.*, 1984) that have a mixture of specificities with a significant trypsin-like activity (Prins *et al.*, 1983; Forsberg *et al.*, 1984). *Dasytricha ruminantium*, *Entodinium* spp., *Isotricha* spp. (Lockwood *et al.*, 1988) are the predominant protozons species that hydrolyse the polypeptides.

Pectinolytic Microorganisms

Pectin is a valuable component of feed of ruminants. Grasses contain approximately 3 to 4% of pectin in the dry matter (Waite and Gorrod, 1959), leguminous plants have 5 to 10% (Van Soest, 1983) and sugar beet pulp consists 25% (Aspinall, 1970) of pectin. The basic building unit of pectin is D-galacturonic acid methoxylated to varying extents at the carboxyl moiety. Other sugars, acetyl groups and calcium ions are present as additional constituents. In mixed cultures of rumen microorganisms pectin produced a metabolite profile, high in acetate and low in butyrate and lactate (Leedle and Hespell, 1983; Marounek *et al.*, 1985). *Treponema saccharophilum* and *Lachnospira multiparus* are the principle pectin degrading bacteria. Pectinolytic activity also arises from the protozoan such as *Eudiplodinium maggii, Ostracodinium dilobum, Epidinium caudatum* and fungi like *Neocallimastix frontalis, Caecomyces communis*. They produce a pectin lyase and a pectin methylesterase (Silley, 1985) enzyme to yield volatile fatty acids from pectin. Pectin lyase generates unsaturated digalacturonate by a non-hydrolytic mechanism (Wojciechowicz 1971; Wojciechowicz *et al.*, 1982). Pectin-grown cells of *Butyrivibrio fibrisolvens* and *Prevotella ruminicola* possess the activity of 2-keto-3-deoxy-6-phosphogluconate aldolase, an enzyme catalysing the pyruvate formation in the Entner–Doudoroff pathway (Wood 1971) from Unsaturated galacturonate (a product of pectin lyase action) (Wood, 1971; Marounek and Duskova 1999).

The pectin fermentation mechanism is proposed as D-galacturonate gets decarboxylated to a pentose and then it is converted to a mixture of 3-7 carbon sugar phosphates by the pentose phosphate pathway (Van Soest 1983). Carbon of this sugar phosphate pool enters the glycolysis (Marounek and Duskova 1999).

Amylolytic Microorganisms

A variety of species of ruminant bacteria are competent enough to grow in starch-containing medium and produce amylase. Of those tested, the highest levels of amylase were produced by *Streptococcus bovis* (Russell and Robinson 1984) and *Ruminobacter amylophilus* (Stackebrandt and Hippe 1986). Other strains that grew well on starch and produced amylase included *Butyrivibrio fibrisolvens* and *Bacteroides ruminicola.* The products of amylolytic attack were the release of oligosaccharides from amylose. With *B. ruminicola*, the products included maltose, maltotriose, maltotetraose, maltopentaose, maltohexaose, and maltoheptaose. Prolonged incubation of enzyme and substrate resulted in a decrease of larger oligosaccharides, with a corresponding increase of smaller oligosaccharides. *R. amylophilus*, and *S. bovis* produced similar patterns of oligosaccharides, with maltose through maltotetraose being the major products detected. *B. fibrisolvens* and *R. amylophilus* also generated a small amount of glucose (Cotta 1988). The major pathway for hexose fermentation in the rumen is the Embden Meyerhof- Paranas (EMP) pathway, the major pathway used by most of the individual species of carbohydrate-fermenting rumen microorganisms.

Tanninolytic Microorganisms

Tannins (hydrolyzable and condensed) are water-soluble polyphenolic compounds that exert anti-nutritional effects on ruminants by forming complexes with dietary proteins. They limit nitrogen supply to animals, besides inhibiting the growth and activity of rumen microflora. However, some ruminant microbes are able to break tannin-protein complexes while preferentially degrading hydrolyzable tannins (Hts) (Goel *et al.*, 2005). *Streptococcus gallolyticus* (Brooker *et al.*, 1994), *Selenomonas* sp. (Odenyo and Osuji 1998) and *Selenomonas ruminantium* (Skene and Brooker 1995) are known to efficiently degrade hydrolyzable tannins. These rumen microbes synthesize tannin degrading enzyme tannase, which has both esterase and depsidase activities and specifically hydrolyzes ester and depside bonds in HTs, releasing glucose and gallic acid. This action diminishes tannin's protein-binding property (Bhat *et al.*, 1998).

Advances in rumen microbiology

The tremendous importance of studying the domestic ruminants is mainly due to their unique quirk of converting the low-quality forages into high- quality, energy-rich, high-protein products (especially milk and meat) consumed by humans worldwide. So it is essential to define the functions of microbial populations within the rumen microbiome to allow directed rumen manipulation strategies and also to explore the vast diversity of rumen microorganisms. With advances in omic technologies, the ability to link host genetics and the rumen microbiome by studying all the biological components became possible in recent times.

Metagenomic approach overcomes the need for culturing, due to which it has become an important tool for understanding the full genomic potential that resides within the rumen microbiome (Denman and McSweeney, 2015). The revolutions in sequencing technology enabled us to achieve greater depths of sequencing, but at the expense of read length, which makes assembly and linking of functional genes to phylogenetic assignment difficult. Bioinformatic algorithms that considered the base composition of genomes have been used to generate k-mer patterns unique to each organism, which has allowed for the sorting of metagenomic fragments into bins of k-mer-like patterns (Tyson *et al.*, 2004; Zhou *et al.*, 2008; Patil *et al.*, 2011; Droge and Mc Hardy 2012; Albertsen *et al.*, 2013). The sequencing projects help in making discoveries of new genes encoding biological degradation of fibre, eliciting animal production or depolymerization of lignocelllosic waste for biofuel production, as well as ruminant methane mitigation via interventions against methanogens. Adopting various advanced approaches like transcriptomics, proteomics metabolomics are being used to decode the functions of annotated genes. These are used to test hypotheses around the role of 'unknown' or 'conserved hypothetical' genes (Morgavi *et al.*, 2012).

The genomic analysis of fibre-degrading bacteria in rumen reveal their strategies for polysaccharide degradation. Diverse bacterial species seem to have different approach of fibre breakdown. *Fibrobacter succinogens* S85 appears to specialise as a cellulose degrader. *F. Succinogenes* lacks the cellulosome components such as dockerins, cohesions and scaffoldins found in *Clostridium thermocellum* and *Ruminococcus flavefaciens* (Doi and Kosugi, 2004). While, *Ruminococcus* spp., encode for variety of enzymes to assemble into a highly organised multi-enzyme scaffold, called a cellulosome, that mediates plant fibre degradation (Berg Miller *et al.*, 2009). *Ruminococcus. albus* has a relatively less developed cellulosome compared with *Ruminococcus flavefaciens.* Recently, the species of *Butyrivibrio* and *Pseudobutyrivibrio* have been recognized due to their great efficiency in xylan- degradation and the report showing presence of rss genes in data set of pyrotag sequence libraries developed from rumen contents (Attwood *et al.*, 2004). The genome sequencing of *B. proteoclasticus* B316 has revealed the presence of genes including glycosyl hydrolases (GHs), carbohydrate esterases attacking ferulic acid and acetylxylan linkages and a variety of pectate lyases (Kelly *et al.*, 2010). Several glycosyl hydrolases of *B. proteoclasticus* B316 are identified as enzymes belonging to various classes of xylanases, xylosidases or arabinosidases, consistent with its ability to grow well on xylan.

The sequence information of rumen methanogen *Methanobrevibacter ruminatium* M1 confirmed it's hydrogenotrophic lifestyle, gene expression (transcriptomic) studies revealed that formate may be an important substrate for methanogenesis during syntrophic growth with other rumen bacteria (Leahy *et al.*, 2010). It has been observed that *M. ruminantium* genome does not have genes encoding methyl coenzyme-M reductase II enzyme system (McrII). McrII is an isozyme of the McrI enzyme and usually found in hydrogenotrophic methanogens. In other methanogens, the Mcr II system is differentially regulated during growth (Reeve *et al.*, 1997; Luo *et al.*, 2002) and is thought to mediate methane formation at high hydrogen partial pressures. A methanogen prophage (φmru) sequence has been discovered from the *M. ruminantium* genome sequence (Attwood *et al.*, 2008; Leahy *et al.*, 2010). The phage analysis has resulted in identification of genes encoding cell lysis function.

Meta-proteomic analysis has potential for complete know-how of the rumen ecosystem than sequencing approaches alone. Meta-proteomics approach helps the researchers to determine if the most abundant proteins in rumen system matched with those foreseen by meta-transcriptomic data and to use this information to identify the most transnationaly active organisms in the rumen microenvironment. This information will assist in the further understanding of the active microbial metabolic pathways in the rumen of cows fed on fresh forage. The abundant taxonomic phyla in the proteome have been identified as *Bacteriodetes, Firmicutes and Proteobacteria,* which corresponds with the most abundant taxonomic phyla as determined by 16S rRNA analysis (Hart *et al.*, 2018).

The Hungate 1000 project is recently been undertaken which aims to produce a reference set of rumen microbial genomes by sequencing the genomes of available cultivable rumen bacteria and methanogenic archaea, together with representative cultures of rumen anaerobic fungi and ciliate protozoa. This led to the creation of a robust database of genomic information, and also provides insight into the life of these microorganisms.

Conclusions

In spite of the potential microbial diversity studied in mammalian rumen, many of them have not yet been cultured. Unless newer application of molecular techniques is implemented, the structural and functional aspect of these microbes may still remain devoid. Recently Metagenomics was introduced to assess the microbial ecosystem of rumen community. Metagenomics perhaps is one of the promptly growing field of molecular biology that focuses on targeting uncultured microorganisms and providing various aspects about the diversity, role and function of the digestive tract of mammalian rumen. It has provided comprehensive database sequence providing in the assessment of genome composition and opportunity to extract the knowledge beyond the catabolic pathways. Rumen microbial community is responsible for the production of volatile fatty acids from lignocellulosic feeds. Thus rumen microbiology represents a vast array of genetic diversity leading to entire metabolic pathways and potentially valuable end products, yet many novel genes or microbes remain as a broad domain of future research.

Acknowledgement

The authors are thankful to Dr. Deep Narayan Prasad, KNIPS, Sultanpur for giving his valuable time in proof reading the chapter.

References

Albertsen M, Hugenholtz P, Skarshewski A, Nielsen KL, Tyson GW, Nielsen PH. 2013. Genome sequences of rare, uncultured bacteria obtained by differential coverage binning of multiple metagenomes. *Nat. Biotechnol.* 31:533-38

Allison, M.J. (1969). Biosynthesis of Amino Acids by Ruminal Microorganisms, *Journal of Animal Science*, 29(5):797-807.

Alzahal O., Kebreab E., France J. & M Cbride B. W. (2007). A mathematical approach to predicting biological values from ruminal pH measurements. *Journal of Dairy Science* 90, 3777–3785.

Alzahal O., Kebreab E., France J., Froetschel M. & Mcbride B. W. (2008). Ruminal temperature may aid in the detection of subacute ruminal acidosis. *Journal of Dairy Science* 91, 202–207.

Aspinall, G.O. (1970). Polysaccharides. Oxford: Pergamon Press.

Attwood, G.T., Cookson, A.C. and Kelly, W.J. (2004). Genome sequencing of *Clostridium proteoclasticum*. *Reproduction Nutrition Development* 44 (Suppl 1), S20.

Attwood, G.T., Kelly, W.J., Altermann, E.H. and Leahy, S.C. (2008). Analysis of the *Methanobrevibacter ruminantium* draft genome: understanding methanogen biology to inhibit their action in the rumen. *Australian Journal of Experimental Agriculture.* 48, 83-88.

Bailey, R. W. and Clarke, R. T. J. (1963). Carbohydrase activity of rumen *Entinodinium* sp. from sheep on a starch-free diet, *Nature*, 198:787.

Barr, D. J. S., Kudo, H., Jakoben, K. D., and Cheng, K.-J. (1989). Morphological development of rumen fungi: *Neocallimastix* sp., *Piromyces communis*, and *Orpinomyces bovis*, gen. nov., *Can. J. Bot.*, 67(9):2815-24.

Behera, B.C., Sethi, B.K., Mishra, R.R., Duttac, S.K. and Thatoi, H.N. (2017). Microbial cellulases – Diversity & biotechnology with reference to mangrove environment: A review. *Journal of Genetic Engineering and Biotechnology*, 15(1):197-210.

Berg Miller, M.E., Antonopoulos, D.A., Rincon, M.T., Band, M., Bari, A., Akraiko, T., Hernandez, A., Thimmapuram, J., Henrissat, B., Coutinho, P.M., Borovok, I., Jindou, S., Lamed, R., Flint, H.J., Bayer, E.A. and White, B.A. (2009). Diversity and strain speciûcity of plant cell wall degrading enzymes revealed by the draft genome of *Ruminococcus flavifaciens* FD-1. PLoS One 4, e6650.

Bhat, T.K., Singh, B. and Sharma OP (1998). Microbial degradation of tannins-a current perspective. *Biodegradation* 9(5):343-357.

Blackburn, T. H., and Hungate, R. E. (1963). Succinic acid turnover and propionate production in the bovine rumen, *Appl. Microbial.* 11: 132-135.

Borneman, W. S., Akin, D. E., and Ljungdahl, L. G. (1989). Fermentation products and plant cell wall degrading enzymes produced by monocentric and polycentric anaerobic ruminal fungi, *Appl. Ewiron. Microbiol.*, 55(5):1066-73.

Breton, A., Bernalier, A., Bonnemoy, F., Fonty, G., Gaillard, B., and Gouet, B. (1989). Morphological and metabolic characterization of a new species of strictly anaerobic rumen fungus, Neocallimastix joy- onii, *FEMS Microbiol. Lett.*, 49(2-3):309-14.

Breton, A., Bernalier, A., Dusser, M., Fonty, G., Gaillard-Martinie, B., and Guillot, J. (1990). *Anaeromyces mucronatus* nov. gen., nov. sp., a new strictly anaerobic rumen fungus with polycentric thallus, *FEMS Microbiol. Lett.*, 70(2):177-182.

Brooker, J.D., O'Donovan, L.A., Skene, I., Clarke, K., Blackall, L. and Muslera, P. (1994). *Streptococcus caprinus* sp nov., a tannin resistant ruminal bacterium from feral goats. *Lett. Appl. Microbiol.* 18:313-318.

Bryant M.P. et al (1977). Growth of Desulfovibrio in lactate or ethanol media low in sulfate in associationwith H_2 utilizing methanogenic bacteria. *Appl. Environ. Microbiol.* 33:1162-1169.

Bryant, M. P. (1970). Normal flora - rumen bacteria, *Am. J. Clin. Nutr.*, 23(11):1440-1450

Bryant, M. P. (1974). Nutritional features and ecology of predominant anaerobic bacteria of the intestinal tract, *Am. J. Clin. Nutr.* 27:1313-1319.

Bryant, M. P., and Robinson, I. M. (1962). Some nutritional characteristics of predominant culturable rumen bacteria, *J. Bacterial* 84:605-614.

Bryant, M.P. (1959). Bacterial species of the rumen. *Bacteriological Reviews*, 23(3):125-153.

Carroll, E. J., and Hungate, R. E. (1955). Formate dissimilation and methane production in bovine rumen contents, *Arch. Biochem. Biophys.* 56:525-536.

Chaucheyras-Durand, F. et al, (1995). Dose effect of live yeasts on rumen microbial communities and fermentations during butyric latent acidosis in sheep: new type of interaction DOI: https://doi.org/10.1017/ASC200693.

Coleman, G. S. (1985). The cellulase content of 15 species of entodinomorphid protozoa, mixed bacteria, and plant debris isolated from the ovine rumen, *J. Agric. Sci.*, 104:349-360.

Cotta, M. A. (1988). Amylolytic activity of selected species of ruminal bacteria. *Appl. Environ. Microbiol.*, **54:**772-6.

Dehority, B. A. (1991). Cellulose digestion in ruminants. In: Haigler, C. H. and Weimer, P. J., (Eds) Biosynthesis and Biodegradation of Cellulose, Marcel Dekker, New York, pp 327.

Demeyer, E. I., and Van Nevel, C. J. (1975). Methanogenesis, an integrated part of carbohydrate fermentation, and its control, In: Digestion and Metabolism in the Ruminant (I. W. McDonald and A. C. I. Warner, eds.) pp. 366-382, University of New England Publishing Unit, Armidale, Australia.

Denman, S. E., and McSweeney, C. S. (2015). *The Early Impact of Genomics and Metagenomics on Ruminal Microbiology. Annual Review of Animal Biosciences,* 3(1), 447–465. doi:10.1146/annurev-animal-022114-110705

Doi, R.H. and Kosugi, A. (2004). Cellulosomes: plant-cell-wall-degrading enzyme complexes. *Nature Reviews Microbiology* 2, 541-551.

Drake H.L., Küsel K., Matthies C. (2006). Acetogenic Prokaryotes. In: The Prokaryotes, Dworkin M., Falkow S., Rosenberg E., Schleifer KH., Stackebrandt E. (eds). Springer, New York NY

Droge, J., Mc Hardy, A.C. (2012). Taxonomic binning of metagenome samples generated by next-generation sequencing technologies. *Brief. Bioinform.* 13:646-55

Ellis, J. E., Williams, A. G., and Lloyd, D. (1989). Oxygen consumption by ruminal microorganisms: protozoal and bacterial contributions, *Appl. Environ. Microbiol.*, 55(10):2583-2587

Ellis, J. L., Kebreab, E., Odongo, N. E., McBride, B. W., Okine, E. K. and France, J. (2007). "Prediction of methane production from dairy and beef cattle," *Journal of Dairy Science* 90(7):3456-3467.

el-Shazly, K. (1952). Degradation of protein in the rumen of the sheep. II. The action of rumen microorganisms on amino acids, *Biochem, J.* 51:647-653.

Forsberg, C.W., Lovelock, L.K.A., Krumholz, L., Buchanan Smith, J.G (1984). Protease activities of rumen protozoa. *Appl. Environ. Microbiol.*, 47:101-110.

Fulghum, R.S. and Moore, W.E.C. (1963). Isolation, enumeration, and characteristics of proteolytic ruminal bacteria. *J. Bact.* 85:808-815.

Goel, G., Puniya, A. K., Aguilar, C. N. and Singh, K. (2005). Interaction of gut microflora with tannins in feeds. Naturwissenschaften, 92(11):497-503.doi:10.1007/s00114-005-0040-7

Hart, E. H., Creevey, C. J., Hitch, T., & Kingston-Smith, A. H. (2018). Meta-proteomics of rumen microbiota indicates niche compartmentalisation and functional dominance in a limited number of metabolic pathways between abundant bacteria. *Scientific Reports*, 8(1). doi:10.1038/s41598-018-28827-7.

Hazlewood, G.P. and Nugent, J.H.A. (1978). Leaf fraction 1 protein as a nitrogen source for the growth of a proteolytic rumen bacterium. *J. Gen. Microbiol.* 106:369-37.

Heredia, A., Jiménez, A. and Guillén, R. (1995). Composition of plant cell walls, *Z. Lebensm.* Unters. *Forsch.* 200:24-31.

Hespell, R.B. (1988). Microbial digestion of hemicelluloses in the rumen. *Microbiological Sciences* 5(12): 362-365.

Ho, Y. W., Bauchop, T., Abdullah, N., and Jalaludin, S. (1990). *Ruminomyces elegans* gen. et sp. nov., a polycentric anaerobic rumen fungus from cattle, *Mycotaxon*, 38:397-405.

Hobson, P.N. and Stewart, C.S. (1997). The Rumen Microbial Ecosystem, Blackie Academic & Professional, London.

Hungate, R. E. (1966). The Rumen and Its Microbes, Academic Press, New York.

Hungate, R. E. (1967). Hydrogen as an intermediate in the rumen fermentation, *Arch. Mikrobiol,* 59:158-165.

Hungate, R. E., Bryant, M. P. and Mah, R. A. (1964). The rumen bacteria and protozoa. *Annual Review of Microbiology*. Vol. 18:pp.131-166.

Janis, C. (1976). The evolutionary strategy of the equidae and the origins of rumen and cecal digestion. *Evolution* 30:757-776.

Joblin, K.N. (1999). Ruminal acetogens and their potential to lower ruminant methane emissions.

Johnson K. A. and Johnson D. E. (1995). Methane emissions from cattle, *Journal of Animal Science*, 73(8):2483-2492.

Johnson, D.E. and Ward, G.M. (1996). Estimate of animal methane emissions. *Environmental Monitoring and Assessment*, 42(1-2):133:141.

Kamra, D.N. (2005). Rumen microbial ecosystem. *Current Science,* 89(1)124-135.

Kelly, W.J., Leahy, S.C., Altermann, E., Yeoman, C.J., Dunne, J.C., Kong, Z., Pacheco, D.M., Li, D., Noel, S.J., Moon, C.D., Cookson, A.L. and Attwood, G.T. (2010). The glycobiome of the rumen bacterium *Butyrivibrio proteoclasticus* B316 Thigh lights adaptation to a polysaccharide-rich environment. *PLoS One* 5, e11942.

Leahy, S.C., Kelly, W.J., Altermann, E., Ronimus, R.S., Yeoman, C.J., Pacheco, D.M., Li, D., Kong, Z., McTavish, S., Sang, C., Lambie, S.C., Janssen, P.H., Dey, D. and Attwood, G.T. (2010). The genome sequence of the rumen methanogen *Methanobrevibacter ruminantium* reveals new possibilities for controlling ruminant methane emissions. *PLoS One* 5, e8926.

Leedle, J.A.Z. and Hespell, R.B. (1983). Brief incubations of mixed ruminal bacteria: effects of anaerobiosis and sources of nitrogen and carbon. *J. Dairy Sci.* 66:1003-1014.

Leng, R.A. and Nolan, J.V. (1984). Nitrogen metabolism in the rumen. *J Dairy Sci.* 67:1072-1089.

Lockwood, B.C., Coombsand, G.H. and Williams, A.G. (1988). Proteinase Activity in Rumen Ciliate Protozoa. *Journal of General Microbiology* 134:2605-2614 .

Luo, H.-W., Zhang, H., Suzuki, T., Hattori, S. and Kamagata, Y. (2002). Differential expression of methanogenesis genes of *Methanothermobacter thermoautotrophicus* (formerly *Methanobacterium thermoautotrophicum*) in pure culture and in cocultures with fatty acid-oxidizing syntrophs. *Applied and Environmental Microbiology*. 68:1173-1179.

Marchessault, R. H. and Sundararajan, P. R. (1983). Cellulose In: The Polysaccharides, Vol. 2, (Ed. Aspinall, G. O.), Academic Press, New York, pp:11.

Marinsek-Logar, R. (1994). Arabinofuranosidases – The debranching enzymes in xylan degradation, Zb. Biotehn. Fak. Univ. Ljubl., Kmet. Zooteh. 64:235–242 (in Slovene).

Marinšek-Logar, R. (1999). Characterization of the xylanolytic enzyme system of rumen bacterium *Prevotella bryantii* B14, PhD Thesis, University of Ljubljana, Dom'ale, Slovenia (in Slovene).

Marounek, M., and Duskova, D. (1999). Metabolism of pectin in rumen bacteria *Butyrivibrio fibrisolvens* and *Prevotella ruminicola*. *Letters in Applied Microbiology*, 29(6):429-433. doi:10.1046/j.1472-765x.1999.00671.x

Marounek, M., Bartos, S. and Brezina. P. (1985). Factors influencing the production of volatile fatty acids from hemicellulose, pectin and starch by mixed culture of rumen microorganisms. Zeitschrift für Tierphysiologie, *Tierernährung und Futtermittelkunde* 53:50-58.

McAllister, T.A., Bae, H. D., Jones, G.A. and Cheng, K.J. (1994). Microbial Attachment and Feed Digestion in the Rumen. *J. him. Sci.* 72:3004-3018.

Montgomery, L., Flesher, B. A., and Stahl, D. (1988). Transfer of Bacteroides succinogenes (Hungate) to Fihrobacter gen. nov. as *Fibrobacter succinogenes* comb. nov. and description of *Fibrobacter intestinalis*, sp. nov., *Int. J. Syst. Bacteriol.*, 38:430.

Morgavi, D. P., Kelly, W. J., Janssen, P. H., & Attwood, G. T. (2012). Rumen microbial (meta)genomics and its application to ruminant production. *Animal*, 7(s1), 184-201.

Morris, E. J. and Cole, O. J. (1987). Relationship between cellulolytic activity and adhesion to cellulose in *R. albus*, *J. Gen. Microbiol.*, 133:1023.

Morvan, B., Bonnemoy, F., Fonty, G. et al. (1996). Quantitative Determination of H_2-Utilizing Acetogenic and Sulfate-Reducing Bacteria and Methanogenic Archaea from Digestive Tract of Different Mammals. *Curr. Microbiol.* 32(3):129-133. https://doi.org/10.1007/s002849900023

Odenyo, A.A. and Osuji, P.O. (1998). Tannin-tolerant ruminal bacteria from East African ruminants. *Can. J. Microbiol.* 44:905-909.

Orpin, C. G. (1975). Studies on the rumen flagellate Neocallimastix frontalis, *J. Gen. Microbiol.* 91(2):249-62.

Patil K.R., Haider P., Pope P.B., Turnbaugh P.J., Morrison M.,et al.(2011).Taxonomic metagenome sequence assignment with structured output models. *Nat. Methods* 8:191–92

Prins, R.A., Van Rheenen, D.L., Van't Dlooster, A.T. (1983). Characterisation of microbial proteolytic enzymes in the rumen. *Ant. Van. Leeuw.* 49:585-595.

Reeve, J.N., Nolling, J., Morgan, R.M. and Smith, D.R. (1997). Methanogenesis: genes, genomes and who's on first? *Journal of Bacteriology*. 179, 5975-5986.

Russell J. B. (1998).The importance of pH in the regulation of ruminal acetate to propionate ratio and methane production in vitro. *Journal of Dairy Science* 81, 3222-3230.

Russell, J.B. and Robinson, P.H. (1984). Compositions and Characteristics of Strains of *Streptococcusbovis*.J. Dairy Sci., 67:1525-1531. https://doi.org/10.3168/jds.S0022-0302(84)81471-X

Scott, T. W., Ward, P. F. V., and Dawson, R. M. C. (1964). The fermentation and metabolism of phenyl-substituted fatty acids in the ruminant, *Biochem.* I. 90:12-24.

Sharp, R., Ziemer, C. J., Stern, M. D. and Stahl, D. A. (1998). Taxon speciûc associations between protozoal and methanogen populations in the rumen and a model rumen system. *FEMS Microbiology Ecology*, 26(1):71-78.

Silley, P. (1985). A note on the pectinolytic enzymes of Lachnospira rnultiparus. *Journal of Applied Bacteriology* 58:145-149.

Skene, I.K. and Brooker, J.D. (1995). Characterization for tannin acyl hydrolase activity in ruminal bacterium, *Selenomonas ruminantium*. *Anaerobe* 1:321-327.

Soliva, C. R., Meile, L., Cie′slak, A., Kreuzer, M. and Machmüller, A. (2004). Rumen simulation technique study on the interactions of dietary lauric and myristic acid supplementation in suppressing ruminal methanogenesis. *British Journal of Nutrition*, 92(4):689-700.

Stackebrandt, E. and Hippe, H. (1986). Transfer of bacteroids-amylopilus to a new genus *Ruminobacter* GEN-NOV, NOV, NOM-REV as *Ruminobacter amylophilus*. *Systematical and Applied Microbiolology*. v.8, p.204-207.

Takenaka, A., Tajima, K., Mitsumori, M. and Kajikawa, H. (2004). Fiber Digestion by Rumen Ciliate Protozoa. *Microbes and Environments*, 19(3), 203–210.

Tyson GW, Chapman J, Hugenholtz P, Allen EE, Ram RJ, et al. 2004. Community structure and metabolism through reconstruction of microbial genomes from the environment. *Nature* 428:37–43

Van Kessel, J. A. S. and Russell, J. B. (1996). The effect of pH on ruminal methanogenesis. *FEMS Microbiology Ecology,* 20(4):205-210.

Van Nevel, C.J. and Demeyer, D.I. (1996). Control of rumen methanogenesis. Environ. Monit. Assess., 42: 73-97. https://doi.org/10.1007/BF00394043

Van Soest, P.J. (1983). Nutritional Ecology of the Ruminant. 2nd Edn. Orvallis, O.R.: O & B Books.

Waite, R. and Gorrod, A.R.N. (1959). The comprehensive analysis of grasses. *J. Sci. Food and Agri.* 10:317-326.

Wallace RJ (1994). Amino acid and protein synthesis, turnover, and breakdown by rumen microorganisms. In: Protein Metabolism in Ruminants (Asplund JM, ed) CRC Press, Boca Raton, Florida, pp:71-111

Wallace, R. J. (1996). The proteolytic systems of ruminal microorganisms. Annales de Zootechnie, INRA/EDP *Sciences*, 45 (Suppl1):301-308.

Wallace, R.J. and Brammall, M.L. (1985). The Role of Different Species of Bacteria in the Hydrolysis of Protein in the Rumen. *J. Gen. Microbiol.*, 131:821-832. https://doi.org/10.1099/00221287-131-4-821.

Wallace, R.J. and Cotta, M.A. (1988). Metabolism of nitrogen containing compounds. In: The Rumen Microbial Ecosystem (Hobson PN, ed) Elsevier Applied Science, London, pp:217-250.

Weimer, P. J. (1992). Cellulose Degradation by Ruminal Microorganisms. *Critical Reviews in Biotechnology*, 12(3), 189–223.doi:10.3109/07388559209069192

Williams, A.G., Coleman, G.S. (1992). The Rumen Protozoa (Ed Brock, T.D.), Springer, New York, USA.

Wojciechowicz, M. (1971). Partial characterization of pectinolytic enzymes of Bacteroides ruminicola isolated from the rumen of a sheep. *Acta Microbiologica Polonica Series* A,3,45-56.

Wojciechowicz, M., Heinrichova, K. and Ziolecki, A. (1982). An exopectate lyase of Butyrivibrio fibrisolvens from the bovine rumen. *Journal of General Microbiology*, 128:2661-2665.

Wolin, M.J. (1979). The Rumen Fermentation: A Model for Microbial Interactions in Anaerobic Ecosystems. In: Alexander M. (Eds) Advances in Microbial Ecology. *Advances in Microbial Ecology*, vol 3. Springer, Boston, MA

Wood, W.A. (1971). Assay of enzymes representative of metabolic pathways. In: Methods in Microbiology, Vol. 6A (Eds.: Norris, J.R. and Ribbons, D.W.) pp. 411-424. London: Academic Press.

Wright, A. D. G., Kennedy, P., O'Neill C. J. et al. (2004). Reducing methane emissions in sheep by immunization against rumen methanogens, *Vaccine*, 22(29-30):3976-3985.

Wright, D. E. (1967). Metabolism of peptides by rumen microorganisms. *Appl. Microbiol.*, 15:547-550.

Zhou F, Olman V, Xu Y. 2008. Barcodes for genomes and applications. *BMC Bioinform*. 9:546

Zorec, M., Vodovnik, M. and Marinšek-Logar, R. (2014). "Potential of Selected Rumen Bacteria for Cellulose and Hemicellulose Degradation", *Food Technology and Biotechnology*, 52(2):210-221.

9

Biohydrogen (Synthesis from Bioprocesses) with Impact of Clean Fuel Development

[a]Sruthy Vineed Nedungadi, [b]Leena Kamlaskar-Kulkarni
[a]Rajesh K. Srivastava*

[a]Department of Biotechnology, GITAM Institute of Technology (GIT), (GITAM Deemed to be University), Gandhi Nagar, Rushikonda, Visakhapatnam-530045 Andhra Pradesh, India
[b]MACS - Agharkar Research Institute, G G Agarkar Road, Pune 411004 Maharashtra, India

Abstract

Hydrogen synthesis from bioprocesses a good approach for generation clean of fuel for world's development. Throughout the ancient and current periods, most of the developmental activities are completed by consumption of non-renewable fossil fuel energy that which has resulted in multiple environmental issues or challenges (Climate change and pollution by toxic gases or byproducts formation) at global level. Biohydrogen synthesis from anaerobic bacterial fermentation could be a good effort for sustainable energy production without any toxic byproduct formation. For production of hydrogen, lignocellulosic material can be utilized as cheap raw substrates. This would help the farmers in providing surplus benefits to farmer for their agriculture residues that normally burnt and cause the serious air pollution. Utilization of agro-waste residues can help in development of sustainable biofuels for world energy need. Burning of agrowaste residues by farmer's can be discouraged via utilization as raw carbon substrates thereby reducing air pollution.

Keywords: Hydrogen; Microbial process; Bacteria; Synthesis; Clean; Energy; Pretreatment; Agroresidues

*Corresponding Author:rajeshksrivastava73@yahoo.co.in

Introduction

Hydrogen is a promising alternative fuel which is clean renewable contains high energy content and does not contribute to greenhouse effect. Hydrogen gas can be produced by many conventional methods such as thermo-chemical gasification, pyrolysis, and solar gasification or by biological methods (bio-hydrogen) (Chookaew et al. 2012). Biological methods are bio-photolysis, dark fermentation or photo fermentation of organic materials. Biohydrogen production has several advantages compared to conventional methods because of low energy requirement and less investment cost. *Clostridium*, *Enterobacter*, *Escherichia coli* are the various anaerobes involved in hydrogen production (Abdeshahian et al. 2014). Inoculum, substrate, pH, temperature, hydraulic retention time (HRT), hydrogen partial pressure (HPP) and various reactor designs used for production. affects hydrogen yield. Hydrogen production yield can be enhanced by pretreatment methods such as heat, alkali, acid, ultrasonic or chemical pretreatment of bacterial inoculums (Bej et al. 2008). Biohydrogen yield can also be increased by over-expression of *hydA* gene or genetically modified bacteria such as *E. coli, Clostridium* sp. and *Enterobacter* sp. Hydrogen is considered an alternate fuel under Energy Policy Act of 1992, United Nations. Energy yield of hydrogen per unit weight is 122 KJ/g which is 2.75 times higher than any conventional fossil fuel (Ainala et al. 2014). During combustion, it produces only water-vapour as exhaust (carbon free fuel). The various sources of Hydrogen production and consumption have been depicted in Figure 1. Hydrogen is used as a raw material in chemical, textile, fiber, glass, electronics, metallurgy industry etc. It is also used as fuel for cryogenic rocket engines and fuel cells. Fuel cells produce electric output and have applications in on-board electric vehicles and in electrifying remote areas cut off from power grid (Chookaew et al. 2012). However, hydrogen has very low volumetric density, which poses specific storage issues. Hydrogen production and storage are costly and energy intensive processes, thereby hindering use of hydrogen as a fuel in large scale commercial applications (Fang et al. 2006).

According to Maximize Market Research Pvt. Ltd, the Global Hydrogen Generation Market was valued at $115.3 billion in 2017 and projected to reach $174.20 billion by 2025, growing at a CAGR (Compound Annual Growth Rate) of 6.0% from 2018 to 2025. The Indian Hydrogen market was valued at $50 million in 2017 and expected to reach $81 million by 2025 at a CAGR of 6.3%. The Banaras Hindu University (BHU), Murrugappa Chettiar Research Centre (MCRC), Chennai and IIT Kharagpur are the leading research groups working on biological, biomass and other renewable energy routes to produce hydrogen (Maximize market Research 2019). With Research and Development support from Ministry of Non-conventional Energy Sources (MNES), the MCRC has demonstrated biological hydrogen production in batch scale from sugar and distillery waste using the effluents at M/s. E.I.D. Parry Ltd., at Nellikuppam, Tamilnadu. The pilot plant is able to produce upto 18,000 liters of hydrogen per hour with about 60% hydrogen mixed largely with CO_2 and CO (Mishra and Das 2014).

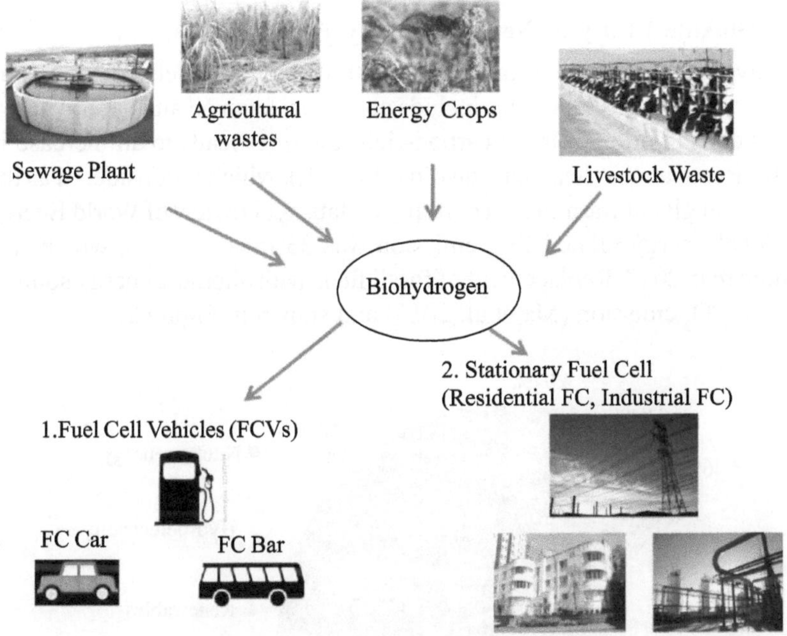

Fig. 1: Biohydrogen Production and Uses

The BHU, IIT Chennai and the National Physical Laboratory are working on the hydrogen storage methods. The BHU has developed various types of metal hydrides with storage capacities of up to 2.4 weight %. The BHU has modified a commercially available motorcycle (100cc, four strokes) and a three- wheeler (175cc, four strokes) to operate on hydrogen as a fuel. IIT Delhi and M/s Mahindra & Mahindra jointly developed hydrogen- fueled three-wheeler. Indian Oil Co-operation is carrying out a pilot project to blend hydrogen into CNG for use in 50 buses in Delhi. They started an H-CNG Dispensing Station at Faridabad in Haryana and Dwarka in New Delhi. By 2020, National Hydrogen Roadmap has projected that one million hydrogen fuelled vehicles would be on the Indian Roads and 1000 MW aggregated hydrogen based power generating capacity be set up in the country (Statistical Reviews of World Energy 2018).

The efforts are being taken by the Agharkar Research Institute, Pune, to develop a process for biohydrogen production by dark fermentation using *Clostridium* sp using wastewaters (Kamalaskar et al. 2010). Not only this but effort full experimentation to achieve improvement in yield by identifying unique genes for biohydrogen production by *C. biohydrogenum* sp. nov.have also been carried out (Kamlaskar et al. 2010;)

Carbon Dioxide Utility in Neutral Energy Production

Our energy requirements are provided almost fully by fossil fuels. These carbon containing traditional fuel sources include coal, petroleum oil and natural gases. The rapid consumption of these carbon-rich resources leads to an increase in concentration of atmospheric carbon dioxide (CO_2), which is considered as the major cause of global warming. According to Statistical review of World Energy 2018, global energy-related CO_2 emission was 33 giga ton (Gt), which was 1.7% more than 2017. Replacement of fossil fuels with alternate energy sources can reduce CO_2 emission (Ma et al. 2015) and shown in Figure 2.

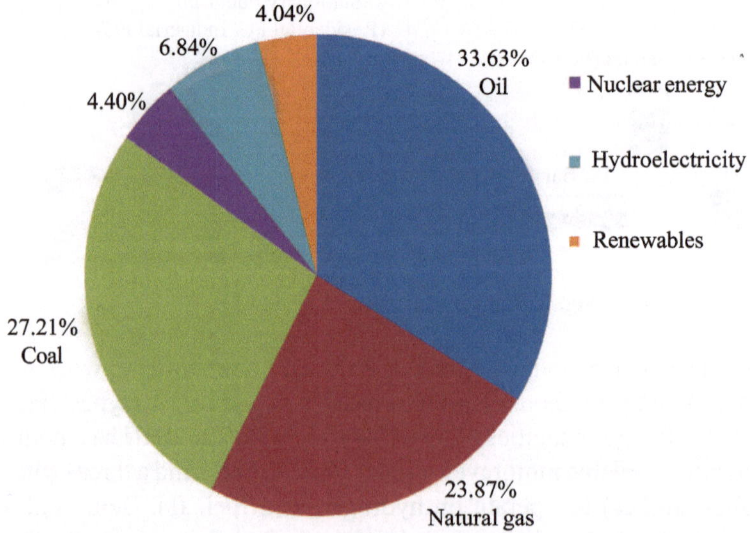

Fig. 2: Classification of world's total primary energy by fuel in year 2018 (Statistical Reviews of World Energy, 2018)

Recent Development on Biohydrogen Production in India

India is one of the few developing countries actively working on the research and development of hydrogen based transportation. Under the Ministry of New and Renewable Energy's Roadmap, various research institutes and vehicle manufactures have been conducting research on developing competitive and safe methods for production, storage and transportation of hydrogen since 2003. Tata Motors developed a hydrogen fuel cell bus. However, a complete transformation to hydrogen is difficult especially for developing countries like India, because of economic and infrastructure challenges (Mahyudin et al. 2010). The present price of hydrogen in India is estimated to be Rs 876 per kg and shown in Figure 3.

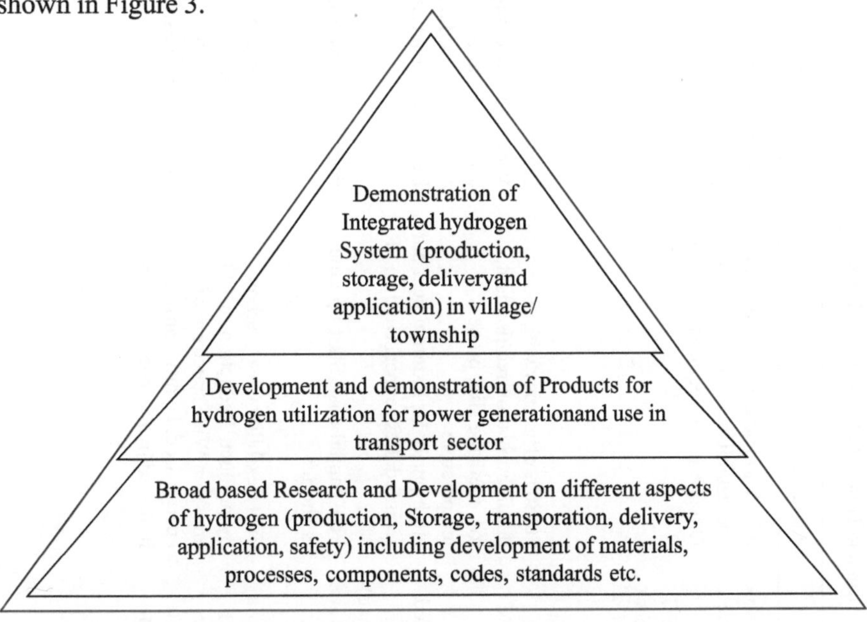

Fig. 3: National scheme of hydrogen energy roadmap (Mahyudin et al. 2010)

Hydrogen Production Methods

Hydrogen can be produced from fossil fuels by steam reforming, thermal reforming or partial oxidation of natural gas. It can be produced from biological methods such as bio-photolysis of water by algae and cyanobacteria, photo-decomposition of organic compounds by photo synthetic bacteria, fermentative hydrogen production from organic compounds and hybrid systems using photosynthetic bacteria. It can also be produced from water by electrolysis, photolysis, thermo-chemical process or thermolysis (Mangayil et al. 2012) and shown in Table 1.

Enzymatic Basis of Microbial Hydrogen Production

Hydrogenase, formate hydrogen lyase and nitrogenase are the enzymes which affects the production of hydrogen. These enzymes catalyze the following reaction:

$2H^+ + 2e^- \leftrightarrow H_2$

Table 1: Methods of Hydrogen Production (Kothari et al. 2017; Lin and Hung 2008; Kim et al 2008)

Method	Process	Advantages	Disadvantages
Steam Reforming (SR)	A mix of steam and hydrocarbons or renewable fuels is heated up to 800-1000! in a catalytic reformer to produce syngas. The gaseous mixture is then cooled down to around 400°C to enter water-shift reactor, where CO reacts with steam to produce further hydrocarbon. $CH_4 + H_2O \rightarrow 3H_2 + CO$ $CO + H_2O \rightarrow H_2 + CO_2$	65-75% efficiency with economical and established infra-structure. Oxygen is not required with lowest	Non-renewable resources
Partial Oxidation (POX)	Hydrocarbon fuel is mixed with oxygen at 1100- 1500°C with no catalyst or 600-900°C with catalyst	70-80% efficiency with decreased desulphurization requirement, low methane slip, process temperature	Expensive Very high processing temperature and low H_2/CO ratio, soot formation, process complexity, produces CO_2
Auto- thermal Reforming (ATR)	Combination of SR and POX. Steam is added in the catalytic partial oxidation to form H_2 and CO. Fuel flexibility (sulphur tolerance)	High purity H_2 External heat not required. Less expensive than SR,	Require oxygen Catalyst deactivation
Pyrolysis	Raw organic material is heated and gasified at a pressure of 0.1-0.5 MPa in the range of 500-900°C	Low capital cost, Start-up time is shorter	Poor light conversion efficiency
Photo-fermentation	Non –sulphur purple photosynthetic bacteria (Rhodobacter *sphaeroids*, *R. capsulates* etc) capture the light energy and convert organic acids into hydrogen and carbon dioxide in the absence of oxygen. $2CH_3COOH + 4H_2O \rightarrow 8H_2 + 4CO_2$.	Fuel flexibility Clean carbon by-product, Reduction in CO_x emission Wide spectrum of light energy used. Hydrogenase is inhibited by presence of oxygen	Require hydrogen impermeable, trans parent bioreactors, nitrogenase requires excess ATP, extensive treatment of ef fluent to remove inhibition necessary

(Contd.)

Dark Fermentation	Microbes like *Clostridium* species act on substrate and undergoes anaerobic fermentation in the absence of light to produce hydrogen and carbon dioxide $C_6H_{12}O_6 + 2H_2O \rightarrow 2CH_3COOH + 2CO_2 + 4H_2$	Light requirement is not mandatory No oxygen limitation Various substrate can be used as carbon source	Thermodynamically unfavorable reaction causes lower hydrogen yield
Direct bio-photolysis	The photosystem PSI and PSII of algae (*Anabaena* sp. ,*Chlamydomanas reinkardtii* etc) convert solar energy into chemical energy, which breaks water molecules into hydrogen	Direct sunlight and water are used Ten–fold conversion of solar energy than crops and trees	Inhibitory effect of hydrogen Low hydrogen productivity Requires high intensity of light
Indirect bio-photolysis	$2H_2O + \text{light} \rightarrow 2H_2 + O_2$ $12H_2O + 6CO \rightarrow C_6H_{12}O_6 + 6O_2$ $C_6H_{12}O_6 + 12H_2O \rightarrow 12H_2 + 6CO_2$ Eg Purple sulphur bacteria	Hydrogen can be produced from blue-green algae. It can fix atmospheric nitrogen	Inhibitory effect of hydrogen on activ ity of nitrogenase

Nitrogenase, Fe-hydrogenase Ni-Fe-Se hydrogenase, Fe-Fe hydrogenase and Ni-Fe hydrogenase are the classes of enzymes which carry out hydrogen production. Microorganisms with Fe hydrogenase and Ni-Fe hydrogenase function as the 'uptake' hydrogenase i.e. that hydrogenase whose normal metabolic function is to derive reductant from hydrogen (Karlsson et al. 2008). Electrons derived from hydrogen are used directly or indirectly through the quinine pool, to reduce NAD (P). The hydrogenase contain complex metallo-clusters as their active sites along with the active enzyme units which are synthesized in a complex processes further involving the auxiliary enzymes and protein maturation steps (Guan et al. 2013). Furthermore, the Ni-Fe hydrogenase are heterodimeric proteins with small (S) and large (L) subunits, where the smaller subunit contains three iron-sulphur clusters, two (4Fe-4S) and one (3Fe-4S) and the larger subunit contains a unique complex nickel iron center with co-ordination to 2CN and one CO, forming a biologically unique metallo-center. The synthesis of Ni-Fe hydrogenase is a complicated process requiring a number of accessory gene products including metal (nickel and iron) capturing, synthesis CO and CN^- as well as cluster insertion and protein maturation (Harris et al 2005). Fe-hydrogenase is reported to be more effective than Ni-Fe hydrogenase and both these hydrogenase are susceptible to deactivation by molecular oxygen. Nitrogenases are the two component protein system that uses Mg, ATP and low potential electron in a variety of substrate. It resides in the photosystems where it reduces nitrogen to ammonia along with evolution of hydrogen (Bruno et al. 2008) and is shown in Figure 4.

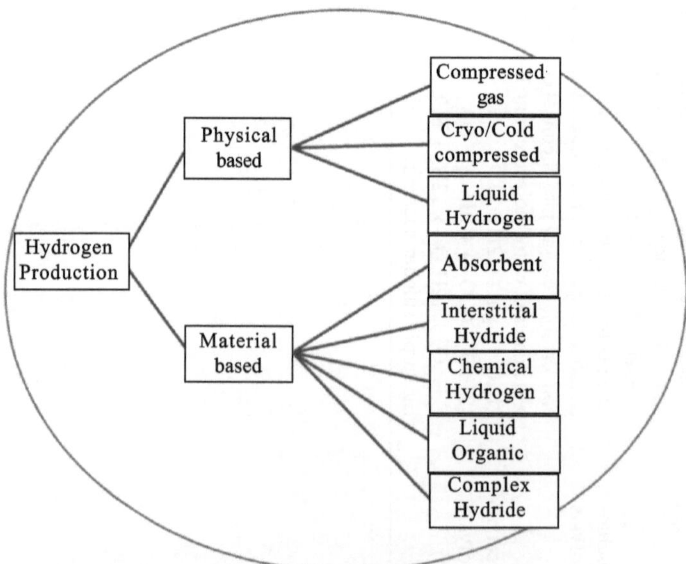

Fig. 4: Hydrogen Storage for transport purposes (mnre.gov report. 2019)

Hydrogen Storage

Gas Storage

Hydrogen is generally stored as gas in compressed form because it is very light with low density of 0.084 kg/m^3. The conventional hydrogen storage tank is significantly heavier than hydrocarbon storage tank for storing the equal amount of energy (Imamura et al. 2005). Hydrogen storage needs special attention due to embrittlement (caused by hydrogen diffusion being smallest molecule in size) of materials of construction of pressure vessels. Compressed gas storage also poses issues of high potential energy and safety hazards due to possibility of explosion of pressure vessels. Currently, hydrogen is being stored in compressed form at 350 bar (5,000 psi) in on-board in demonstration vehicles and 700 bar (10,000 psi) in Type IV carbon composite cylinders (Kabbour et al. 2006).

Liquid Storage

The cryogenic hydrogen is to be stored in specially insulated vessels at - 252.88!. The energy required to liquefy hydrogen is about 47 MJ / kg of hydrogen. The liquid hydrogen has tendency to diffuse into the material of construction at high pressures and make them brittle. Therefore, the storage vessels may be made up of FCC with special insulation, comprising double walled with vacuum in between, opacifiers and multi-layer insulations. Till now, liquid hydrogen is only used in space application (Karlsson et al. 2008).

Solid State Storage

Solid state storage of hydrogen is a much safer and efficient method than pressurized or cryogenic storage of hydrogen. The basic parameters of hydrogen storage materials as targeted by Department of Energy (DoE), USA are given in Table 2.

Table 2: Technical system targets for Onboard hydrogen storage for light-duty fuel cell vehicles (Energy gov reports-2015)

Storage Parameter	2020	2025	Ultimate
System Gravimetric	(1.5 kWh/kg)	1.8 kWh/kg	2.2 kWh/kg
Capacity Specific-energy	(0.045 kgH_2/kg)	0.055 kg H_2/kg	0.065 kg H_2/kg
from H_2 (net useful energy/max system mass)			
System Volumentric Capacity	1.0 kWH/kg	1.3 kWh/kg	1.7 kWh/kg
Energy density from H_2 (net useful energy/max system mass)	0.030 kg H_2/kg	0.040 kg H_2/kg	0.050 kg H_2/kg
System Storage Cost,	\$10/kWh net	\$9/kWh net	\$8/kWh net
(Storage system cost	\$333/kg H_2	\$300/ kg H_2	\$266/ kg H_2
System Storage Cost (Fuel Cost)	\$4/gge at pump	\$4/gge at pump	\$4/gge at pump

(1 kg $H_2 \approx$ 1 gal gasoline equivalent (gge) on energy basis, Lower heating value for H_2: 33.3 kWh/kg H_2),

Carbon Nanotubes and Other Carbon Structures

Single wall carbon nanotubes (SWNTs), multiwall nanotubes, graphitic nanofibers (GNFs), carbon aerogels (CAs) and carbon nanorods are studied for hydrogen storage (Prachi et al. 2016). Zhou et al. 2004 reported that activated carbons and activated carbon fibers can store 5 wt% of hydrogen at low temperature (77K) and high pressure (30-60 bar). Recent studies on carbon aerogels shows 5 wt% of hydrogen adsorption for surface area of 3200 m^2/g at 77K and pressure 20-30 bar (Kabbour et al. 2006).

Zeolites

Zeolites are three dimensional alumino-silicate structures made up of TO_4 tetrahedrons sharing all four corners. They are crystalline and their adsorption characteristics are considered for gas storage. Their gravimetric hydrogen storage capacity is very less. Langmi et al. 2003 had shown that NaY zeolite maximum gravimetric hydrogen capacity of 1.81wt% can be obtained at pressure and temperature of 15bar and 77K respectively. Harris et al. 2005 had shown that CaX zeolite gravimetric hydrogen capacity of 2.19 wt% was obtained at 15bar pressure and 77K temperature (Harris et al. 2005).

Metal Organic Frameworks (MOFS)

MOFs are synthetic nanoporus materials consisting of organic ligands connecting metal ions or clusters that form cage structure. The large surface area of MOFs make them best candidate for hydrogen storage. Rosewell et al. 2004 observed maximum hydrogen uptake of MOF-5 to be 1.32wt% at 1 bar and 77K (Rosewell et al. 2004).

Metal Hydrides

Metal hydrides can be defined as a concentrated single phase compound between a host metal and hydrogen. Metal hydrides have high hydrogen storage density than hydrogen gas or liquid hydrogen and they can absorb and desorb hydrogen with a small change in hydrogen pressure. Magnesium hydride is most widely studied material as it combines a high H_2 storage capacity of 7.7 wt% with the benefit of the low cost of the abundantly available magnesium with good reversibility (Zaluska et al. 2001; Imamura et al. 2005; Zhu et al. 2006).

Hydrogen Production from Anaerobes

The hydrogen production occurs in anaerobic conditions are known as fermentative hydrogen production in which bacteria degrade organic substrate by oxidation to provide metabolic energy to the cell by release of ATP molecules (Rai et al. 2012) and is shown in Figure 5.

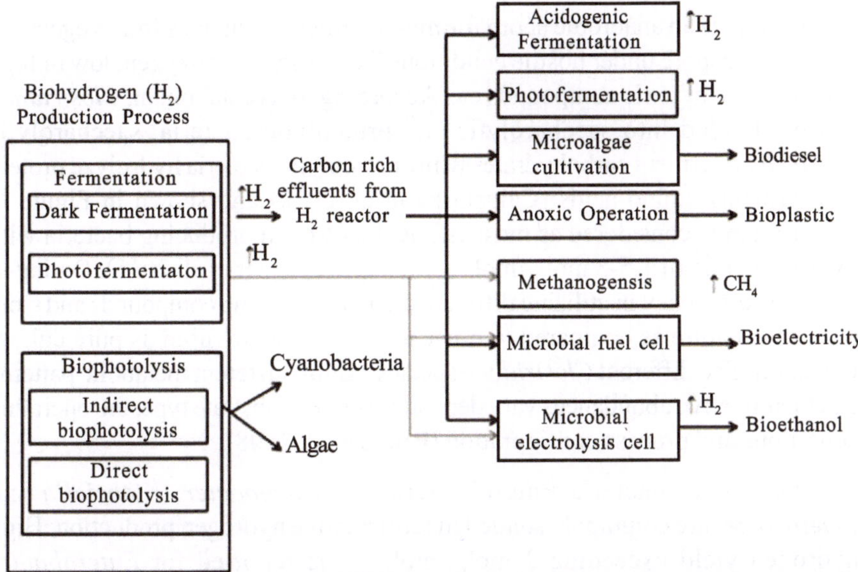

Fig. 5: Schematic representation of the primary biological routes integrated with various secondary processes for effective Hydrogen production (Chandrasekhar et al. 2015)

Obligate anaerobes such as *Clostridium*, *Ethanoligenes*, *Desulfovibrio* species and facultative anaerobes such as *Enterobacter*, *Klebsiella*, *Escherichia coli*, *and Bacillus* species are the various anaerobes involved in hydrogen production.

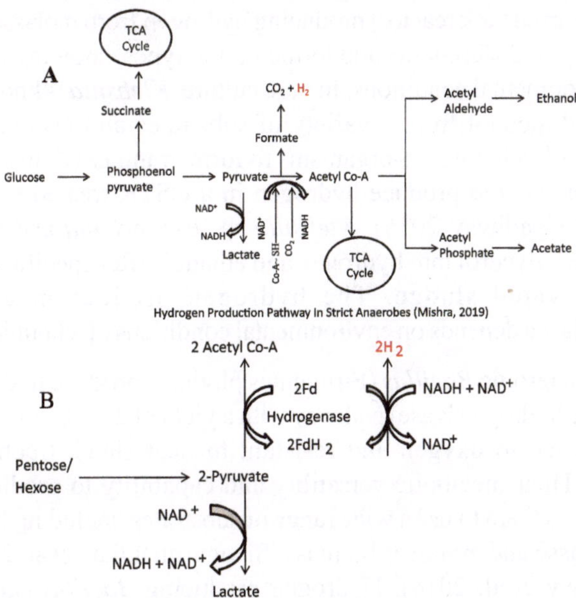

Fig. 6: Hydrogen Production Pathway (A) strict anaerobe (B) facultative anaerobe (Mishra 2019)

Clostridium is an anaerobic spore forming microbe. It changes from vegetative state to endospore under hostile conditions like presence of oxygen, low or high pH, presence of toxic compounds etc. According to its catabolism, *Clostridium* can be divided into saccharolytic and proteolytic bacteria. saccharolytic *Clostridium* ferment carbohydrates while proteolytic bacteria hydrolyze protein and ferments amino acids (Calusinska et al. 2010) and shown in Figure 6. Clostridium is considered as most efficient hydrogen producing bacteria with hydrogen yield of 1.5-3 $mol_{H2}/mol_{hexoses}$ (Chandrasekhar et al. 2015). Owing to their high efficiency in utilizing different carbon containing compounds and strict anaerobic nature as compared to other bacteria, they are used as pure culture inoculum. The different *Clostridium* species exhibit different metabolic patterns and their relative abundances vary depending on the substrate type, the operating conditions and process configuration (Bruno et al. 2008).

Amongst other bacteria Enterobacteriacea, *Citrobacter*, *Kiebsiella* and *Enterobacter* are commonly studied in fermentative hydrogen production. High hydrogen yield exceeding 2 $mol_{H2}/mol_{hexoses}$ is reported for *Enterobacter aerogenes* which exhibit the highest hydrogen production within the Enterobacteriacea family compared to *Citrobacter amalonaticus* and *E. coli* and also when compared to bacterial isolates from diverse environmental sources (Chenlin and Fang 2007).

Citrobacter spp. was identified as one of the hydrogen producers in CSTR (continuous stirred tank reactor) producing hydrogen from molasses wastewater, independently of different pH and fermentation types, showing its tolerance to various environmental conditions. In pure culture, *Klebsilla* is known to produce hydrogen and alcohol from a variety of substrates and also has capacity to agglutinate with other micro-organisms to form granules (Zaluska et al. 2001). *Klebsiella pneumonia* produce hydrogen in a CSTR treated sugarcane juice (Parihar and Upadhyay 2015). *Klebsilla* sp, *Escherichia* and *Shigella* were able to convert glycerol into hydrogen and ethanol, after specific enrichment of aerobic activated sludge. The hydrogen production efficiency of Enterobacteriacea depends on environmental conditions (Trchounian et al. 2014).

Facultative anaerobic *Bacillus* (Firmicutes phylum) produce hydrogen through FHL (formate hydrogen lyase) pathway with a yield of 2 $mol_{H2}/mol_{hexoses}$. *Bacillus* spp. is sensitive to oxygen and resistant to heat shock treatment through sporulation. Their metabolic versatility and capability to produce hydrolytic enzymes allows them to use a wide range of substrates, including lignocellulosic biomass bagasse and molasses biomass (Sivagurunanthan et al. 2014), proteins etc (Chookaew et al. 2014). Hydrogen producing *Bacillus* can be found in different environment such as sewage sludge aerobic digester, corn-stalk biohydrogen reactor (Sivagurunanthan et al. 2014), marine intertidal sludge or cattle dung. Facultative anaerobic Gamma Proteobacteria such as *Shewanella oneidensis* and *Pseudomonas stutzeri* (Sanchez-Torres et al. 2013) were characterized for the ability to produce hydrogen at high yields in pure culture and shown in Table 3.

Factors Affecting Hydrogen Production

Culture of Microbes

A number of pure cultures of bacteria have been used to produce hydrogen from various substrates. *Clostridium* and *Enterobacter* were most widely used as inoculum for hydrogen production. *Clostridium* species are Gram stain-positive, rod-shaped, strict anaerobes and endospore formers, whereas *Enterobacter* are Gram stain-negative, rod-shaped, and facultative anaerobes. Most of the studies using pure cultures of bacteria were conducted in batch mode and used glucose as substrate (Prahukhot et al. 2016).

The bacteria capable of producing hydrogen widely exist in natural environments such as soil, wastewater sludge, compost. Thus these materials can be used as inoculum for hydrogen production. At present, the mixed cultures of bacteria from anaerobic sludge, municipal sewage sludge, compost and soil have been widely used as inoculum. Mixed cultures are more preferred than pure cultures, because the former are simpler to operate and easier to control, and may have a broader source of feedstock (Pandu and Joseph 2012). But the hydrogen produced by hydrogen producing bacteria in mixed culture may be consumed by hydrogen consuming bacteria. In addition, when mixed cultures are treated under harsh conditions, hydrogen-producing bacteria would have a better chance than some hydrogen-consuming bacteria to survive (Wu and Lin 2004). Thus, in order to harness hydrogen from a fermentative hydrogen production process, the mixed cultures can be pretreated by certain methods to suppress as much hydrogen-consuming bacterial activity as possible while still preserving the activity of the hydrogen-producing bacteria. The pretreatment methods reported for enriching hydrogen-producing bacteria from mixed cultures includes heat-shock, acid, base, aeration, freezing and thawing, chloroform, sodium 2-bromoethanesulfonate or 2-bromoethanesulfonic acid and iodopropan (Singh et al. 2013).

Table 3: Hydrogen Yield by Different Strict Anaerobes and Facultative Anaerobes

Inoculum	Substrate	Maximum Hydrogen yield	Reference
Clostridium acetobutylicum	Glucose	2.0 mol/mol glucose	Chenlin and Fang, 2007
Clostridium acetobutylicum ATCC824	Glucose	1.08 mol/mol glucose	Zhang et al. 2007
Clostridium butyrium CGS5	Xylose	0.73 mol/mol xylose	Leena et al. 2016
Clostridium butyrium CGS2	Starch	9.95 mmol/g COD	Chen et al. 2007
Clostridium pasteurianum CH_4	Sucrose	2.07 mol/mol hexose	Lin and Hung 2008
Clostridium paraputrifcum M-21	Chitinuous waste	2.2 mol/mol substrate	Fang et al. 2006
Clostridium thermocellum 27405	Cellulosic biomass	2.3 mol/mol glucose	Leena et al. 2016
Clostridium thermolacticum	Lactose	3.0 mol/mol lactose	Calusinska et al. 2010
Clostridium sp. strain no. 2	Cellulose	0.3 mol/mol glucose	Trchounian et al. 2014
Clostridium sp. Fan P2	Glucose	0.05 mol/L medium	Pandu and Joseph 2012
Clostridium sp YN1	Glucose	1.7 mol/ mol glucose	Abdeshahian et al. 2014
Clostridium sp	Glucose	3.31 mol/kg COD_R	Goud et al. 2014
Clostridium butyricum CWBI1009	Glucose	2.4 mol/mol glucose	Hiligsmann et al. 2014
Clostridium perfringens ATCC13124	Glucose	490.6 ml/g glucose	Wang et al. 2014
Clostridium butyricum EB6	Palm Oil Mill Effluent (POME)	5.35 L/L-Pome	Singh et al. 2013
Clostridium butyricum	*Scenedesmus* Obliquus biomass	113.1 ml/ g VS_{alg}	Batista et al. 2014
Clostridium sp	Glycerol	1.1±0.1 mol/mol glycerol	Mendes et al. 2005
Citrobacter freundii H3	Glycerol	3545ml/glycerol	Maru et al. 2013
Enterobacter aurogenes HO-39	Glucose	1.0 mol/mol glucose	Zaluska et al. 2001
*Enterobacter aurogenes*NBRC 13534	Glucose	0.05 mol/L medium	Okamoto et al. 2000
Enterobacter aurogenes HU-101	Glycerol	0.6 mol/mol glycerol	Nath and Das 2011
Enterobacter aurogenes	Starch	1.09 mol/mol starch	Goud et al. 2014
Enterobacter aurogenes E82005	Molasses	3.5 mol/mol sugar	Trchounian et al. 2014
Enterobacter cloacae IIT-BT-08	Sucrose	6 mol/mol sucrose	Langmi et al. 2003
Enterobacter cloacae IIT-BT-08	Cellobiose	5.4 mol/mol cellobiose	Langmi et al. 2003
Enterobacter cloacae IIT-BT-08	Distillery effulent	7.38±0.24 mol/kg COD_R	Mishra et al. 2014
Enterobacter sp	Glycerol	3506 mL/L glycerol	Maru et al. 2013

(Contd.)

Enterobacter aerogenes	*Scenedesmus* Obliquus biomass	57.6 mL/ g VS_{alg}	Batista et al. 2014
Enterobacter aerogenes MTCC 2822	Cheese whey	2.04 mol/mol lactose	Rai et al. 2012
Escherichia coli MC13-4	Glucose	1.2 mol/mol glucose	Calusinska 2010
Escherichia coli	Glucose	2.0 mol/mol glucose	Bisaillon et al. 2006
Escherichia coli	Glucose	2.0 mol/mol glucose	Bisaillon et al. 2008
Escherichia coli XL-1 Blue	Fructose	1.17 mol/mol fructose	Sivagurunanthan et al 2014
Escherichia coli XL-1 Blue	Glucose	0.96 mol/mol glucose	Sivagurunanthan et al 2014
Escherichia coli XL-1 Blue	Xylose	0.69 mol/mol xylose	Sivagurunanthan et al. 2014
Klebsiella pneumonia TR17	Glycerol	0.25 mol/mol glycerol	Chookaew et. al. 2012
Klebsiella sp HE1	Glycerol	0.345 mol/mol glycerol	Wu et al 2011
Pseudomonas sp. GZI	Waste sludge	0.07 mol/g TCOD	Guan et al. 2013
Thermoanaerobacterium thermosaccharolyticum KU001	Glucose	2.4 mol/mol glucose	Ueno et al. 1995
Thermotoga elfii	Glucose	84.9 m mol/L medium	Nath and Das 2011
Hydrogen producing bacteria B49	Glucose	0.1 ml/L culture	Wang et al. 2014
Ruminococcus albus	Glucose	2.52 mol/mol glucose	Okamoto et al. 2000
Citrobacter amalonaticus Y19	Glucose	8.7 mol/mol glucose	Wang et al. 2003
Ethanoligenes harbinese YUAN-3	Glucose	1.93 mol/mol glucose	Yanga et al. 2006

Substrate for Hydrogen

Biological hydrogen can be produced from various substrates like glucose, sucrose (Mizuno et al. 2000; Van Ginkel et al. 2001; Duangmanee et al. 2002; Khanal et al. 2004), food and food processing wastes (Shin et al 2004; Van Ginkel et al 2005; Wu and Lin 2004), cellulose containing waste (Okamoto et al. 2000), municipal solid waste, activated sludge (Wang et al. 2003), agricultural waste such as wheat straw, corn straw (Fang et al. 2006), sugarcane molasses (Li and Chen 2007).

Carbohydrate rich substrates are the primary choice hydrogen production. Waste from food processing industry such as potato processing industry, rice slurry, sugar and distillery industry and used in anaerobic fermentation (Karlsson et al. 2008; Mishra et al. 2014; Parihar and Upadhyay 2015). During hydrolysis of carbohydrates rich substrates, hydrolytic bacteria produce simple sugars such as sucrose, glucose, xylose and hexose (Aceves) and these sugars are consumed by anaerobic bacteria to produce biohydrogen (Van Niel et al. 2002).

Food waste, food processing waste, oils and dairy products are also used for hydrogen production. Lipase enzyme hydrolysis lipid which result in generation of fatty acids and glycerol that can be hydrolyzed to acetyl- CoA, acetate and hydrogen from NADH oxidation during β-oxidation pathway (Karadag et al. 2015; Mendes et al. 2005) and shown in Figure 7.

Food waste, food processing waste from cheese whey, casein, fish, meat, egg etc are the protein substrate for biohydrogen production. Bacteria convert protein into polypeptides and amino acids by protease, which is further broken down to volatile fatty acid, carbon-dioxide and hydrogen (Wu et al. 2007). Plants and agricultural biomass are good sources of cellulose, hemicellulose and lignocelluloses materials. They contain different sugars which can be converted to hydrogen (Xiao et al. 2013).

It compared hydrogen production from carbohydrate-rich, protein-rich and fat-rich organic solid waste. It was observed that hydrogen producing microbes could evolve much more hydrogen from carbohydrate rich organic waste (Xiao et al. 2010) and is shown in table 4.

Enzymes

In most of the facultative anaerobic bacteria, hydrogen production is catalyzed by the enzyme formate hydrogen lyase (FHL). In anaerobic condition under acidic environment, formic acid is converted into hydrogen. This reaction is catalyzed by the formate hydrogen lyase. The FHL consists of formate dehydrogenase, hydrogenase and electron transfer carrier, responsible for formic acid oxidation to CO_2 and H_2. During the reaction, formic acid acts as an electron

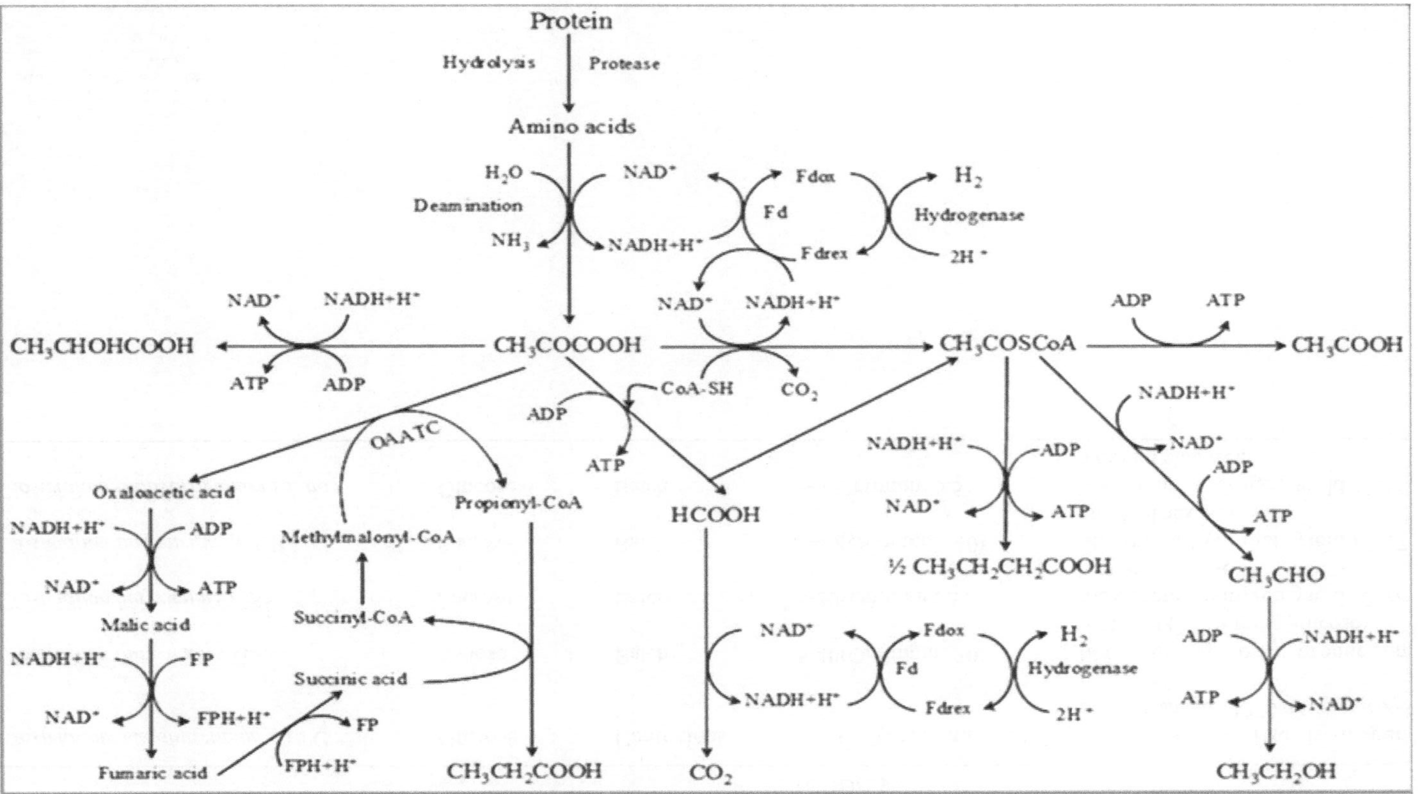

Fig.7: Proposed metabolic pathway for anaerobic bio-hydrogen production from protein (Wang and Wan 2008)

Table 4. The comparison of various substrates for hydrogen production by *Clostridium* (Wang et al. 2014; Kamlaskar et al. 2010)

Inoculum	Substrates	Reactor- type	Substrate Concentration g (COD/L)	Optimal Index Value
Clostridium acetobutylicum ATCC 824	Glucose	Continuous	1.1-11.2(optimum:11.2)	Maximum specific hydrogen production rate (1270 mL/g glucose L-reactor)
Clostridium butyricum CGS5	Xylose	Batch	5-40(Optimum: 20)	Maximum hydrogen production potential (172.9 ml/ g glucose)
Clostridium butyricum CGS5	Sucrose	Batch	5-30(Optimum:20)	Maximum hydrogen yield (2.78 mol/mol sucrose)
Clostridium pasteurianum CH_4	Sucrose	Batch	5-40(Optimum: 40)	Maximum hydrogen yield (2.07 mol/mol hexose)
Clostridium biohydrogenum sp. nov.	Glucose	Batch	4-8(Optimum:5.5)	Maximum hydrogen yield (3.35 mol/mol glucose),

donor and proton are the only electron acceptor and thus leads to the formation of H_2 (Xiao et al. 2013).

Hydrogenase enzymes are classified into three groups based on the number and identity of the metal in their active sites such as (Ni-Fe), (Fe-Fe) and Fe-hydrogenase. On the basis of their active sites they all contain Fe and Co as a ligand to the Fe atom (Zhu and Béland 2006). Among all these, Fe-Fe hydrogenase is known to be the most potent in terms of fermentative hydrogen production. These enzymes are monomeric as in *Clostridium* or multimeric as in *Thermotogamaritima* and *Thermoanaerobactertengcon-genesis* that consist of three and four subunits, respectively. The (Fe-Fe) hydrogenases are organized into modular domains (Yanga et al. 2006). The accessory cluster known as the F cluster, functions inter and intra molecular electron transfer centers. The accessory cluster is linked electronically to the catalytic cluster known as the H cluster. The (Fe-Fe) hydrogenase have the activity about 10-100 times higher than others hydrogenase (Zhou et al. 2004). Genomic analysis shows that there is a great deal of varieties in (Fe-Fe) hydrogenase with some *Clostridia* containing a large number of hydrogenase with different modular structure. *Clostridium biohydrogenum* sp. is reported to have novel hydrogenases with such high efficiency producing continuous biohydrogen from glucose and organic wastewaters (Kamalaskar et al. 2010).

Temperature

Anaerobic bacteria are more sensitive to temperature regime. The anaerobic fermentation process lies among four temperature ranges, ambient (15-30!), mesophilic (30-39!), thermophilic (50-64!) and hyperthermophilic (>64!) (Parihar and Upadhyay 2015). Change in temperature ranges affects the consumption of substrate, biohydrogen yield, formation of metabolites in the form of volatile fatty acids and presence of microbes in the system. Most of the hydrogen production studies at the laboratory scale were performed using mesophilic microorganisms because of the ease of operation and maximum specific growth rates (Lic et al. 2007). Recently, many studies are done on thermophilies. Lower viscosity, better mixing, less risk of contamination, higher reaction rates and no need of cooling bioreactor are the major advantages of thermophilies. The yield of hydrogen using thermophilies could reach maximum theoretical value of 4 mol H_2/ mol glucose, which is much higher than mesophiles that reached the maximum yield of less than 2 mol H_2/ mol glucose (Maru et al. 2013).

pH

pH influences hydrogenase activity and metabolism pathway of hydrogen producing bacteria. Changes in external pH value affect several physiological parameters such as the internal pH, proton motive force and membrane potential.

In low pH, pyruvate is converted to volatile fatty acids which concomitant with hydrogen production, whereas a neutral pH leads to methane production by methanogenic microorganism (Nath and Das 2011). In an alkaline environment, fermentative pathways are prone to solventogenesis. The H^+ that shuttles among metabolic intermediates causes the formation of reduced compounds such as aldehydes, alcohols and altered membrane potentials, which result in decreased cellular growth rates (Ozgur 2010; Kapdan and Kargi 2006). The accumulation of volatile fatty acids during the fermentation process leads to a decrease in the system pH, which decrease the system buffering capacity and ultimately inhibits hydrogen production. Ma et al (2015) reported that the optimal range of pH for the hydrogen production is 5.5-6.5 and at this pH maximum gas production and least solvent production occurs (Ma et al. 2015) and is shown in table 5.

Table 5: The effect pH on hydrogen production (Zhang et al. 2007; Yanga et al. 2006; Van Ginkel and Lay 2001; Trchounian et al. 2014)

Inoculum	Substrates	Optimal pH	Optimal index value
Anaerobic digester sludge	Rice slurry	4.5	Maximum hydrogen yield (346 mL/g starch)
Anaerobic sludge	Sucrose	5.5	Maximum hydrogen yield (3.7 mol/mol sucrose)
Anaerobic sludge	Glucose	5.5	Maximum hydrogen yield (2.1 mol/mol glucose)
Enterobacter cloacae IIT-BT-08	Sucrose	6.0	Maximum hydrogen production rate (29.63 mmol/g dry cell)
Clostridium biohydrogenumspnov.	Glucose	5.5	Maximum hydrogen yield (3.35mol/mol glucose)

Volatile Fatty Acids

Volatile fatty acids (VFA) are produced in the form of different solvents in hydrogen production process. Most of the fatty acids are produced by the hydrolysis process in the acidogenic phase. These acids include acetic acid, propionic acid, i-butyric acid, n-butyric acid and lactic acid. In anaerobic treatment process, the drop in pH occurs due to either accumulation of VFA or excessive generation of carbon dioxide or both. The theoretical H_2 yield is expected to be 4 moles of H_2 per mole of glucose via acetate pathway and 2 moles of H_2 per mole of glucose via butyrate pathway (Van Ginkel et al. 2005).

The VFA generation in hydrogen production is also affected by temperature, as at higher temperature (45!), acetate and butyrate concentration is higher (26-30%) than the mesophilic temperature (30-35!) where the concentration of acetate, propionate and butyrate is comparatively low (20-25%) (Van Ginkel et al. 2002).

Alcohol Production

Production of solvents has been reported during the production of hydrogen as the pH is below than optimal owing to very high volatile fatty acid production. Higher ethanol concentration also reduces hydrogen production. Ethanol production consumes electron and favours propionate formation by directly utilizing hydrogen, which decreases the yield of hydrogen (Lin and Hung 2008). Clostridium is known to produce butanol along with ethanol. The yield of hydrogen is lower when butanol is produced as a byproduct along with volatile fatty acids as the surge of electrons is diverted towards the alcohol production as against the hydrogen production to maintain the eco-physiological environment within the bacterial cell (Singh et al. 2013).

Hydraulic Retention Time (HRT)

HRT is an important factor in the selection of microorganisms is required with growth rates that can withstand the mechanical dilution caused by continuous volumetric circulation. A shorter HRT would restrict the growth rate of methanogenic microorganisms. For satisfactory hydrogen yields, the optimum HRTs were between 8 and 14 hour for a wide variety of substrates. The optimum hydrogen yield, the HRT is influenced by several factors, including the type and composition of the substrate, the type of microorganisms, the organic loading rate and the system redox condition (Nath and Das 2011).

Reactor Type

Continuous Stirred Tank Reactors (CSTR) and up-flow anaerobic sludge blanket (UASB) are two mostly used reactors for biohydrogen production. In a conventional CSTR, biomass is well suspended in the mixed liquor, which has the same composition as the effluent. Since biomass has the same retention time as the HRT, washout of biomass may occur at shorter HRT. In addition, biomass concentration in the mixed liquor and the hydrogen production is limited. Immobilized-cell reactors provide an alternative to a conventional CSTR, because they are capable of maintaining higher biomass concentrations and could operate at shorter HRT without biomass washout. Biomass immobilization can be achieved through forming granules, biofilm, or gel-entrapped bioparticles. Zhang et al. found that the formation of granular sludge facilitated biomass concentration up to 32.2 g VSS/L and enhanced hydrogen production (Mu et al. 2006; Mangayil et al. 2012) and is shown in table 6.

Table 6: Different reactors available for biohydrogen production (Kothari et al. 2017)

Reactor type	Substrate used	OLR	Yield (H_2)	HRT
CSTR	Kitchen waste	20 g COD/ld	2 mmol/g(COD)	4d
	Rice winery wastewater,	34 g COD/ld	2.14mol/mol(hexose)	1d
	Corn starch	26.7 g COD/ld	0.92 mol H2/glucose	18h
	Fruit waste water	5-15 kgCOD/m^{3}d	5.4-4.2mol/kg(COD)	15-5h
UASB	Citric acid wastewater	10.0 to 75.0 kg COD/m^3 d	0.84 molH2/mol hexose	12h
	Sludge of coffee drinking wastewater	-	1.78 mol H_2 to 2.76 l H_2/l	12h
	Coffee drink manufacturing wastewater	20 g COD/ld	1.29 mol H_2/mol hexose	6h
	Cheese whey	5- 20 g COD/ld	0.38 and 0.36 l H_2/l	6h

Metal Ion

Growth of microbial cell, enzymatic activity and metabolic pathway involved in hydrogen producing bacteria is highly influenced by low concentration of trace metals. Fe and Mg is essential for hydrogenase enzyme functioning. In hydrogenase catalyzed hydrogen production process, electrons are transported through intra-molecular electron transport chain to the active site where proton is reduced and hydrogen is produced (Woodward 2000). Higher concentration of Fe was observed to form cell clumps, which limits the mass transfer. Limited iron concentration (< 10 mmol /L) for *Clostridium pasteurianum* was found to change the pattern of product during glucose fermentation and lactate. On the other side with iron concentration up to 25mmol/L, metabolism of *C. acetobutyricum* was acidogenic and hydrogen was formed as the major metabolite (Langmi et al. 2003). Trace concentration of Ni is required to activate the function of (Ni-Fe) hydrogenase, thus it is conducive for fermentation of hydrogen production (Kim et al. 2008). Wang and Wan (2008) studied Influence of Ni^{2+} concentration on biohydrogen production and found that maximum hydrogen production (296.1ml/g-glucose) at 0.1. mg/L of Ni^{2+} concentration (Wang and Wan 2008; Kothari et al. 2017).

Hydrogen Partial Pressure (HPP)

Hydrogen partial pressure (HPP) is the concentration of hydrogen gas produced within the reactor during the biohydrogen production process and in excess it inhibits the process. The hydrogen partial pressure beyond 60 Pa does not favour the gas production and leads to the synthesis of alcohol. Increase in HPP leads to the lowering of H^+/H_2 ratio and inhibits the flow of electron so that the electrons from reduced ferredoxin to molecular H_2 via the hydrogenases system also get inhibited (Kotay and Das 2009). Several studies have shown the significant relation between the reactor temperature and hydrogen partial pressure. Favorable pressure reported so far for biohydrogen production are 50kPa at 60 °C, 20kPa at 70 °C 2kPa at 98 °C (Calusinska 2010). The effect of hydrogen partial pressure on biohydrogen production can be decreased by sparging of inert gas like nitrogen in the reactor (Van Niel et al. 2002)

Strategies for Enhancement of Hydrogen Production

Pretreatment of Bacterial Inoculums

It involves pretreatment of hydrogen producing bacteria, which posses the ability to sporulate when passed through stress conditions such as pH, temperature, radiation and their resulting spore should be more resistant, so that they can survive in severe conditions (Li and Chen 2007). To remove the hydrogen consuming bacteria from the process, various pre-treatments of bacterial inoculums are employed such as heat pretreatment, alkali pretreatment, acid pretreatment, ultrasonic, chloroform, chemical treatment and combined treatment which increase hydrogen production. Kotay and Das 2009 studied various pretreatment methods on *Enterobacter cloacae* IIT-BT 08, *Citrobacter freundii* IIT-BT L139 and *Bacillus coagulans* IIT-BT S1 and observed that pretreatment reduce competitive microbial load by 98%, improved the nutrient solublization and 1.5-4 times increase in hydrogen yield (Kotay and Das 2009). It compared acid, base and heat shock pretreatment and found heat shock was the best one with a hydrogen yield of 2.00 mol-H_2/mol-glucose. Wang et al. (2003) compared ionizing irradiation, heat-shock, acid and base pretreatment and observed the seed sludge pretreated by ionizing irradiation achieved the best hydrogen yield of 2.15 mol/mol glucose with a substrate degradation rate 98.9% (Wang et al. 2003).

Co-culture

Co-cultures provide better anaerobic conditions for strictly anaerobic hydrogen producers and eliminate the need of an expensive reducing agent to remove O_2 present in the reactor. Presence of single strain of hydrogen producing bacteria, can only occupy the single feature such as strictly anaerobic bacteria cannot survive in the presence of even at slight amount of oxygen while facultative can do. However, a combination of different bacteria that can retain several hydrolytic enzymes and the co-culture of bacteria, which can exist in wide range of acidic and alkaline conditions, give the feature of enhanced hydrogen yield in co-culture system (Mahyudin et al. 2010; Maru et al. 2013) and is shown in table 7.

Table 7: Various pure strains bacterial co-cultures for biohydrogen production (Kothari et al. 2017; Kotay and Das 2009)

Cultures	T (°C)	Substrate	Hydrogen Yield (mol/mol)
C.butyricum and *E.coli*	37	Glucose	2.09
Enterobacter aerogens and *C.butyricum*	37	Starch	2.0
Enterobacter aerogens and C. butyricum	37	Sweet potato	2.7
B.thermoamylovorans and C. beijerinckii L9.	40	Brewery yeast waste	91.6 (ml H2) from a 80-ml co-culture
Thermoanaerobacteriumthermosacchrolyticum GD17 and *C. thermocellum* JN4	60	Cellulose	0.8
Clostridium butyricum-NRRL 1024 and Clostridium pasteurianum-NRRL B-598	30	Wheat starch	109 ml H_2 g TS
C. acetobutylicum x9 and *Ethaniolegenes herbinese*	37	Microcrystalline cellulose	1.32
C. thermocellum DSM1237 and *C. thermopalmarium* DSM 5974	55	Cellulose	1.36
Enterobacter aerogens W23 and *Candida matosa* HY 35	35	Glucose	2.59

Biochemical Engineering

Biohydrogen production pathway is reported via modification in enzyme hydrogenase and genetic modifications in microbes that can enhance biohydrogen production in bioprocesses.

In Pathway

In the hydrogen production, glucose is oxidized in two steps:glyceraldehydes 3-phosphate to 1,3-biphosphoglycerate and Pyruvate to acetyl-Co A. The modification at pyruvate may increase yield of Hydrogen. Three types of biochemical reactions are involved in the generation of H_2 biologically. First one is found in the family of *Enterobactereacae* where it employs two major enzymes viz. Pyruvate formate lyase (PFL) and Formate hydrogen lyase (FHL) to mediate biohydrogen production (Mendes and de Castro 2005). PFL acts upon splitting of Pyruvate into acetyl-Co A and formate in anaerobic condition whereas FHL cleaves formate to H_2 and CO_2.The second type of H_2 producing reaction involves pyruvate ferredoxin oxidoreductase (PFOR) and Fd-dependent hydrogenase (hyd-A). In third type reaction of hydrogen production, NAD (P) H is utilized by bacteria to evolve H_2. This reaction is catalyzed by two major enzymes: NAD (P)H-ferredoxin oxido-reductase (NFOR) and HydA (Mishra et al. 2019).

$$\text{Pyruvate} + \text{COA} \underset{PFL}{\rightarrow} \text{Acetyl} - \text{CoA} + \text{Formate}$$

$$\text{Formate} + \underset{FHL}{\rightarrow} H_2 + CO_2 \quad (1)$$

$$\text{Pyruvate} + \text{CoA} + 2\text{Fd}_{ox} \underset{PFOR}{\leftrightarrow} \text{Acetyl} - \text{CoA} + CO_2 + 2\text{Fd}$$

$$2\text{Fd}_{rd} + 2H^+ \underset{Hyda}{\rightarrow} 2\text{Fd}_{ox} + H_2 \quad (2)$$

$$\text{Glucose} + 2\text{NAD}^+ \underset{\text{EMP}}{\rightarrow} 2\ \text{Pyruvate} + 2\text{NADH}$$

$$2\text{NADH} + 4\text{Fd}_{ox} \underset{\text{NFOR}}{\rightarrow} + 2\text{NAD}^+\ 4\text{Fd}_{rd}$$

$$4\text{Fd}_{rd} + 4H^+ \underset{\text{HydA}}{\rightarrow} 4\text{Fd}_{ox} + 2H_2 \quad (3)$$

The process discussed in equation 1-3 gives the maximum theoretical yield of 2 or 4 mole H_2 as per the presence of facultative and strictly anaerobic bacteria. But in several extreme thermophiles 3.3- 4 mol H_2/ mol of glucose can be achieved naturally. These bacteria utilize both NFOR and PFOR for H_2 production (Mu et al. 2006).

Enzyme Hydrogenase

Many studies are going on to express (Fe-Fe) hydrogenases in *E. coli* by over expression of hydA from an organism such as *Clostridium* but it is observed that in order to co-express maturation gene hyd E, hyd F and hyd G that are required for H-cluster maturation, insertion of the organism does not possess these enzyme (Trchounian et al. 2014).

In microbes

Genetically modified bacteria such as *E. coli, Clostridium* sp. and some species of *Enterobacter* were successfully used for high yield of biohydrogen. The genetic expression in *C. acetobutylicum* to increase the H_2 production is regulated by antisense RNA. Bacterial strains of *Clostridium* are found to possess great potential for breaking cellulose in to hydrogen such as *C. cellulolyticum* and *C. populeti* (Mishra and Das 2014). The property of these strains i.e. cellulose degrading pathway can be expressed in *C. acetobutylicum* to achieve highest hydrogen yield. The heterogonous expression of pyruvate decarboxylase and alcohol dehydrogenase from *Zymomonas mobilis* can be used to increase the cellulose degrading ability of *C.*

cellulolyticum. Thus the approach of metabolic engineering enables the researchers to develop efficient strains to improve the biohydrogen production (Zulkhairi et al. 2013; Wang et al. 2003).

Conclusions and future prospectus

Hydrogen is clean energy fuel without carbon emission in environment and it is sustainable mode of energy and its development can be achieved from utilization of efficient microbial processes and sustainable nature of abundant carbon substrates (lignocellulosic matter). These substrates need efficient pretreatment processes with modified microbial strains that can express their capability for consumption of various nature of monomer sugars (D- glucose or xylose or others) in effective ways with least byproducts formation. Researchers are involved in creation of effective microbial strains (engineered strain or pathways) for consumption of various nature of raw and plentiful amount of waste residue (good sources of organic matters). We can promote hydrogen synthesis from different approaches such as thermo-chemical gasification, pyrolysis, and solar gasification or by biological methods. But biological mode of hydrogen generation can help to biofuel development in sustainable mode for world societies for their need. It can maintain the current and future securing of energy stocks for people. We need to develop the cost effective processes with validation at pilot or industrial scale of processes for commercialization of fuel sources. Now-days other approaches are used for hydrogen and microbial fuel cell or microbial electrolysis cell can help the hydrogen biogeneration via helping in wastewater treatment.

References

Abdeshahian P, Al-Shorgani NKN, Salih NKM, Shukor H, Kadier A, Hamid AA, Kalil MS. 2014. The production of biohydrogen by a novel strain *Clostridium* sp. YM1 in dark fermentation process. *International Journal Hydrogen Energy,* 39: 12524-12531

Aceves-Lara C, Latrille E, Bernet N, Buffière P, Steyer J. 2008. A pseudo-stoichiometric dynamic model of anaerobic hydrogen production from molasses. *Water Research* 42: 2539-2550

Ainala SK, Seol E, Sekar BS, Park S. 2014. Improvement of carbon monoxide-dependent hydrogen production activity in *Citrobacter amalonaticus* Y19 by over-expressing the CO sensing transcriptional activator, CoA. *International Journal Hydrogen Energy* 39: 10417-10425

Allied Market Research, 2018. India Hydrogen Market Is Expected to Reach $81 Million by 2025. https://www.alliedmarketresearch.com/press-release/india-hydrogen-market.html

Batista AP, Moura P, Marques P, Ortigueira J, Alves L, Gouveia L. 2014. *Scenedesmus obliquus* as feedstock for biohydrogen production by *Enterobacter aerogenes* and *Clostridium butyricum*. *Fuel* 117: 537–543

Bej B, Basu RK, Ash SN. 2008. Kinetic studies on acid catalyzed hydrolysis of starch. *Journal of Scientific and Industrial Reasearch* 67: 295-298

Bruno LM, Filho JL, Castro HF. 2008. Comparative performance of microbial lipases immobilized on magnetic polysiloxane polyvinyl alcohol particles. *Brazilian Archives of Biology and Technology* 51: 889-896

Calusinska M, Happe T, Joris B, Wilmotte A. 2010. The surprising diversity of clostridial hydrogenases: a comparative genomic perspective. *Microbiology* 156: 1575–1588

Chandrasekhar K, Lee YJ, Lee DW. 2015. Biohydrogen Production: Strategies to Improve Process Efficiency through Microbial Routes. *International Journal of Molecular Science 16*(4): 8266-8293

Chenlin L, Fang HHP. 2007. Fermentative hydrogen production from wastewater and solid wastes by mixed cultures. *Critical Reviews in Environmental Science and Technology* 37, 1–39

Chookaew T, Thong OS, Prasertsan P. 2012. Fermentative production of hydrogen and soluble metabolites from crude glycerol of biodiesel plant by the newly isolated thermotolerant *Klebsiella pneumoniae* TR17. *International Journal Hydrogen Energy* 37, 13314-13322.

Chookaew T, Thong OS, Prasertsan P. 2014. Biohydrogen production from crude glycerol by immobilized *Klebsiella* sp. TR17 in a UASB reactor and bacterial quantification under non-sterile conditions. *International Journal Hydrogen Energy* 39: 9580-9587

Duangmanee T, Chyi Y, Sung S. 2002. Biohydrogen production in mixed culture anaerobic fermentation. In: Proc. 14th World Hydrogen Energy Conference, Québec, Canada.

EAI, 2017. India Hydrogen Energy. EAI. Catalyzing CleanTech and Sustainability. http://www .eai.in / ref/ae/hyn/hyn.html. 2017.

Energy gov.reports. 2015. Hydrogen Storage, DOE Technical Targets for Onboard Hydrogen Storage for Light-Duty Vehicles. https://www.energy.gov/sites/prod /files/2015/05 /f22/ fcto_

Fang HHP, Li C, Zhang T. 2006. Acidophilic biohydrogen production from rice slurry. *International Journal Hydrogen Energy* 31: 683-692

Goud RK, Sarkar O, and Mohan SV. 2014. Regulation of biohydrogen production by heat-shock pretreatment facilitates selective enrichment of *Clostridium* sp, *International Journal Hydrogen Energy* 39: 7572-7586

Guan W, Fan X, Yan R. 2013. Effect of combination of ultraviolet light and hydrogen peroxide on inactivation of *Escherichia coli* O157:H7, native microbial loads, and quality of button mushrooms. *Food Control* 34, 554-559

Harris IR, Walton A, AL-Mamouri MM. Johnson SR, Book D, Speight JD, Edwards PP, Gameson I, Andreson PA, Langmi HM. 2005, Hydrogen storage in ion exchange zeolites. *Journal of Alloys and Compounds* 404-406: 637-642

Hiligsmann S, Beckers L, Masset J, Hamilton C, Thonart P. 2014. Improvement of fermentative biohydrogen production by *Clostridium butyricum* CWBI1009 in sequenced-batch, horizontal fixed bed and bio disc like anaerobic reactors with biomass retention, *International Journal Hydrogen Energy* 39: 6899-6901

Imamura H, Masanari K, Kusuhara M, Katsumoto H, Sumi T, Sakata Y. 2005. High hydrogen storage capacity of nanosized magnesium synthesized by high energy ball miling. *Journal of Alloys and Compounds* 86: 211-216

Kabbour H, Baumann TF, Satcher JH, Sauliner A, Ahn CC. 2006. Toward new candidate for hydrogen storage: high surface area carbon aerogels. *Chemistry Material* 18: 6085-6087

Kamalaskar LB, Dhakephalkar PK, Meher KK, Ranade DR. 2010. High biohydrogen yielding *Clostridium* sp. DMHC-10 isolated from sludge of distillery waste treatment plant. *Inter national Journal of Hydrogen Energy.* 35: 1063-10644.

Kapdan IK, Kargi F. 2006. Bio-hydrogen production from waste materials. Enzyme and *Microbial Technology* 38:569–582.

Karadag D, Köroðlu OE, Ozkaya B, Cakmakci M. 2015. A review on anaerobic biofilm reactors for the treatment of dairy industry wastewater. *Process Biochemistry* 50:262-271

Karlsson A, Vallin L, Ejlertsson J. 2008. Effects of temperature hydraulic retention time and hydrogen extraction rate on hydrogen production from the fermentation of food industry residues and manure, *International Journal Hydrogen Energy* 33: 953-962

Khanal SK, Chen WH, Li L, Sung S. 2004. Biological hydrogen production: effects of pH and intermediate products. *International Journal Hydrogen Energy* 29: 1123-1131

Kim EJ, Kim MS, Lee JK. 2008. Hydrogen evaluation under photoheterotrophic and dark fermentative condition by recombinant *Rhodobacter spharoides* containing the gene for fermentative pyruvate metabolism of *Rhodospirillium rubrum. International Journal Hydrogen Energy* 33: 5131-5136

Kotay SM, Das D. 2009. Novel dark fermentation involving bioaugmentation with constructed bacterial consortium for enhanced biohydrogen production from pretreated sewage sludge. *International Journal Hydrogen Energy* 34: 7489–7496

Kothari R, Kumar V, Pathak VV, Ahmad S, Aoyi O, Tyagi VV. 2017. A critical review on factors influencing fermentative hydrogen production. *Frontiers in Bioscience*, Landmark (Ed) 22: 1195-1220

Langmi HM, Walton A, Al-Mamouri MM, Johnson SR, Book D, Speight JD, Edwards PP Gameson I, Andreson PA, Harris IR. 2003. Hydrodgen adsorption in zeolites A,X,Y and RHO. *Journal of Alloys and Compounds* 356-357: 710-715

Leena K, Neelam K, Soham P, Anita PD, Ranade DR, Dhakephalkar PK. 2016. Genome sequence and gene expression studies reveal novel hydrogenases mediated hydrogen production by *Clostridium biohydrogenum* sp. nov., MCM B-509T. *International Journal of Hydrogen Energy* 41(28):11990-11999.

Li D, Chen H. 2007. Biological hydrogen production from steam exploded straw by simultaneous sachariûcation and fermentation. *International Journal Hydrogen Energy* 32: 1742-1750

Lin CY, Hung WC. 2008. Enhancement of fermentative hydrogen/ ethanol production from cellulose using mixed anaerobic cultures. *International Journal of Hydrogen Energy* 33 3660–67

Ma J, Zhao QB, Laurens LLM, Jarvis EE, Nagle NJ, Chen S, Frear CS. 2015. Mechanism, kinetics and microbiology of inhibition caused by long-chain fatty acids in anaerobic digestion of algal biomass. *Biotechnology and Biofuels*. 8: 141

Mahyudin AR, Zulaicha DH, Eniya LD, Syaftik N, Adelia N. 2010. Biohydrogen Production in 30L of BSTR Using Biomass Waste by *Enterobacter aerogens* and Its Simultaneous Utilization for Fuel Cell, 7th Biomass Asia Workshop, Jakarta, Indonesia

Mangayil R, Karp M, Santala V. 2012. Bioconversion of crude glycerol from biodiesel production to hydrogen, *International Journal Hydrogen Energy* 37: 12198-12204

Maru BT, Bielen AAM, Constantí M, Medina F, Kengen SWM. 2013. Glycerol fermentation to hydrogen by *Thermotoga maritima*: Proposed pathway and bioenergetic considerations. *International Journal of Hydrogen Energy* 38: 5563–5572.

Maru BT, Constanti M, Stchigel AM, Medina F, Sueiras JE. 2013. Biohydrogen production by dark fermentation of glycerol using *Enterobacter* and *Citrobacter* Sp. *Biotechnology. Progress* 9: 31-38

Maximize market Research. 2019. Global Hydrogen Storage Market : Industry Analysis & Forecast (2019-2026)-By Storage form, Storage type, Application and Geography. https://www.maximizemarketresearch.com/market-report/global-hydrogen-storage-market/ 149 20/ Pvt. Ltd.

Mendes AA, de Castro, HF. 2005. Effect on the enzymatic hydrolysis of lipids from dairy wastewater by replacing gum Arabic emulsifier for sodium chloride. *Brazilian Archives of Biology and Technology* 48: 135-142

Mishra P, Das D. 2014. Biohydrogen production from *Enterobacter cloacae* IIT-BT 08 using distillery effluent. *International Journal Hydrogen Energy* 39: 7496–7507

Mishra P, Krishnana S, Rana S, Singha L, Sakinaha M, Wahida ZA. 2019. Outlook of fermentative hydrogen production techniques: An overview of dark, photo and integrated dark-photo fermentative approach to biomass. *Energy Strategy Reviews* 27-37

Mizuno O, Dinsdale R, Hawkes FR, Hawkes DL, Noike T. 2000. Enhancement of hydrogen production from glucose by nitrogen gas sparging. *Bioresource Technology* 73: 59-65.

Mnre.gov report 2019. Hydrogen Energy. https://mnre.gov.in/hydrogen-energy

Monlau F, Barakat A, Trably, Dumas C, Steyer JP, Carrere H. 2013. Lignocellulosic Materials Into Biohydrogen and Biomethane: Impact of Structural Features and Pre-treatment. *Critical Review of Environmental Science Technology* 43: 260-322

Mu Y, Yu HQ, Wang G. 2006. Evaluation of three methods for enriching H_2-producing cultures from anaerobic sludge. *Enzyme and Microbe Technology* 40(4): 947–953

Nath K, Das D. 2011. Modeling and optimization of fermentative hydrogen production. *Bioresource Technology* 102(18): 8569-8581

Okamoto M, Miyahara T, Mizuno O, Noike T. 2000. Biological hydrogen potential of materials characteristic of the organic fraction of municipal solid wastes. *Water Science and Technology* 41: 25-32.

Özgür E, Afsar N, de Vrije T, Yücel M, Gündüz U, Claassen PAM, Eroglu I. 2010. Potential use of thermophilic dark fermentation effluents in photofermentative hydrogen production by *Rhodobacter capsulatus*. *Journal of Cleaner Production 18*: S23–S28

Pandu K, Joseph S. 2012. Comparisons and Limitations of Biohydrogen Production Processes. *Research Journal of Biotechnology* 7 (2): 59-71

Parihar R. K., K. Upadhyay. 2015. Production of Bio-Hydrogen Gas from Wastewater by Anaerobic Fermentation Process: A Review. *International Journal Chemistry and.* Studies 3: 07-14

Prahukhot Prachi R, Wagh Mahesh M, Gangal Aneesh C. 2016. A review on Solid State Hydrogen Storage Material. *Advances in Energy and Power* 4(2): 11-22

Rai PK, Singh SP, Asthana RK. 2012. Biohydrogen production from cheese whey wastewater in a two-step anaerobic process. *Applied Biochemistry and Biotechnology* 167: 1540-1549

Roswell JLC, Millward AR, Park KS, Yaghi OM. 2004. Hydrogen adsorption in functionalized metal-organic frameworks. *Journal of American Chemical Society* 126:5666-5667

Sanchez-Torres V, Zulkhairi M, Yusoff M, Nakano C, Maeda T, Ogawa HI, Wood TK. 2013. Influence of *Escherichia coli* hydrogenases on hydrogen fermentation from glycerol, *International Journal Hydrogen Energy* 38: 3905-391

Shin HS, Youn JH, Kim SH. 2004. Hydrogen production from food waste in anaerobic mesophilic and thermophilic acidogenesis. *International Journal Hydrogen Energy* 29: 1355-1363.

Singh L, Wahid ZA, Siddiqui MF, Ahmad A, Ab MH, Sakinah RM. 2013. Biohydrogen production from palm oil mill effluent using immobilized *Clostridium butyricum* EB6 in polyethylene glycol. *Process Biochemistry* 48: 294-298

Sivagurunathan P, Kumar G, Lin C. Y. 2014. Hydrogen and ethanol fermentation of various carbon sources by immobilized *Escherichia coli* (XL1-Blue), *International Journal Hydrogen Energy*, 6881-6888.

Statistical Reviews of World Energy. 2018. BP Statistical Review of World Energy, 2019 | 68th edition. https://www.bp.com/content/dam/bp/business-sites/en/global/corporate/pdfs/energy -economics/statistical-review/bp-stats-review-2019-full-report.pdf

Trchounian K, Sargsyan H, Trchounian A. 2014. Hydrogen production by *Escherichia coli* depends on glucose concentration and its combination with glycerol at different pH. *International Journal Hydrogen Energy* 39: 6419-6423

Ueno Y, Kawai T, Sato S, Otsuka S, Morimoto M. 1995. Biological production of hydrogen from cellulose by natural anaerobic microflora. *Journal of Fermentation Bioengineering* 79: 395-397

Van Ginkel S, Oh SE, Logan BE. 2005. Biohydrogen gas production from food processing and domestic wastewaters. *International Journal Hydrogen Energy* 30: 1535-1542

Van Ginkel SS, Lay S. 2001. Biohydrogen production as a function of pH and substrate concentration. *Environmental Science and Technology* 35: 4726-4730

Van Niel EWJ, Budde MAW, de Haas GG, Van der Wal FJ, Claassen PAM. 2002. Stams AJ M. Distinctive properties of high hydrogen producing extreme thermophiles, *Caldicellulosiruptor saccharolyticus* and *Thermotoga elfii*. *International Journal Hydrogen Energy* 27: 1391-1398

Wang CC, Chang CW, Chu CP, Lee DJ, Chang BV, Liao CS. 2003. Producing hydrogen from wastewater sludge by *Clostridium bifermentans*. *Journal of Biotechnology* 102: 83-92.

Wang H, Ma S, Bu H. 2014. Fermentative hydrogen production by newly isolated *Clostridium perfringens* ATCC 13124, *Journal of Sustainable and renewable Energy* 6: 013130.

Wang J, Wan W. 2008. Comparison of different pretreatment methods for enriching hydrogen producing bacteria from digested sludge. *International Journal Hydrogen Energy* 33: 2934–2941

Woodward J, Orr M, Cordray K, Greenbaung E. 2000. Biotechnology: Enzymatic Production of Biohydrogen. *Nature* 405:1014

Wu J, Wang Z, Xu S. 2007. Preparation and characterization of sericin powder extracted from silk industry wastewater. *Food Chemistry*103:1255-1262

Wu JH, Lin CY. 2004. Biohydrogen production by nesophilic fermentation of food wastewater. *Water Science and Technology* 49: 223-228.

Wu KJ, Lin YH, Lo YC, Chen CY, Chen WM, Chang JS. 2011. Converting glycerol into hydrogen, ethanol, and diols with a *Klebsiella* sp. HE1 strain via anaerobic fermentation, *Journal of the Taiwan Institute of Chemical Engineers* 42: 20-25

Xiao B, Han Y, Liu J. 2010. Evaluation of biohydrogen production from glucose and protein at neutral initial pH. *International Journal Hydrogen Energy* 35: 6152-6160

Xiao Y, Zhang X, Zhu M, Tan W. 2013. Effect of the culture media optimization, pH and temperature on the biohydrogen production and the hydrogenase activities by *Klebsiella pneumoniae* ECU-15. *Bioresource Technology*. 137: 9-17

Yanga H, Shaoc P, Lub T, Shena J, Wangb D, Xu BZ. 2006. Continuous biohydrogen production from citric acid wastewater via facultative anaerobic bacteria. *International Journal Hydrogen Energy* 6: 1306-1313

Zaluska A, Zaluska L, Strom-Oslen JO. 2001. Structure, catalysis and atomic reactions on the nano-scale: a systematic approach to metal hydrides for hydrogen storage. *Applied Physics A* 72: 157-165

Zhang ZP, Show KY, Tay JH, Liang DT, Lee DJ, Jiang WJ. 2007. Rapid formation of hydrogen-producing granules in an anaerobic continuous stirred tank reactor induced by acid incubation. *Biotechnology and Bioengineering* 1040–50.

Zhou L, Zhou YP, Sun Y. 2004. Enhanced storage of hydrogen at temperature of liquid nitrogen. *International Journal of Hydrgen Energy* 9: 319-322

Zhu H, Béland M. 2006. Evaluation of alternative methods of preparing hydrogen producing seeds from digested wastewater sludge. *International Journal of Hydrogen Energy* 31: 1980–1988.

Zhu M, Wang H, Ouyang LZ, Zeng MQ. 2006. Composite structure and hydrogen storage properties in Mg-based alloys. *International Journal of Hydrogen Energy* 31: 251-257

Zulkhairi M, Yusoff M, Hashiguchi Y, Maeda T, Wood TK. 2013. Four products from *Escherichia coli* pseudogenes increase hydrogen production. *Biochemical and Biophysical Research Communication* 439: 576 579

10

Anaerobic Process of Bio-butanol Production and Its Use as Biofuel

Kajal Gopal Singh and D.R. Ranade*

A2-991, Saarrthi Souvenir, Mhalunge, Pune-411045, Maharashtra, India

Abstract

Rise in global energy demand has led to an increase in consumption of fossil fuels. In recent years, there has been an increasing awareness about the consequences of pollution and global warming caused by burning of fossil fuels. This has led to an increased interest in the search for sustainable energy. Transportation sector is a major consumer of world's energy resources. n-Butanol is one of the most promising alternative bio-fuels which has various advantages over other biofuels. The microbial n-butanol is produced by anaerobic conversion of the substrate to n-butyric acid and subsequently utilized to form n-butanol. The present chapter explores the scope of bacterial butanol production as need of the hour and its use as an alternative biofuel.

Keywords: Biofuel, *Clostridium*, Ethanol, Butanol, Fermentation, Optimization.

Introduction

In recent times fuels and petro-chemicals from renewable sources have gained global interest due to global warming, volatility of oil prices and increasing politico-legal restrictions on non-renewable energy sources. Bio-butanol is a promising alternative biofuel due to its advantages over the more established biofuels like ethanol and methanol. The chemical properties such as higher specific heat, lower volatility, lower polarity, lower corrosivity and lower heat of vaporization makes bio-butanol less prone to ignition problems, as compared to other biofuels. Louis Pasteur first reported microbial butanol production in 1862. Later in 1915,

Corresponding Author: kajalgopalsingh@gmail.com

Weizmann patented industrial acetone-butanol-ethanol (ABE) production process. Many methods have been explored for butanol production but anaerobic fermentation method of butanol production has been proven advantageous as compared to other production methods. This chapter deals with methods of butanol production and its use as biofuel.

Need for Biofuels

Rise in global energy demand (Figure 1.1a) has led to an increase in consumption and thereby to rapid depletion of finite fossil fuel. As a consequence, uncertainty in oil prices are increasing and affecting the world economy. The use of fossil fuels over a long period of time has led to climate change and global warming due to emission of greenhouse gases coupled with environmental pollution. The global prosperity of a country is dependent on the energy production. This and increasing awareness about the consequences of pollution and global warming altogether have led to an increased interest in the search of sustainable renewable energy sources. Transportation sector is a major consumer of the world's energy resources. Bio-fuels offer an attractive option to stop the dependence on petroleum-based fuels. n-Butanol is one of the most promising alternative bio-fuel which has various advantages over ethanol (Liu et al., 2012). Currently, most of the butanol produced (at a cost of 7.0 - 8.4 billion dollars per year) is synthesized through chemical processes based on oxo-synthesis, Reppe synthesis or crotonaldehyde hydrogenation (Kolesinsk et al., 2019).

Why n-Butanol?

Among the various available bio-fuels, bio-diesel, bio-ethanol and bio-butanol have gained the attention of commercial sectors. n-Butanol (C_4H_9OH) is an alcohol which can be used as a fuel, because of its versatile characteristics such as octane number close to gasoline, low water miscibility, low vapor pressure and high energy content. n-Butanol is also used as a raw material for the chemical industry (Figure 1.1 b).

n-Butanol has multipurpose use in the manufacture of butyl acrylate, butyl phthalate, glycerol ethers and as an industrial solvent in the production of plastic, medical extractant, hydraulic fluid and detergents (Wypych, 2001). As a bio-fuel, n-Butanol is found to be superior to ethanol. It has many advantages over ethanol like higher energy content, higher blending rate with gasoline, no engine modification, easy distribution using current infrastructure, and a better auto emission performance. It can not only be used as a fuel but also as a fuel additive. When butanol is used as fuel in a car, no modifications in engine are required upto 40% butanol concentration of the total fuel contents (Suib, 2013). It is assumed that in near future, bio-butanol has the potential to substitute for both bio-ethanol and bio-diesel in the biofuel market, making it worth \$247 billion (Green, 2011). The current global demand for butanol is exceeding 1.2 billion gallons per year and it has been valued over USD 6 billion annually. The Asia-Pacific region was the largest consumer of bio-butanol in 2018. Growing production of coatings, adhesives, resin and textile in countries, such as China, India and Japan, are driving the market growth (https://www.researchandmarkets.com). Inspite of the advantages, bio-butanol being an alcohol-based fuel is not compatible with some fuel system components. It can reduce the carbon footprint, mitigate the supply and price fluctuation during the transportation sectors, relieve the energy security problem, and provide related job opportunities to improve social equality (Yue et al., 2014). Therefore, with further improvements or upgrade, the bio-butanol is believed to be the next generation biofuel.

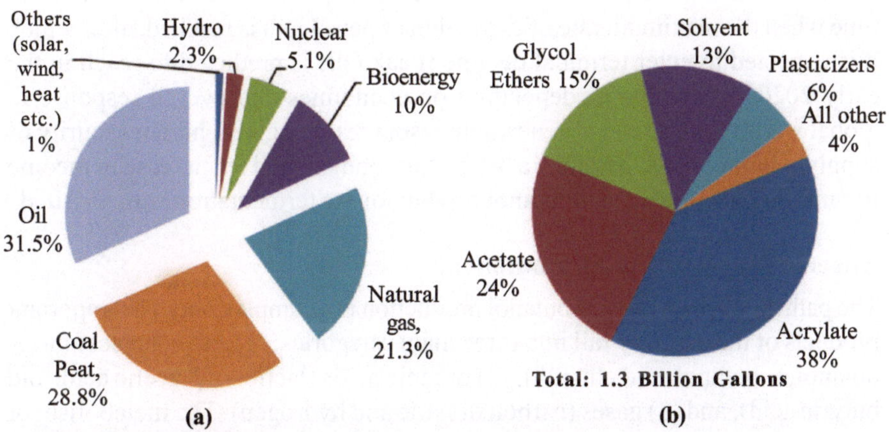

Fig. 1.1: a) World primary energy demand (figure based on Ho et al., 2014) **b)** Use of butanol as an industrial feed stock (http://greenchemicalsblog.com)

History of n-butanol production

Butanol fermentation in form of Acetone-Butanol-Ethanol (ABE) fermentation is considered as one of the oldest industrial fermentation processes. It is used since 1911, when amyl alcohol, butanol and acetone were produced as precursors to synthetic rubber (Dürre, 2008). During World War- I, ABE fermentation was further improved to produce acetone for its use in smokeless munitions powder production. Until 1920's, n-butanol was considered as an unwanted by-product of the ABE fermentation process. This perspective changed with the discovery of n-butanol and butyl acetate. These were used as solvents for nitrocellulose lacquer which is used in the automotive industry (Jones and Woods, 1986). Chaim Weizmann (1916) pioneered the commercial use of n-butanol. He invented the historic industrial ABE fermentation using *Clostridium acetobutylicum* to convert fermented corn starches into acetone, which could then be used to make dynamite. Till date, *Clostridium acetobutylicum*ATCC824 remains the best studied and manipulated strain (Green, 2011). Solvent fermentation continued until 1950's. However the scenario changed in the western countries since 1954. In western countries the price of petroleum became lower than that of sugar and at same time due to high demand for n-butanol fermentative production got replaced rapidly by the chemical process. After this,n-butanol production via hydrolysis of halo-alkanes or hydration of alkenes obtained from petroleum was encouraged. This process mostly included aldol-condensation of acetaldehydes, followed by the dehydration and then hydrogenation of croton aldehyde. The industrial discovery of oxo synthesis boosted the chemical synthesis of n-butanol (Ndaba et al., 2015).This growing demand for crude oil led to the "peak oil" scenario which was met in 2010. The world was facing the problem of oil reserves depletion. As of 2019, point in time when the maximum rate of extraction of petroleum is reached, after which it is expected to enter terminal decline (Peak Oil) is predicted to reach in the early 2020s to the 2040s, depending on economics and overall response to global warming. The use of renewable resources will give a higher security of supply, a higher national value, a better environment and an increase in income in rural regions. Thus the n-butanol production by fermentation was revisited.

Anaerobic production of n-butanol

The pathway of anaerobic n-butanol production is a complex one. The important products of the pathway fall into three main categories: (1) solvents (acetone, i-butanol, n-butanol, and ethanol), (2) organic acids (lactic acid, acetic acid, and butyric acid), and (3) gases (carbon dioxide and hydrogen). The metabolism of n-butanol fermentation is a unique feature of the members of genus *Clostridium* and can be divided into two distinct phases. In the first phase, called acidogenesis, sugars are metabolized to produce acetate and butyrate which accumulate and reduce the pH of the fermentation broth. The second phase is termed as solventogenesis wherein the acids produced in the first phase are utilized to produce acetone, butanol and ethanol with simultaneous increase in pH. The shift is initiated when the n-butyric concentration increases to more than 2 g/l and pH decreases to 5. An anaerobic microorganism has to maintain a balance of NADH and NAD+ because of its inability to utilize the electron transport system to generate ATP. The conversion of glucose to ABE is the only way to regenerate NAD+ because all other end products yield NADH. The aceto-acetyl-CoA is converted to acetoacetate, then to acetone and later to i-propanol. This conversion takes place during solventogenesis phase of fermentation where reduction of acetate and butyrate occurs. Majority of the acetyl-CoA is converted to acids and solvents whereas some are used for the lipid synthesis and growth (Figure 1.2).

Acetone is considered as an undesirable co-product of the ABE process because of its poor fuel properties and corrosiveness to rubber engine parts. Iso-propanol-Butanol-Ethanol (IBE), an alcohol mixture is considered as a green solvent and biofuel. It is the characteristic of some Clostridial strains that produce iso-propanol instead of acetone together with butanol and ethanol (George et al., 1983). Ethanol, iso-propanol and n-butanol have been suggested as feedstocks (with 2,3-butanediol) that can support a major fraction of the chemical industry. They are also used as co-solvents and octane number enhancers in blends of methanol and gasoline as transportation fuels. The strains of iso-propanol producing *C. beijerinckii,* accumulate acetone rarely and even if they do it is in very low concentration.

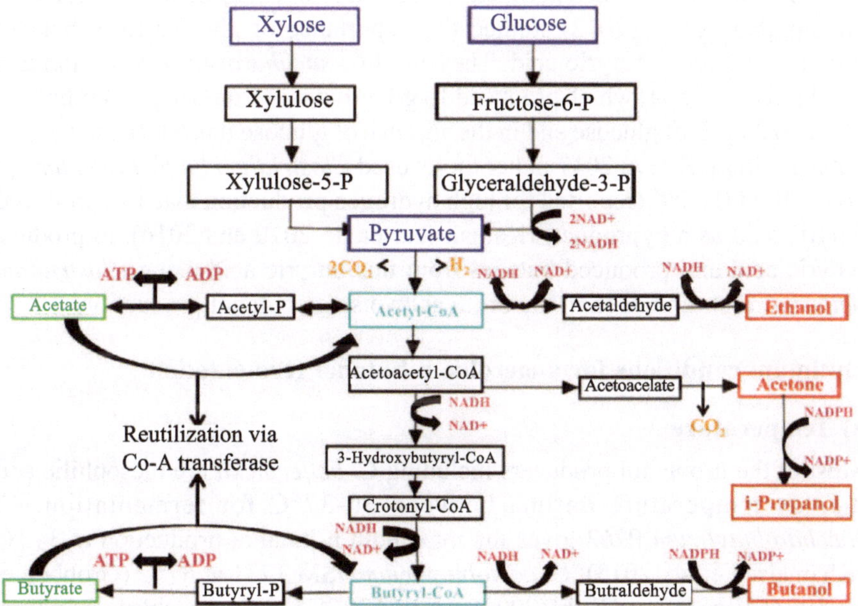

Fig. 1.2: Molecules in green–acids, molecules in red - the solvents, Gases produced- orange, Energy generated or utilized in pink

Solvent-producing Clostridia reutilize acetate or butyrate catalyzed by CoA-transferase to form solvents. This uptake is a de-toxic bio-reaction to prevent the inhibitory effect of the organic acids on cell growth (Gheshlaghi et al., 2009). However, this uptake does not take place in the absence of sugars because the reducing power is not sufficient to drive the uptake mechanism (Tashiro et al. 2007). The high sugar concentration is an important parameter to ensure a successful ABE fermentation (Karakashev et al. 2007). Thus butyrate cannot solely be used as a carbon source for n-butanol production. It needs presence of fermentable sugars in a fermentation broth. Studies have shown promising results on n-butanol production from the mixture of butyrate and glucose fermentation (Tashiro et al., 2007).

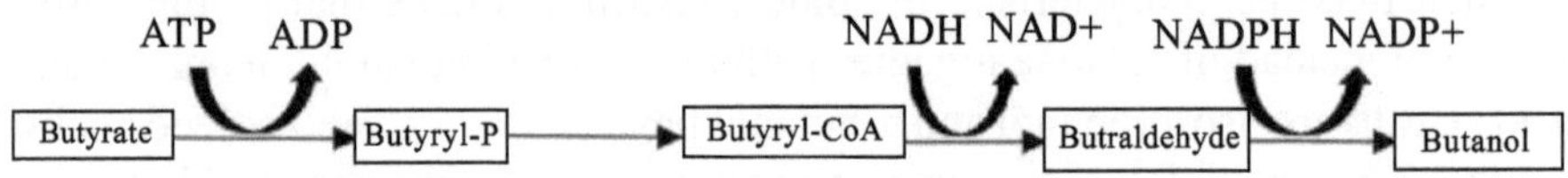

Fig. 1.3: Mechanism to convert butyrate to butanol (Al-Shorgani et al., 2012)

In stoichiometric analysis, the supplement of n-butyric acid in the medium was shown to benefit n-butanol production from ABE fermentation by reducing the required energy to produce n-butanol (NADH) (Figure 1.3) and increasing butyrate phosphate production which up-regulates solventogenesis (Chang, 2010). Al shorghani et al. (2012) studied the importance of glucose for n-butanol production from n-butyric acid. They used *C. saccharoperbuty-lacetonicum* N1-4 (ATCC 13564) which produced 13 g/L n-butanol from 10 g/L of n-butyric acid and 20 g/L of glucose and in the absence of glucose it produced 0.7 g/L n-butanol. Singh K.G in 2017 in her thesis used *Clostridium biohydrogenum* sp. nov., MCM B-509^T (reported for high hydrogen production that also produced butyric acid as a byproduct) (Kamalaskar et.al., 2010 and 2016), to produce butyric acid and produced butanol from this butyric acid using *Clostridium beijerinckii* strain CHTa in the broth as two stage butanol production.

Optimum conditions for anaerobic n-butanol fermentation

a) Temperature

Most of the n-butanol producers including *C. beijerinckii* are mesophilic and display temperature optima between 30-37°C for fermentation. *C. saccharobutylicum* P262 gives the maximum n-butanol production at 34 °C (Khamaiseh et al., 2013), *C. acetobutylicum* DSM 1731 at 37°C (Grobben et al., 1993). Elgadafi and Kalil (2009) reported 30 °C was the suitable temperature for maximum n-butanol production by *C. acetobutylicum* NCIMB 13357. There are few thermophilic strains which show butanol production at the higher temperature. *C. thermocellum* has been reported to produce n-butanol from cellulose at 60 °C. (Nakayama et al, 2011). *Thermoanaerobacterium saccharolyticum* is another thermophilic anaerobic Clostridium that has the ability to directly ferment primary sugars into n-butanol (Bhandiwad et al., 2014).

b) pH

pH of the fermentation medium plays a very important role in n-butanol production. The optimum pH is between pH 6-7. n-Butanol production at pH 5 or below is poor and this is due to the formation of acidic metabolites which destroy cell's ability to maintain internal pH (Bowles and Ellefson, 1985). The medium having high (6-7) initial pH can attain faster equilibrium pH level in the medium as compared to low (4-5) initial pH in the medium. Jones and Woods (1986) have shown that the pH range over which the solvent synthesis occurs, varies depending on the strain and culture conditions used. With the cultures *Clostridium acetobutylicum,* the solvent production was observed to commence at acidic pH between pH 3.8 to 5.5 (Bahl et al, 1982; Nischio et al, 1983). According to Gottschalk and Morris, 1981, lowering of pH is not always the pre-requisite trigger for solventogenesis. In fact, *C.acetobutylicum P62* and *C. beijerinckii* ATCC 10132 have the optimum pH of 6.5 (Isar and Rangaswamy., 2012). *Clostridium acetobutylicum* B18 produced a large amount of n-butanol over a wide range of pH (4.5–6.0) (Geng and Park., 1993). Hartmanis et al (1984) co-relates the concentration of n-butyric acid which thereby decreases pH of the medium to switch metabolism to solventogensis.

c) Substrate concentration

The initial sugar concentration plays an important role in ABE fermentation. Sugar concentrations lower than 20 g/L leads to the acidic condition with little solvent formation whereas at higher concentrations more acid is produced and then the process progresses towards solventogenesis (Monot et al, 1982). Razak et al (2013) have reported optimum n-butanol production (8.17 g/L n-butanol) by *C. acetobutylicum* ATCC 824 to occur at 70 g/L glucose concentration.

d) Other essential factors

In microbial fermentation, the product concentration depends on different conditions. Components of the media play a significant role in metabolism for n-butanol production. Singh et al, 2016 studied the effect of media components on n-butanol production using *C. beijerinckii* strain CHTa. The orders of significance of the nutrients were in the following order: yeast extract > $MgSO_{4.}7H_2O$ > glucose> cysteine HCl > pH> $FeSO_{4.}7H_2O$> K_2HPO_4> $CaCO_3$> $(NH_4)_2SO_4$. Cysteine hydrochloride is a reducing agent essential for maintaining anaerobiosis, but its level is not increased because of toxic effects of reducing agents at higher concentration (Fukushima et al., 2002).

Yeast extract, is a nitrogen source which provides minerals, proteins, growth factors etc. thereby promoting growth and enhancing n-butanol production (Al-Shorgani et al. 2013). Singh et al. (2016) found that yeast extract enhanced the n-butanol yield from 7.6 to 9.9 g L^{-1} when yeast extract concentration was increased from 5.1 to 7.5 g L^{-1}. In 2002, Canganella et al. also concluded that the increase in yeast extract concentration from 0.05 to 2% enhanced the glucose tolerance but glucose concentrations above 10% caused an increase in doubling time even in the presence of 2% yeast extract.

High glucose concentrations have an inhibitory effect on the n-butanol production because of inhibition of glucose permease at n-butanol concentration of 12gL^{-1}(Ounine 1985). Mermelstein et al., 1993 suggested that initial glucose concentration should not exceed 90 g L^{-1} because it negatively affects the growth. Further similar results were obtained by Singh et al., 2016, where the increase in n-butanol production was observed with the increase in glucose concentration upto 100 g L^{-1}. He and Chen, 2013, showed that at the glucose concentration of 100 g L^{-1}, highest total solvent production was found, which was 1.39 times of that achieved at 60 g L^{-1} glucose. The solvent yield (g/g) however decreased with the increase in substrate concentration. $MgSO_{4.}7H_2O$ is reported to increase nutrient uptake from the media. Pasternak et al., 2010, reported Mg^{2+} to be essential for the activation and production of the majority of enzymes like ribozymes, endonucleases etc. Suzuki et al., 1985, showed that Mg^{2+} acts as an activator of enzymes like transferases and decarboxylases that plays an important role in the biochemistry of alcohol production.

Thus yeast extract, $MgSO_{4.}7H_2O$ and glucose concentration were found significant for n-butanol production only when all the three were in the optimum concentration. Using the optimal conditions, the n-butanol concentration was 10.08 g L^{-1} and productivity of 0.45g L^{-1} h^{-1} which was 36.5% and 40% higher as compared to that obtained with un-optimized medium (6.4 g L^{-1}, 0.27 g L^{-1} h^{-1}) in 24h. (Singh et al, 2016).

Bacteria reported for fermentation

Anaerobic bacteria from the genus *Clostridium* are reported to produce n-butanol via ABE fermentation processes. In 1862, the production of a C4 alcohol was described by Louis Pasteur using "*Vibrionbutyrique*". It was a mixed culture of solventogenic *Clostridium* species. *Bacillus butylicus* was documented as the n-butanol producing species for the first time by Albert Fitz. Later Martinus Beijerinck isolated and described a similar strain of solventogenic bacterium called *Granulobacter saccharobutyricum* which in 1926 was classified as *Clostridium acetobutylicum* (Dürre, 2008).

Clostridium is an omnipresent group of bacteria found in soil, sewage, vegetation, plant and animal products and the digestive tracts of many animals. The species of Clostridia that produces n-butanol are *C. acetobutylicum, C. beijerinckii, C. saccharoperbutylacetonicum, C. pasteurianum* (Keis et al. 1995, Dhamole et al. 2012).*Clostridium beijerinckii,* was originally classified as *C. acetobutylicum.* Many of the strains from each of the above species were later found to be *C. saccharoperbutylacetonicum* and *C. saccharobutylicum* (Shaheen et al. 2000) including *Clostridium saccharoperbutylacetonicum* strains N1-4 and N1-504.

*Clostridium beijerinckii BA101*is a mutant strain of *Clostridium beijerinckii NCIMB* 8052 which produces n-butanol up to 21 g/L and is considered as a hypern-butanol producing strain (Chen and Blaschek., 1999). Numerous attempts have been made to engineer the *Clostridium* n-butanol pathway in controllable hosts, such as *Escherichia coli*(0.55–1.2 g/L) (Atsumi et al., 2008; Inui et al., 2008), *Saccharomyces cerevisiae* (2.5 mg/L) (Steen at al., 2008), and *Bacillus subtilis* (120 mg/L) and *Pseudomonas putida* (500mg/L), (Nielsen at al., 2009) *Lactobacillus brevis*(300 mg/L), for efficient n-butanol fermentation. Liao and coworkers in 2011 reported that the high titer of 15 g/L of n-butanol in engineered *E. coli* where butyl-CoA dehydrogenase gene of *Clostridia* was replaced by trans-enoyl-CoA reductase from *Treponema denticola.* Recently under aerobic conditions, high n-butanol titer of 10.38 g/L was achieved using *Bacillus* sp.15 (Ng et al., 2016).

Clostridium, the native n-butanol producer, is still the best n-butanol producer whereas the non-native producers, engineered strains, are not competitive enough. Few non-native bacteria are very specific for their substrates e. g., lactic acid bacteria and yeasts can produce n-butanol only from 2, 3-butanediol (Speranza et al. 1997).

Substrates for n-Butanol production

Despite of superior fuel properties, n-butanol has to compete in production costs with ethanol in order to dominate in biofuel markets. Medium composition and substrate concentrations are known to affect the yield, productivity and final concentration of n-butanol in the process (Kheyrandish et al., 2015).One of the challenges for ABE fermentation is the substrate cost which greatly contributes to the overall cost of low value-added fermentation products. The reduction in feedstock cost offers the best opportunity to decrease the cost of production. Thus cheaper, abundant, and sustainable feedstocks such as wastes and agricultural residues should be used to improve the cost of the process (Van der Merwe et al., 2013).

Solventogenic Clostridia can utilize a wide range of substrates viz. mono, di, oligo and polysaccharides like glucose, fructose, xylose, arabinose, lactose, saccharose, starch, pectin, inulin etc. (Patakova et al., 2011). A number of studies have reported bio-butanol production from lignocellulosic biomass including wheat straw, barley straw(Qureshi et al., 2010a), corn fiber(Qureshi et al., 2007), corn stover, switchgrass (Qureshi et al., 2010b) and dried distilled grains and soluble (DDGS) (Ezeji and Blaschek, 2008). However, the biomass required pretreatment by methods using dilute sulfuric acid and the fermentation results varied significantly depending on the biomass species. Biobutanol can also be produced from algae (called Solalgal Fuel) or diatoms (Vinod et al., 2014).

Furthermore, there are many reports for the use of various biomass substrates such as a hardwood (Shah and Lee., 1994), domestic organic waste (Claassen et al., 2000), agricultural waste (Zverlov et al., 2006), palm oil waste (Tiam mun et al., 1995), whey (Raganati et al., 2013)and sago starch (Madihah et al., 2001) for ABE fermentation by different strains. The substrates like whole grains or lingo-cellulosic biomass unlike simple sugars require pretreatment either in form of hydrolysis using enzymes, acid treatment or gasification. After this treatment, the lysate is used for ABE fermentation (Figure 1.4).

Considering the economics of n-butyric acid and n-butanol fermentation, agricultural products would not be feasible substrates due to its use as food and feed which makes it a high cost commodity. Use of low cost substrate can be one way of reducing the cost of fermentation. Considering the number of distilleries in the country and the wastewater generated therein, development of a bio-process to n-butyric acid and further to n-butanol while treating the distillery waste will be very beneficial to the distilleries to increase the revenue.

Fermentation technologies used for n-Butanol production

There are several studies that has been focused on improving the concentration, productivity, and yield of n-butanol production by using different culture modes under the various conditions. Batch culture, fed-batch culture, continuous culture and these processes integrated with recovery process are the various operation modes used for fermentation. Batch reactors are mostly preferred in the majority of studies on n-butanol fermentation because of its simplicity, easy manipulation, and low risk of contamination (Chang, 2010). This reactor setup requires minor or no modification depending upon the feeding material used for the fermentation. The batch mode fermentation is mostly used for the study of various parameters like pH control, carbon to nitrogen sources ratio (C/N ratio), the addition of electron carriers and the partial pressure of gasses like hydrogen or carbon monoxide in the fermentor headspace. However, lag phase in the beginning and the end product inhibition at the end; affects the productivity in a batch reactor. In batch fermentation, the average butanol productivity was 0.2 g/L/h, which was not economical enough for large scale production (Tashiro et al., 2013). Use of fed-batch techniques or continuous culture techniques integrated with product removal techniques can eliminate problems related to batch fermentation.

The fed-batch fermentation is commonly used in the industrial production with its advantage of easier maintenance over batch reactors. At the same time, fed-batch fermentation faces problems of toxicity due to acid accumulation and thereby inspite of the increase in substrate input the production of desired end-product is blocked. This can be overcome by coupling the fed-batch process with some removal technique. Besides batch and fed-batch fermentations, continuous fermentation was also studied for n-butanol production. This process had the advantage of the continuous removal of toxic end-product. With the continuous feeding of medium and removal of the product, a high n-butanol productivity of 1.47 g/L/h can be achieved which is high when compared to batch fermentation (Chang, 2010). However, the continuous fermentation for n-butanol production was not operated at large scale because of the physiological complexity of then-butanol producing Clostridia. When n-butanol concentration is too high in the medium, the bacteria stops producing n-butanol. Even though high productivity is obtained in a single-stage continuous system; the concentration of n-butanol is low when compared to batch fermentations (Cheng, J. ed., 2009). The complexity of n-butanol fermentation makes it almost impractical to use a single-stage continuous reactor on an industrial scale. The problem of solvent concentration can be solved by using two or more multistage continuous fermentation systems (Ni and Sun, 2009)

In order to further improve the productivity, techniques like high cell density technique were commonly integrated into fermentation processes. While higher cell density gave higher productivity, the built up of cells in the reactor could also produce problems such as clogging by cell mass and mass transfer limitation of nutrient and products.

Biphasic nature of ABE fermentation and solvent toxicity makes the classical fed-batch and continuous cultivation process incompetent. To overcome this problem, the coupling of in situ recovery process to the fed-batch culture (Ezeji et al., 2004), and two/multistage continuous fermentation has been conducted (Godin and Engasser, 1990).Table 1.1 shows the list of fermentation processes reported using Clostridial strains.

Various methods are reported to reduce the butanol toxicity which includes gas stripping evaporation, membrane assisted solvent extraction, adsorption; liquid-liquid extraction. However, most of these methods focus on the separation of n-butanol as it is formed and not on increasing the initial concentration of n-butanol in the broth which would reduce the downstream processing cost. Recently, Dhamole et al., (2012) reported the extractive fermentation with non-ionic surfactants for n-butanol production. It addresses two interlinked problems with n-butanol fermentation, first it relives n-butanol toxicity to the microbes thus enhancing the butanol production (and hence titer), and second, it concentrates the n-butanol in downstream processing. The surfactants self-assemble into micelles above their critical micelle concentration and entrap (hydrophobic) butanol into the micelles, thus relieving the butanol toxicity to the microorganisms. Surfactant reduces the substrate or product toxicity by entrapping the substrate/ product. Surfactant based cloud point extraction significantly reduces the process volume and concentrates the product in the surfactant rich phase consequently into reduction in downstream processing cost.

Also, surfactant recovery (>95%) and reuse is possible with surfactants (Dhamole et al., 2012). According to Singh et al., 2017 found that surfactant addition doubled the productivity of both, butanol and total solvent. Butanol productivity obtained by them was 2-3 times higher than similar studies on extractive fermentation with non-ionic surfactants.

Problems related to the biotechnological production such as solvent toxicity can also be solved by classical but powerful ways like random mutagenesis and process optimization. The *C. beijerinckii* strain BA101, one of the highest producers of n-butanol was created by mutagenesis. This was done by treating *C. beijerinckii* NCIMB 8052 with a mutagen N-methyl-N9-nitro-N-nitrosoguanidine and then the mutants were selectively enriched on 2-deoxyglucose (Formanek et al., 1997).An antisense RNA has also been used to improve n-butanol tolerance. Desai and Papoutsakis (1999) studied the use of an anti-sense RNA to reduce the activities of butyrate forming enzymes in *C. acetobutylicum.* Despite many efforts including genetic manipulation of strains, the best results ever obtained for ABE fermentations till now is less than 2% n-butanol concentration and a yield of less than 25% of the theoretical yield from glucose. Lack of an efficient gene knock-out system makes metabolic engineering of Clostridia difficult and costly affair (Lee at al., 2008). Optimizing the ABE fermentation process is a necessity of the industry since long time.

Use of co-culture in fermentation has also been reported by many scientists (Table1.2). Fermentation of the complex biomass requires hydrolysis step prior to the actual fermentation to n-butanol. Soni et al. in 1982 used consortia of fungi *Trichoderma reesei* and *Aspergillus wentii* to treat agricultural wastes such as bagasse, rice straw, and wheat straw and obtain hydrolysates which were then fermented with the cultures of *C. saccharoperbutylacetonicum* to obtain 16 g/L of n-butanol. Similarly, Petitdemange et al. in 1983 carried out the direct fermentation of cellulose using co-cultures of cellulolytic (*C. cellulolyticum* H10) and glycolytic (*C. acetobutylicum*) Clostridial strains. However, the major end product of fermentation was n-butyric acid along with small amounts of acetone, ethanol, and n-butanol.

Conversion of biomass to butyric acid and then to n-butanol can reduce the toxicity problems during fermentation (Dwidar et al., 2012). To avoid ancillary products, the cultures can be added sequentially where n-butyric acid

Table 1.1ABE fermentation processes using different clostridial strains

Strain/ fermentation mode	Substrate	ABE (g/L)	Reference
a) Batch fermentation			
1. *C. acetobutylicum* ATCC824^T	Glucose (80 g/L)	15.3	Ventura et al., 2013
2. *C. beijerinckii* BA101	Glucose(80 g/L)	32.6	Chen andBlascheck., 1999
3. *C. saccharoper-butylacetonicum*N1-4	Glucose (66 g/L)	24.2	Thang et al.,2010
4. *C. bejerinckii* strain CHTa	Glucose (90 g/l)	n-Butanol- 11.8 Acetone - 0.16 i-Propanol -1.5	Singh et al.,2016
b) Continuous fermentation			
1. *C. beijerinckii*DSM 6423	Glucose(58.3 g/L)	Butanol-4.48 iPropanol-3.40	Survaseet al., 2011
2. *C.beijerinckii*NCIMB8052	Glucose (30 g/L)	9.4	Lee et al., 2008
3. *C.acetobutylicum*	Glucose (60 g/L)	19	Chen et al 2013
4. *C.saccharobutylicum*	Molasses(40 g/L Total sugar)	15.27	Ni et al2013
c) Fed-batch fermentation			
1. Clostridium acetobutylicum.	Baggase (65 g/L total sugars)	21.11	Pang et al .,2016
1. C.acetobutylicum (repeated batch mode)	Glucose (60 g/L)	14.53	Dolejš et al., 2014
3. C.beijerinckii BA 101 Integrated fed batch fermentation	Glucose (500 g/L)	232	Ezeji et al.,2004 a
d) C. beijerinckii BA *Continuous with immobilization*	Glucose (60 g/L)	7.9	Qureshi et al. 2000
e) C. beijerinckii BA 101 *Continuous with gas stripping*	Glucose (1163 g/L)	460	Ezeji et al.,2004 b

Table1.2: ABE fermentation process using Co-cultures

Co- culture	Strategy	Substrate	Butanol (g/L)	ABE (g/L)	ABE productivity (g/L/h)	Reference
Trichoderma reesei and E. coli	Co-culture of T. reesei and an E. coli strain metabolically engineered to produce isobutanol	Pretreated cornstover	iB-1.88	ND	ND	Minty et al.,2013
C. thermocellum and C. acetobutylicum	Sequential co-culture and feeding butyrate	Cellulose solka floc	2.4	3.9	0.023	Yu et al., 1985
C. thermocellum and C.acetobutylicum	Sequential co-culture	crystalline cellulose	0.6	1.4	0.005	Nakayama et al., 2011
C. thermocellum and C. saccharoperbutylacet onicum strain N1-4	Sequential co-culture	crystalline cellulose	7.9	9.9	0.038	Nakayama et al., 2011
C. thermocellum and C. beijerinckii	Sequential co-culture	crystalline cellulose	2.1	3	0.011	Nakayama et al.,2011
C. cellulolyticum and C.acetobutylicum	co-culture	Cellulose solka floc	0.8	1.1	0.005	Petitdemange., 1983
Bacillus cellulolyticus and C. acetobutylicum	co-culture	Pretreated palmpressed fiber	0.49	ND	ND	Ponthein and Cheirsilp, B., 2011
C. thermocellum and C. beijerinckii	Sequential co-culture	AECC*	8.75	16	0.089	Lin et al., 2013
C. cellulovorans and C. beijerinckii	co-culture	AECC*	8.3	11.8	0.148	Wen et al.,2014
C. hydrogenum MCMB 509T and C. beijerinckii MCM B581	Sequential co-culture	Distillery waste	5.96	8.04	0.08	Thesis Singh K.G, 2017

*AECC: Alkali extracted deshelled corn cobs

concentration is dominated in the first stage and n-butanol concentration dominates the second stage. Singh et al., in 2016 isolated a bacterium that produced n-butanol from n-butyric acid. This isolate was identified as *Clostridium beijerinckii* strain that not only produced butanol in high concentration but also in less time. The strain produced Butanol yield of 10.08 g L^{-1} and productivity of 0.45 g L^{-1} $h^{-1}$was obtained with optimum operational and medium conditions. The same strain was tested for butanol production from butyric acid produced from distillery waste (Thesis Singh K.G, 2017).

It was observed that yield of n-butanol produced was 0.42 g/g of carbohydrates utilized. The study proved that co-culture of different microorganisms with specific metabolic capacities is an advantage which improves the substrate conversion and the product yield. A similar experiment was carried out by Li et al, 2013. Using co-cultures of *Clostridium beijerinckii* and *Clostridium tyrobutyricum* in immobilized-cell fermentation mode for butanol production from cassava starch, the n-butanol production (6.66 g/L), yield (0.18 g/g), and productivity (0.96 g/L/h) were obtained with cassava starch. Nakayama et al in 2011 used co-culture of a thermophilic cellulolytic strain *C. thermocellum* JN4 and a butanol-producing strain *C. saccharoperbutylacetonicum* N1-4.

Recent studies on the process development of fermentation led to a novel continuous two-stage, dual-path fermentation where two different strains were used and reported to give a high n-butanol productivity of 8 g/L/h in a stable process (Ramey, 1998). David Ramey was the first person to report this two stage ABE fermentation reaction where n-butanol production was carried out by first producing n-butyric acid and then induction of metabolic shift to n-butanol. The US patent 5753474 (1998) describes a process that manufactures n-butanol and volatile organic acids by fermenting carbohydrates via a multistage fermentation process. A patent filed by Shang-Tian Yang, (2008) on the method of producing n-butanol by using a feedstock comprising of a carbohydrate source which on fermentation produces n-butyric acid. This acid was then hydrogenated with metal oxides catalysts under elevated pressures and temperatures to produce n-butanol. Huang et al. (2004) reported a two step process wherecultures of *Clostridium tyrobutyricum* were immobilized in a continuous process to maximize the production of hydrogen and n-butyric acid followed by *Clostridium acetobutylicum*that converts this n-butyric acid to n-butanol. This led to n-butanol production of 5.1 g/L with the productivity of 4.64 g/L/hr utilizing 42% glucose, in comparison to the up to 25% glucose utilization in traditional ABE fermentation by *Clostridium acetobutylicum* alone. Richter et al in 2012 carried out optimization of the process of conversion of n-butyic acid to n-butanol along with co-fermentation of glucose in a two-stage continuous fermentation system by *C. saccharoperbutylacetonicum* strain N1-4. This process had the molar yields, Y (n-butanol/n-butyrate) and Y (n-butanol/glucose) of 2.0, and 0.718, respectively and a production rate of 0.39 g/L/h.

Inhibitions and other problems associated with ABE fermentation

A number of feedstock and microorganisms have successfully been utilized to produce bio-butanol but there are numerous drawbacks associated with ABE fermentation which hinders it to compete economically with petrochemical synthesis. Some of the identified limitations of the ABE based on the process aspect are as follows:

1. Low final butanol concentration (<20 g /L) caused by inhibition during fermentation
2. Low yield of butanol due to hetero-fermentation (0.28–0.33 g of n-Butanol / g of substrate)
3. Low productivity <0.5 g/L/h
4. High substrate costs
5. Carbohydrates are not utilized at its best

Not only inhibition is a major challenge for the ABE fermentation, research has also shown that lack of feedstock biomass alternatives is widespread. Current ABE fermentation depends on very expensive feedstock and efforts should be directed towards decreasing the substrate costs and several other drawbacks.

Thus, rational approaches and advanced knowledge is necessary for the effective strategies to limit these problems to improve the yields of the products of ABE fermentation.

Conclusion

Butanol is a very promising biofuel. This chapter summarizes the methods, substrates and fermentation conditions necessary for obtaining high n-butanol production from anaerobic fermentation. Low final butanol concentration (<20 g /L) caused by inhibition during fermentation can be improved by stripping or removal of n- butanol simultaneously during downstream fermentation. The problem of low yield and low productivity of butanol is due to hetero-fermentation which can be solved by using strain improvement methods or by using two stage fermentation methods thereby eliminating the production of other solvents. Using these combinations along with strain improvement techniques n-butanol production can be a game changing fuel.

Acknowledgements

We are thankful to Agharkar Research Institute, Pune and to entire Bioenergy Team of the institute for support.

References

Al-Shorgani, N.K.N., Ali, E., Kalil, M.S. and Yusoff, W.M.W., 2012. Bioconversion of butyric acid to butanol by Clostridium saccharoperbutylacetonicum N1-4 (ATCC 13564) in a limited nutrient medium. *BioEnergy Research*, 5(2), pp.287-293

Atsumi, S., Cann, A.F., Connor, M.R., Shen, C.R., Smith, K.M., Brynildsen, M.P., Chou, K.J., Hanai, T. and Liao, J.C., 2008. Metabolic engineering of *Escherichia coli* for 1-butanol production. *Metabolic Engineering*, 10(6), pp.305-311

Bahl, H., Andersch, W., Braun, K. and Gottschalk, G., 1982. Effect of pH and butyrate concentration on the production of acetone and butanol by Clostridium acetobutylicum grown in continuous culture. *European Journal of Applied Microbiology and Biotechnology*, 14(1), pp.17-20.

Bhandiwad, A., Shaw, A.J., Guss, A., Guseva, A., Bahl, H. and Lynd, L.R., 2014. Metabolic engineering of Thermoanaerobacterium saccharolyticum for n-butanol production. *Metabolic Engineering,* 21, pp.17-25.

Bowles, L.K. and Ellefson, W.L., 1985. Effects of butanol on Clostridium acetobutylicum. *Applied and Environmental Microbiology*, 50(5), pp.1165-1170.

Canganella, F., Kuk, S.U., Morgan, H. and Wiegel, J., 2002. Clostridium thermobutyricum: growth studies and stimulation of butyrate formation by acetate supplementation. *Microbiological Research*, 157(2), pp.149-156.

Chang, W.L., 2010. Acetone-butanol-ethanol fermentation by engineered Clostridium beijerinckii and Clostridium tyrobutyricum (Doctoral dissertation, The Ohio State University)

Chen, C.K. and Blaschek, H.P., 1999. Acetate enhances solvent production and prevents degeneration in Clostridium beijerinckii BA101. *Applied Microbiology and Biotechnology*, 52(2), pp.170-173.

Cheng, J. ed., 2009. Biomass to renewable energy processes. CRC Press.

Claassen, P.A., Budde, M.A. and López-Contreras, A.M., 2000. Acetone, butanol and ethanol production from domestic organic waste by solventogenic clostridia. *Journal of Molecular Microbiology and Biotechnology*, 2(1), pp.39-44.

Desai, R.P. and Papoutsakis, E.T., 1999. Antisense RNA strategies for metabolic engineering of Clostridium acetobutylicum. *Applied and Environmental Microbiology*, 65(3), pp.936-945.

Dhamole, P.B., Wang, Z., Liu, Y., Wang, B. and Feng, H., 2012. Extractive fermentation with non-ionic surfactants to enhance butanol production. *Biomass and Bioenergy*, 40, pp.112-119.

Dhamole, P.B., Mane, R.G. and Hao.F., 2015. Screening of non-ionic surfactant for enhancing biobutanol production. *Applied Biochemistry and Biotechnology*, 177(6), pp. 1272-1281.

Dolejš, I., Krasòan, V., Stloukal, R., Rosenberg, M. and Rebroš, M., 2014. Butanol production by immobilised Clostridium acetobutylicum in repeated batch, fed-batch, and continuous modes of fermentation. *Bioresource Technology*, 169, pp.723-730.

Dürre, P., 2008. Fermentative butanol production. *Annals of the New York Academy of Sciences*, 1125(1), pp.353-362.

Dwidar, M., Park, J.Y., Mitchell, R.J. and Sang, B.I., 2012. The future of butyric acid in industry. *The Scientific World Journal*, 2012. Article ID 471417, 10.

Elgadafi, R.A. and Kalil, M.S., 2009. Production of butanol in batch culture by using Clostridium acetobutylicum NCIMB 13357. M. Eng (Doctoral dissertation, Thesis, Universiti Kebangsaan Malaysia, Malaysia).

Ezeji, T.C., Qureshi, N. and Blaschek, H.P., 2004 a. Acetone butanol ethanol (ABE) production from concentrated substrate: reduction in substrate inhibition by fed-batch technique and product inhibition by gas stripping. *Applied Microbiology and Biotechnology*, 63(6), pp.653-658.

Ezeji, T. and Blaschek, H.P., 2008. Fermentation of dried distillers' grains and solubles (DDGS) hydrolysates to solvents and value-added products by solventogenic clostridia. *Bioresource Technology*, 99(12), pp.5232-5242.

Formanek, J., Mackie, R. and Blaschek, H.P., 1997. Enhanced Butanol Production by Clostridium beijerinckii BA101 Grown in Semidefined P2 Medium Containing 6 Percent Maltodextrin or Glucose. *Applied and Environmental Microbiology*, 63(6), pp.2306-2310.

Fukushima, R.S., Weimer, P.J. and Kunz, D.A.,2002. Photocatalytic interaction of resazurin N-oxide with cysteine optimizes preparation of anaerobic culture media. *Anaerobe*, 8:29-34.

Geng, Q. and Park, C.H., 1993. Controlled-pH batch butanol-acetone fermentation by low acid producing Clostridium acetobutylicum B18. *Biotechnology letters*, 15(4):421-426.

George, H.A., Johnson, J.L., Moore, W.E.C., Holdeman, L.V. and Chen, J.S., 1983. Acetone, isopropanol, and butanol production by Clostridium beijerinckii (syn. Clostridium butylicum) and Clostridium aurantibutyricum.*Applied and Environmental Microbiology,* 45(3):1160-1163.

Gheshlaghi, R.E.Z.A., Scharer, J.M., Moo-Young, M. and Chou, C.P., 2009. Metabolic pathways of clostridia for producing butanol. *Biotechnology Advances*, 27(6):764-781.

Godin, C. and Engasser, J.M., 1990. Two-stage continuous fermentation of Clostridium acetobutylicum: effects of pH and dilution rate. *Applied Microbiology and Biotechnology*, 33(3), pp.269-273.

Gottschal, J.C. and Morris, J.G., 1981. Non-production of acetone and butanol by Clostridium acetobutylicum during glucose-and ammonium-limitation in continuous culture. *Biotechnology Letters*, 3(9):525-530.

Green, E.M., 2011. Fermentative production of butanol—the industrial perspective. *Current Opinion in Biotechnology*, 22(3):.337-343.

Grobben, N.G., Eggink, G., Cuperus, F.P. and Huizing, H.J., 1993. Production of acetone, butanol and ethanol (ABE) from potato wastes: fermentation with integrated membrane extraction. *Applied Microbiology and Biotechnology*, 39(4-5):494-498.

Hartmanis, M.G., Klason, T. and Gatenbeck, S., 1984. Uptake and activation of acetate and butyrate in Clostridium acetobutylicum. *Applied Microbiology and Biotechnology*, 20(1), pp.66-71.

He, Q. and Chen, H.Z., 2013. Improved efficiency of butanol production by absorbent fermentation with a renewable carrier. *Biotechnology for Biofuels*, 6(1), p.1.

Huang, W.C., Ramey, D.E. and Yang, S.T., 2004. Continuous production of butanol by Clostridium acetobutylicum immobilized in a fibrous bed bioreactor. In Proceedings of the Twenty-Fifth Symposium on Biotechnology for Fuels and Chemicals Held May 4–7, 2003, in Breckenridge, CO (pp. 887-898). Humana Press.

Inui, M., Suda, M., Kimura, S., Yasuda, K., Suzuki, H., Toda, H., Yamamoto, S., Okino, S., Suzuki, N. and Yukawa, H., 2008. Expression of Clostridium acetobutylicum butanol synthetic genes in *Escherichia coli. Applied Microbiology and Biotechnology*, 77(6), pp.1305-1316

Isar, J. and Rangaswamy, V., 2012. Improved n-butanol production by solvent tolerant *Clostridium beijerinckii. Biomass and Bioenergy*, 37, pp.9-15.

Jones, D.T. and Woods, D.R., 1986. Acetone-butanol fermentation revisited. *Microbiological Reviews*, 50(4), pp.484.

Kamalaskar, L.B., Dhakephalkar, P.K., Meher, K.K. and Ranade, D.R., 2010. High biohydrogen yielding Clostridium sp. DMHC-10 isolated from sludge of distillery waste treatment plant. International *Journal of Hydrogen Energy*, 35(19), pp.10639-10644.

Kamalaskar, L., Kapse, N., Pore, S., Dhakephalkar, A.P., Ranade, D.R. and Dhakephalkar,P.K., 2016. Genome sequence and gene expression studies reveal novel hydrogenases mediated hydrogen production by *Clostridium Biohydrogenum* sp. nov., MCM B-509T. *International Journal of Hydrogen Energy*, 41(28) ,p.p 11990-11999.

Karakashev, D., Thomsen, A.B. and Angelidaki, I., 2007. Anaerobic biotechnological approaches for production of liquid energy carriers from biomass. *Biotechnology Letters*, 29(7), pp.1005-1012.

Keis, S., Bennett, C.F., Ward, V.K. and Jones, D.T., 1995. Taxonomy and phylogeny of industrial solvent-producing clostridia. *International Journal of Systematic and Evolutionary Microbiology*, 45(4), pp.693-705.

Khamaiseh, E.I.S., Hamid, A.A., Yusoff, W.M.W. and Kalil, M.S., 2013. Effect of some environmental parameters on biobutanol production by clostridium acetobutylicum ncimb 13357 in date fruit medium. *Pakistan Journal of Biological Sciences*, 16(20), p.1145.

Kheyrandish, M., Asadollahi, M.A., Jeihanipour, A., Doostmohammadi, M., Rismani-Yazdi, H. and Karimi, K., 2015. Direct production of acetone–butanol–ethanol from waste starch by free and immobilized *Clostridium acetobutylicum. Fuel*, 142, pp.129-133.

Kolesinska, B., Fraczyk, J., Binczarski,M., Modelska,M., Berlowska,J., Dziugan,P., Antolak,H., Kaminski, Z.J., Witonska, I. A. and Kregiel, D., 2019.Butanol Synthesis Routes for Biofuel Production:Trends and Perspectives. Material. 12, 350.

Lee, S.Y., Park, J.H., Jang, S.H., Nielsen, L.K., Kim, J. and Jung, K.S., 2008. Fermentative butanol production by Clostridia. *Biotechnology and Bioengineering,* 101(2), pp.209-228.

Lee, S.M., Cho, M.O., Park, C.H., Chung, Y.C., Kim, J.H., Sang, B.I. and Um, Y., 2008. Continuous butanol production using suspended and immobilized Clostridium beijerinckii NCIMB 8052 with supplementary butyrate. *Energy & Fuels,* 22(5), pp.3459-3464.

Lin, Y.J., Wen, Z.Q., Zhu, L., Lin, J.P. and Cen, P.L., 2013. Butanol production from corncob in the sequential co-culture of Clostridium thermocellum and Clostridium beijerinckii. *Journal of Chemical Engineering of Chinese Universities*, 3, p.p 016.

Madihah, M.S., Ariff, A.B., Sahaid, K.M., Suraini, A.A. and Karim, M.I.A., 2001. Direct fermentation of gelatinized sago starch to acetone–butanol–ethanol by Clostridium acetobutylicum. *World Journal of Microbiology and Biotechnology*, 17(6), pp.567-576.

Mermelstein, L.D., Papoutsakis, E.T., Petersen, D.J. and Bennett, G.N., 1993. Metabolic engineering of Clostridium acetobutylicum ATCC 824 for increased solvent production by enhancement of acetone formation enzyme activities using a synthetic acetone operon. *Biotechnology and Bioengineering,* 42(9), pp.1053-1060.

Minty, J.J., Singer, M.E., Scholz, S.A., Bae, C.H., Ahn, J.H., Foster, C.E., Liao, J.C. and Lin, X.N., 2013. Design and characterization of synthetic fungal-bacterial consortia for direct production of isobutanol from cellulosic biomass. *Proceedings of the National Academy of Sciences,* 110(36), pp.14592-14597.

Monot, F., Martin, J.R., Petitdemange, H. and Gay, R., 1982. Acetone and butanol production by Clostridium acetobutylicum in a synthetic medium. *Applied and Environmental Microbiology*, 44(6):1318-1324.

Nakayama, S., Kiyoshi, K., Kadokura, T. and Nakazato, A., 2011. Butanol production from crystalline cellulose by cocultured *Clostridium thermocellum* and *Clostridium saccharoperbutylacetonicum* N1-4. *Applied and Environmental Microbiology,* 77(18), pp.6470-6475.

Ndaba, B., Chiyanzu, I. and Marx, S., 2015. n-Butanol derived from biochemical and chemical routes: A review. *Biotechnology Reports,* 8, pp.1-9.

Ng, C.Y.C., Takahashi, K. and Liu, Z., 2016. Isolation, characterization, and optimization of an aerobic butanol producing bacterium from Singapore. *Biotechnology and Applied Biochemistry,* 63(1), pp.86-91.

Ni, Y. and Sun, Z., 2009. Recent progress on industrial fermentative production of acetone–butanol–ethanol by Clostridium acetobutylicum in China. *Applied Microbiology and Biotechnology*, 83(3):415-423.

Nielsen, D.R., Leonard, E., Yoon, S.H., Tseng, H.C., Yuan, C. and Prather, K.L.J., 2009. Engineering alternative butanol production platforms in heterologous bacteria. *Metabolic Engineering*, 11(4):262-273.

Nishio, N., Biebl, H. and Meiners, M., 1983. Effect of pH on the production of acetone and butanol by Clostridium acetobutylicum in a minimum medium. *Journal of Fermentation Technology*, 61(1):101-104.

Ounine, K., Petitdemange, H., Raval, G. and Gay, R., 1985. Regulation and butanol inhibition of D-xylose and D-glucose uptake in Clostridium acetobutylicum. *Applied and Environmental microbiology*, 49(4):874-878.

Pang, Z.W., Lu, W., Zhang, H., Liang, Z.W., Liang, J.J., Du, L.W., Duan, C.J. and Feng, J.X., 2016. Butanol production employing fed-batch fermentation by Clostridium acetobutylicum GX01 using alkali-pretreated sugarcane bagasse hydrolysed by enzymes from Thermoascus aurantiacus QS 7-2-4. *Bioresource Technology*, 212, pp.82-91.

Pasternak, K., Kocot, J. and Horecka, A., 2010. Biochemistry of magnesium. *Journal of Elementology*, 15(3):601-616.

Petitdemange, E., Fond, O., Caillet, F., Petitdemange, H. and Gay, R., 1983. A novel one step process for cellulose fermentation using mesophilic cellulolytic and glycolytic clostridia. *Biotechnology Letters*, 5(2):119-124.

Ponthein, W. and Cheirsilp, B., 2011. Development of Acetone Butanol Ethanol (ABE) Production from Palm Pressed Fiber by Mixed Culture of *Clostridium* sp. and *Bacillus* sp. *Energy Procedia*, 9:459-467.

Qureshi, N., Schripsema, J., Lienhardt, J. and Blaschek, H.P., 2000. Continuous solvent production by Clostridium beijerinckii BA101 immobilized by adsorption onto brick. *World Journal of Microbiology and Biotechnology*, 16(4), pp.377-382.

Qureshi, N., Ezeji, T.C., Ebener, J., Dien, B.S., Cotta, M.A. and Blaschek, H.P., 2008. Butanol production by Clostridium beijerinckii. Part I: use of acid and enzyme hydrolyzed corn fiber. *Bioresource Technology*, 99(13):5915-5922.

Qureshi, N., Saha, B.C., Dien, B., Hector, R.E. and Cotta, M.A., 2010. Production of butanol (a biofuel) from agricultural residues: Part I–Use of barley straw hydrolysate. *Biomass and Bioenergy*, 34(4):559-565.

Qureshi, N., Saha, B.C., Hector, R.E., Dien, B., Hughes, S., Liu, S., Iten, L., Bowman, M.J., Sarath, G. and Cotta, M.A., 2010. Production of butanol (a biofuel) from agricultural residues: Part II–Use of corn stover and switchgrass hydrolysates. *Biomass and Bioenergy*, 34(4):566-571.

Raganati, F., Olivieri, G., Procentese, A., Russo, M.E., Salatino, P. and Marzocchella, A., 2013. Butanol production by bioconversion of cheese whey in a continuous packed bed reactor. *Bioresource Technology*, 138:259-265.

Ramey, D.E., Environmental Energy, Inc., 1998. Continuous two stage, dual path anaerobic fermentation of butanol and other organic solvents using two different strains of bacteria. U.S. Patent 5,753,474.

Razak, M.N.A., Ibrahim, M.F., Yee, P.L., Hassan, M.A. and Abd-Aziz, S., 2013. Statistical optimization of biobutanol production from oil palm decanter cake hydrolysate by Clostridium acetobutylicum ATCC 824. *BioResources*, 8(2):1758-1770.

Richter, H., Qureshi, N., Heger, S., Dien, B., Cotta, M.A. and Angenent, L.T., 2012. Prolonged conversion of n butyrate to n butanol with Clostridium saccharoperbutylacetonicum in a two stage continuous culture with in situ product removal. *Biotechnology and Bioengineering*, 109(4), pp.913-921.

Shah, M.M. and Lee, Y.Y., 1994. Process improvement in acetone-butanol production from hardwood by simultaneous saccharification and extractive fermentation. *Applied Biochemistry and Biotechnology*, 45(1), pp.585-597.

Shaheen, R., Shirley, M. and Jones, D.T., 2000. Comparative fermentation studies of industrial strains belonging to four species of solvent-producing clostridia. *Journal of Molecular Microbiology and Biotechnology*, 2(1), pp.115-124.

Shen, C.R., Lan, E.I., Dekishima, Y., Baez, A., Cho, K.M. and Liao, J.C., 2011. Driving forces enable high-titer anaerobic 1-butanol synthesis in Escherichia coli. *Applied and Environmental Microbiology,* 77(9):2905-2915.

Singh K.G, 2017. Studies on anaerobic bacteria producing butyric acid and butanol from distillery waste, Savitribai Phule Pune University, Pune, India.

Singh, K., Lapsiya, K., Gophane, R. and Ranade, D.R., 2016. Optimization for butanol production using Plackett-Burman Design coupled with Central Composite Design by *Clostridiumbeijerinckii* strain CHTa isolated from distillery waste manure, *Journal of Biochemical Technology*., 7(1):1063-1068.

Singh, K., Gedam, P., Raut, A., Dhamole, P., Dhakephalkar, P. and Ranade D.R., 2017. Enhanced n-butanol production by *Clostridium beijerinckii* MCMB 581 in presence of selected surfactant, 3 *Biotech*, 7(3):161.

Soni, B.K., Das, K. and Ghose, T.K., 1982. Bioconversion of agro-wastes into acetone butanol. *Biotechnology Letters*, 4(1), pp.19-22.

Speranza, G., Corti, S., Fontana, G., Manitto, P., Galli, A., Scarpellini, M. and Chialva, F., 1997. Conversion of meso-2, 3-butanediol into 2-butanol by lactobacilli. Stereochemical and enzymatic aspects. *Journal of Agricultural and Food Chemistry*, 45(9), pp.3476-3480.

Steen, E.J., Chan, R., Prasad, N., Myers, S., Petzold, C.J., Redding, A., Ouellet, M. and Keasling, J.D., 2008. Metabolic engineering of Saccharomyces cerevisiae for the production of n-butanol. *Microbial Cell Factories*, 7(1), p.1.

Suib, S.L.ed. 2013. New and future developments in catalysis: catalytic biomass conversion. Newnes.

Survase, S.A., Jurgens, G., Van Heiningen, A. and Granström, T., 2011. Continuous production of isopropanol and butanol using Clostridium beijerinckii DSM 6423. *Applied Microbiology and Biotechnology,* 91(5):1305-1313.

Suzuki, T., Mori, H., Yamané, T. and Shimizu, S., 1985. Automatic supplementation of minerals in fed batch culture to high cell mass concentration. *Biotechnology and Bioengineering,* 27(2):192-201.

Tashiro, Y., Shinto, H., Hayashi, M., Baba, S.I., Kobayashi, G. and Sonomoto, K., 2007. Novel high-efficient butanol production from butyrate by non-growing Clostridium saccharoperbutylacetonicum N1-4 (ATCC 13564) with methyl viologen. *Journal of bioscience and bioengineering,* 104(3), pp.238-240.

Thang, V.H., Kanda, K. and Kobayashi, G., 2010. Production of acetone–butanol–ethanol (ABE) in direct fermentation of cassava by *Clostridium saccharoperbutylacetonicum* N1-4. *Applied Biochemistry and Biotechnology*, 161(1-8), pp.157-170.

Tiam mun, L., Ishizaki, A., Yoshino, S. and Furukawa, K., 1995. Production of acetone, butanol and ethanol from palm oil waste by *Clostridium saccharoperbutylacetonicum* N1-4. *Biotechnology Letters,* 17(6):649-654.

Ventura, J.R.S., Hu, H. and Jahng, D., 2013. Enhanced butanol production in Clostridium acetobutylicum ATCC 824 by double overexpression of 6-phosphofructokinase and pyruvate kinase genes. *Applied Microbiology and Biotechnology*, 97(16):7505-7516.

Vinod, K.U., Harvey, K.S., and Anyanwu, M.Z., 2014. Algal biomass as a global source of transport fuels: Overview and development perspectives. *Progress in Natural Science*, 4, p.006.

Wen, Z., Wu, M., lin, Y., Yang, L., Lin ,J. and Cen, P., 2014. Artificial symbiosis for acetone-butanol-ethanol fermentation from alkali extracted deshelled corn cobs by co-culture of *Clostridium beijerinckii* and *Clostridium cellulovorans*. *Microbial Cell Factories*, 13(1):92.

Wypych, G., 2001. Solvent Use in Various Industries. Handbook of Solvents. Ontario, ChemTec Publishing, 2001., pp.847-922.

Yang, S.T., The Ohio State University, 2008. Methods of producing butanol. U.S. Patent Application 12/062,091.

Yu, E.K.C., Chan, M.H. and Saddler, J.N., 1985. Butanol production from cellulosic substrates by sequential co-culture ofClostridium thermocellum andC. acetobutylicum. *Biotechnology Letters*, 7(7), pp.509-514.

Yue, D; Slivinsky, M; Sumpter, J; You, F., 2014. Sustainable design and operation of cellulosic bioelectricity supply chain networks with life cycle economic, environmental, and social optimization. *Industrial and Engineering Chemistry Research*, 53, 4008–4029.

Zverlov, V.V., Berezina, O., Velikodvorskaya, G.A. and Schwarz, W.H., 2006. Bacterial acetone and butanol production by industrial fermentation in the Soviet Union: use of hydrolyzed agricultural waste for biorefinery. *Applied Microbiology and Biotechnology*, 71(5):587-597.

11

Anaerobic Process of Solid-Waste Management and Bio-composting

Shraddha Shaligram Vajjhala[1], Vikas Patil[1], Ashish Polkade[2*]*

[1]National Centre for Microbial Resource (NCMR) –National Centre for Cell Science (NCCS), Pune-21, Maharashtra, India
[2]Bharat Eco Solutions and Technologies, A19, Meera Classics, Aundh Ravet BRTS Road, Kalewadi, Pune-57, Maharashtra, India

Abstract

Solid Waste Management (SWM) has become a significant global concern, specifically for developing countries like India because of increasing population, rapid urbanization and high rate of municipal solid waste generation. Bio-composting and anaerobic digestion are the methods majorly used for solid waste management. Present chapter comprises detailed information on methods, microbiology involved and pros and cons of these methods to provide the detail information of solid waste management processes to the readers. Variable nature of substrates are available for bio-composting process. It includes sludge from effluent treatment plant (ETP), sewage treatment plant (STP) and bio-toilets, municipal solid waste, food waste and garden waste etc. Nutritional content, carbon /nitrogen (C/N) ratio of the substrate, available space and cost of the process are the main factors that affect selection method for waste management. Anaerobic digestion is the preferred method of solid waste management in comparison to other processes of composting due to methane generation as biogas and bioconversion of organic compounds to manure.

Keywords: Waste, Anaerobic digestion, Composting, Sludge, Biogas

**Corresponding Authors: shraddhashaligram1386@gmail.com; director@visionecologica.in*

Introduction

Almost every human activity leaves behind some or the other kind of waste. In layman language solid waste can be defined as "Garbage" or "Refuse" generated by various human activities. The Resource Conservation and Recovery Act (RCRA) defines solid waste as waste generated from sludge of wastewater treatment plant, water supply treatment plant, or facilities for air pollution control and other rejected material from industrial, commercial, mining, food industries and agricultural processes, and mainly from community activities. It has been estimated that the human population will be around 9 billion by 2050 (Lin et al. 2018) which will increase the demand for energy, water and food. India is a diverse country with world's second largest population with different religions, culture and traditions, developing day and night therefor demand for resources required for living are high. This in turn is inversely proportional to high rate of solid waste generation. Rapid growth in population and continuous improvement in standard of living makes it more difficult to attain the concept of sustainable development in this scenario. Waste are disposed negligently in open dumps which causes serious environmental as well as health issues due to the presence of contaminants and pathogens. This kind of unplanned approach of waste management which leads to health and environmental issues needs immediate attention (Ali et al. 1999)

To combat this issue there is an urgent need to manage our waste and recycle it for conserving our "Mother Nature". Managing solid waste is the biggest problem for India as the urbanization and industrialization is increasing day to day (Kumar et al. 2017). Metro cities or cities with elevated population are a major task for operative SWM. If there are no strict measures or actions taken to control the generation of organic waste it will lead to deterioration of soil, water and air which will directly have a destructive impact on food, energy and water supplies. Thus, use of specific low cost methods for reuse and treatment of waste can lend a helping hand towards this disturbing situation. We need to reduce the volumes of waste or come up with a strategy for environmental disposal of waste to avoid further impacts on the environment and public health.

The contents of municipal solid waste are refuse generated by household activities, industries, agriculture and sewage etc. It has waste, compostable and recyclable materials, with the municipality overseeing its disposal. Classically, this garbage is collected, separated and sent to either a landfill or a municipal recycling centre for processing. With megacities spurting a growth of 30.47% (Census 2011), India's basic necessities have sometimes been ignored. With an increasing focus towards services such as water, electricity and food for the growing population, the Indian administration has unfortunately ignored another major public service: waste management. According to the Press Information

Bureau, India generates 62 million tonnes of waste (mixed waste containing both recyclable and non-recyclable waste) every year, with an average annual growth rate of 4% (PIB 2016). This generated waste can be divided into three major categories: Organic (all kinds of biodegradable waste), dry (or recyclable waste) and biomedical (or sanitary and hazardous waste). Out of this 62 million tonnes waste less than 60% is collected and around 15% processed see Figure 2.

Types of waste

MSW generally falls in following five major classes.

1. **Biodegradable Waste:** These types of waste are biodegradable and majorly include green waste like vegetables, flowers, leaves, fruits, food and kitchen waste.
2. **Recyclable Waste:** It includes waste which can be recycled and reused. Examples of this kind of wastes are paper, glass, metals and recyclable grade plastic etc.
3. **Inert waste:** This includes waste which is inert in nature like dirt and debris.
4. **Composite waste:** This includes tetra packs, waste clothing etc.
5. **Domestic hazardous/toxic waste:** This kind of waste generally considered as hazardous to the life forms and needs special treatment plants. It includes waste medicines, e-waste, paints, polishes, lights fertilizers, pesticides and many others.

The per capita MSW generation in India ranges from 0.17 kg to 0.62 kg per day (Kumar et al. 2017). According to an estimate India generates about 133760 tonnes of MSW per day. Only 68% of the generated waste (around 91152 tonnes) is collected and only 19.35% of collected waste subject to treatment process. Rest are unmanaged and leave as such which creates health and environmental issues. Indian MSW approximately contains 40–60% compostable, 30–50% inert and 10 -30% recyclable waste. Analysis carried out by National Environmental Engineering Research Institute (NEERI) reveals that in totality Indian waste consists of nitrogen content (0.64 ± 0.8) %, phosphorus (0.67 ± 0.15)%, potassium (0.68 ± 0.15)%, and C/N ratio (26 ± 5) % (Joshi and Ahmed 2016).

Methods of management of solid waste

Major portion of solid waste contains bio-degradable contents in the form of domestic, food, agricultural, industrial and hotel-kitchen waste etc. There are several treatment methods of waste management which has been developed but currently two major processes anaerobic digestion (AD) and composting are employed for management of biodegradable solid waste.

Thermal Treatment: This method uses heat to treat the waste. Different forms of thermal treatment includes

a. **Incineration:** This approach allows combustion of waste material in the presence of oxygen.

b. **Gasification and Pyrolysis:** These are almost similar methods with minor variations. Gasification involves decomposition of organic components of waste by exposing them to high temperature and low oxygen. Pyrolysis uses absolutely no oxygen as compared to gasification which permits a very low amount of oxygen in the process. Gasification is more profitable as it allows the burning process to recover energy without causing air pollution.

c. **Open Burning:** This type of treatment is the traditional method of solid waste management but harmful for the environment. Incinerators lack pollution regulator devices to monitor the degree of environmental pollutants they release. They release hazardous substances such as hex-chlorobenzene, dioxins, carbon monoxide, particulate matter, volatile organic compounds, polycyclic aromatic compounds, and ash. Regrettably this procedure is still practiced by many local authorities due its inexpensive solution to solid waste (Nandan et al. 2017).

Dumps and landfills: A landfill is a site for the disposal of waste materials by burial. Landfilling involves management of waste on small or large land area away from society.

a. **Sanitary landfills** are the most commonly used solution for solid waste treatment. These landfills are designed to either remove or reduce the risk of hazard to environment or public. Landfills sites are situated where landscapes work as natural buffers between the environment and the landfill.

b. **Bioreactor landfills:** This type of landfills is the result of recent research in solid waste management technology.

Bioreactor landfills employ microorganisms to speed up waste decomposition. It is based on continuous addition of liquid to the waste materials to provide optimum moisture conditions for rapid microbial actions. For this the landfill leachate is re-circulated and whenever the amount of leachate is not adequate, liquid waste such as sewage sludge is used to maintain the moisture level(Nandan et al. 2017).

Biological Waste Treatment

Anaerobic digestion and composting are the two plausible approaches for organic waste management. Fig. 1 shows detailed composite of both AD and composting. Composting is most frequently used waste disposal or treatment method under controlled aerobic decomposition of organic waste materials by exploiting of small invertebrates and microorganisms. The most common composting techniques include static pile composting, vermin-composting, windrow composting (mostly used in agricultural field) and in-vessel composting. Anaerobic Digestion uses biological processes to decompose organic materials in absence of oxygen. There is no supply of oxygen or air in this kind of digestion process and anaerobic bacteria plays a key role in digestion of waste. Anaerobic digestion and composting techniques can be combined to get Biogas and compost making this process sustainable. The AD requires water and solid feedstock as an input and generates biogas which comprises of the major component as methane along with CO_2 and H_2S. Biogas needs to be purified through H_2S and CO_2 scrubbers. Pure methane can be used to generate energy and use as electrical power or it can also be used as Bio -CNG for fuel. Solids can be separated using solid-liquid separators or screw press and liquid can be recirculated with water. Normally 4-6% solids can be recovered through this process other feedstock can be used in the composting process. Effluent after solid separation can also be converted to organic fertilizer after some value addition (Mollazadeh 2015).

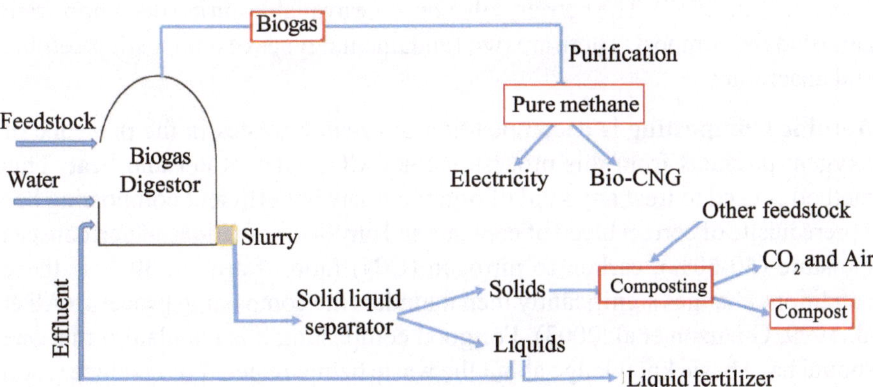

Fig. 1: Schematic diagram showing Input and output system of Anaerobic Digestion and composting system

Composting

Composting is traditional practice of waste management which has been commenced anciently from small scale e.g. home to big scale *viz.* industries.

The rapid development in urbanisation and rise in amount of generated wastes causes landfills to reach their capacity and reject organic wastes.

Composting is a viable solution of this issue of organic waste treatment. Additionally the final product is compost which is valuable for soil resource as it has uses in agriculture, horticulture etc. MSW has almost 40% organic waste, hence composting is the best way to reduce the load of waste management along with generation of nutrient rich compost for soil. If we discuss in detail about composting and its benefits, composting converts organic waste to humus like material which augments physical, chemical and biological properties of soil. Composting of MSW has many advantages which includes reduction in the volume of waste, generation of high temperature in this process allows killing of pathogens which may be present in the waste. In addition it discourages germination of weeds in agricultural fields and degrades putrid compounds. Now a day's people are attracted towards organic farming products. Hence production of organic grade MSW compost is gaining popularity due to its key role in improving the biological, physical and chemical properties of soil.

Another term for composting is "Microbial farming". Basically microorganisms require carbon as energy source and nitrogen for building proteins. Microorganisms secrete enzymes to break down complex carbohydrates into simpler forms, and use them as food. Microorganisms are so combust that they continue this process until the remaining nutrients are consumed by the last microorganisms. Most of the carbon is converted into carbon dioxide and water (Nandan et al. 2017). The nutrients that become available during decomposition persist in the compost. There are two fundamental types of composting aerobic and anaerobic:

Aerobic Composting is decomposition of organic wastes in the presence of oxygen; products from this process include CO_2, NH_3, water and heat. This method is used to treat any kind of organic waste but efficient composting has a prerequisite of correct blend of contents and physical conditions which consists moisture (40-60%), carbon to nitrogen (C/N) ratio of around 30:1. If these conditions changes significantly then it hinders the composting process. (Ali et al. 1999, Guanzon et al. 2003). For good composting it is mandatory that one should have basic knowledge about the waste being treated. For instance wood and paper serve as carbon source while sewage and food waste provides rich source of nitrogen for microbes involved in this process. Another important aspect is the reasonable supply of oxygen throughout the process which is obligatory for the good composting.

Anaerobic Composting is the decomposition of organic wastes in the absence of oxygen. It generates methane CH_4, CO_2, NH_3 and trace amounts of other gases and organic acids. Anaerobic composting traditionally used to treat the animal manure and human sewage sludge. Recently it gained popularity for municipal solid waste (MSW) and green waste management. It has been noted that anaerobic composting requires an entirely different set of microorganisms and conditions than aerobic composting process.

Both anaerobic and aerobic decay produce heat as a by-product. The temperatures in an aerobic system can become high enough to kill pathogens or weed seeds. This phenomenon is absent in a digester system. However, the digester's acidic environment itself eventually does the trick. The inhospitable environment takes six months to a year to kill off dangerous microorganisms (Mehta and Kanak 2018).

Why composting?

If waste or organic matter is directly added to the soil without composting, it results in a change in the ecosystem (Kononova et al. 1966). If organic matters of the waste are not sufficiently humidified and comes in contact with soil it lead to degradation of the precious microflora resulting in production of transitional metabolites which are not companionable with normal plant growth (Zucconi et al. 1981). In addition if the decomposition procedure is not done completely there will be a fight for nutrients especially nitrogen between microorganisms and plant roots which leads to high C: N ratio with production of ammonia in soil (Golueke 1977).

Therefore, composting is a way out from all these issues which provides a stable product from biological oxidative transformation. Composting, if controlled properly, will give safe and agriculturally useful product. Industrial methods have been developed which produce compost in a short time that is compatible with agricultural use. The main prerequisite for compost is that it should be appropriate for agricultural use as an organic soil improvement, for further explanation physical, chemical and biological stability, non-phytotoxicity and balance among mineral elements are the essential characteristics for compost to be useful for the soil and for crops (Loehr 1977, Sopper and Kerr 1979 Brinton 1979, de Bertoldi et al. 1983).

Anaerobic digestion (AD)

AD as name suggests is a biological process that converts organic matter to biogas and digestate in absence of oxygen. AD serves as method for degradation of variety of organic waste such as lignocellulose biomass, food waste, animal waste and manure. Sewage sludge also work as one of the important substrate for this process. AD digestion has two end products one is gaseous i.e. biogas mainly methane which is well known source of renewable energy and other is the digestate which is rich in nutrients and can be used for soil amendment (Lin et al. 2018). Hence using AD can not only alleviate the risks associated with environment health but it also nourishes the concept of food, energy and water security. AD allows for energy recovery through the formation of biogas. The latter can be locally exploited by a combined heat and power (CHP) generator, to cover the thermal energy requirements for AD and produce electric energy for distribution within the electricity network.

Steps of Anaerobic Digestion

There are three major steps for anaerobic digestion

Hydrolysis: The first step of the AD is the hydrolysis of complex organic material to its monomers by the hydrolytic enzymes. Amylases, cellulases, hemicelluloses, proteases and lipases are important enzymes involved in the hydrolysis process. The improvement of biogas production from lignocellulolytic materials and improved methane yield from lipid-rich waste are due to pre-treatment with cellulase and lipase-producing microorganisms respectively.

Bacterial community in hydrolysis comprises strict anaerobes such as *Bacteriocides, Clostridia* and *Bifidobacteria* with some facultative anaerobes such as *Streptococci* and *Enterobactericeae* (Parawira 2012).

Acidogenesis: The simpler organics obtained from hydrolytic step are then fermented to organic acids and hydrogen by the fermentative bacteria which are commonly known as acidogens. The volatile organic acids are converted into acetate and hydrogen by the acetogenic bacteria like *Dendrosporobacter quercicolus* (Dar et al. 2008). The volatile fatty acids involved in process are acetic acid, propionic acid, isobutyric acid, butyric acid, isovaleric acid, and valeric acid (Lee et al. 2015).

Methanogenesis

Archael methanogens use hydrogen and acetic acid to convert them into methane. Methane production from acetic acid is done by acetoclastic methanogens and hydrogen and carbon dioxide is carried out by hydrogenotrophic methanogens. Appropriate environmental conditions play key role in acetogensis and methanogenesis simultaneously which ultimately leads to methane formation.

Types and role of Microbes in both AD and composting

AD is carried out by four different types of microorganisms. These include hydrolytic, fermentative, acetogenic and methanogenic (Braber 1995). Depolymerzation take place in the first step of AD. Polymeric solids are depolymerised into smaller molecules by hydrolytic bacteria such as *Bacteroides succinogenes, Clostridium lochhadii, Clostridium cellobioporus, Ruminococcus flavefaciens, Rumminococcus albus, Butyrivibrio fibrisolvens, Clostridium thermocellum, Clostridium stercorarium,* and *Micromonospora bispora .Clostridia* are the predominant bacteria in AD of MSW. It can be stated that 50% of organic matter are degraded in this stage. Intermediary reactions also happen: for example, products of depolymerization reactions are altered to fermentation products (Chynoweth and Pullammanappallil 1996). The concluding step is methanogenesis. Methanogens are slow-growing anaerobe that has the talent to degrade acetate, methanol, carbon dioxide, formate, methyl mercaptans ,carbon monoxide, methylamines and also reduced metals (Chynoweth and Pullammanappallil 1996). To operate anaerobic digesters effectively in acidic conditions, methanogens that has ability to utilize hydrogen are desirable as they can grow proficiently at acidic pH and can stand high volatile fatty acids (VFA). An acid tolerant hydrogenotrophic methanogen viz. *Methanobrevibacter acididurans* is an acid tolerant hydrogenotrophic methanogen who's isolation source is anaerobic digester running on alcohol distillery wastewater has been described earlier by (Savant and Ranade 2004). This organism may propagate optimally at pH 6.0. In the experiments reported herein, *M. acididurans* showed better methanogenesis under acidic environment with high VFA, particularly acetate, than *Methanobacterium bryantii*, a common hydrogenotrophic inhabitant of anaerobic digesters. Addition of *M. acididurans* culture to digesting slurry of acidogenic as well as methanogenic digesters running on distillery wastewater presented escalation in methane production and reduction in accumulation of volatile fatty acids. The results verified the practicability of application of *M. acididurans* in anaerobic digesters.

Microbiology of Composting

A plethora of microorganisms exists in a composting system. Though the microbial population is immense, bacteria and fungi are the most important microorganisms for composting (Haug 1993). To further discuss about the thermodynamics of composting, there are three classes of microorganisms which are involved, cryophiles also known as psychrofiles, mesophiles and thermophiles (Stoffela and Kahn 2001). For cryophilic bacteria the optimum temperature is around 13 °C. Mesophilic bacteria are found to be predominant between 0 °C to 40°C. Above this temperature, thermophilic bacteria work rapidly. Mesophilic bacteria again come in action when the temperature of compost reduces (Wassenaar 2003). Considering the structure of rows of long piles called as windrow's the role of mesophilic and thermophilic microorganisms is central. The function of mesophiles and thermophiles can be discussed in four stages. In the first step the substrate for mesophile are abundant and hence they are predominant and active at this stage. Heat generation in this stage increases the temperature of the compost pile. Mesophiles are known for temperature range between 35 ºC and 45 ºC (Miller 1996, Stofella and Kahn 2001). As the process continues the rise in temperature makes environment suitable for thermophilic microbes. Once the compost pile reaches about 65 ºC to 70 ºC the food sources reduces for microorganisms eventually the temperature falls which results in additional mesophilic stage. Temperature at this stage falls to ambient temperature again (Miller 1996, Stofella and Kahn 2001). Articles suggest that almost 80 to 90% of the microbial activity during composting is due to bacteria (Haug 1993, Stofella and Kahn 2001). These compose of the following species *Bacillus, Pseudomonas, Arthrobacter, Alicaligenes* (Stofella and Kahn 2001), as well as *Staphilococci* (Hassen et al. 2001) which is active in the mesophilic stage. In thermophilic stage, bacteria are predominantly of the *Bacillus sp*., as *B. subtilis, B. stearothermophilic*, and *B. licheniformis*. Temperature above 65°C is handled by of *B. stearothermophilus* (Miller 1996).

Substrate variations

Bio-degradable waste which is known as substrates for the waste treatment process can be categorised in various categories on the basis of their source of origin, characteristics, contents and composition. Identification or selection of method to treat this type of waste is to be finalized on the basis of factors like space availability, economy, C/N ratio (Table number. 2) or other chemical characters. Substrates and their C/N ratio is mentioned in the Table 2.

Anaerobic digestion of Human Excreta and Bio-toilets

Human waste treatment and disposal has been one of the major sanitation problems in developing countries like India. In India, under the small budget sanitation programme, latrines are connected to biogas plants running on cattle dung to achieve hygienic disposal of human night soil (HNS). Anaerobic digestion offers a good alternative for human waste treatment but the presence of enteric bacterial pathogens existing in Human Night Soil (HNS) is still a major concern considering the hygienic safety . A two-stage anaerobic digestion process which consists of two separate digesters (acidogenic and methanogenic) was designed. Efficacy of this two stage AD process to inactivate *Salmonella typhi* was compared with a single stage digestion process. The two-stage process was efficient in biogas generation from HNS. Thus, the two-stage process ensures complete hygienic safety in anaerobic digestion of human night soil (Kunte et al. 2004). There are researchers who studied anaerobic digestion process of HNS with addition of cattle dung slurry which is promoted in villages of Indian. This is widely used for disposal of human excreta as an alternative to conventional sanitary systems. It is thought that intestinal pathogens may get eliminated during AD however there are no conclusive reports for the same. The saprophytic organisms which are present in the fermenting mass mark it impossible to detect the residual pathogens. Use of an antibiotic-resistant strain of *Salmonella typhimurium* as a test organism to learn its persistence during AD displayed that the organism can be entirely eradicated in nine days(Gadre et al. 1986).

Bio toilets are a type of waste management unit which completely manages human waste (mainly human faeces) and converts in into biogas and water, with the help of bacterial activities. Bio-digester toilets are designed in order to convert human waste into gases and manure. There is technology called zero waste bio-digester technology which uses psychrotrophic bacteria like *Clostridium* and *Methanosarcina* capable of surviving in cold or hot environment and dwell on human excreta by breaking it down to gas and water (Ali et al. 2013). Best practical example of bio toilet is Indian railways (Sharma et al. 2013, Salunke et al. 2013). In these bio-toilets, human waste is treated by bacteria which are benign to the humans. The gases generated in this escape's to atmosphere and treated waste water is discharged after chlorination (Naik et al. 2016).

Garden waste / lignocellulosic waste

This category of waste includes agricultural residues, energy crops, forestry residues, garden trimmings, cellulosic remains at green vegetable market etc. This kind of waste has high C/N ratio and its biodegradability depends on the composition and nature of waste. Organic components present in this kind of waste are cellulose, hemicellulose and lignin. There are reports which stated that the degradation of cellulose and hemicellulose in anaerobic digestion is lower than composting (Lin et al. 2014). It is also noticed that lignin is barely degraded in AD due to absence of aerobic microbes which are mainly participate in the process of lignin degradation, if one want to degrade lignin in AD then pretreatment is prerequisite. These kind of substrates are also been used for ethanol production after pre-treatment with dilute acid or acidophiles. Market waste consisting of rotten vegetables including potatoes , onions and fruit skin etc. are subjected to anaerobic digestion with different parameters to increase the biogas production and the slurry formed is used as a manure (Ranade et al.1987). Food waste includes waste from industrial sector as well as waste from municipal and all the other food sectors. Food waste has low C/N ratio also the pH of food waste is around 4-5. It has high amount of soluble organic waste which can be easily degraded and can be converted into intermediates. The composition of food waste is carbohydrates, fats and proteins. It is well known fact that fats has a highest methane potential which is followed by carbohydrates and proteins (Muzenda 2014). Overall it can be said that food waste is a good substrate for AD. Composting of food waste needs complete attention and monitoring since there are chances in developing bad odour if composting process is inclined towards anaerobic due to generation of ammonia, sulphur and some volatile components. In India, nearly 34 food parks are approved where commercial production of juices and other food items are been produced at large scale. Which generates tons of waste with varied compositions, in such cases research base decision is required wherein substrates, their available quantum, season of generation, available storage place is to be taken in consideration. After study of its biogas potential and other available nutritional parameters selection of method to process it can be advised. Vision Ecologica Pvt Ltd, Pune had successfully processed 90 tonnes/ day waste from the part of north India with anaerobic digestion. Ratio of mixing of different waste generated in Food Park can be decided on the basis of its characteristics.

Sewage waste

Sewage waste is also known as bio-solids and it is the by-product of municipal wastewater treatment plants (Kumar et al. 2017). Sewage waste is somewhat similar to food waste as it also has low C/N ratio. Degradation or composting of this waste also causes odour generation. It was observed that the presence of polymers in the sewage waste makes it recalcitrant in both AD and composting (Raman et al. 2017). Sewage waste has a very serious issue of human pathogens present in faeces which is also leading to spread of antimicrobial resistance (AMR) along with spread of contagious diseases (Taneja and Sharma 2019, Hendriksen et al. 2019). Hence proper treatment strategy with respect to AMR should also be considered for sewage waste. Transmission of resistance genes to human commensal and pathogenic bacteria are encouraged by the environmental antibiotic pollution (Rutgersson et al. 2014). In particular, waste water treatment plants helping facilities which produce antibiotics have been involved in the transferral of antibiotic resistance genes into human microbiota which is serious threat to antibiotic effectiveness given the size of India's pharmaceutical sector (Johnning et al. 2013). There are no regulations governing the discharge of antimicrobial waste into the environment, and these are needed (Laxminarayan and Chaudhury 2016)

Current reports have shown that pharmaceutical industries also add to this issue of AMR. A major portion of antibiotic need is globally produced by countries such as India and China. Reports also suggest that 90 per cent of pharmaceutical wastewater treatment plants release production residues into either water or soil or even both. This in turn provides ideal breeding grounds for drug-resistant bacteria. The situation now is crucial and AMR researchers should work with scientists and industries to catch out the mode of AMR transmission and help learn the control mechanisms. This approach will surely help to overcome AMR.There are few solutions for this problem which are discussed below:

Microbial enzyme based treatment option

Conventional methods if compared with advanced methods for waste water treatment such as reverse osmosis, membrane bioreactors and nano filteration are found to be more capable processes for removal of micro pollutants with nanogram per litre (ng/L) concentration levels (Huang and Lee 2015).

Since last twenty years researchers has extensively studied about how enzymes can contribute for bioremediation of various pollutants (Pandey et al. 2017, Aitken 1993).Various types of bacteria and fungi produce active enzymes. These enzymes are grafted in membrane filters or used in membrane bioreactor to eliminate toxic pharmaceutical pollutants which results in purification of wastewater.Enzymes such as laccase, lipase, ligninnases, and cellulose are widely used for inactivation of antibiotics in effluents which consequently prevents the pollution of our environment. (https://www.teriin.org/blog/drying-antimicrobial-resistance-innovative-wastewater-treatment)

The Energy and Resources Institute's (TERI) method

The Energy and Resources Institute (TERI), in association with The University of Pannonia, Hungary, is on the brink of developing a simple treatment system to remediate pharmaceuticals from wastewater. According to preliminary results, immobilisation and/or binding of these enzymes to membranes can effectively break down/eliminate micro-pollutants. In this innovation, the bacterial enzymes embedded in the membrane reactor help degrade antibiotics (de Cazes et al. 2014). Nanomaterials are employed for thorough removal of bio-transferred intermediants and for elimination of antimicrobial resistant bacteria from wastewater (Amin et al. 2014). These rapid and nifty enzymes has capacity to treat wastewater without toxic by-products or generating biomass. They also require low energy and operational costs. This treatment strategy led to effective decontamination of antibiotic residues. These enzymatic membrane reactor structure will be made more advance for commercial use by eliminating existing process limitations. The projected enzymatic technology leads to sustainable and greener processes, while treating emerging pharmaceuticals in wastewater (de Cazes et al. 2014).

Municipal Solid Waste (MSW)

Biodegradable MSW includes food and kitchen waste, leaves, flowers, vegetables and non-vegetarian food scraps. It also includes inert waste, electronic waste, and toxic waste. AD as well as composting both the methods has been successfully implemented for management of biodegradable waste from MSW (Joshi and Ahmed 2016). There are methods which are using integrated AD with composting to manage MSW with biogas and manure/ compost as valuable end products.

Possible Public consequences for use of composting processes

It is well known fact that urban solid waste (mostly sewage) contains pathogenic microorganisms which create risk for public and environmental health (Gaby 1975, Alderslade 1980, Dean and Lund 1981). Sludge (faecal origin) is richer in pathogens than any other municipal solid waste. Studies have listed the pathogenic microorganisms isolated from solid urban waste and sewage sludge. It includes bacteria, viruses, filamentous fungi, yeast and parasites. It is advisable that for free usage of compost, the end product should have low concentration of pathogen. It is also necessary that there should be some assurance that that there should be no regrowth of pathogens (Kawata et al. 1977, Burge et al. 1978, Connery et al. 1979, Dean & Lund 1981, World Health Organization 1981). Temperature plays a very important role in decreasing the pathogenic organisms during the complete process. As the temperature rises up to 70°C for 30 min or 65°C for several hours it can be presumed that a sufficiently germ-free end product is obtained (Kawata et al. 1977, Connery et al. 1979).There are reports which have mentioned the presence of antibiotic resistance (AMR) bacteria in waste, which is another serious matter to be considered. Many articles stated that AMR will the leading cause for human death in near future (Cassini et al. 2018) hence there is a need to treat and monitor waste for AMR. Another way to handle this situation would be sensible use of antibiotics by us and knowledge of effluent treatment or disposal method by the industries.

Co-digestion

Anaerobic Co-digestion of Sewage Sludge

Sewage sludge becomes harder to hydrolyse and digest due to the presence of complex microstructures which includes extracellular polymeric substances. These substances have low volatile solid degradation (30–50%) even at a long retention time (20–30 day) this in turn causes low methane yield (Appels et al. 2008, Pereira et al. 2015). To resolve this problem for better hydrolysis and methane production pre-treatment can be employed. Variety of pre- treatment are available such as mechanical, thermal, chemical, biological process or integration of these, have been developed at laboratory or pilot level recently (Kim et al. 2010, Houtmeyers et al. 2014, Zhen et al. 2017). But this comes with few disadvantages as well. Pre-treatment increases the yield of methane but the energy demand and capital cost is high. Moreover this process is complex to operate and maintain which makes it uneconomical. Hence these kinds of methods mostly exist in research stage but hard to employ practically.

Simultaneous digestion in absence of air of a homogenous mixture of substrates is called as anaerobic co-digestion. The combination of two or multiple substrates or any kind of biomass resource improves the economic viability of AD as high amount of methane is produced. (Alvarez et al. 2000), and hence is an eminent and realistic route to overcome the drawbacks of a traditional digestion process. C/N ratio can be improved in co-digestion which in turn enhances biogas production, manages nutrient imbalance of substrates and also creates a stable condition for AD (Kumaran et al. 2016). An example of AD co-digestion is the digestion of sewage sludge with MSW, livestock manure, industrial waste etc. as co substrates. There are reports where it has been demonstrated that improving C/N ratio by anaerobically co-digesting sewage sludge with waste which have higher organic content such as food waste leads to increase in biogas production (Tanimu et al. 2014).

Organic fraction of municipal solid waste (OFMSW) with sewage sludge is the most reported research example of co-digestion (Alvarez et al. 2014). As compared to OFMSW composting or traditional method one study has reported that OFMSW used as co-substrates in waste water treatment plant (WWTPs) had economic and environmental benefits (Krupp et al. 2005). Hence introducing anaerobic co-digestion can be an intelligent move to address current situation of OFMSW. Partial amount of the municipal waste is made up of food waste, and by 2020 the amount of municipal waste will increase beyond expectation (Tarmudi et al. 2009). Traditionally dumping of solid waste on landfill is used as cheap way for its management but the increasing load will need more dumpsites in future. The extraordinary amount of food waste at landfill would lead to concerns such as foul odour, toxic leachate and vermin infestation (Lee et al. 2007). Currently there is no thumb rule to tackle this issue. The only possibility is co-digestion of sewage sludge with food waste.

AD/ Bio-methanation as most preferred way for solid waste management for technology reasons

Man has made an artificial system called as biogas reactors for production of energy which exploits anaerobic microorganism with ability to decompose complex organic matter. This has been in trend since centuries. The biogas (methane) produced in this process has high economic value as it is an alternate for fossil fuel. AD is perfect blend of production of renewable energy with environmental benefits, such as decline of greenhouse emissions, reduction of odour's and organised way of discarding waste completely (Labatut and Pronto 2018). AD caters for recycling of nutrients such as nitrogen and phosphorous as the end products can be useful on agricultural land as bio-fertilizer, which has the potential to substitute artificial fertilizers (Campos et al. 2019). AD allows conserving energy in the form of reduced products like methane, ethanol, H_2, and volatile fatty acids. As compared to aerobic digestion AD benefits environment by low cell biomass production (Meegoda et al. 2018). In aerobic treatment plants to treat wastewater, maximum economy is invested on power supplies for aeration while on the other hand anaerobic wastewater treatment has low power requirements (Gebrezgabher et al. 2010). Anaerobic digestion of solid waste produces slurry base output which can be converted to excellent liquid manure and such successful efforts are been made in India (Tampio et al. 2016). This is another benefit with AD technology to sustain in the field.

Table 1: Use of anaerobic digestion over aerobic composting (Pros/ Cons)

	Advantages	Disadvantages
Anaerobic digestion	Small area Reduced odor Net energy producers	Relative slow degradation System instability Large investment Post treatment of digestate
Composting	Fast degradation Small investments Compost as soil	Large area Odor pollution Greenhouse gas emissions Leachate production Net energy consumer

Table 2: Different substrates and its C/N ratio and preferred method used to manage waste

Sr.No	Substrates	C/N Content	Preferred method of management	References
1.	Bio toilets	25:1	Composting	Sharma et al. 2016.
2.	Garden waste/ Ligno-cellulosic waste	>50	Composting	Brown et al. 2012, Bohacz 2018, Tong et al. 2018, Kuhad et al. 2011.
3.	Canteen/ hotel Food waste	10-20	AD	Brown et al. 2012, Karimi and Karimi 2018, Kuczman et al. 2018, Yadav et al. 2016, Lou et al.2018
4.	ETP/STP Sludge	10-20	AD and composting both	Zhang et al. 2014, de Mes et al. 2003, van et al. 2008, Davis and Hall 1997.
5.	Animal waste	5-30	AD	Sanchez et al.2015, Liu et al. 2011, Khoiyangbam et al. 2011.

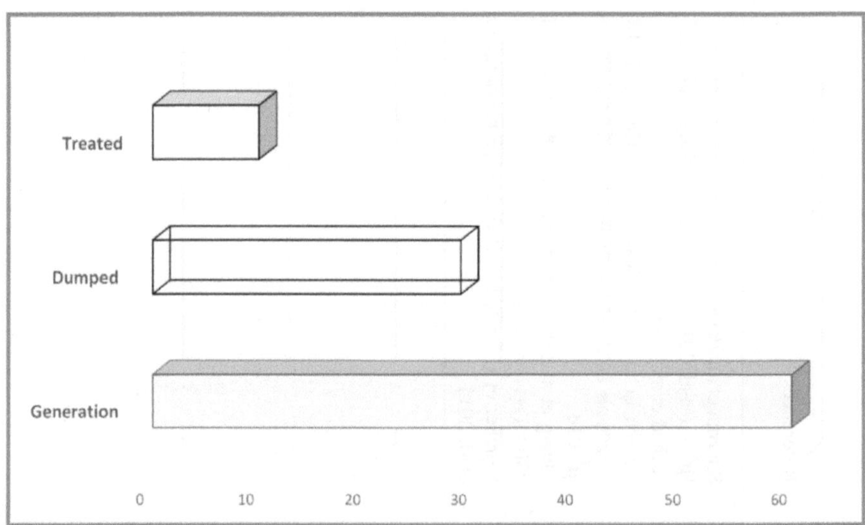

Fig. 2: Collection –Dumped- Treated waste statistics (Numbers in million MT per annum)

Conclusion

A well-organized system or structure that includes proper procedure for collection and separation and disposal of solid waste would certainly lead to a more feasible way for managing solid waste. In order to be successful in waste management process social/ community initiative, government initiative, corporate support and individual efforts are equally important factors along with technical knowledge of the subject. One should always think twice before starting any activity, about the amount of waste that will be generated at the end of the action and ways to manage the same. Also we should start managing biodegradable waste (domestic waste) at home with the aid of pilot models available in market through which you can create your own manure / compost. It is rightly said by someone that reduce, reuse or recycle is the smartest approach to save our environment.

Acknowledgements

We would like to thank Dr. Om Prakash Sharma for giving us this opportunity to contribute for this topic. We would also like to thank Department of Biotechnology (DBT), Government of India (Grant No. BT/Coord.II/01/03/2016) 'Establishment of National Centre for Microbial Resource' for the constant support. We gratefully thank Dr. Ranade and Dr.Om Prakash Sharma for their expert advice and revision of this chapter.

References

Aitken MD. 1993. Waste treatment applications of enzymes: opportunities and obstacles.*Chem. Eng. J.* 52: 49– 58.

Alderslade R. 1980. The problem of assessing possible hazards to the public health associated with the disposal of sewage sludge to land: recent experience in the United Kingdom. Proceedings of the second European Symposium "Characterization Treatment and Use of Sewage Sludge", Vienna, 21-23 October (P. L'Hermite & H. Ott, Eds), p. 372. D. Reidel Publishing Co., Dordrecht, Holland.

Ali M, Bosse T, Hindriks KV, Hoogendoorn M, Jonker CM, Treur J. 2013. Recent Trends in Applied Artificial Intelligence: 26th International Conference on Industrial Engineering and Other Applications of Applied …/Lecture Notes in Artificial Intelligence.

Ali SM, Cotton AP, Westlake K. 1999. Down to earth -Solid waste disposal for low-income Countries. Water Engineering and Development Centre, Lougborough University, UK.

Alvarez JM, Dosta J, Güiza MSR, Fonoll X, Pece, M, Astals S. 2014.A critical review on anaerobic co-digestion achievements between 2010-2013.*Renew. Sustain. Energy Rev.* 36: 412–427.

Alvarez JM, Macé S, Llabrés P. 2000. Anaerobic digestion of organicsolid wastes. an overview of research achievements and perspectives. *Bioresour. Technol.* 74:3–16.

Amin MT, Alazba AA, Manzoor U. 2014. "A Review of Removal of Pollutants from Water/ Wastewater Using Different Types of Nanomaterials," *Advances in Materials Science and Engineering*, vol. 2014, Article ID 825910, 24 pages,

Appels L, Baeyen J, Degrève J, Dewil R. 2008. Principles and Potential of the Anaerobic Digestion of Waste-Activated Sludge. *Progress in Energy and Combustion Science* 34: 755-781.

Bohacz J. 2018. Composts and Water Extracts of Lignocellulosic Composts in the Aspect of Fertilization, Humus-Forming, Sanitary, Phytosanitary and Phytotoxicity Value Assessment *Waste and Biomass Valorization* 1-14.

Braber K. 1995. Anaerobic digestion of municipal solid waste: a modern waste disposal option on the verge of breakthrough. Biomass and Bioenergy 9(1-5): 365-376.

Brinton WF. 1979. The effect of different fertilizer treatments on humus quality, *Compost Science and Land Utilization*, 20 (5): 3841.

Brown D, Shi J, Li Y. 2012. Comparison of solid-state to liquid anaerobic digestion of lignocellulosic feedstocks for biogas production. *Bioresour Technol* 124: 379-386.

Burge W D, Marsh PB. Millner PD. 1978. Occurrence of pathogens and microbial allergents. In Composting of Municipal Residues and Sludges, p. 128. Information Transfer Inc. and Hazardous Material Control Research Institute, Rockville, MD, U.S.A.

Campos J L, Crutchik D, Franchi O, Pavissich J P,Belmonte M, Pedrouso A, Mosquera-Corral A, Val del R A. 2019. Nitrogen and Phosphorus Recovery From Anaerobically Pretreated Agro-Food Wastes: A Review. *Frontiers in Sustainable Food Systems*, 2:91.

Cassini A, Högberg LD, Plachouras D, Quattrocchi A, Hoxha A, Simonsen GS, Colomb M, Kretzschmar ME, Devleesschauwer B, Cecchin M, Ouakrim D, OliveiraT, Struelens J, Suetens C, Monnet D. 2018. Attributable deaths and disability-adjusted life-years caused by infections with antibiotic-resistant bacteria in the EU and the European Economic Area in 2015: a population-level modelling analysis. *Lancet Infect Dis.*

Chynoweth DP, Pullammanappallil P. 1996. Anaerobic digestion of municipal solid wastes. In A.C. Palmisano and M. A. Barlaz (eds): Microbiology of solid waste, Boca Raton, FL, USA: CRC Press.

Connery J, Epstein E, Guymont F, Marcus S, Passman F, Post J, Taffel W. 1979. Workshop on the Health and Legal Implications of Sewage Sludge Composting. Energy Resources Co. Inc., Cambridge, MA, U.S.A

Dar SA, Kleerebezem R, Stams AJM, Kuenen JG, Muyzer G. 2008. Competition and coexistence of sulfatereducing bacteria, acetogens and methanogens in a lab-scale anaerobic bioreactor as affected by changing substrate to sulfate ratio. *Appl Microbiol Biotechnol* 78(6): 1045–1055.

Davis RD, Hall JE. 1997. Production, treatment and disposal of wastewater sludge in Europe form a UK perspective. *Eur Water Pollut Control* 7:2

de Bertoldi M, Vallini G, Pera A. 1983. The biology of composting: A review. Waste Manag. Res 1: 157-176.

de Cazes M, Abejón R, Belleville MP, Sanchez-Marcano J. 2014. Membrane bioprocesses for pharmaceutical micropollutant removal from waters. *Membranes*, *4*(4), 692–729.

de Mes TZD, Stams AJM, Reith JH, Zeeman G. 2003. Methane production by anaerobic digestion of wastewater and solid wastes. In: Reith JH, Wijffels RH, Barten H (eds) Bio-methane & biohydrogen: status and perspectives of biological methane and hydrogen production. Dutch Biological Hydrogen Foundation, Petten, The Netherlands, pp 58–102

Dean RB, Lund E.1981. Water Reuse; Problems and Solutions. Academic Press, London,

Gaby W L. 1975. Evaluation of health hazards associated with solid waste/sewage sludge mixtures. USEPA 670/2-75-023, Cincinnati, Ohio, U.S.A.

Gadre RV, Ranade DR. Godbole SH. 1986. A note on survival of salmonellas during anaerobic digestion of cattle dung. *Journal of Applied Bacte- riolog* 60: 93-96.

Gebrezgabher S, Meuwissen M, Prins B, Lansink A. 2010. "Economic analysis of anaerobic digestion—A case of Green power biogas plant in The Netherlands". *NJAS - Wageningen Journal of Life Sciences*, 57(2):109-115.

Golueke CG. 1977. Biological Reclamation of Solid Wastes. Emmaus, PA: Rodale Press.

Guanzon YB, Holmer RJ. 2003. Composting of organic wastes: A main component for successful integrated solid waste management in Philippine cities, Paper presented during the National Eco-Waste Multisectoral Conference and Techno Fair at Pryce Plaza Hotel, Cagayan de Oro City, Philippines, 16-18 July, pp. 4-5.

Hassen A, Belguith K, Jedidi N, Cherif A, Cherif M, Boudabous A. 2001. Microbial characterization during composting of municipal solid waste, *Bioresource Technology* 80: 217- 225.

Haug RT. 1993. The practical handbook of compost engineering. Boca Raton, FL, USA: Lewis Publishers.

Hendriksen RS, Munk P, Njage P, Bunnik B van, McNally L, Lukjancenko O, et al. 2019. Global monitoring of antimicrobial resistance based on metagenomics analyses of urban sewage. *Nat Commun.* 10*(1)*:1124

Houtmeyers S, Degreve J, Willems K, Dewil R, Appels L. 2014. Comparingthe inûuence of low power ultrasonic and microwave pre-treatments on thesolubilisation and semi-continuous anaerobic digestion of waste activatedsludge. *Bioresour. Technol.* 171: 44–49.

Huang L, Lee D.J. 2015. "Membrane bioreactor: A mini review on recent R&D works." *Bioresource Technology*. 194: 383-388.

Johnning A, Moore ER, Svensson-Stadler L, Shouche YS, Larsson DG, Kristiansson E. 2013. Acquired genetic mechanisms of a multiresistant bacterium isolated from a treatment plant receiving wastewater from antibiotic production. *Appl Environ Microbiol.* 79(23):7256–63.

Joshi R, Ahmed S. 2016. Status and challenges of municipal solid waste management in India: a review. *Cogent Environ Sci* 2:1139434

Karimi S, Karimi K 2018. Efficient ethanol production from kitchen and garden wastes and biogas from the residues. *J Cleaner Prod* 187:37–45.

Kawata K, Cramer WN. Burge WD. 1977. Composting destroys pathogens in solid wastes, *Water and Sewage Works* 124 (4): 76-79.

Khoiyangbam RS, Gupta N, Kumar S.2011. *Biogas Technology: Towards Sustainable Development* (TERI Press, New Delhi, 2011).

Kim DH, Jeong E, Oh SE, Shin HS. 2010. Combined (alkaline+ultrasonic) pretreatment eûect on sewage sludge disintegration. *Water Res*. 44: 3093–3400.

Kononova MM. 1966. Soil Organic Matter. Pergamon Press, London, UK, 544 pp.

Krupp M. Schubert J, Widmann R. 2005.Feasibility study for co-digestionof sewage with OFMSW on two wastewater treatment plants in Germany.*Waste Manage*. 25: 393–399.

Kuczman O, Gueri MVD, De Souza SNM, Schirmer WN, Alves HJ, Secco D, BurattoW G, Ribeiro CB, Hernandes FB. 2018. Food Waste Anaerobic Digestion of a Popular Restaurant in Southern Brazil. *J. Cleaner Prod.* , 196:382– 389.

Kuhad RC, Lata PC, Singh A.2011.Composting of lignocellulosic waste material for soil amendment. In: Singh, A., Parmar, N., Kuhad, R.C. (eds.) Bioaugmentation, Biostimulation and Biocontrol, pp. 107–128. Springer, Heidelberg (2011).

Kumar S, Smith SR, Fowler G, Velis C, Kumar SJ, Arya S, Rena, Kumar R, Cheeseman C. 2017. Challenges and opportunities associated with waste management in India. *R Soc Open Sci* 4(3):160764.

Kumar V, Chopra AK, Kumar A. 2017A review on sewage sludge (Biosolids) a resource for sustainable agriculture. *Arch Agr Environ Sci.* 2: 340-347.

Kumaran P, Hephzibah D, Sivasankari R, Saifuddin N, Shamsuddin AH. 2016. A review on industrial scale anaerobic digestion system deploymentin Malaysia: opportunities and challenges. Renew. *Sustain.Energy Rev*. 56:929–940.

Kunte D, Yeole TY, Ranade D. 2004.Two-stage anaerobic digestion process for complete inactivation of enteric bacterial pathogens in human night soil. *Water science and technology: a journal of the International Association on Water Pollution Research*. 50: 103-8.

Labatut RA, Pronto JL. 2018. Sustainable Waste-to-Energy Technologies: Anaerobic Digestion. Sustainable Food Waste-To-energy Systems, - Elsevier Pages 47-67

Laxminarayan R, Chaudhury RR. 2016. Antibiotic Resistance in India: Drivers and Opportunities for Action. *PLoS Med* 13(3): e1001974

Lee DJ, Lee SY, Bae JS, Kang JG, Kim KH, Rhee SS, Park JH, Cho JS, Chung J, Seo DC. 2015. "Effect of volatile fatty acid concentration on anaerobic degradation rate from field anaerobic digestion facilities treating food waste leachate in South Korea," *Journal of Chemistry*, 1-9.

Lee SH, Choi KI, Osako M. Dong JI. 2007. Evaluation of environmental burdens caused by changes of food waste management in Seoul,Korea. *Sci. Total Environ*. 387: 117–120.

Lin L, Xu F, Ge X, Li Y. 2018.Improving the sustainability of organic waste management practices in the food-energy-water nexus: A comparative review of anaerobic digestion and composting. *Renew Sust.Energy Rev* 89: 151–167.

Lin L, Yang L, Xu F, Michel FC, Li Y. 2014.Comparison of solid-state anaerobic digestion and composting of yard trimmings with effluent from liquid anaerobic digestion. *Bioresour. Technol*. 169: 439–446.

Liu D, Zhang R, Wu H, Xu D, Tang Z, Yu G, Xu Z, Shen, Q. 2011.Changes in biochemical and microbiological parameters during the period of rapid composting of dairy manure with rice chaff. Bioresource technology 102(19): 9040–9049. Loehr RC. 1977. Land as a Waste Management Alternative. *Ann Arbor Science Publ. Inc., Ann Arbor,* MI, U.S.A

Lou XF, Nair J, Ho G. 2013.Potential for energy generation from anaerobic digestion of food waste in Australia. *Waste Manag Res*. 31:283.

Meegoda JN, Li B, Patel K, Wang LB. 2018. A Review of the Processes, Parameters, and Optimization of Anaerobic Digestion. *Int J Environ Res Public Health*. 15(10):2224.

Mehta CM, Kanak S. 2018. Comparative study of aerobic and anaerobic composting for better understanding of organic waste management: a mini review. *Plant Archives* 18 (1): 44-48.

Miller FC. 1996. Composting of municipal solid waste and its components. In A. C. Palmisano, and M. A. Barlaz (Eds): *Microbiology of solid waste*: 115-154, Boca Raton, FL, USA: CRC Press.

Mollazadeh N. 2014. Composting: a new method for reduction of solid waste and wastewater. *International Research Journal of Applied and Basic Sciences* 8: 311–317.

Muzenda E. 2014. Biomethane generation from organic waste: a review. In: Proceedings of the world congress on engineering and computer science, vol. II, WCECS, 22–24 October, 2014, San Francisco, USA

Naik N, Shettar M, Bhat R, Kowshik S. 2016.Waste Management Analysis in Trains Using Anaerobic Bio Digester. *International Journal of Advances in Science Engineering and Technology*, Volume- 4, Issue-2.

Nandan A, Yadav BP, Baksi S, Bose D. 2017. Recent Scenario of Solid Waste Management in India. *World Scientific News* 66: 56-74.

Pandey K, Singh B, Pandey A, Badruddin IJ, Pandey S, Mishra VK, Jain PA. 2017. Application of microbial enzymes in industrial waste water treatment. *Int J Curr Microbiol Appl Sci* 6(8):1243–1254.

Parawira W. 2012. Enzyme research and applications in biotechnological intensification of biogas production, *Critical Reviews in Biotechnology*, 32:2, 172-186.

Pereira AJM, Elvira PSI, Oneto SJ, Cruz D LR, Portela JR, Nebot E. 2015. Enhancement of methane production in mesophilic anaerobic digestion of secondary sewage sludge by advanced thermal hydrolysis pretreatment. *Water Res.* 71:330–340.

Raman AAA, Asghar A, Buthiyappan A, Daud, W. 2017.16 –Treatment of Recalcitrant Waste. Current Developments in Biotechnology and Bioengineering, Elsevier, 10.1016/B978-0-444-63665-2.00016-3.

Ranade DR, Yeole TY, Godbole SH. 1987. Production of biogas from market waste. *Biomass* 13:147–153.

Rutgersson C, Fick J, Marathe N, Kristiansson E, Janzon A, Angelin M, Johansson A, Shouche Y, Flach CF, Joakim D. G.2014. Fluoroquinolones and qnr genes in sediment, water, soil, and human fecal flora in an environment polluted by manufacturing discharges. *Environmental science & technology* 48(14):7825–32.

Salunkhe AR, Chavan ND, Wadar AA, Patil DS, Karande PA. 2018. Bio-Toilet IJEECS Volume 7, Issue 3.

Sanchez-Garcia M, Alburquerque JA, SanchezMonedero MA, et al. 2015.Biochar accelerates organic matter degradation and enhances N mineralisation during composting of poultry manure without a relevant impact on gas emissions. *Bioresour Technol*192: 272–279.

Savant D, Ranade D. 2004. Application of Methanobrevibacter acididurans in anaerobic digestion. *Water science and technology: a journal of the International Association on Water Pollution Research.* 50: 109-14.

Sharma M, Neog K, Sugam RK, Ramji A. 2016. Ceew Decentralised Waste Management in Indian Railways A Preliminary Analysis Issue Brief June ceew.in

Sopper WE, Kerr SN. 1979. Utilization of Municipal Sewage Effluent and Sludge on Forest and Disturbed Land. The Pennsylvania State University Press, University Park, U.S.A.

Stofella PJ, Kahn BA. 2001. Compost utilization in horticultural cropping systems. Boca Raton, FL, USA: Lewis Publishers.

Tampio E, Marttinen S, Rintala J. 2016. Liquid fertilizer products from anaerobic digestion of food waste: mass, nutrient and energy balance of four digestate liquid treatment systems. *J. Cleaner Prod.* 125: 22–32.

Taneja N, Sharma M. 2019. Antimicrobial resistance in the environment: The Indian scenario. *Indian Journal of Medical Research*, 149:2 119-128.

Tanimu MS, Tinia IMG, Razif MH, Idris A. 2014. Eûect of carbonto nitrogen ratio of food waste on biogas methane production in a batchmesophilic anaerobic digester. *Int. J. Innovat. Manage. Technol.* 5:116–119.

Tarmudi Z, Abdullah ML, Tap AOM. 2009. An overview of municipal solid waste generation in Malaysia. *Jurnal Teknologi* 51: 1–15.

Tong J, Sun X, Li S, Qu B, Wan L. 2018.Reutilization of Green Waste as Compost for Soil Improvement in the Afforested Land of the Beijing Plain. *Sustainability 10:* 2376.

Van Lier JB, Mahmoud N, Zeeman G. 2008. Anaerobic wastewater treatment. In: Henze M, Van Loosdrecht MCM, Ekama GA, Brdjanovic D (eds) Biological watewater treatment principles, modelling and design. IWA Publishing, London, pp 401–442.

Wassenaar,T. M. 2003. Actinomycetes pp. <http://www.bacteriamuseum.org/species/actinomycetes.html> (March 21, 2004).

World Health Organization. 1981. The Risk to Health of Microbes in Sewage Sludge Applied to Land, Stevenage, 69 January. EURO Reports and Studies, N.54, Copenhagen, Denmark.

Yadav D, Barbora L, Rangan L, Mahanta P. 2016. Tea waste and food waste as a potential feedstock for biogas production. *Environ Progress Sustain Energy* 35(5):1247–1253.

Zhang JN, Lü F, Shao LM, & He PJ. 2014. *The use of biochar-amended composting to improve the humification and degradation of sewage sludge. Bioresource Technol.* 168: 252–258.

Zhen G, Lu X, Kato H, Zhao Y, Li YY. 2017. Overview of pretreatment strategies for enhancing sewage sludge disintegration and subsequent anaerobic digestion: current advances, full-scale application and future perspectives. *Renew. Sustain. Energy Rev*. 69: 559–577.

Zucconi F, Forte M, Monaco A, De Bertoldi M. 1981. Biological evaluation of compost maturity. *BioCycle* 22 (4): 27-9.

12

Role of Anaerobes in Wastewater Treatment and Nitrogen Cycling

Rakesh Kumar Gupta[1,2], *Suraj P. Nakhate*[1,2], *Ashish K. Singh*[1,2]
Atul R. Chavan[1,2], *Bhagyashri Poddar*[1,2], *Anshuman A*. Khardenavis*[1,2], *Hemant J. Purohit*[2]

[1]*Academy of Scientific and Innovative Research (AcSIR), Ghaziabad Uttar Pradesh, India*
[2]*Environmental Biotechnology and Genomics Division (EBGD) CSIR–National Environmental Engineering Research Institute (NEERI) Nehru Marg, Nagpur, Maharashtra, India*

Abstract

Anaerobes are a diverse groups of microorganisms which do not require oxygen or energy for respiration/metabolism and are one of the important components of nitrogen cycle. They are capable of using various other organic and even inorganic materials as electron acceptors during respiration. Anaerobic based wastewater treatment is a biological approach where microorganisms degrade organic contaminants (i.e. rich in carbon and nitrogen) in the absence of oxygen. In this chapter, the role of anaerobes in wastewater treatment has been discussed as it offers an alternative, cheap and eco friendly treatment technology among different biological treatment methods. Anaerobic biological treatment of wastewater is well developed, understood, and frequently used in anaerobic digesters to treat complex organic carbon and nitrogen compounds such as primary and secondary wastewater sludge. The anaerobic sludge contains diverse groups of anaerobic microorganisms that work together in association via a multistage processe consisting of the following biochemical reactions like hydrolysis, acidification, acetogenesis, and methanogenesis thereby converting the organic matter to biogas. The composition of biogas includes 70% methane (CH_4) and 30% carbon dioxide (CO_2) in addition to traces

**Corresponding Author: aa_khardenavis@neeri.res.in*

of other gases (e.g., H_2 and H_2S). The methane produced as a by-product can be used as an energy source. The potentiality of anaerobes in anaerobic based bioreactors for its maximum efficiency can be implemented in a variety of ways. The chapter discusses about the working principle, performance, treatment potential, and cost implication in the application of anaerobic treatment technology. The main bioprocess parameters such as pH, temperature, mixing, organic loading rate (OLR), and hydraulic retention time (HRT) have also been discussed in brief. Anaerobic treatment offers several benefits over aerobic treatment, including lower energy requirements, less chemicals, low capital, operation, and maintenance costs, and less sludge production. Hence, it saves money and the environment and overall simplifies the whole process.

Keywords: Anaerobes, Anaerobic Technology, Nitrogen Cycle, Wastewater Treatment

Introduction

Globally, over 80% of all the wastewater is released to the natural water streams like the river, lake, sea, and coastal areas without treatment (UN-WWDR 2017). Wastewater is roughly composed of 99% water with the remaining 1% consisting of suspended, colloidal, and dissolved solids (UN-Water 2015). The remaining 1% portion of wastewater contains a variety of pollutants such as organics, carbon, phosphates, nitrogen, and other contaminants in significant amount (Sengupta et al. 2015). The presence of organic matter, nitrogen, and phosphorus in excess concentration in wastewater may pose a serious threat to human health and the environment (Akpor et al. 2014, Yamashita and Yamamoto-Ikemoto 2014). In the present scenario, wastewater generation and its treatments are global problems since this contaminated water can't be ignored and discharged into natural water streams without adequate treatment.

Since a long time, wastewater has been treated by various technological approaches based on various physical, chemical, and biological methods for nitrogen removal (Capodaglio et al. 2015). Biological wastewater treatment is based on the ability of the microorganisms to feed on complex nutrients and organic matter, converting them into simpler compounds and substances. This type of treatment can be divided into two categories, aerobic and anaerobic treatment (Zheng et al. 2013). In the past, aerobic based removal bio-processes were favoured for wastewater treatment because they are reliable, fast, simple to operate, and can abide to greater extent of process changes or variances. In spite of these advantages, aerobic processes require higher capital and operating costs (electrical energy for aeration), and special monitoring and maintenance for system operation (Mittal 2011), which makes it an overall costly treatment.

The anaerobic bio-processes for the treatment have several benefits over aerobic treatment, including lower energy requirements, less chemicals, and less sludge production (Aziz et al. 2019). Hence in the last few decades, this has led to an emergence of anaerobic treatment as a cheap and cost effective option to aerobic treatment.

In this chapter, the degradation of organic matter under anaerobic condition, process microbiology and biochemistry, and environmental factors affecting anaerobic treatment are discussed. Besides, the different forms of the anaerobic reactors presently in use for wastewater treatment are discussed.

Major Sources of Nitrogen Pollution

Since the industrial revolution, agricultural and industrial activities such as production and application of nitrogen-based fertilizers for enhancement of food production, and burning of conventional fuels are the cause of nitrogen pollution (Kuypers et al. 2018). Besides, poorly managed animal wastes generated from various sources such as a slaughterhouse, poultry, livestock, and fisheries operation cause high-strength of organic and inorganic nitrogenous pollution (Driscoll et al. 2003, Manuel 2014, Ghaly and Ramakrishnan 2015).

In domestic wastewater, urine and faeces contribute about 75% and 20% of the nitrogen respectively. It was reported that 30% of the ammonia in fertilizers is used by the crops in the agricultural processing (Vinnerås 2001, Wilsenach and Van Loosdrecht 2003, Tjandraatmadja and Diaper 2006) while the remaining 70% is wasted. There are a lot of wastes in agricultural residue burnings that also contribute to increasing ammonia emissions resulting in atmospheric precipitation into water (Menon 2017). The anthropogenic activities add additional nitrogen to the atmosphere in the form of nitrogen oxide gases (Ghaly and Ramakrishnan 2015).

Nitrogen Toxicity and Its Consequences

The nitrogen is an important vital element for all organisms and it is key com-ponent for the biosynthesis of many organic compounds such as amino acids, nucleic acids (DNA & RNA) and proteins (Ohyama 2010). In small amounts, nitrogen is an essential nutrient that plants need to grow and they are beneficial to many ecosystems (Kuypers et al. 2018). Nitrogen has versatile chemical variability, but mostly it exists naturally as unreactive gaseous nitrogen (N_2) in the atmosphere (House and House 2015, Menon 2017). On the other hand, some others fixed forms of nitrogen driven compounds such as ammonia/ammonium, nitrite, nitrate, and nitrous oxide are reactive in nature and harmful in the environment. Hence these compounds are of primary concern in the environment and pose a risk to living organisms (Philips et al. 2002, Chislock et al. 2013).

Effect on Human Health

Excess nitrogen in wastewater, i.e., ammonia (NH_3) is toxic when combined with industrial pollution of acids due to the formation of aerosols that damage the human lungs and respiratory tract (Menon 2017). Nitrogen compounds in water can cause health issues to the human population, mainly small children and women (Manassaram et al. 2005, Ward et al. 2018). Elevated concentrations of nitrate (NO_3^-) can cause infant methemoglobinemia (also called blue baby syndrome). It is a health condition of infant where the skin turns blue (cyanotic) resulting in reduction in the amount of haemoglobin due to which blood is not able to carry the oxygen to baby (Greer and Shannon 2005).The maximum permissible limit of nitrate is 50 mg/L while ammonium is 0.05 mg/L that are acceptable in drinking water (WHO, 2011, Schullehner et al. 2017). Long-term hazard to NO_3 and NO_2 at upper level of Maximum Contamination Level (MCL) can potentially cause many diseases such as haemorrhaging of the spleen, increased starchy deposits, and dieresis (USEPA 1992, Hoon 2013).

Effect on Environment

The other major problem associated with excess nitrogen is eutrophication, a type of pollution caused by nitrates (NO_3^-) (Canfield et al. 2010, Jacobson et al. 2017). Eutrophication promotes an explosive growth of algae, called algal blooms that deplete the oxygen (O_2), and when the algae die, the dead algal biomass is reduced by heterotrophic bacteria. The nitrogen-rich aquatic system may become hypoxic (oxygen-poor) or anoxic (wholly depleted of oxygen) for the fish and other aquatic organisms. The consequences include fish mortality, decline in biodiversity, deterioration of the water quality due to increased toxicity, and foul-smell (Bernhard 2010, Chislock et al. 2013, Erisman et al. 2013).

Problem of Nitrogen Pollution in Fish Aquaculture

Nitrogen is a basic supplement in aquaculture practice which has sustained aquatic food production (Luo et al. 2018, Ngatia et al. 2019). The aquaculture farming is practiced by two ways, firstly in closed and secondly in open systems. Open aquaculture systems often involve a large cage in the ocean where fish are reared, fed, and then caught for processing. On the other hand, closed aquaculture system is a land-based approach utilizing filtration and recirculation systems. There is an evidence that impact of nitrogen in aquaculture during nitrogen loading from feeding and metabolic activities of fish can cause environmental problems (Chen 1991, Martinez-Porchas et al. 2012, Suantika et al. 2017).

Aquaculture Pollution

The source of nitrogen pollution in cultivation of fish, marine fish, and shellfish has been the nitrogen-containing unused feed and generated faecal waste matter which can lead to several environmental problems (Chen 1991). The emissions of these wastes from aquaculture facilities into the ecosystem will not only affect other fish, but will also result in nutrient pollution. It causes eutrophication, global acidification, oxygen depletion, and loss of biodiversity and productivity in aquatic and marine systems (Gowen 1994, Porrello et al. 2003, Zhang et al. 2006, Burkholder and Shumway 2011, Bergström et al. 2015). Also, the utilization of anti-microbials, antifoulants, and pesticides closely associated with aquaculture can also be introduced into the water bodies which pollutes the open water and makes it unsafe for human drinking, recreational use, and for other wildlife (Healey et al. 2016).

Habitat Destruction

Practically all aquaculture practices are performed as often as possible inside the confines of aquatic and coastal regions and even near to it. The release of pollutants from aquaculture practices can accumulate in the ponds and other water bodies over a period of time. These pollutants in addition to the incidences of illnesses, for example, the bacterially-caused vibriosis and the viral "white spot" infection have resulted in the devastation of many water bodies and others territories of biological systems like mangrove ecosystem leading to the abandonment of water bodies (Beveridge et al. 1994, Costa-Pierce 2002, Naylor et al. 2000, Greenberg 2014, Beardmore et al. 1997).

Disease Transfer

Aquaculture waste can be separated into solid and dissolved waste of which the solid waste is derived from uneaten part or spilled feed and from fish faeces (Amirkolaie 2011). Dissolved waste comes mostly from metabolites excreted by the fish particularly carbon, nitrogen, and phosphorous. The concentration of the two types of pollutants increases within a location and eventually reduces the water quality of that particular system while also leading to an influx of disease-carrying fish. Improper handling of fish including transport of fishing baskets and containers that are repeatedly used are a source of contamination. Ice used for temporary preservation is also often a source of contamination by human pathogens (Ayyappan and Jena 1999). The cultivation of species in wild conditions can serve as a vector for expansion of zoonotic diseases such as that observed in case of disease transfer across countries which was speeded up during the worldwide transfer of salmon hatchlings (Greenberg 2014).

Aerobic and Anaerobic Microorganisms

Based on their ability or inability for utilizing oxygen as a terminal electron acceptor in oxidation/reductions during metabolic processes, microorganisms are categorized into various types.

Obligate Aerobes

The group of microbes that can use oxygen as the terminal electron acceptor are named as obligate aerobes, for example, *Mycobacterium tuberculosis* and *Nocardia asteroides* (Ryan and Ray 2004, Levinson 2010).When oxygen presence in the wastewater system can support obligate aerobes, it is called the aerobic system and their application to treat wastewater is called aerobic treatment.

Obligate Anaerobes

Obligate anaerobes are group of microorganisms that cannot utilize oxygen as a terminal electron acceptor and hence they cannot survive in the presence of oxygen. Obligate anaerobic bacteria include members of the genera *Fusobacterium*, *Actinomyces*, *Clostridium*, *Bacte-roides*, *Prevotella*, *Peptostreptococcus*, *Propionibacterium*, *Porphyromonas* and *Veillonella* (Kim and Gadd 2008). The application of such anaerobic microorganisms in treatment of wastewater is called anaerobic treatment (Hogg 2005, Cyprowski et al. 2018*)*.

Facultative Anaerobes

Facultative anaerobes can utilize oxygen if it is present as the terminal electron acceptor but can grow without oxygen under certain conditions. Some facultative anaerobic bacteria include *Staphylococcus* sp., *Streptococcus* sp., *Escherichia coli, Salmonella, Listeria* sp., *Shewanella oneidensis, and Yersinia pestis* (Ryan and Ray 2004*)*. Denitrifying facultative anaerobes possess the capacity to use chemically bound oxygen such as in nitrites (NO_2^-) and nitrates (NO_3^-) as the terminal electron acceptor under anoxic conditions. This way of transformation of nitrate (NO_3^-) to nitrogen gas (N_2) without oxygen is called anoxic denitrification (Hogg 2005).

Anaerobes and Denitrification Process

Microbes can transform reactive and toxic nitrogen compounds into inert and harmless dinitrogen gas. Biological nitrogen treatment mainly relies on nitrification and denitrification processes through the activity of key microorganisms (Yang and Zhang, 1995). Microbial transformations of organic form of nitrogen into ammonia is called ammonification, while subsequently the ammonia is oxidized to nitrate via nitrification processes ($NH_4^+ \rightarrow NO^-_2 \rightarrow NO^-_3$). In the last conversion process, nitrate is eventually converted to dinitrogen gas through denitrification process ($NO_3^- \rightarrow NO_2^- \rightarrow NO \rightarrow N_2O \rightarrow N_2$) or via anaerobic ammonium oxidation (anammox; $NO_2^- + NH_4^+ \rightarrow N_2$) (Holló et al. 1990, Halling-Sørensen 1993, Kuypers et al. 2018, Farazaki and Gikas 2019).

Denitrification is a microbial process which is one of the important parts of nitrogen cycle (Falkowski et al. 2000, Galloway et al. 2004), wherein the nitrate and nitrite are converted to gaseous forms of nitrogen like nitrous oxide (N_2O) and nitrogen (N_2) as depicted in Fig. 1 (Roberson et al. 2016). Denitrification is favoured under limited oxygen concentration (Zafarzadeh et al. 2011) and a diverse group of microorganisms have been studied for their potential to denitrify owing to their ability to switch their metabolism from aerobic respiration to anaerobic respiration under oxygen depletion conditions. They include a large group of heterotrophic microorganisms viz., *Bacillus, Paracoccus, Pseudomonas, Propionibacterium,* and *Thiobacillus* species which require an external supply of carbon source for growth (Skiba 2008).

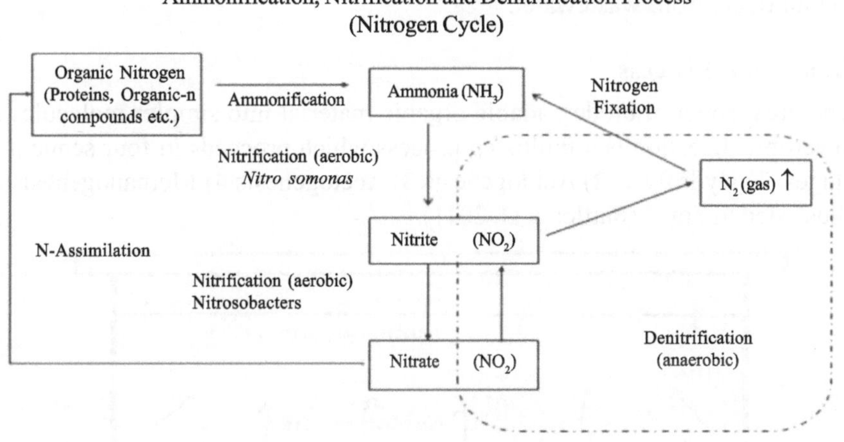

Fig. 1: Flow diagram showing the different reactions in Nitrogen Cycle (dotted line shows the denitrification process)

To control the denitrification process, pH, temperature, and oxygen perform a major role (Saleh-Lakha et al. 2009, Rodziewicz et al. 2019). For optimum denitrification process, the pH should be between 7-8 and an increase in the pH can lead to a decrease in N_2O/N_2 ratio (Nommik 1956, Van Cleemput and Patrick 1974). Temperature also plays a major role in the denitrification process which shows optimum efficiency at 25–30°C with a decrease in activity at lower or higher temperatures (Šimek et al. 2002, Miao et al. 2019).

Intense aeration conditions facilitate the ammonium conversion via oxidation to nitrate. Subsequently, purposeful addition of methanol as 'carbon source' induces heterotrophic denitrification leading to reduction of nitrate to dinitrogen (Megraw and Knowles 1989a,b). However, there are a few drawbacks in the conventional nitrogen removal methods such as the requirement of large amount energy, resources, and investment which makes it a very resource-intensive strategy. Tis strategy has another drawback since it also produces a greenhouse gas in the form of 'nitrous oxide' thereby contributing to global warming (Pai et al. 1999, Campos et al. 2016). In view of above problem, anaerobic wastewater treatment processes assume significance since such processes also leads to decline in external carbon source supplement and cost expensive aeration provision (Ahn et al. 2014).

For example, in many treatment systems, influent/raw wastewater (rich in organic carbon and nitrogen) is directly projected to the anaerobic conditions for denitrification step (Chiu et al. 2007, Morita et al. 2008, Chen et al. 2009, Paetkau and Cicek 2011) or, in others, an anoxic or discontinuous air circulation approach is utilized to advance the nitrification and denitrification for expelling the nutrients from wastewater.

Anaerobic Process

The breakdown of biodegradable organic material into simpler molecules in anaerobic digestion is a multistep process which proceeds in four sequential stages: 1) Hydrolysis 2) Acidogenesis 3) Acetogenesis 4) Methanogenesis as illustrated in Fig. 2 (Sattler et al. 2011).

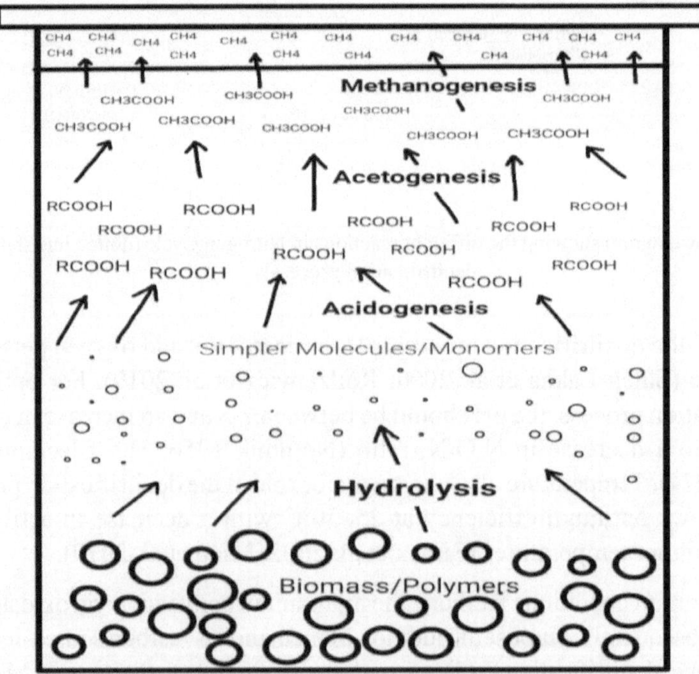

Fig. 2: Various steps in anaerobic digestion process

Hydrolysis

The hydrolysis step refers to the breakdown of large or complex polymeric molecules like carbohydrates, lipids, proteins into their simple soluble monomer molecules by the action of hydrolytic exo-enzymes secreted by the anaerobic bacteria. The products of this step include lactic acid, alcohol, carbon dioxide, volatile fatty acids, ammonia, and hydrogen, etc. which can pass through cell membrane of anaerobic bacteria. (Chernicharo, 2007).

Acidogenesis

In the second stage, microorganisms convert monomeric molecules produced during hydrolysis into small-chain fatty acids, named as Volatile Fatty Acids (VFAs) such as Propionate, Butyrate, and Valerate. Hydrogen gas is also produced during this process (Lyberatos and Skiadas 1999, Sattler et al. 2011).

Acetogenesis

In the third stage, the volatile organic acids are converted into acetic acid and hydrogen by a transitional metabolic group of microorganisms called the acetogenic microorganisms, which serve as substrate for methanogenic bacteria (Lyberatos and Skiadas 1999, Chernicharo and Augusto 2007, Sattler et al. 2011).

Methanogenesis

This is the final phase of the anaerobic process. Hydrogen and acetic acid produced during acetogenesis are utilized by archeal methanogens and converted into methane and carbon dioxide. Methanogenic microorganisms are categorized into two main groups; one group of methanogenic microorganisms forming methane from acetic acid or methanol are called acetoclastic methanogens, whereas second group, the hydrogenotrophic methanogens utilize hydrogen to reduce carbon dioxide to methane (Lyberatos and Skiadas 1999, Chernicharo and Augusto 2007, Sattler et al. 2011).

Multiple Factors Affecting Anaerobic Bio-processes

The stability and susceptibility of the anaerobic bio-process is dependent on multiple environmental factors/conditions (Lay et al. 1997, Karakashev et al. 2005). For designing of effective process strategies, it is mandatory to know the factors that influence the overall process stability. The efficiency of anaerobic digestion can be optimized by altering several factors such as organic matter, specific growth rate (Moletta et al. 1986), concentration of volatile acids (Park et al. 2018), decay rate (Angelidaki et al. 1993, Bareha et al. 2019), pH (Liu et al. 2008, Zhou et al. 2018), substrate utilization, fluctuations in temperature (Shi et al. 2019) and solubility start-up and reactor operation condition (Cristancho et al. 2006) and flow rate (Pauss et al. 1993, Gustin et al. 2011).

pH, Acidity and Alkalinity

Anaerobic bio-processes are strongly vulnerable to pH change (Mamo et al. 2019, Wang et al. 2019, Mamo et al. 2019). The methanogens and even their enzyme systems are more vulnerable to pH changes. The most suitable "Optimal," pH range for the methane producing bacteria is 6.6 - 7.5 (Gulhane et al. 2016). Whereas non-methanogens work at wide pH range from 5 to 8.5. The pH is maintained within the bioreactor during anaerobic digestion by means of the good buffering system resulting from the carbon dioxide (CO_2)–bicarbonate and volatile fatty acid – ammonia formed during the process. Routinely added calcium hydroxide [Ca $(OH)_2$], calcium or sodium carbonate-bicarbonate provides the buffering capacity to the reactor (Dahiya and Mohan 2019, Sheng et al. 2019). Presence of plentiful quantity of ammonia (NH_4^+-N) leads to accelerated hydrolysis of macromolecules to micromolecules like amino acids, simple sugars, and fatty acids which are transformed to VFAs rapidly (Fig. 3). Increase in the reactor acidity can cause inhibitory action on protein hydrolyzing enzymes like peptidase and protease (Wang et al. 2014). At pH 5 and 6, there is a linear increase in the concentration of ionised ammonia (NH_4^+) transported with help of transporter e.g. Amt (Nikol et al. 2019). Inside the cell, pH is quite high and leads to deprotonation where ammonia possessing no charge behaves as a "protonophore" and diffuses out passively through plasma membrane (Weiner et al. 2019). This may lead to the disruption of proton gradient across the plasma membrane (Fig.3) (Angelidaki and Ahring 1993, Hansen et al. 1998).

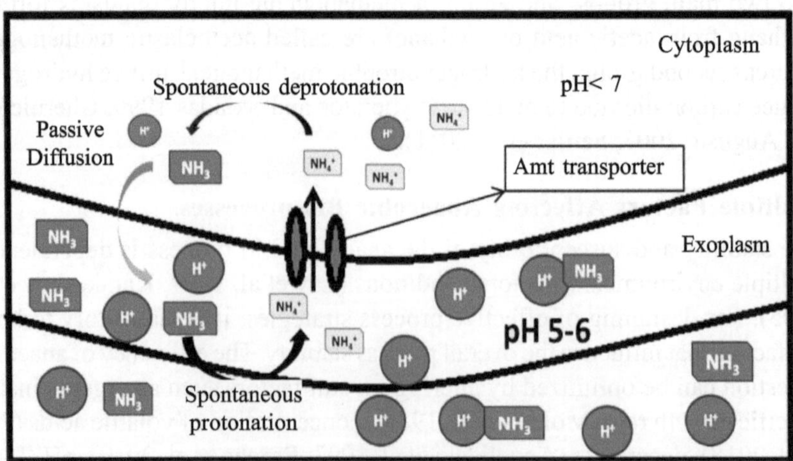

Fig. 3: Passive diffusion of unionized VFA through plasma membrane and subsequent ammonia transport across cell

Temperature

A microbial system based anaerobic bio-process is affected by temperature (Feng et al. 2019). It is observed that higher temperature leads to increased microbial metabolic activity until an optimum temperature is reached (Lin et al. 2016). The biogas production through sludge digestion in the anaerobic process can occur over range of temperatures from 25 to 60°C (Lin et al. 2016), although most of the anaerobic processes happen efficiently at mesophilic condition (30-40°C). Hence stabilization time of biomass has been shown to vary with the temperature. Similarly, substrate removal rate and methane production rate are also temperature dependent processes as has been depicted by the equation-I which is valid over a wide range of mesophilic temperatures (4 to 30 °C), mostly suitable for the aerobic systems (Von Sperling 2007; Goswami et al. 2016).

$$K_T = K_{20}.\ \theta^{(T-20)} \quad (1)$$

Where:

K_T = reaction coefficient at a temperature T (d^{-1})

K_{20} = reaction coefficient at a standard temperature of 20 °C (d^{-1})

θ = temperature coefficient (–)

T = temperature of the medium (°C)

Inhibitory Substances

The anaerobic bio-process have been used for treating multiple types of organic waste feeds each of which may consist of one or more inhibitory components added from the influent as it is or may be a substrate converted to a toxic form during digestion (Vijayaraghavan and Murthy 1997, Chen 2014). Lignin has been reported to be more hazardous in the anaerobic digestion of lignocellulosic material (den Camp 1988, Yin et al. 2000). Many wastewater treatment plants engaged in the treatment of effluents from petroleum refinery, tanning industries, and food industries possess a large number of sulfates, ammonia (Omil et al. 1995, Milán et al. 2001). In addition, alkaline earth metals and heavy metals have also been demonstrated for their inhibitory effect on anaerobic digestion (Oleszkiewicz and Sharma 1990).

Ammonia – Nitrogen Inhibition

Ammonia provides shielding activity to pH change at a lower temperature but at a higher temperature and at lower pH it is inhibitory to the anaerobic digestion process. Total ammonia nitrogen (TAN) comprises of protonated, positively charged ammonium (NH_4^+) and unionized ammonia or free ammonia (Akindele and Sartaj 2018). Their relative percentage at a given time varies with temperature and pH. Since free ammonia nitrogen (FAN) does not possess any charge; it can easily pass out of the hydrophobic core of microbial plasma membrane leading to proton imbalance which may be fatal for some transporters situated on the membrane surface. An altered intracellular pH leads to halting of enzyme activities (Kayhanian 1994). It is reported that 50% of the methanogenic activity occurs at 365 mg/L NH_3-N and inhibition of anaerobic bio-process starts at the level of 1500 mg- N/L while 100% seizing of the methanogenic activity could be found at 6000-13000 mg N/L (Zhang and Angelidaki 2015).

Heavy Metals Inhibition

Among the microbial communities involved in anaerobic digestion, acidogens are comparatively less affected by the presence of metal ions than methanogens (Lin, 1993). The effluent generated from both domestic and industrial sources like leather tanneries, coal mines etc. contain an array of heavy metals like Nickel, Copper, Manganese, Ferrous, Mercury, Lead, Chromium, and Cadmium (Abdel-Shafy and Mansour 2014). Ni, Cu, and Fe serve as cofactors for some enzymes and enhance the anaerobic bio-process, which are supplemented in the form of trace element solution to enhance the efficiency of anaerobic digestion (Nelson 2008, Mudhoo and Kumar 2013). On the other hand, the presence of heavy metals like Hg, Cd, and Cr show inhibitory action to enzyme activity resulting in reactor failure. Similarly, Cu, As, and Pb are also reported to have an inhibitory action on methanogens (Abdel-Shafy and Mansour 2014).

Anaerobic Processes Based Wastewater Treatment Technologies

Anaerobic Ammonium Oxidation (Anammox)

In the previous decade, Anaerobic Ammonia Oxidation (ANAMMOX) has been developed as an emerging substitute and low-cost novel approach for expelling nitrogen contamination from wastewater (Van deGraaf et al. 1995, Strous et al. 1999, Fux et al. 2002, Zhu et al. 2008). In the ANAMMOX process, transformation of ammonia to nitrogen occurs through oxidation reaction catalysed by the anaerobic ammonia oxidizing bacteria (AOB) which simultaneously utilize nitrite as electron acceptor under anaerobic conditions.Here carbon dioxide serves as the carbon source for anaerobic AOB, hence there is no need to add external carbon source (Van et al. 1996, Zhu et al. 2008).

In the 'partial nitritation', the process happens under oxygen limitation condition by anaerobic AOB viz., *Nitrosomonas europea* oxidize half of the ammonia to nitrite. Further nitrite and residual unoxidized ammonia is converted to dinitrogen gas by ANNAMOX process by bacteria such as 'Candidatus *Kuenenia stuttgartiensis*'. Hence the activity of *Nitrospira* sp. or *Nitrobacter* sp. in the partial nitritation-anammox systems is undesired owing to production of nitrate thereby reducing the efficiency of nitrogen removal. Benefits of oxygen-limited partial nitritation-anammox reactors are to minimize the provision for external carbon source, lower the generation of nitrous oxide, and negligible requirement for aeration facility compared to the conventional nitrogen removal reactor system. Nowadays, partial nitritation-anammox technology is progressively being applied for ammonium-rich wastewaters such as effluents from anaerobic sludge digesters. Execution of these frameworks in full-scale domestic wastewater treatment systems with lower concentrations of ammonium could prove to be a more practical solution for sustainable sewage treatment.

Complete Ammonia Oxidizer (Comammox)

Complete ammonia oxidizers (COMAMMOX) is a term used for complete nitrification, which was proposed in 2006 for the study of competent microorganisms in microbial community and biofilm formation under nutrient-limited condition (Martens Habbena et al. 2009, Schleper and Nicol 2010, Gubry-Rangin 2011). Comammox bacteria have the ability to grow under nutrient limiting conditions and in spite of their slow growing characteristic they are known for their high yield. Enrichment culture of comammox was first detected in biofilm samples wherein metagenomic study revealed a gene set for complete nitrification. This set of genes was detected in *Nitrospora*, a key member belonging to nitrogen oxidizing bacteria (NOB) in various natural and synthetic systems (Daims et al. 2015, Pinto et al. 2015, Van Kessel et al. 2015, Wang et al. 2015, Daims et al. 2016, Palomo et al. 2016, Bartelme et al. 2017, Wang et al. 2017). *Nitrospora* have been observed in agriculture soils and have also been found in aquatic habitats such as wastewater treatment plants (WWTPs). Two species of *Nitrospora* (*N. nitrosa* and *N. nitrificans*) had the complete set of enzymes for nitrification in its genome ((Daims et al. 2001, Van Kessel et al. 2015, Pinto et al. 2016).

Aerobic Deammonitrification

The conversion of ammonia (NH_4^+) to nitrogen (N_2) is a one step process (Zhu et al. 2008).Two distinctive approaches have been proposed for aerobic deammonitrification: the simultaneous or concurrent nitrification and denitrification and the isolated nitrification and denitrification.In the primary approach, aerobic deammonification is accomplished by oxygen consuming nitrifiers and anaerobic ANAMMOX AOB. An aerobic ammonia oxidizing bacteria (AOB), *Nitrosomonas*, was reported on the surface layer of microbial flock or biofilm which transformed ammonia from wastewater to nitrite under aerobic conditions and later these all shifted to the inner side of the anoxic zone/layer where rest of the conversion took place by the anaerobic AOB. Alternate cycles of aerobic/anoxic flock/biofilm system have been made for aerobic deammonification (Zhu et al. 2008). In the second approach, aerobic deammonification is accomplished by AOB nitrifiers. The conversion of NH_4^+ to NO_2^- occurs on the surface of the floc/biofilm under aerobic conditions while NO_2^- conversion occurs on the internal layer of the floc/biofilm which is reduced into inert N_2gas (Zhu et al. 2008).

Simultaneous Nitrification and Denitrification (SND)

Simultaneous nitrification and denitrification (SND) implies that nitrification and denitrification both happen simultaneously in the single reactor (Keller et al. 1997, Helmer and Kunst 1998). Two approaches of SND systems have been accounted for: (i) physical and (ii) biological. In the conventional physical system, SND happens as an effect of dissolved oxygen (DO) concentration gradient within sludge flocs/biofilm due to diffusional constraint. The nitrifiers are present in aerobic zone with dissolved oxygen higher than 1–2 mg/L, while the denitrifiers develop in anoxic zones with DO under 0.5 mg/L.

Oxygen Limited Autotrophic Nitrification-Denitrification (OLAND)

In OLAND process, a blended bacterial network which involves aerobic ammonium-oxidizing bacteria (AOB) and anoxic ammonium-oxidizing or anammox bacteria, proliferate well in a single reactor such as rotating biological contactor (RBC) (Kuai and Verstraete 1998). Ammonia oxidising bacteria (AOB) incompletely convert ammonium (NH_4^+) to nitrite (NO_2^-) at low DO as the electron acceptor (partial or halfway nitritation), and anammox microbes further utilize the residual ammonium combined with the produced nitrite (NO_2^-) as the electron acceptor (Hien et al. 2017, Windey et al. 2005).

Aerobic Denitrification

The name suggests that the microorganisms utilize the nitrogen under aerobic conditions. Aerobic nitrifying bacteria like *Thiosphaera pantotropha* can remove the nitrogen by nitrification and denitrification simultaneously. Aerobic denitrifying bacteria are Gram-negative bacteria belonging to the phylum proteobacteria (Ji et al. 2015). Various factors can affect the aerobic denitrification process like DO concentration, temperature, pH, and carbon to nitrogen load ratio (C/N load ratio). Nitrate and oxygen both compete as electron acceptor, however, in contrast to the O_2 respiration, denitrification requires input of energy (Baek and Shapleigh, 2005; Bergaust et al. 2008). Denitrification processes are considered to occur under anoxic conditions, since expression of genes *Nir* and *Nor* need to low oxygen (Baek and Shapleigh2005). Nitrate (NO_3-) and oxygen (O_2) may compete for the electron flux facilitated by organic substrate however denitrification process is not compatible with oxygen (O_2) respiration owing to its higher energy requirement (Bergaust et al. 2008). Aerobic denitrifying bacteria grow and demonstrate optimal activity at a temperature range between 25ºC and 37°C (Robertson et al. 1989) with a drastic reduction in aerobic denitrification when the temperature goes down below 10°C (Lukow and Diekmann 1997).

Dissimilatory Nitrate Reduction to Ammonium (DNRA)

This process is also known as nitrate/nitrite ammonification which is performed under anaerobic respiration conditions by chemoorgano heterotrophic microorganisms. This group of microorganisms utilize nitrate as an electron acceptor (rather than oxygen) which is reduced to nitrite, and further converted to ammonium ($NO^-_3 \rightarrow NO_2^- \rightarrow NH_4^+$) as depicted in Fig. 4 (Seitz et al. 1986, Brunet et al. 1996, Lam at al. 2011, Ettwig et al. 2016, Robertson et al. 2016). DNRA process converts the nitrate to ammonium within the system, but the denitrification process removes the reactive nitrogen from the system. DNRA and denitrification can happen simultaneously (Andalib et al. 2011). DNRA accounts for15% of total nitrogen reduction. Dissimilatory nitrate reduction to ammonium, unlike denitrification, which eliminates reactive nitrogen from the system, retains nitrogen within the system. It does not contain N_2 or N_2O because DNRA takes nitrate and transforms it into ammonium. DNRA thus recycles nitrogen instead of causing N-loss, resulting in more sustainable primary production and nitrification (Christensen et al. 2003, Nizzoli et al. 2006, Burgin and Hamilton 2007, Marchant et al. 2014). In biological systems, denitrification and DNRA can happen simultaneously. Normally DNRA is responsible for about 15% of the total nitrate reduction rate, which incorporates both DNRA and denitrification (Marchant et al. 2014). However, the overall significance of each procedure is affected by natural factors.

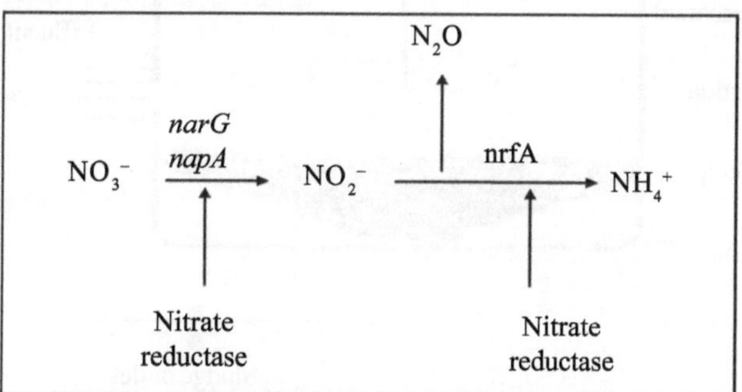

Fig. 4: Dissimilatory nitrate reduction to ammonium

Bioreactors in Anaerobic Bioprocesses

Single-Stage Anaerobic Reactors

The anaerobic reactor equipped with only a single vessel to carry out all the four stages of anaerobic digestion is called single-stage anaerobic reactor (Micolucci et al. 2018, Van et al. 2019). Organic acids produced by acidogens leads to a lowering of pH. On the contrary, methanogens perform best at higher pH values thus affecting the overall efficiency of the digestion process (Zhou et al. 2018). Therefore, a balanced activity of acidogens and methanogens is crucial to keep the reactor operative. The miniature version of a single-stage reactor can be a single conical flask fitted with airtight cork while the large scale version can consists of a thousand-gallon capacity tank producing biogas for a small community (Fig. 5) (Wu et al. 2017). Anaerobic lagoons are ponds constructed to digest livestock manure, industrial wastewater, and municipal wastewater (Gloyna and WHO 1971, Nizami et al. 2013). Though low on construction cost, it loses control on reactor performance. Depending upon the rates of digestion, facilities can be considered as standard rate or high rate anaerobic digesters. These digesters suffer from the limitation of the organic loading rate (OLR) which is still unsatisfactory at 1- 4 kg VS/L/day because of acidification in single-stage anaerobic reactors (Verrier et al. 1987, Zhang et al. 2007).

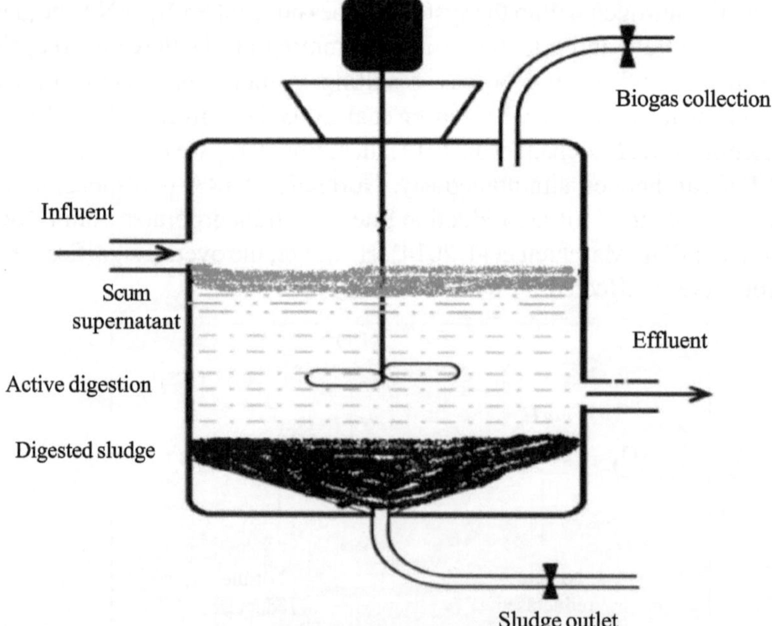

Fig. 5: Schematic diagram of Single-stage Anaerobic Continuous Stirred Tank Reactor (CSTR)

Anaerobic Filter (AF)

The anaerobic filter (AF) consists of a perforated scaffold to aid the development and growth of microbial consortia. There are two variants of this type of reactor system; up flow (Tilley 2014) and the down flow stationary fixed film (DSFF) reactor (Young and Yang 1989). Generally, biomass grows in suspended form and/or gets entangled because of the development of biofilm inside the empty pore surface. Development of biofilm inside the hollow pore has its own advantage since it prevents the biofilm from being washed off with the flow of sludge current. The nurturing effect of small pores on adhered slow growing microbes to form a biofilm results in relatively more solid retention time (SRT) than hydraulic retention time (HRT) (Bodik et al. 2000).

The direction of the flow of reactor contents makes the major difference between the two reactors. In AF, the feed starts its journey from the bottom of reactor moving upward through packing media (Fig 6). In the DSFF reactor, influent enters on top of the packing media and starts to seep down through it. A key factor affecting reactor efficiency is that in AF, most of the biomass is present in suspended form leading to a risk of wash off (Yousefzadeh et al. 2017). On the contrary, DSFF possesses the majority of biomass anchored to support medium. Strong media like needle punched polyester (NPP), red channel tile clay, PVC, or glass can be utilized for this purpose. The longer retention time signifies efficient removal of moderately low-strength chemical oxygen demand (2000 to 20000 ppm) of the soluble organics in industrial wastewater (Tilley et al. 2014, Ladu and Lu 2014).

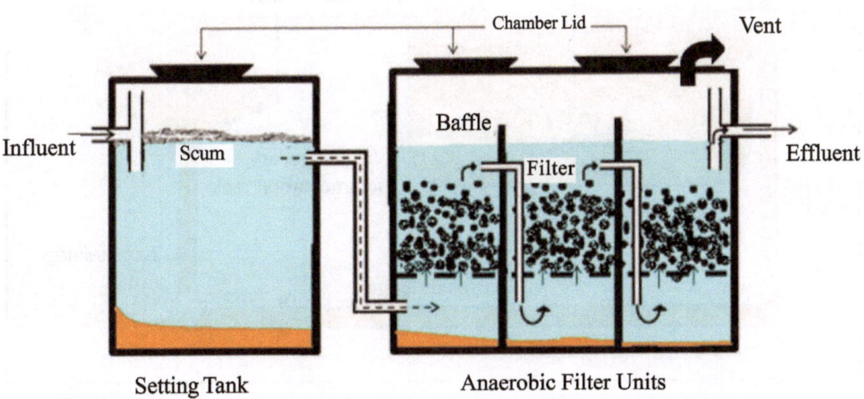

Fig. 6: Schematic diagram of Anaerobic Filter (AF) system

Anaerobic Contact Processes (ACP)

As the name suggests, the anaerobic contact process (ACP) utilizes the recirculation of activated sludge to enhance the contact time between high solid containing influent and activated biomass. The simple tandem arrangement of the primary anaerobic digester with adjacent settling tank provides opportunity to grab the metabolically active sludge and recirculate it into main digester tank thus increasing solids retention time (SRT) and reducing the hydraulic retention time (HRT) (Basitere et al. 2019). The benefit of this is reflected in gaining remarkable increase in treatment efficiency. So ACP provides advantage of high biomass concentration, better efficiency, and small reactor size (Fig. 7). The reactor architecture includes a large primary digester and a settling tank which offers low hydraulic and organic loading rates. The vigorous gas generation hinders the settling of biomass thereby preventing effective flocculation. This drawback can be overcome by rapid vacuum degasification which can recover to some extent the gas formation problem. Incorporating inclined plates into the settler design helps in the settlement of biomass. In the primary digester tank, the bottom area possesses digestion zone, middle area possesses clearing zone, and the top area possesses gas formation zone while in the settling tank, the settlement is achieved by gravity itself (Ripley 2000, Upendrakumar et al. 2006, Hassan et al. 2013).

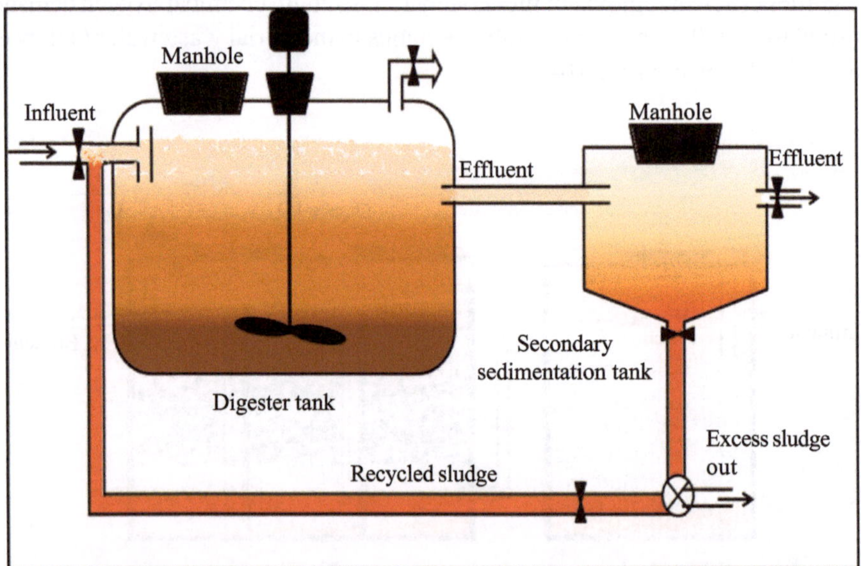

Fig. 7: Schematic diagram of Schematic diagram of Anaerobic Contact Process (ACP)

Anaerobic Side-Stream Reactor (ASSR)

Like the anaerobic contact process in which presence and availability of activated sludge in the main digesting tank increases by recirculating the effluent emerging out from main digestion tank, ASSR utilizes recirculation of some part of effluent activated sludge but with the additional anaerobic side-stream reactor (Fig. 8). DO has a significant impact on the side-stream reactor system. It is reported that ASSR (sludge reduction efficiency 61.1%) is five times more efficient than Aerobic SSR (sludge reduction efficiency of 7.9%) (Pang et al. 2018).

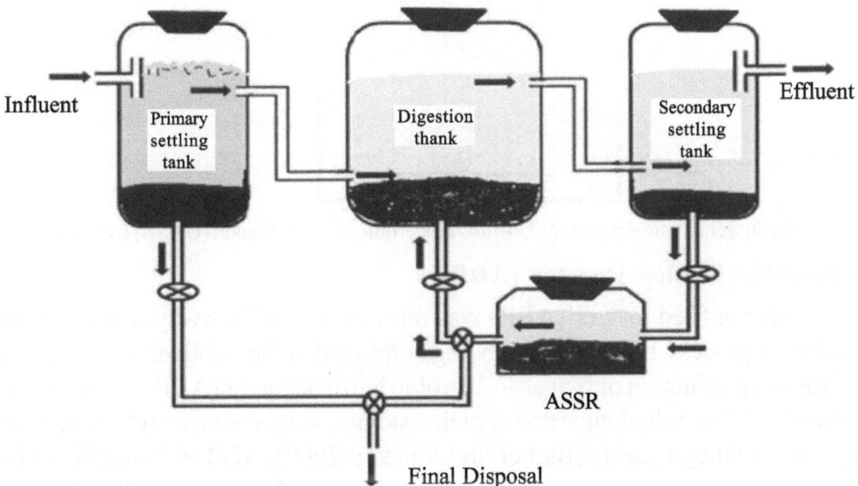

Fig. 8: Schematic diagram of Anaerobic Side-stream Reactor (ASSR)

Upflow Anaerobic Sludge Blanket (Usab) Reactor

Upflow Anaerobic Sludge Blanket (UASB) technology has been well accepted and successfully applied to many wastewater treatments in the United States, Canada, Europe, etc. The reactor configurations and model may be rectangular or cylindrical with full-scale reactor volumes varying from 30 to 5500 m^3 (Fig. 9) (Antwi et al. 2017). This is a methane-producing (methanogenic) technology that evolved from the anaerobic processes. UASB technology works on the principle of anaerobic digestion wherein the granular sludge is suspended in the tank. Wastewater moves upwards through the blanket and the degradation process occurs by the action of anaerobic microorganisms generating a high concentration of methane which is a major constituent of biogas produced as a by-product (Banu and Kaliappan 2007). This methane (CH_4) gas can be captured and utilized as an energy source to generate electricity. For successful operation of UASB, the essential pre-requisite is the occurrence of granulation of the biomass and a highly efficient solids-liquid-gas separator (Christensen et al. 1984, Ke et al. 2008).

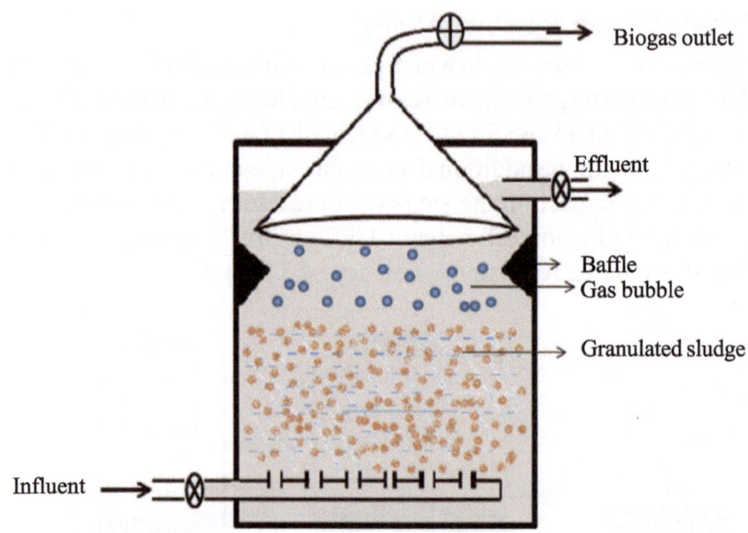

Fig. 9: Schematic diagram of Up-flow Anaerobic Sludge Blanket (UASB) Reactor

Anaerobic Baffled Reactor (ABR)

Anaerobic baffled reactor (ABR) was initially used effectively in wastewater treatment process, nowadays it is being employed in the stabilization of organic waste and production of biogas. Anaerobic baffled reactor (ABR) demonstrates better flexibility to hydraulic and organic loadings, longer biomass retention times, and lower sludge yields (Barber and Stuckey 2000). ABR is basically a plug flow reactor in which the reactor contents move without mixing (Fig. 10).

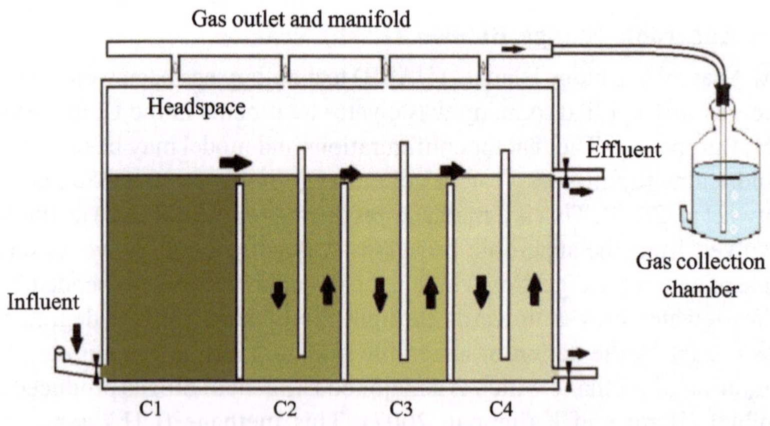

Fig. 10: Schematic diagram of Anaerobic Baffled Reactor (ABR)

The serpentine movement of the reactor content exhibits increased solid and hydraulic retention time with stratification of microbial community being observed over a period of operation of the digester (Wang et al. 2004). The first chamber where influent enters the reactor is reported to be dominated by the hydrolytic community followed by abundance of acidogens producing volatile organic acids in the 2nd chamber. Subsequently in the next chamber acetogens are abundant which produce acetate, as food for the methanogens. The last chamber has been reported to be dominated by methanogens. Relative pH in all the chambers has been shown to form a gradient over the length of the reactor with the first chamber characterized by acidic effluent followed by increase in pH across the remaining chambers to achieve neutral or more basic pH in the fourth chamber. In ABR, effluent recycling has been reported to provide pH maintaining effect thus preventing souring of the reactor (Banu et al. 2007, Liu et al. 2010). The best achievement of the reactor has been observed in the form of extended sludge retention time achieved owing to recirculation of effluent resulting in the improved biogas yield and carbon recovery factor of 96.7% of input carbon (Boopathy and Tilche 1991, Gulhane et al. 2016).

Conclusion

Notably, till date, anaerobic wastewater treatment has emerged as a competitive wastewater treatment technology. Many polluted wastewater containing organic and nitrogenous components and even those that were earlier accepted not to be too good for anaerobic treatment are now being treated by anaerobic high rate conversion processes with high efficiency and relatively low costs. Anaerobic technologies are more efficient than other treatments in terms of operational complexity, time, space, and cost. With this wide application, it may be possible in reducing the toxic components such as ammonia (NH_3), ammonium (NH_4^+), Nitrite (NO_2^-), Nitrate (NO_3^-), and Nitrous oxide (N_2O). These processes can be applied for treating various wastewaters with variable C:N content with little or no need for adding external carbon source. Further, this treatment application does not degrade the water quality, minimizes human health risk, and saves the environmental integrity by mitigating global warming and climate change by reducing nitrous oxide (N_2O). The main purpose of this chapter is to explore the potentiality of anaerobes and the anaerobic bio-processes for achieving treatment of wastewater in a single step and at the same time reducing the operational complexity, cost of operation, increasing the stability, and optimizing conditions for operating reactors.

Acknowledgement

All authors have contributed equally to the manuscript.

The authors are thankful to Director, CSIR-National Environmental Engineering Research Institute for providing necessary facilities for this work [KRC Manuscript No.: CSIR-NEERI/KRC/2019/DEC/EBGD/3]. Mr. Rakesh K. Gupta and Mr. Suraj P. Nakhate acknowledge the Junior Research Fellowship (JRF) from University Grants Commission (UGC), Government of India, India for carrying out this work. Mr. Ashish K. Singh and Mr. Atul R. Chavan acknowledge the Junior Research Fellowship (JRF) from Council of Scientific and Industrial Research (CSIR), Government of India, India for carrying out this work. Funds from Science and Engineering Research Board (SERB), New Delhi, for project - EMR/2016/006589 are gratefully acknowledged.

References

Abdel-Shafy., H. I., & Mansour, M. S.M. (2014). Biogas production as affected by heavy metals in the anaerobic digestion of sludge. *Egyptian Journal of Petroleum*, *23*(4), 409-417.

Ahn, Y. M., Wi, J., Park, J. K., Higuchi, S., & Lee, N. H. (2014). Effects of Pre-aeration on the anaerobic digestion of sewage sludge. *Environmental Engineering Research*, *19*(1), 59-66.

Akindele, A.A. & Sartaj, M. (2018). "The toxicity effects of ammonia on anaerobic digestion of organic fraction of municipal solid waste. *Waste Management*: 757-766.

Akpor, O. B., Otohinoyi, D. A., Olaolu, D. T., & Aderiye, B. I. (2014). Pollutants in wastewater effluents: impacts and remediation processes. *International Journal of Environmental Research and Earth Science*, *3*(3), 050-059.

Amirkolaie, A. K. (2011). Reduction in the environmental impact of waste discharged by fish farms through feed and feeding. *Reviews in Aquaculture*, 3(1), 19–26.

Andalib, M., Nakhla, G., McIntee, E., & Zhu, J. (2011). Simultaneous denitrification and methanogenesis (SDM): Review of two decades of research. *Desalination*, *279*(1-3), 1-14.

Angelidaki, I., & Ahring, B. K. (1993). Thermophilic anaerobic digestion of livestock waste: the effect of ammonia. *Applied Microbiology and Biotechnology*, 38(4), 560-564.

Antwi, P., Li, J., Boadi, P. O., Meng, J., Quashie, F. K., Wang, X.,&Buelna, G. (2017). Efficiency of an upflow anaerobic sludge blanket reactor treating potato starch processing wastewater and related process kinetics, functional microbial community and sludge morphology. *Bioresource Technology*, *239*, 105-116.

Ayyappan, S., & Jena, J. K. (1999).Environmental Issues in Indian Freshwater Aquaculture.

Aziz, A., Basheer, F., Sengar, A., Khan, S. U., & Farooqi, I. H. (2019).Biological wastewater treatment (anaerobic-aerobic) technologies for safe discharge of treated slaughterhouse and meat processing wastewater. *Science of The Total Environment, 686(10), 681-708..*

Baek, S. H., and Shapleigh, J. P. (2005). Expression of nitrite and nitric oxide reductases in free-living and plant-associated Agrobacterium tumefaciens C58 cells. *Applied and Environmental Microbiology*, 71: 4427-4436.

Banu, J. R., Kaliappan, S., & Yeom, I. T. (2007). Treatment of domestic wastewater using upflow anaerobic sludge blanket reactor. *International Journal of Environmental Science & Technology*, *4*(3), 363-370.

Banu, J. R., & Kaliappan, S. (2007). Treatment of tannery wastewater using hybrid upflow anaerobic sludge blanket reactor. *Journal of Environmental Engineering and Science*, 6(4),415-421.

Barber, W. P., & Stuckey, D. C. (2000). Nitrogen removal in a modified anaerobic baffled reactor (ABR): 2, nitrification. *Water Research*, 34(9), 2423-2432.

Bareha, Y., Girault, R., Guezel, S., Chaker, J., & Trémier, A. (2019). Modeling the fate of organic nitrogen during anaerobic digestion: Development of a bioaccessibility based ADM1. *Water Research, 154*, 298-315.

Bartelme, R. P., McLellan, S. L., & Newton, R. J. (2017). Freshwater recirculating aquaculture system operations drive biofilter bacterial community shifts around a stable nitrifying consortium of ammonia-oxidizing archaea and comammox Nitrospira. *Frontiers in Microbiology, 8*, 101.

Basitere, M., Njoya, M., Rinquest, Z., Ntwampe, S. K., & Sheldon, M. S. (2019). Performance evaluation and kinetic parameter analysis for static granular bed reactor (SGBR) for treating poultry slaughterhouse wastewater at mesophilic condition. *Water Practice and Technology, 14*(2), 259-268.

Beardmore, J. A., Mair, G. C., & Lewis, R. I. (1997). Biodiversity in aquatic systems in relation to aquaculture. *Aquaculture Research, 28*(10), 829-839.

Bergaust, L., Shapleigh, J., Frostegard, A. and Bakken, L. (2008) Transcription and activities of NOx reductases in Agrobacterium tumefaciens: The influence of nitrate, nitrite and oxygen availability. *Environmental Microbiology*, 10: 3070-3081.

Bergström, P., Lindegarth, S., & Lindegarth, M. (2015). Modeling and predicting the growth of the mussel, Mytilus edulis: implications for planning of aquaculture and eutrophication mitigation. *Ecology and Evolution, 5*(24), 5920-5933.

Bernhard, A. (2010). The nitrogen cycle: Processes, *Players, and Human Impac*

Beveridge, M. C., Ross, L. G. & Kelly, L. A. (1994).Aquaculture and biodiversity. *Ambio, 23*(8) 497-502.

Bodik, I., Herdova, B. and Kratochvil, K.(2000) "The application of anaerobic filter for municipal wastewater treatment." *Chemical Papers-Slovak Academy Of Sciences,* 54(3): 159-164.

Boopathy, R., & Tilche, A. (1991). Anaerobic digestion of high strength molasses wastewater using hybrid anaerobic baffled reactor. *Water Research, 25*(7), 785-790.

Brunet, R. C., & Garcia-Gil, L. J. (1996). Sulfide-induced dissimilatory nitrate reduction to ammonia in anaerobic freshwater sediments. *FEMS Microbiology Ecology*, 21(2), 131-138.

Burgin, A.J., Hamilton, S.K., 2007. Have we overemphasized the role of denitrification in aquatic ecosystems? A review of nitrate removal pathways. *Frontiers in Ecology and the Environment*, 5: 89-96.

Burkholder, J.M., & Shumway, S.E. (2011).Bivalve shellfish aquaculture and eutrophication. *Shellfish Aquaculture and the Environment*, 155-215.

Campos, J. L., Valenzuela-Heredia, D., Pedrouso, A., Val del Rio, A., Belmonte, M., & Mosquera-Corral, A. (2016). Greenhouse gases emissions from wastewater treatment plants: minimization, treatment, and prevention. *Journal of Chemistry*, Article ID 3796352

Canfield, D. E., Glazer, A. N., & Falkowski, P. G. (2010). The evolution and future of Earth's nitrogen cycle. *Science*, 330(6001), 192-196.

Capodaglio, A. G., Hlavínek, P., & Raboni, M. (2015). Physico-chemical technologies for nitrogen removal from wastewaters: a review. *Revista Ambiente & Agua*, 10(3):481-498.

Chen, H., Liu, S., Yang, F., Xue, Y., & Wang, T. (2009).The development of simultaneous partial nitrification, ANAMMOX and denitrification (SNAD) process in a single reactor for nitrogen removal. *Bioresource Technology*, 100(4), 1548-1554.

Chen, J. L., Ortiz, R., Steele, T. W., & Stuckey, D. C. (2014). Toxicants inhibiting anaerobic digestion: a review. *Biotechnology Advances*, 32(8), 1523-1534.

Chen, S. N. (1991). Environmental problems of aquaculture in Asia and their solutions. *Rev. Sci. Tech*, 10(3), 609-627.

Chislock, M. F., Doster, E., Zitomer, R. A., & Wilson, A. E. (2013). Eutrophication: causes, consequences, and controls in aquatic ecosystems. *Nature Education Knowledge*, *4*(4), 10.

Chiu, Y. C., Lee, L. L., Chang, C. N., & Chao, A. C. (2007). Control of carbon and ammonium ratio for simultaneous nitrification and denitrification in a sequencing batch bioreactor. *International Biodeterioration & Biodegradation*, 59(1), 1-7.

Christensen, D. R., Gerick, J. A., & Eblen, J. E. (1984). Design and operation of an upflow anaerobic sludge blanket reactor. *Journal (Water Pollution Control Federation)*, 1059-1062.

Costa-Pierce, B. A. (2002). The'Blue Revolution'- Aquaculture must go green. *World Aquaculture*, *33*(4), 4-5.

Cristancho, D. E., & Arellano, A. V. (2006). Study of the operational conditions for anaerobic digestion of urban solid wastes. *Waste Management*, 26(5), 546-556.

Cyprowski, M., Stobnicka-Kupiec, A., £awniczek-Wa³czyk, A., Bakal-Kijek, A., Go³ofit-Szymczak, M., & Górny, R. L. (2018). Anaerobic bacteria in wastewater treatment plant. *International Archives of Occupational and Environmental Health*, *91*(5), 571-579.

Dahiya, S., & Mohan, S. V. (2019). Selective control of volatile fatty acids production from food waste by regulating biosystem buffering: A comprehensive study. *Chemical Engineering Journal*, *357*, 787-801.

Daims H, Lucker S, Wagner M (2016) A new perspective on microbes formerly known as nitrite-oxidizing bacteria. *Trends in Microbiology*, 24, 699–712

den Camp, H. J. O., Verhagen, F. J., Kivaisi, A. K., & de Windt, F. E. (1988).Effects of lignin on the anaerobic degradation of (ligno) cellulosic wastes by rumen microorganisms. *Applied Microbiology and Biotechnology*, *29*(4), 408-412.

Driscoll, C., Whitall, D., Aber, J., Boyer, E., Castro, M., Cronan, C. & Lawrence, G. (2003). Nitrogen pollution: Sources and consequences in the US northeast. *Environment: Science and Policy for Sustainable Development*, 45(7), 8-22.

Erisman, J. W., Galloway, J. N., Seitzinger, S., Bleeker, A., Dise, N. B., Petrescu, A. R., ...& de Vries, W. (2013). Consequences of human modification of the global nitrogen cycle. *Philosophical Transactions of the Royal Society B: Biological Sciences*, 368(1621), 20130116.

Ettwig, K. F., Zhu, B., Speth, D., Keltjens, J. T., Jetten, M. S., & Kartal, B. (2016). Archaea catalyze iron-dependent anaerobic oxidation of methane. *Proceedings of the National Academy of Sciences*, *113*(45), 12792-12796.

Falkowski, P., Scholes, R. J., Boyle, E. E. A., Canadell, J., Canfield, D., Elser, J., ... & Mackenzie, F. T. (2000). The global carbon cycle: a test of our knowledge of earth as a system. *Science*, *290*(5490), 291-296.

Farazaki, M., & Gikas, P. (2019). Nitrification-denitrification of municipal wastewater without recirculation, using encapsulated microorganisms. *Journal of Environmental Management*, 242, 258-265.

Feng, Q., Song, Y. C., Kim, D. H., Kim, M. S., & Kim, D. H. (2019). Influence of the temperature and hydraulic retention time in bioelectrochemical anaerobic digestion of sewage sludge. *International Journal of Hydrogen Energy*, 44(4), 2170-2179.

Fux, C., Boehler, M., Huber, P., Brunner, I., & Siegrist, H. (2002). Biological treatment of ammonium-rich wastewater by partial nitritation and subsequent anaerobic ammonium oxidation (anammox) in a pilot plant. *Journal of Biotechnology*, 99(3), 295-306.

Galloway, J. N., Dentener, F. J., Capone, D. G., Boyer, E. W., Howarth, R. W., Seitzinger, S. P. & Karl, D. M. (2004). Nitrogen cycles: past, present, and future. *Biogeochemistry*, 70(2) 153-226.

Ghaly, A. E., & Ramakrishnan, V. V. (2015). Nitrogen sources and cycling in the ecosystem and its role in air, water and soil pollution: a critical review. *Journal of Pollution Effects & Control*, 1-26.

Gloyna, E. F., & World Health Organization. (1971). Waste stabilization ponds. Vol. 7, nos. 24-25, 175 p.

Goswami, R., Chattopadhyay, P., Shome, A., Banerjee, S. N., Chakraborty, A. K., Mathew, A. K., & Chaudhury, S. (2016). An overview of physico-chemical mechanisms of biogas production by microbial communities: a step towards sustainable waste management. *3 Biotech*, 6(1), 72.

Gowen, R. J. (1994). Managing eutrophication associated with aquaculture development. *Journal of Applied Ichthyology*, 10(4), 242-257.

Greer, F. R., & Shannon, M. (2005). Infant methemoglobinemia: The role of dietary nitrate in food and water. *Pediatrics*, *116*(3), 784-786.

Gubry-Rangin, C., Hai, B., Quince, C., Engel, M., Thomson, B. C., James, P., & Nicol, G. W. (2011). Niche specialization of terrestrial archaeal ammonia oxidizers. *Proceedings of the National Academy of Sciences*, 108(52), 21206-21211.

Gulhane, M., Khardenavis, A. A., Karia, S., Pandit, P., Kanade, G. S., Lokhande, S. &Purohit, H. J. (2016). Biomethanation of vegetable market waste in an anaerobic baffled reactor: Effect of effluent recirculation and carbon mass balance analysis. *Bioresource Technology* 215, 100-109.

Guštin, S., & Marinšek-Logar, R. (2011).Effect of pH, temperature and air flow rate on the continuous ammonia stripping of the anaerobic digestion effluent. *Process Safety and Environmental Protection*, 89(1), 61-66.

Halling-Sørensen, B. (1993). Biological nitrification and denitrification. *Studies in Environmental Science*, 54, 43-43.

Hansen, K. H., Angelidaki, I., & Ahring, B. K. (1998). Anaerobic digestion of swine manure: inhibition by ammonia. *Water Research*, 32(1), 5-12.

Hassan, S. R., Zwain, H. M., & Dahlan, I. (2013). Development of Anaerobic Reactor for Industrial Wastewater Treatment: An Overview, Present Stage and Future Prospects. *Journal of Advanced Scientific Research*, 4(1).

Healey, B., Erba, W. D., & Leavitt, K. (2016). *Aquaculture and Its Impact on The Environment.* https://blogs.umass.edu/natsci397a-eross/aquaculture-and-its-impact-on-the-environment. Accessed on 28th Decmeber, 2019.

Helmer, C., & Kunst, S. (1998). Simultaneous nitrification/denitrification in an aerobic biofilm system. *Water Science and Technology*, 37(4-5), 183-187.

Hien, N. N., Van Tuan, D., Nhat, P. T., Van, T. T. T., Van Tam, N., Que, V. N. X., & Dan, N. P. (2017). Application of oxygen limited autotrophic nitritation/denitrification (OLAND) for anaerobic latex processing wastewater treatment. *International Biodeterioration & Biodegradation*, 124, 45-55.

Hogg, S. (2005). *Essential Microbiology (1st ed.)*. Wiley. pp. 99–100. ISBN 978-0-471-49754-7.

Holló, J., Czakó, L., & Mihaáltz, P. (1990). Biological nitrification and denitrification. In *Proceedings of the 5th European Congress on Biotechnology, Copenhagen* (pp. 8-13).

Hoon, C. H. (2013). The removal methods of phosphorus/phosphate and nitrogen/nitrate from water and wasterwater, Ph.D. Thesis.

House, J. E., & House, K. A. (2015). *Descriptive inorganic chemistry*. 3rd Edition Academic Press.

Jacobson, P. C., Hansen, G. J., Bethke, B. J., & Cross, T. K. (2017). Disentangling the effects of a century of eutrophication and climate warming on freshwater lake fish assemblages. *PloS one*, *12*(8), e0182667.

Ji, B., Yang, K., Zhu, L., Jiang, Y., Wang, H., Zhou, J., & Zhang, H. (2015). Aerobic denitrification: a review of important advances of the last 30 years. *Biotechnology and Bioprocess Engineering*, 20 (4), 643-651.

Karakashev, D., Batstone, D. J., & Angelidaki, I. (2005). Influence of environmental conditions on methanogenic compositions in anaerobic biogas reactors. *Applied and Environmental Microbiology*, 71(1), 331-338.

Kayhanian, M., 1994.Performance of a high-solids anaerobic digestion process under various ammonia concentrations. *J. Chem. Technol. Biotechnol.* 59, 349– 352

Ke, S., Zhang, M., Shi, Z., Zhang, T., & Fang, H. H. (2008). Phenol degradation and microbial characteristics in upflow anaerobic sludge blanket reactors at ambient and mesophilic temperatures. *International Journal of Environment and Pollution*, 32(1), 68-77.

Keller, J., Subramaniam, K., Gösswein, J., & Greenfield, P. F. (1997).Nutrient removal from industrial wastewater using single tank sequencing batch reactors. *Water Science and Technology*, *35*(6), 137.

Kuai, L., &Verstraete, W. (1998). Ammonium removal by the oxygen-limited autotrophic nitrification-denitrification system. *Applied and Environmental Microbiology,* 64 (11), 4500-4506.

Kuypers, M. M., Marchant, H. K., & Kartal, B. (2018).The microbial nitrogen-cycling network. *Nature Reviews Microbiology*, *16*(5), 263.

Ladu, J. L. C., & Lü, X. W. (2014). Effects of hydraulic retention time, temperature, and effluent recycling on efficiency of anaerobic filter in treating rural domestic wastewater. *Water Science and Engineering*, 7(2), 168-182.

Lam, P., &Kuypers, M. M. (2011). Microbial nitrogen cycling processes in oxygen minimum zones. *Annual Review of Marine Science*, 3, 317-345.

Lay, J. J., Li, Y. Y., Noike, T., Endo, J., & Ishimoto, S. (1997). Analysis of environmental factors affecting methane production from high-solids organic waste. *Water Science and Technology*, 36(6-7), 493-500.

Levinson, W. (2010). *Review of Medical Microbiology and Immunology* (11th ed.). McGraw-Hill. pp. 150–157. ISBN 978-0-07-174268-9.

Lin, C. Y. (1993). Effect of heavy metals on acidogenesis in anaerobic digestion. *Water Research*, 27(1), 147-152.

Lin, Q., He, G., Rui, J., Fang, X., Tao, Y., Li, J., & Li, X. (2016). Microorganism-regulated mechanisms of temperature effects on the performance of anaerobic digestion. *Microbial Cell Factories*, 15(1), 96.

Liu, C. F., Yuan, X. Z., Zeng, G. M., Li, W. W., & Li, J. (2008). Prediction of methane yield at optimum pH for anaerobic digestion of organic fraction of municipal solid waste. *Bioresource Technology*, 99(4), 882-888.

Liu, R. R., Tian, Q., Yang, B., & Chen, J. H. (2010). Hybrid anaerobic baffled reactor for treatment of desizing wastewater. *International Journal of Environmental Science & Technology*, 7(1), 111-118.

Lukow, T. & Diekmann, H. (1997) Aerobic denitrification by a newly isolated heterotrophic bacterium strain TL1. *Biotechnology Letters,* 19: 1157-1159

Luo, Z., Hu, S., & Chen, D. (2018). The trends of aquacultural nitrogen budget and its environmental implications in China. *Scientific Reports*, *8*., 10877

Lyberatos, G., & Skiadas, I. V. (1999). Modelling of anaerobic digestion–a review. *Global Nest Int J*, *1*(2), 63-76.

Mamo, T. Z., Dutta, A., & Jabasingh, S. A. (2019). Start-up of a pilot scale anaerobic reactor for the biogas production from the pineapple processing industries of Belgium. *Renewable Energy*, *134*, 241-246.

Manassaram, D. M., Backer, L. C., & Moll, D. M. (2005). A review of nitrates in drinking water: maternal exposure and adverse reproductive and developmental outcomes. *Environmental Health Perspectives*, 114(3), 320-327.

Manuel, J. (2014). Nutrient pollution: A persistent threat to waterways. *Environmental Health Perspectives*, 122(11), A304-A309.

Marchant, H. K., Lavik, G., Holtappels, M., & Kuypers, M. M. (2014). The fate of nitrate in intertidal permeable sediments. PloS one, 9(8), e104517.

Martens-Habbena, W., Berube, P. M., Urakawa, H., José, R., & Stahl, D. A. (2009). Ammonia oxidation kinetics determine niche separation of nitrifying Archaea and Bacteria. *Nature*, 461(7266), 976-979.

Martinez-Porchas, M., & Martinez-Cordova, L. R. (2012). World aquaculture: environmental impacts and troubleshooting alternatives. *The Scientific World Journal*, *2012*.

Megraw, S. R., & Knowles, R. (1989b). Methane-dependent nitrate production by a microbial consortium enriched from a cultivated humisol. *FEMS Microbiology Ecology*, *5*(6):359-365.

Megraw, S. R., & Knowles, R. (1989). Isolation, characterization, and nitrification potential of a methylotroph and two heterotrophic bacteria from a consortium showing methane-dependent nitrification. *FEMS Microbiology Ecology*, *5*(6), 367-374.

Menon, (2017). India Has an Ammonia Problem but No Policy to Deal With It. On line available at https://thewire.in/129196/ammonia-india-haze-pollution-fossil-fuels.

Miao, L., Yang, G., Tao, T., & Peng, Y. (2019). Recent advances in nitrogen removal from landfill leachate using biological treatments–a review. *Journal of environmental management*, 235, 178-185.

Micolucci, F., Gottardo, M., Pavan, P., Cavinato, C., & Bolzonella, D. (2018). Pilot scale comparison of single and double-stage thermophilic anaerobic digestion of food waste. *Journal of Cleaner Production*, 171, 1376-1385.

Milán, Z., Sánchez, E., Weiland, P., Borja, R., Martýn, A., & Ilangovan, K. (2001). Influence of different natural zeolite concentrations on the anaerobic digestion of piggery waste. *Bioresource Technology*, 80(1), 37-43.

Mittal, A. (2011).Biological wastewater treatment. *Water Today*, 1, 32-44.

Moletta, R., Verrier, D., & Albagnac, G. (1986). Dynamic modelling of anaerobic digestion. *Water Research*, 20(4), 427-434.

Morita, M., Uemoto, H. & Watanabe, A. (2008). Nitrogen-removal bioreactor capable of simultaneous nitrification and denitrification for application to industrial wastewater treatment. *Biochemical Engineering Journal*, *41*(1), 59-66.

Mudhoo, A., & Kumar, S. (2013). Effects of heavy metals as stress factors on anaerobic digestion processes and biogas production from biomass. *International Journal of Environmental Science and Technology*, *10*(6), 1383-1398.

Naylor, R. L., Goldburg, R. J., Primavera, J. H., Kautsky, N., Beveridge, M. C., Clay, J., ... & Troell, M. (2000). Effect of aquaculture on world fish supplies. *Nature*, 405(6790), 1017-1024.

Nelson, D. L., Lehninger, A. L., & Cox, M. M. (2008). *Lehninger Principles of Biochemistry*. Macmillan.

Ngatia, L., Grace III, J. M., Moriasi, D., & Taylor, R. (2019). Nitrogen and Phosphorus Eutrophication in Marine Ecosystems.In *Monitoring of Marine Pollution*. Intech Open.

Nicol, G. W., Hink, L., Gubry-Rangin, C., Prosser, J. I., & Lehtovirta-Morley, L. E. (2019). Genome Sequence of "Candidatus Nitrosocosmicus franklandus" C13, a Terrestrial Ammonia-Oxidizing Archaeon. *Microbiology Resource Announcements*, 8(40), e00435-19.

Nizami, A. S., Saville, B. A., & Maclean, H. L. (2013). Anaerobic digesters: perspectives and challenges. In *Bioenergy Production by Anaerobic Digestion* pp. 169-181, Routledge.

Nizzoli, D., Welsh, D.T., Fano, E.A., Viaroli, P., 2006. Impact of clam and mussel farming on benthic metabolism and nitrogen cycling, with emphasis on nitrate reduction pathways. *Marine Ecology Progress Series,* 315, 151-165

Nömmik, H. (1956). Investigations on denitrification in soil. *Acta Agriculturae Scandinavica, 6*(2), 195-228.

Ohyama, T. (2010). Nitrogen as a major essential element of plants. *Nitrogen assimilation in plants. Signpost, Trivandrum, Kerala, India.*

Oleszkiewicz, J. A., & Sharma, V. K. (1990). Stimulation and inhibition of anaerobic processes by heavy metals—a review. *Biological Wastes, 31*(1), 45-67.

Omil, F., Méndez, R., & Lema, J. M. (1995). Anaerobic treatment of saline wastewaters under high sulphide and ammonia content. *Bioresource Technology, 54*(3), 269-278.

Paetkau, M., & Cicek, N. (2011). Comparison of nitrogen removal and sludge characteristics between a conventional and a simultaneous nitrification–denitrification membrane bioreactor. *Desalination, 283,* 165-168.

Pai, S. L., Chong, N. M., & Chen, C. H. (1999). Potential applications of aerobic denitrifying bacteria as bioagents in wastewater treatment. *Bioresource Technology, 68*(2), 179-185.

Palomo, A., Fowler, S. J., Gülay, A., Rasmussen, S., Sicheritz-Ponten, T., & Smets, B. F. (2016). Metagenomic analysis of rapid gravity sand filter microbial communities suggests novel physiology of Nitrospira spp. *The ISME journal*, 10(11), 2569.

Pang, H., Zhou, Z., Niu, T., Jiang, L. M., Chen, G., Xu, B., & Qiu, Z. (2018). Sludge reduction and microbial structures of aerobic, micro-aerobic and anaerobic side-stream reactor coupled membrane bioreactors. *Bioresource Technology, 268,* 36-44.

Park, J. G., Lee, B., Jo, S. Y., Lee, J. S., & Jun, H. B. (2018). Control of accumulated volatile fatty acids by recycling nitrified effluent. *Journal of Environmental Health Science and Engineering*, 16(1), 19.

Pauss, A., & Guiot, S. R. (1993). Hydrogen monitoring in anaerobic sludge bed reactors at various hydraulic regimes and loading rates. *Water Environment Research, 65*(3), 276-280.

Philips, S., Laanbroek, H. J., & Verstraete, W. (2002). Origin, causes and effects of increased nitrite concentrations in aquatic environments. *Reviews in Environmental Science and Biotechnology, 1*(2), 115-141.

Pinto, A. J., Marcus, D. N., Ijaz, U. Z., Bautista-de lose Santos, Q. M., Dick, G. J., & Raskin, L. (2016). Metagenomic evidence for the presence of comammoxNitrospira-like bacteria in a drinking water system. *Msphere, 1*(1), e00054-15.

Porrello, S., Lenzi, M., Persia, E., Tomassetti, P., & Finoia, M. G. (2003). Reduction of aquaculture wastewater eutrophication by phytotreatment ponds system: I. Dissolved and particulate nitrogen and phosphorus. *Aquaculture, 219*(1-4), 515-529.

Ripley, L. E. (2000). *U.S. Patent No. 6,096,214.* Washington, DC: U.S. Patent and Trademark Office.

Robertson, E. K., Roberts, K. L., Burdorf, L. D., Cook, P., & Thamdrup, B. (2016). Dissimilatory nitrate reduction to ammonium coupled to Fe (II) oxidation in sediments of a periodically hypoxic estuary. *Limnology and Oceanography*, 61(1), 365-381.

Rodziewicz, J., Ostrowska, K., Janczukowicz, W., & Mielcarek, A. (2019). Effectiveness of Nitrification and Denitrification Processes in Biofilters Treating Wastewater from De-Icing Airport Runways. *Water*, 11(3), 630.

Ryan KJ; Ray CG, eds. (2004). *Sherris Medical Microbiology* (4th ed.). McGraw Hill. pp. 261–271, 273–296. ISBN 0-8385-8529-9.

Saleh-Lakha, S., Shannon, K. E., Henderson, S. L., Goyer, C., Trevors, J. T., Zebarth, B. J., & Burton, D. L. (2009). Effect of pH and temperature on denitrification gene expression and activity in Pseudomonas mandelii. *Appl. Environ. Microbiol., 75*(12), 3903-3911.

Sattler, M. L. (2011). Anaerobic Processes for Waste Treatment and Energy Generation, Integrated Waste Management - Volume II, Sunil Kumar, IntechOpen, DOI: 10.5772/17731

Schleper, C., & Nicol, G. W. (2010).Ammonia-oxidising archaea–physiology, ecology and evolution. In *Advances in Microbial Physiology* 57,1-41. Academic Press.

Schullehner, J., Stayner, L., & Hansen, B. (2017). Nitrate, nitrite, and ammonium variability in drinking water distribution systems. *International Journal of Environmental Research and Public Health*, 14(3), 276.

Seitz, H. J., & Cypionka, H. (1986). Chemolithotrophic growth of Desulfovibriodesulfuricans with hydrogen coupled to ammonification of nitrate or nitrite. *Archives of Microbiology*, *146*(1), 63-67.

Sengupta, S., Nawaz, T., & Beaudry, J. (2015). Nitrogen and phosphorus recovery from wastewater. *Current Pollution Reports*, *1*(3), 155-166.

Sheng, L., Liu, J., Zhang, C., Zou, L., Li, Y. Y., & Xu, Z. P. (2019). Pretreating anaerobic fermentation liquid with calcium addition to improve short chain fatty acids extraction via in situ synthesis of layered double hydroxides. *Bioresource Technology*, *271*, 190-195.

Shi, X., Zhao, J., Chen, L., Zuo, J., Yang, Y., Zhang, Q., & Zhou, J. (2019). Genomic dynamics of full-scale temperature-phased anaerobic digestion treating waste activated sludge: Focusing on temperature differentiation. *Waste Management*, 87, 621-628.

Šimek, M., Jýìšová, L., & Hopkins, D. W. (2002). What is the so-called optimum pH for denitrification in soil?. *Soil Biology and Biochemistry*, 34(9), 1227-1234.

Strous, M., Kuenen, J. G., &Jetten, M. S. (1999). Key physiology of anaerobic ammonium oxidation. *Applied and Environmental Microbiology*, 65(7), 3248-3250.

Suantika, G., Situmorang, M. L., Aditiawati, P., IndrianiAstuti, D., Azizah, F. F. N., & Muhammad, H. (2017).Closed Aquaculture System: Zero Water Discharge for Shrimp and Prawn Farming in Indonesia.In *Biological Resources of Water*. Intech Open.

Tilley, E. (2014). *Compendium of sanitation systems and technologies*. 2nd Edition, Eawag.

Tjandraatmadja, G., & Diaper, C. (2006). Sources of Critical Contaminants in domestic wastewater: a literature review. CSIRO.

U.S. Environmental Protection Agency (U.S. EPA). (1992). Framework for ecological risk assessment. U.S. Environmental Protection Agency, Washington, D.C. EPA/630/R-92/001.

UN-Water. 2015a. Wastewater Management: A UN-Water Analytical Brief. UN-Water. www.unwater.org/fileadmin/user_upload/unwater_new/docs/UNWater_Analytical_Brief_Wastewater_Management.pdf

UN-WWDR 2017.United Nation World Water Development Report 2017. New York, United Nations. www. https://unesdoc.unesco.org/ark:/48223/pf0000247153

Upendra Kumar, K. C., Bachman, T., & Williams, E. (2006). *U.S. Patent Application No. 11/347,621*.

Van Cleemput, O., & Patrick Jr, W. H. (1974). Nitrate and nitrite reduction in flooded gamma-irradiated soil under controlled pH and redox potential conditions. *Soil Biology and Biochemistry*, 6 (2), 85-88.

Van de Graaf, A. A., de Bruijn, P., Robertson, L. A., Jetten, M. S., & Kuenen, J. G. (1996). Autotrophic growth of anaerobic ammonium-oxidizing micro-organisms in a fluidized bed reactor. *Microbiology*, 142(8), 2187-2196.

Van de Graaf, A. A., Mulder, de Bruijn, Jetten, M. S., Robertson, L. A., & Kuenen, J. G. (1995). Anaerobic oxidation of ammonium is a biologically mediated process. *Applied and Environmental Microbiology* 61(4), 1246-1251.

Van Kessel, M.A., Speth, D.R., Albertsen, M., Nielsen, P.H., den Camp, H.J.O., Kartal, B. et al (2015) Complete nitrification by a single microorganism. *Nature* 528:555–559

Van, D. P., Fujiwara, T., Leu Tho, B., Toan, S., Phu, P., Hoang Minh, G., & Minh, G. H. (2019). A review of anaerobic digestion systems for biodegradable waste: Configurations, operating parameters, and current trends. *Environmental Engineering Research, 25*(1), 1-17.

Verrier, D., F. Roy, & Albagnac, G. "Two-phase methanization of solid vegetable wastes." *Biological Wastes*22.3 (1987): 163-177.

Vijayaraghavan, K., & Murthy, D. V. S. (1997). Effect of toxic substances in anaerobic treatment of tannery wastewaters. *Bioprocess Engineering*, 16(3), 151-155.

Vinnerås, B. (2001). Faecal separation and urine diversion for nutrient management of household biodegradable waste and wastewater. Ph.D. Thesis, Swedish University of Agricultural Sciences, Uppasala.

Von Sperling, M. (2007). Activated sludge and aerobic biofilm reactors. IWA publishing.

Wang, B.Z., Zhao, J., Guo, Z., Ma, J., Xu, H., Jia, Z. (2015). Differential contributions of ammonia oxidizers and nitrite oxidizers to nitrification in four paddy soils. ISME J 9:1062–1075

Wang, J., Huang, Y., & Zhao, X. (2004).Performance and characteristics of an anaerobic baffled reactor. *Bioresource Technology, 93*(2), 205-208.

Wang, K., Yin, J., Shen, D., & Li, N. (2014). Anaerobic digestion of food waste for volatile fatty acids (VFAs) production with different types of inoculum: effect of pH. *Bioresource Technology, 161*, 395-401.

Wang, X., Li, Y., Zhang, Y., Pan, Y. R., Li, L., Liu, J., & Butler, D. (2019). Stepwise pH control to promote synergy of chemical and biological processes for augmenting short-chain fatty acid production from anaerobic sludge fermentation. *Water Research, 155*, 193-203.

Wang, Y., Ma, L., Mao, Y., Jiang, X., Xia, Y., Yu, K., & Zhang, T. (2017). Comammox in drinking water systems. *Water Research, 116*, 332-341.

Ward, M. H., Jones, R. R., Brender, J. D., De Kok, T. M., Weyer, P. J., Nolan, B. T., & Van Breda, S. G. (2018). Drinking water nitrate and human health: an updated review. *International Journal of Environmental Research and Public Health, 15*(7), 1557.

Weiner, I. D., & Verlander, J. W. (2019). Emerging Features of Ammonia Metabolism and Transport in Acid-Base Balance. In *Seminars in Nephrology,* 39(4): 394-405. WB Saunders.

WHO (World Health Organisation), (2011). Guidelines for drinking water quality.Fourth edition. World Health Organization, Geneva, Switzerland, 541 pp. Available from http://whqlibdoc.who.int/publications/2011/9789241548151_eng.pdf

Wilsenach, J., & Van Loosdrecht, M. (2003). Impact of separate urine collection on wastewater treatment systems. *Water Science and Technology*, 48(1): 103-110.

Windey, K., De Bo, I., &Verstraete, W. (2005). Oxygen-limited autotrophic nitrification–denitrification (OLAND) in a rotating biological contactor treating high-salinity wastewater. *Water Research*, 39(18): 4512-4520.

Wu, B., Li, Y., Lim, W., Lee, S. L., Guo, Q., Fane, A. G., & Liu, Y. (2017). Single-stage versus two-stage anaerobic fluidized bed bioreactors in treating municipal wastewater: Performance, foulant characteristics, and microbial community. *Chemosphere*, 171, 158-167.

Yamashita, T., & Yamamoto-Ikemoto, R. (2014). Nitrogen and phosphorus removal from wastewater treatment plant effluent via bacterial sulfate reduction in an anoxic bioreactor packed with wood and iron. *International Journal of Environmental Research and Public Health*, 11(9): 9835-9853.

Yang, P. Y., & Zhang, Z. (1995). Nitrification and denitrification in the wastewater treatment system. In Proceedings of the UNESCO—University of Tsukuba International Seminar on Traditional Technology for Environmental Conservation and Sustainable Development in the Asian-Pacific Region, Tsukuba Science City, *Japan* (p 11-14).

Yin, C. R., Seo, D. I., Kim, M. K., & Lee, S. T. (2000). Inhibitory effect of hardwood lignin on acetate-utilizing methanogens in anaerobic digester sludge. *Biotechnology Letters*, 22(19).

Young, J. C., & Yang, B. S. (1989). Design considerations for full-scale anaerobic filters. *Research Journal of the Water Pollution Control Federation*, 1576-1587.

Yousefzadeh, S., Ahmadi, E., Gholami, M., Ghaffari, H. R., Azari, A., Ansari, M., & Rezaei, S. (2017). A comparative study of anaerobic fixed film baffled reactor and up-flow anaerobic fixed film fixed bed reactor for biological removal of diethyl phthalate from wastewater: a performance, kinetic, biogas, and metabolic pathway study. *Biotechnology for Biofuels*, 10(1), 139.

Zafarzadeh, A., Bina, B., Nikaeen, M., Movahedian, A.H., & Haji, K.M. (2011). Effect of dissolved oxygen and chemical oxygen demand to nitrogen ratios on the partial nitrification/ denitrification process in moving bed biofilm reactors.

Zhang, Ruihong, et al. (2007) "Characterization of food waste as feedstock for anaerobic digestion." *Bioresource Technology*, 98.4: 929-935.

Zhang, S., Liu, J., Wei, S., Gao, J., Wang, D., & Zhang, K. (2006). Impact of aquaculture on eutrophication in Changshou Reservoir. *Chinese Journal of Geochemistry*, 25(1), 90-96

Zhang, Y., Angelidaki, I., (2015). Recovery of ammonia and sulfate from waste streams and bioenergy production via bipolar bioelectrodialysis. *Water Research*, 85, 177–184.

Zheng, C., Zhao, L., Zhou, X., Fu, Z., & Li, A. (2013). Treatment technologies for organic wastewater. *Water Treatment*, 249-286.

Zhou, M., Yan, B., Wong, J. W., & Zhang, Y. (2018). Enhanced volatile fatty acids production from anaerobic fermentation of food waste: a mini-review focusing on acidogenic metabolic pathways. *Bioresource Technology*, 248, 68-78.

Zhu, G., Peng, Y., Li, B., Guo, J., Yang, Q., & Wang, S. (2008). Biological removal of nitrogen from wastewater.In *Reviews of Environmental Contamination and Toxicology*, 159-195. Springer, New York, NY.

13

The Emerging Threat of Anaerobic Infections

*Aruna Poojary and Anurag Kumar Bari

Department of Pathology & Microbiology, Breach Candy Hospital Trust Mumbai, Maharashtra, India

Abstract

The normal human microbial flora comprises of both aerobes and anaerobes at various sites. Anaerobes are recognised as pathogens causing simple abscesses to life threatening human infections like gas gangrene. They are also incriminated in a large number of mixed bacterial infections and their diagnosis and specific therapy is important for better patient outcomes. The emergence of Clostridioides difficile *as an important pathogen causing antibiotic associated diarrhoeas and its emerging resistance to antimicrobial agents has brought anaerobic microbiology into the forefront. Unfortunately, anaerobic clinical microbiology is difficult to practice with long turnaround time, difficult to grow pathogens, new genera and species being discovered based on 16S rRNA sequencing methods and increasing incidence of antimicrobial resistance being reported in anaerobes. This chapter delves into the important aspects required to improve utilization of anaerobic microbiology services and the variety of human infections we encounter in clinical practice due to anaerobic bacteria.*

Introduction

Anaerobes play a very important role in health and disease of human beings by maintaining a balance between the host and the colonizing flora (Park et al.2009). They are found as commensal flora in many parts of the human body and have a symbiotic role along with the aerobic commensal flora. Most human infections with anaerobes are often mixed with aerobic bacteria. Conventional methods of detecting anaerobes usually take time and reduce the sensitivity of detecting

**Corresponding Author: arunapoojary@gmail.com*

them. Molecular methods based on protein analysis or DNA detection have made rapid diagnosis of anaerobes possible. These methods also help in rapid detection of drug resistance as well which play a role in selection of appropriate antibiotic therapy (Johannson et al.2014). Approximately two-thirds of anaerobic infections are caused by the following anaerobes i.e. *Bacteroides fragilis, Prevotella, Porphyromonas, Fusobacterium, Peptostreptococcus, Clostridium ramosum, Clostridium perfringens & Clostridioides difficile* (Finegold et al.1995). Better methods of detection and new species being identified; anaerobes are emerging as important causes of community and hospital acquired infections. Drug resistance is another important concern which we are faced with in today's time.

Predisposing factors for anaerobic infections (Finegold et al.1995, Finegold et al.2002)

- trauma leading to contaminated wounds
- diabetes mellitus
- neutropenia (congenital or acquired)
- colonic malignancy
- abdominal surgeries

Pathogenesis of anaerobic infections

Anaerobic infections occur when the anatomical barriers are disrupted and these bacteria enter into otherwise sterile spaces. They may also occur if a disturbance is induced in the ratio of aerobes and anaerobes like in the gut after taking antibiotics causing antibiotic associated diarrhoea (AAD). Most anaerobic infections are derived from the commensals of the oropharynx, gastrointestinal tract and vagina. Anaerobic infections especially involving the skin and soft tissue can also arise in trauma associated wounds occurring due to road traffic accidents and indwelling implants like intrauterine devices. Often these are mixed infections comprising of aerobes and anaerobes (Bowler et al.2001). Another important aspect of pathogenesis of spore bearing anaerobes is that they can persist in the environment for a long time by forming endospores only to germinate when conditions are conducive for growth. This has an important role to play in the causation of Clostridial myonecrosis in contaminated trauma wounds like road traffic accidents. These infections can become life threatening if not intervened by a surgical procedure immediately (Nichols et al.2001). Similarly, spores of *Clostridioides difficile* may persist in the healthcare environment and unless thorough terminal disinfection is carried out, they can cause infection in the subsequent patients entering the same clinical area by horizontal transmission. (Barra-Carrasco et al.2014).

Virulence factors: The specific properties that help anaerobes cause disease are called as virulence factors. To establish an infection the anaerobe should be able to attach itself to cells, invade them, destroy tissues, prevent its own destruction by host defences and multiply to spread the infection and cause disease.

- Adherence to tissues – Several gram-negative anaerobic bacteria have the ability to attach to human cells. The most common target cells studied so far are the epithelial cells and erythrocytes (RBCs).
- Destruction of tissues – destruction of tissues can be caused by enzymes and toxins produced by the bacteria.
- Protection from host immune responses like oxygen toxicity and protection against phagocytosis is brought about the capsule, metabolic cell products like volatile fatty acids and enzymes like proteases.

Laboratory Diagnosis of anaerobic infections

Most clinical microbiology laboratories do not perform anaerobic cultures as they are technically demanding, financially draining and also have delayed turnaround time (TAT). Yet for those labs which venture into anaerobic microbiology, the knowledge of the clinical team about role of anaerobes in causing infections and the laboratory's policy to reject inappropriate samples helps in improving diagnosis and also brings in cost reduction (Nagy et al.2018) (Table 1).

The tests that a routine clinical microbiology laboratory can offer for diagnosis for anaerobic infections include anaerobic cultures, identification of organisms and their drug susceptibility testing using standard methods. Molecular methods for detecting *Clostridioides difficile* and enzyme immunoassays for the same are also readily available. Cultures for *C.difficile* have research significance and do not play an important role in managing *C.difficile* associated infections due to the long TAT.

Microbiology sampling for diagnosis of anaerobic infections: Training clinical teams on appropriate specimen collection, handling and storage of specimens goes a long way in better utilization of anaerobic microbiology services. Microbiology sampling for anaerobes begins from suspecting anaerobic infection, then collecting the appropriate specimen, sending it to the laboratory in appropriate transport medium as early as is possible. In the lab, the process is completed by inoculating the specimen on appropriate media and incubating in

Table 1: Specimen acceptance & rejection criteria for anaerobes

Specimen source	Acceptable specimen	Unacceptable specimens
Deep seated organ space abscesses	Pus aspirate, usually invasive and received as intraoperative specimens	Pus swabs without transport media
Skin, soft tissue, decubitus ulcers, diabetic foot infections	Deep biopsy specimens, needle aspirates of pus, bone curettes, advancing wound edge specimens can be taken after debridement and cleaning with sterile saline	No superficial swabs or dry swabs
Muscle	Deep biopsy or intraoperative tissue specimen	No superficial swabs
Bone and joint	Aspirates, bone curettes and bone biopsies. Biopsies should be taken from multiple sites. Joint fluids may be collected in anaerobic blood culture bottles	No superficial swabs or dry swabs
Ear	Aspirates from tympanocentesis	No superficial swabs or dry swabs
Eye	Tissues, anterior chamber, vitreous aspirates and conjunctival biopsies	No superficial swabs or dry swabs
Head & Neck	Pus aspirates, intraoperative biopsy specimens	No superficial swabs or dry swabs
Periodontal	Pus aspirates, subgingival pocket biopsies and aspirates	Surface gingival swabs and oral swabs
Central nervous system	Pus aspirates especially from contiguous sites like middle ear, cribriform plate or teeth	Routine CSF
Blood	Anaerobic blood cultures	Blood collected in vacutainers, intravascular catheter tips
Pulmonary	Transthoracic aspirate of lung abscess, lung biopsy, bronchoscopic aspirate of lung abscess or tissue biopsy	Sputum, endotracheal or tracheal secretions, bronchoalveolar lavage
Abdominal cavity	Fluid, pus aspirates, tissue biopsies	No superficial swabs or dry swabs.
Urine	Suprapubic aspirate of urine, intraoperative specimens from kidney or blocked ureters	Voided or catheterised urine specimens
Female genital tract	Endometrial specimens taken by protected brush, tissue from septic abortion, tissue biopsies, pelvic pus aspirates by culdocentesis, peritoneal fluid and surgical specimens	Avoid cervical and vaginal swabs except when bacterial vaginosis is suspected
Gastrointestinal tract	Tissue biopsies	Stomach, small, large bowel, colostomy contents, colocutaneous fistula specimens
Stool	Unformed or liquid stool specimens for *Clostridiodes difficile* cultures or PCR or immunoassay for toxin detection	Formed or solid stool samples for *C.difficile*. Stool culture for other anaerobes

specific anaerobic conditions for the desired time which is usually at least 48 hours.

Clinical clues when anaerobic infections are suspected and thereby collecting specimens and requesting for anaerobic cultures are given below in Table 2 (Bergan et al.1983, Brook et al.2007)

Table 2: Clinical & microbiology clues to Anaerobic infection

Foul or putrid smelling discharge from a wound
Penetrating trauma injuring mucosal tracts where anaerobes exist as commensal flora
Abdominal surgeries with manipulation of the gastrointestinal tract
Presence of gas in tissue or crepitus felt on palpation
Deep seated visceral abscesses
Septic thrombophlebitis
Preceding aminoglycoside therapy
Lack of inflammatory cells but plenty of gram negative or gram-positive bacilli on gram stain smears of wound, pus or tissue specimens
Lack of aerobic growth in pus or tissue specimens obtained from the site of infection
Blood cultures flashing positive with wet mount and gram stain showing bacteria which do not grow on aerobic media

Collection of specimens : Clinical microbiology labs must have defined policy for sample collection as this is the most crucial pre analytical step for microbiology results to co- relate with the patient's clinical condition (Brook et al.2007, Nagy et al. 2018). Some important points to note when collecting specimens for anaerobic cultures are as follows:

- Anaerobes easily loose viability, hence all specimens should be collected in transport media or anaerobic broth media
- Avoid collecting pus swabs as they are prone to desiccation
- Avoid collecting superficial wound swabs as they are often colonized with aerobic flora
- Avoid using antiseptic agents for skin antisepsis as they destroy viability of anaerobes, use saline instead to clean the area before sampling is performed
- These media should be readily available on site in surgical wards, operating rooms and emergency rooms, so that specimens collected are transported quickly to the microbiology laboratory
- No anaerobic specimens must be stored in the refrigerator at 2 to 8° C as they tend to loose viability

- Processing should be performed as early as possible inside an anaerobic work station.

Transportation & handling: Since anaerobes easily loose viability it is extremely important that collected samples should be transported and handled correctly. All specimens should be immediately placed into anaerobic transport media (liquid/tissue) or into pre-reduced anaerobically sterilized semisolid transport media (Stuart's or Amie's transport media) for swabs. Specimens should be sent to the laboratory immediately. As far as possible samples should be immediately inoculated into pre-reduced anaerobic media, if not store samples at room temperature especially large volume specimens (Isenberg et al.2007). When the transport and processing time is >24 hours, one can use anaerobic specimen collector or pouches which are rendered anaerobic before use.

Direct examination of specimens: Gram stain of specimens often provides important clues to presence of anaerobes. Multiple species, pleomorphic pale staining (*Bacteroides spp*), Fusiform shape (*Fusobacterium spp*), gram negative cocci (*Veillonella*), box car appearance (*Clostridium spp*) are typical morphologies seen on Gram stain. The number of inflammatory cells are often few to nil in the presence anaerobes. Sterile aerobic culture results with plenty of organisms on smear also suggest the presence of anaerobic infections (Brook et al.2017).

Culture methods available: Since anaerobes are often part of polymicrobial infections, it is essential to use non selective and selective aerobic and anaerobic media for primary isolation. Initial basic media required for anaerobes includes non-selective supplemented (Vitamin K & hemin) blood agar, selective media allowing growth of *Bacteroides and Prevotella* species along with liquid anaerobic broth (thioglycolate broth), enriched and differential media. (Nagy et al. 2018, Isenberg et al. 2007 and Brook et al. 2017). These media may be freshly prepared or commercially available. (Table 3)

Table 3: Media for recovery and isolation of anaerobes from clinical specimens (Isenberg et al. 2007, Brook et al. 2017)

Media	Utility
Enriched Non selective media	
Brain Heart Infusion blood agar, Brucella blood agar, CDC anaerobe agar, Columbia blood agar, TSA blood agar, Schaedler blood agar	General purpose media, allows growth of all significant anaerobes
Selective media	
KVLB or LKV Kanamycin -Vancomycin Laked blood agar	Selects growth of Bacteroides, Fusobacterium and Prevotella species, enhances the pigment producing *P.melaninogenica* and

(Contd.)

	Porphyromonas does not grow
Bacteroides bile esculin agar	Allows growth of *Bacteroides fragilis* group & also *Bilophila wadsworthia*
Phenyl ethyl alcohol sheep blood agar	Allow growth of most gram positive and gram-negative anaerobes and inhibits proteus
Colistin Nalidixic acid Blood agar	Inhibits gram negative bacteria, allows growth of gram-positive bacteria
Cycloserine Cefoxitin fructose agar (CCFA)	Selects the growth of *C.difficile*
Egg Yolk agar	Useful when *Clostridium* species especially those producing lecithinase and lipase are suspected
THIO supplemented with vitamin K1, hemin, $NaHCO_3$, chopped meat carbohydrate & chopped meat-glucose	Supports the growth of all anaerobes. Serves as a back-up enrichment broth if primary culture media fail to grow. Can be held at 35°C for 7 days

Blood culture collection methods: Blood culture is an important specimen for detecting anaerobic bacteraemia. Patients at risk are usually those undergoing gastrointestinal surgeries, pelvic abscesses or other deep-seated organ space infections. Most laboratory protocols contain commercially available anaerobic bottles as a part of their set of blood culture bottles. Some important issues when collecting blood cultures are as follows:

- At least 8 to 10 cc of blood must be collected in each bottle
- The volume is more important than the timing of blood cultures
- Two sets are preferable over one set as it increases the yield
- Blood cultures must be collected either prior to antimicrobial therapy administration or before change of antibiotics
- Skin antisepsis is very important to ensure contaminants don't grow in the culture
- When only an anaerobic bottle flashes positive, primary subculture must be done on both aerobic and anaerobic media

Incubation methods for anaerobes: Incubating inoculated cultures in appropriate anaerobic conditions (gas mixture containing 80 to 90% nitrogen (N_2), 5% hydrogen (H_2), and 5 to 10% carbon dioxide (CO_2) and temperature is an additional crucial step to ensure recovery of anaerobes. This is achieved by use of anaerobic chambers, anaerobic pouches or bags and anaerobic jars. (Isenberg HD 2007) The cultures must be evaluated at 48 hours and worked up to ensure absence of facultative aerobes (aerotolerance test)

Identification of anaerobes: While conventional methods can be used to identify anaerobes, the TAT of these methods reduces their clinical utility. An automated identification system (e.g. ANC card of Vitek 2 Compact -

BioMérieux) is a rapid method for identification of most clinically significant anaerobes (Rennie et al.2008). Another important advancement in identification of anaerobes has been the Matrix Assisted Laser Desorption Ionization – Time of Flight (MALDI-TOF) technology that uses proteomics to identify a variety of microorganism's including anaerobes (Shannon et al. 2018).

Table 4: Common anaerobes causing human infections

Gram positive bacilli	Gram negative bacilli
Spore bearers • *Clostridium species* Non spore bearing • *Bifidobacterium* • *Arcanobacterium* • *Lactobacilli* • *Propionibacterium*	• *Bacteroides fragilis complex* • *Fusobacterium* • *Porphyromonas species* • *Prevotella species* • *Bilophila wadsworthia*
Gram positive cocci • *Peptostreptococci* • *Peptococcus* • *Gemella* • *Parvimonas* • *Finegoldia magna*	**Gram negative cocci** • *Veillonella*

Drug susceptibility testing for anaerobes: Recommended drug susceptibility testing methods for anaerobes include microbroth dilution (BMD) for *Bacteroides species*, agar dilution method for other anaerobes. Gradient diffusion methods (e.g. E-test) are also acceptable for anaerobic drug susceptibility testing (Brook et al. 2013).

Infections syndromes caused by spore bearing organisms

Clostridium species are the spore bearing anaerobes and they are incriminated for a large variety of infection syndromes. Many a times these can be life threatening and early diagnosis is the key to preventing mortality.

- Traumatic gas gangrene (Stevens et al.2002) - Gas gangrene caused by *C.perfringens* is characterized by extensive local destruction of muscle (myonecrosis), rapid destruction of viable tissue, shock, and ultimately death. *C.perfringens* is divided into different types from A – O based on the spectrum of toxins it produces. These organisms produce large quantities of exotoxins like alpha (phospholipase C, PLC) and theta (perfringolysin O, PFO) toxins which are its major virulence factors. It also produces abundant amount of gas which gives rise to the typical "crepitus" which is the hall mark of the disease. An immediate surgical intervention is mandatory in such cases to save the life of the patient.

- Spontaneous non traumatic gas gangrene – *Clostridium septicum* has typically been associated with spontaneous non traumatic gas gangrene (Stevens et al. 2012). It starts abruptly with confusion, excruciating pain followed by rapidly progressing tissue destruction and demonstration of " crepitus" due to gas present in the tissue plane. (Johnson et al.1994). The disease progresses fast with bullae formation filled with cloudy, haemorrhagic, purplish fluid and the skin showing signs of compromised vascularity (Stevens et al.1990). The alpha (α) toxin is the most potent and lethal toxin that has both haemolytic and necrotizing activities. It is responsible for the rapid decline in the general condition of the patient. Other toxins like DNase(β toxin), hyaluronidase (γ toxin) and the septicolysin (δ toxin) are also responsible for pathogenic effect (Yiru et al. 2014). Important predisposing factors for this rare fatal disease are as follows: leukaemia, colonic malignancies, lymphoproliferative disorder, gastrointestinal surgery, proton pump inhibitor therapy, diverticulitis, chemotherapy and radiation therapy. Neutropenia which may be congenital or acquired has been associated with increasing incidence of this disease with finding of necrotizing enterocolitis, typhlitis and distal ileitis (Stevens et al. 2012). With the gastrointestinal mucosal barrier compromise associated with these conditions, *C. septicum* hematogenously spreads to tissues causing myonecrosis and other features of the disease (Smith et al.1994). Although rare, the incidence of spontaneous non traumatic gas gangrene is increasing due to the various iatrogenic interventions being done with the advancement of medicine (Yiru et al.2014). A high degree of clinical suspicion in high risk patients as mentioned above along with typical features of the disease and rapid microbiology diagnosis in the form of blood culture and tissue microscopy is the key to improving clinical prognosis. While antibiotics play a very important role in management of spontaneous gas gangrene, surgical debridement must be performed at the earliest to reduce the burden of the diseased tissue and organisms and to improve survival outcomes (Yiru et al.2014).
- Tetanus (Pavani et al. 2010): Tetanus is a disease of the nervous system which is characterised by persistent tonic spasm with brief violent exacerbations. The pathogen responsible for the disease is *C.tetani*, a typical racket shaped gram-positive bacilli with terminal spores. It produces two toxins i.e. tetanospasmin and tetanolysin. Tetanospasmin is a nerve toxin which enters the lower motor neurons through the presynaptic terminals. It then travels the retrograde axonal transport system and reaches the cell bodies of neurons in the brain stem and the spinal cord

where it exerts its action. Tetanospasmin leaves the motor neurons without inhibition and therefore leads to a hyper spasmodic state. Tetanus is a disease of high mortality but easily preventable due to the toxoid vaccine which is both safe and efficacious. Due to the widespread use of tetanus toxoid vaccine, the disease is fortunately rare.

- Cellulitis – Cellulitis caused by *Clostridia* is also called as " anaerobic cellulitis" or "crepitant cellulitis" (Stevens et al.2012). This disease is often seen in diabetics involving the subcutaneous or retroperitoneal tissue and can progress rapidly into fulminating systemic disease and death. Typically, the muscle and fascia are not involved which segregates it from the necrotising fasciitis often caused by *Clostridia*, *C.histolyticum*, *C.sordelli* and *C.perfringens* have been involved in infection of the skin and subcutaneous tissues leading to cellulitis, abscess formation, endophthalmitis and endocarditis in patient undergoing surgical procedures, injuries or in injection drug abusers (Zink et al.2004, Hsu et al.2008 and Assadian et al.2004).
- Bacteraemia – Clostridial bacteraemia account for a very small but significant percentage of bacteraemia seen among both adults and children (Rechner et al.2002). Often it occurs in late stages of invasive Clostridial infections like gas gangrene due to traumatic or non-traumatic causes (Finegold et al.2002). Other risk factors for Clostridial bacteraemia are diabetes, colonic malignancies, female genital tract infections chemotherapy, radiation therapy, after surgeries performed on the gastrointestinal tract, perforations of the bowel or stomach, leukaemia, lymphomas, other lymphoproliferative disorders and rarely even urinary tract infections causing emphysematous cystitis (Millard et al.2016). A review of cases from a rural area in the United States of America for Clostridial bacteraemia showed that *C.perfringens* (21.7%) was the most common species isolates from blood cultures followed by *C.septicum* (19.6%) (Rechner et al.2002). Patients with clostridial bacteraemia are likely to have more than 50% mortality if managed medically versus surgical management (Rechner et al.2002).
- Enteritis (Finegold et al. 2002): *C. perfringens* type C is known to cause necrotising jejunitis in patients who consumed rabbit, fish paste or tinned meat. The patients present with severe abdominal pain and diarrhoea a few hours after consuming these foods. The physiological events that follow are circulatory collapse with intestinal obstruction due to jejunal necrosis and mucosal oedema. This syndrome affects children and is also called as "pigbel" due to its association with pork feasting in New Guinea. The organisms multiply in the small intestine and produces large

amounts of beta toxin (β) toxin. β toxin is normally destroyed by the intestinal proteinases. But when consumed with sweet potato, the trypsin in sweet potato inhibits the destruction of β toxin and causes intestinal mucosal oedema and necrosis. The disease can be prevented successfully by active immunisation with *C. perfringens* Type C toxoid.

- Botulism (Pavani et al.2010): Botulism results from a protein intoxication produced mostly by *C. botulinum* with subterminal spores. There are varying types of botulinum toxins based on antigenic differences named from type A to G. Types A, B, E & F are specifically associated with human disease while type C & D exclusively affect animals. The toxin primarily affects the peripheral neuromuscular junction and the autonomic synapses causing muscle weakness. Botulism presents in many clinical varieties namely food botulism, infant botulism, wound botulism, adult botulism and rarely inhalational botulism has also been described. The cardinal features of botulism clinical presentation as per Centers for Disease Control & Prevention are as follow: a) No fever unless a complicating infection is present b) symmetrical neurologic manifestations comprising of acute bilateral cranial neuropathies associated with symmetrical descending weakness c) patient remains responsive d) heart rate is normal or slow in the absence of hypotension e) other than for blurred vision, sensory deficits do not occur. Infant botulism is acquired from ingestion of spores from high burden environment rather than ingestion of preformed toxins. It presents with constipation, feeding difficulties, drooling, hypotonia and weak cry. Wound botulism typically lacks gastrointestinal symptoms but present with neurological symptoms. Botulism toxin has been used to treat chronic pain syndromes, blepharospasm, achalasia and anal fissures. It is also widely used for many cosmetic procedures.

- Food poisoning : Food borne outbreaks of *Clostridia* are mainly due to *Clostridium perfringens* and are attributable to usually a single food type like meat or poultry (Aregawi et al.2018, Shandera et al.1983). The spores germinate in raw or cooked foods under anaerobic conditions. When they are ingested, they germinate into vegetative forms producing large number of enterotoxins which cause loose motions.

- Antibiotic associated diarrhoea (AAD): AAD is defined as otherwise unexplained diarrhoea associated with administration of antibiotics (Bartlett et al.2002). Almost all antibiotics are incriminated in causing AAD especially clindamycin, cephalosporins, fluoroquinolones, azithromycin, tetracycline etc (Bartlett et al.1992). Severity of AAD can range from mild diarrhoea to colitis to the most severe form of disease called pseudomembrane colitis. Apart from *Clostridiodes difficile*, other enteric pathogens also cause AAD like *Salmonella*, *S. aureus*, *Clostridium perfringens* type A and rarely *Candida albicans* (Bartlett et al.1992). *C.difficile* is the most common cause of colitis and is fast emerging as the most important pathogen causing AAD both in hospital and community settings. The management of *C.difficile* associated diarrhoea (CDAD) includes a high degree of suspicion in high risk patients, early identification using sensitive laboratory methods like a molecular method, isolation of this patient into a single room and treatment with either oral metronidazole or vancomycin. An important feature of CDAD is that patients may not have a previous history of diarrhoea with antibiotic therapy and the response to therapy is prompt (Bartlett et al.2002). *C.difficile* has become the most common cause of hospital acquired infection in the United States and the most common healthcare associated intestinal infection in Europe and worldwide (O'Donoghue et al.2011). *C. difficile* ribotype O27 strain is a hypervirulent strain that has caused many outbreaks in the United States & Europe resulting in higher morbidity and mortality among patients (McGlone et al.2012, Kwon et al.2015).

Infection syndromes caused by non-spore bearing anaerobes

Infection syndromes caused by non-spore bearing anaerobes range from superficial skin and soft tissue infections and bite wounds to deep invasive infections like organ space abscesses, bacteraemia, endocarditis, septic arthritis etc (Finegold et al.2002, Wexler et al.2007). Most of these infections are caused by mixed organisms comprising of aerobes and anaerobes and serve as excellent examples of synergism exists between these two types of bacteria (Bartlett et al.1972). The presence of a foul and putrid odour which is very much suggestive of the presence of anaerobes.

- Pleuropulmonary infections - The role of anaerobes in causing pleuropulmonary infections is being increasingly recognised (Bartlett et al.1972, Brook et al.1993, Ayyagiri et al.1981). Anaerobes have been reported in empyema, lung abscesses and septic emboli in the lung. Some of the predisposing factors are aspiration of oral, oropharyngeal or gastrointestinal secretions, primary preceding anaerobic infection at another site and obstruction, stasis or tissue necrosis in the lung (Bartlett et al.1972). In a study conducted by A.De et al showed that 72% of patients with pleuropulmonary infections had anaerobes isolated. 58% cases had mixed aerobic and anaerobic bacteria while 14% cases had only anaerobes isolated for various pleuropulmonary infections. (A.De et al 2002) Gram negative anaerobic bacilli comprising of *Prevotella melaninogenicus*, *Fusobacterium species* and *Bacteroides* fragilis were common pathogens recovered. *Peptostreptococci* species were the most common gram positive anaerobic species isolated in this study comprising of 100 patients with pleuropulmonary infections (A. De et al 2002). In order to diagnose pleuropulmonary infections it appropriate samples must be collected. Expectorated sputum is not the correct specimen for anaerobic culture (Bartlett et al.1972). The appropriate specimens to be sent for culture are pulmonary tissue, pleural fluids, per cutaneous transthoracic lung aspirates and blood culture. Management of these infections include surgical drainage and antimicrobial therapy.

- Intra-abdominal infections – Intra abdominal infections occur due to a wide variety of conditions like primary or secondary peritonitis, diverticulitis, appendicitis and intra-abdominal abscesses. They are the second most common infections causing mortality in the intensive care unit (Marcus et al.2016). The most serious infections occur when there is inflammation of the viscus and perforation. This usually involves mixed bacterial flora comprising of aerobic and anaerobic infections. The colon contains approximately 10^{12} bacteria per gram of faeces. Anaerobes predominate among the colonic flora and outnumber aerobic bacteria by 1000 to 1 (Goldstein et al.2002). The most common anaerobe isolated from intra-abdominal infections is *Bacteroides fragilis* (Edmiston et al.2002). This despite the fact that it accounts for only 0.5% of the normal colonic flora (Goldstein et al.2002) *B.fragilis* belongs to the *B.fragilis* group of organisms which also have other virulent species like *B.thetaiotaomicron*, *B.distasonis*, *B.vulgatus*, *B.ovatus* and *B.uniformis* (Goldstein et al.2002). Other anaerobes commonly seen with intra-abdominal infections are *Prevotella*, *Bilophila wadsworthia* and *Peptostreptococcus* group. About 4% of anaerobic intraabdominal infections are complicated with associated bacteraemia (Goldstein et al.2002). The management of intra-abdominal infections includes source control and antimicrobial therapy with broad spectrum agents to cover for both aerobes and anaerobes.

- Female genital tract – Anaerobes present in the female genital tract play a very important role in maintaining health and also causing disease (Hill et al.1980). It has been shown that in women of reproductive age group, anaerobic bacteria were in a ratio of 10:1 with aerobes present (Gorbach et al.1973). 50 to 90% of female genital tract infections comprise of anaerobic bacteria (Finegold et al. 1995). Like other sites of infections, most female genital infections are also mixed in nature comprising of aerobes and anaerobes. Chorioamnionitis is an ascending infection occurring during pregnancy which has been associated with bacterial vaginosis. Anaerobes also play an important role in causing endometritis, adnexitis, pelvioperitonitis more often referred to as Pelvic Inflammatory Disease (PID). Anaerobic Actinomyces species are known to be typically associated with intra uterine devices (IUDs) (Mikamo et al.2011). Infections of the external genitalia include Bartholin's cyst abscess, vaginal abscess, labial cyst abscess and labial wounds. The most common anaerobes isolated from these infections are *Bacteroides fragilis complex* and *Prevotella* species (Brook et al.2004).

- Skin and soft tissue: Anaerobes other than the spore bearing species can also be involved in skin and soft tissue infections especially in children causing perirectal and facial abscesses. *Bacteroides fragilis* is the most common pathogen involved in this (Elliott et al.2000). Most of these infections are polymicrobial in nature. Diabetic foot wound infections (DFI) are an important group of chronic non healing skin and soft tissue infections affecting particularly the foot, characterised by mixed aerobic and anaerobic flora. Presence of anaerobes in the in DFI has been largely underestimated (Percival et al.2018). Therefore, proper management of DFI wounds requires that pus aspirate or tissue be sent to the laboratory for aerobic and anaerobic cultures. Superficial swabs from such sites should be strongly discouraged as they are not representative of true pathogens and grow only colonizers. In addition, DFI management includes tight control of blood sugars and surgical intervention to remove dead necrotic tissue. *Bacteroides fragilis complex*, *Peptostreptococcus species*, *Eubacterium*, *Prevotella*, *Porphyromonas*, *Fusobacterium* etc are the common organism isolated from DFIs (Garg et al.2017, Sasikumar et al.2018).

- Bite wounds: Human bite wounds also give rise to wound infections due to anaerobes. The flora mainly comprises of oral anaerobes like *Prevotella, Fusobacterium* and *Peptostreptococcus* species (Merriam et al.2003). Dog and cat bite wounds are also predominantly mixed with the anaerobes like *Prevotella, Porphyromonas, Fusobacterium, Bacteroides fragilis* group, *Propionibacterium* and *Peptostreptococcus* being isolated (Talan et al.1999).

- Brain abscesses : Brain abscess is a life threatening neurosurgical emergency with an incidence of about 2% in developed countries and 8% in developing countries (Vishwanath et al.2016) Approximately, 13.7% to 17% of brain abscesses may be caused by anaerobes (Lakshmi et al.2011, Helweg-Larsen et al.2012). The most common source for anaerobic brain abscess is infections from contiguous sites like the paranasal sinuses, dental and middle ear infections (Heineman et al.1963). Often polymicrobial in nature, anaerobes comprise of a significant percentage of pathogens causing brain abscesses. Correct identification of the pathogen plays a crucial role in the management of patients with brain abscesses. The most common organisms isolated are *Fusobacterium*, *Prevotella* and *Bacteroides fragilis* group.

- Bacteraemia due to non-spore bearing anaerobes: Anaerobes play an important role in causing bacteraemia accounting for 1 to 17% of positive blood cultures depending on the institution (Brook et al.2010). The most common sources of infection for anaerobic bacteraemia (AB) are abdomen (50 to 70%), pelvic infections (5% to 10%) and skin and soft tissue infections (5% to 20%) (Lassmann et al.2007). Among the non-sporing organisms causing bacteraemia *Bacteroides fragilis* group is the most common (65%) followed by *Peptostreptococcus* and *Fusobacterium species* (Lassmann et al.2007). A study performed by Kim et al showed that despite AB being infrequent (2.3%), it had high mortality (21.4%) and survival rate was worse when patient did not receive appropriate therapy (Kim et al.2016). This makes it important that in high risk group patients like those with abdominal malignancies or abdominal surgeries, anaerobic blood cultures be included in the paired blood culture set.

Emerging threat of drug resistance in anaerobes

Drug resistance among anaerobes is an emerging problem that has been reported consistently from various parts of the world over the last decade (Boyanova et al.2015). This is more often for the non sporing gram negative bacilli *Bacteroides fragilis* group and gram positive spore bearer *Clostridioides difficile*. There are many hurdles to better understanding of resistance trends among anaerobes, some of which are as follows:

- Not suspecting anaerobes as causative agents of infection
- Not requesting and not performing anaerobic cultures especially for deep seated infections
- Not growing anaerobes thereby not having information on empiric choices of therapy
- Labour intensive drug susceptibility processes like agar dilution or broth dilution methods for anaerobes
- Absence of data on changing drug resistance patterns except from a few reference centre's around the world.

Empiric therapy of anaerobic bacteraemia and sepsis can cause treatment failures which in turn impacts morbidity & mortality in patients (Kim et al.2016). In a study conducted by Salonen et al on the clinical significance of anaerobic bacteraemia, they found that antibiotic treatment in the ICU was the only significant factor affecting mortality of patients. They also found that in 32% of their patients empiric therapy had to be change because it was ineffective (Salonen et al.2007). Another study by Nguyen et al showed that with monomicrobial bacteraemia caused by *Bacteroides fragilis group*, the mortality was 45% if the patient's empiric therapy antibiotics were resistant (Nguyen et al. 2002).

Drug resistance in *Bacteroides fragilis* group – *B.fragilis* is the most common anaerobe isolated from clinical specimens. The common antibiotics that are used in the treatment of these infections are β lactam agents (with and without inhibitors), carbapenems, clindamycin and metronidazole (Wexler et al.2007). Drug resistance in *Bacteroides* is a well described phenomenon which may vary geographically. Drug resistance towards β lactam agents is mediated by production of âlactamase enzyme (chromosomally encoded Class 2e Cephalosporinase) which is found in almost all *Bacteroides species* (Hecht et al.2006, Rogers et al. 1993). Other mechanisms like change in porins, efflux mechanisms, methylation of the ribosomal targets (clindamycin), alteration of DNA targets (fluroquinolones) have also been reported (Wexler et al. 2007).

The emerging threat with *Bacteroides* species is the resistance to some of the most active antimicrobial agents carbapenems & 5-Nitroimidazoles appearing in the early 21st century (Sadarangani et al.2015). An international clone of MDR *B.fragilis* has also been identified and characterised (Sóki et al.2016) Carbapenem resistance is mediated by Class B mettalo βlactamases encoded by cfiA & ccrA genes while 5-Nitroimidazole resistance is mediated by the nim genes through nitroimidazole reductase (Livermore et al.2000, Löfmark et al.2005). This has led to multidrug resistant (MDR) strains i.e. resistant to >3 antimicrobial agents which have been associated with mortality in many cases (Sherwood et al.2011, Hartmeyer et al.2012, Meggersee et al.2015, Urba2 n et al.2015, Vishwanath et al.2019, Shafquat et al.2019). A study carried out by Nagy et al from 10 European countries showed that 66.3% of the Bacteroides strains were MDR (Nagy et al.2013).

Drug resistance in *Prevotella* group – *Prevotella species* is a common human commensal anaerobe at various site in the human body. In the last decade, there is growing concern about drug resistance observed in *Prevotella* especially *P.bivia*. Most of them produce beta lactamase that hydrolyse penicillins and cephalosporins. Alauzet et al studied *Prevotella* species using data compiled from 13 studies comprising of 2541 isolates from 15 countries. (Alauzet et al

2010) They showed that overall â lactamase production was about 13 to 84%, clindamycin resistance ranged from 0 to 31% (Greece), moxifloxacin resistance was 16 to 38% (Greece), Metronidazole was 0 to 10.8 % (France) suggesting that drug resistance could become problem with this group of anaerobes also (Corentine et al.2010). More recently, Sherrard et al have described *Prevotella species* producing extended spectrum âlactamases (ESBL) from the respiratory tract of patients with cystic fibrosis. (Sherrad et al 2013) The clinical implication the authors noted of this important finding was that these ESBL producing *Prevotella* species would hydrolyse Ceftazidime used for the treatment of *Pseudomonas* infections in cystic fibrosis patients leading to persistence of lung infections (Laura et al.2016).

Drug resistance in *Clostridiodes difficile* (CD) – In the last decade, surveillance performed at many centre's worldwide has shown the emergence of CD resistant to multiple antimicrobial agents (Freeman et al.2005, Spigaglia et al.2016, Syndman et al.2015, Spigaglia et al.2011). CD is resistant to antimicrobial agents that are commonly used in treating bacterial infections like Penicillin's, cephalosporins, fluoroquinolones, clindamycin, erythromycin, tetracyclines, lincosamides and aminoglycosides (Spigaglia et al.2016, Johanesen et al.2015). This leads to occurrence and recurrence of CD infection (CDI) and also drives the emergence of new strain types (Johanesen et al.2015). While metronidazole and vancomycin still remain primary agents for treatment of CDI, reports of decreased susceptibility have increased in the past few years leading to treatment failures (Leffler et al.2015, Goudarzi et al.2013, Adler et al.2015). A study conducted in Israel showed that vancomycin resistance was up to 47% in CD (Adler et al.2015). In addition, resistance has also been reported against other therapeutic options like rifamycin's, fidaxomicin, tetracycline and chloramphenicol (Kim et al.2016). In a study conducted by Peng et al in Florida with 139 CD isolates showed 28.06% strains resistant to 3 antibiotics, 12.24% strains resistant to four types of antibiotic and 14.39% strains resistant to 5 types of antibiotics (Peng et al.2017). An outbreak of CD ribotype O17 investigated in Portugal showed the strain to be resistant to more than five types of antibiotics namely : tetracyclines, fluoroquinolones, rifampicin, macrolides /streptogrammins, aminoglycosides and carbapenems (Isidro et al.2018). Thus antibiotic selection pressure continues to be an important driver of resistance in CD.

Overcoming emerging threats due to anaerobic infections

- A high degree of clinical suspicion is a must when deep seated infections or sterile site infections are being managed

- Request for anaerobic cultures, this will only help diagnostic microbiology laboratories to scale up and upgrade their services to offer anaerobic cultures and identification methods
- Validate drug susceptibility methods which are easy to perform and interpret in routine diagnostic settings
- Similar to aerobes, antibiogram for anaerobes should also be prepared by microbiology laboratories to guide empiric therapy and understand drug resistance trends
- Strive to comply with the institute antimicrobial stewardship program as it will help reduce antibiotic selection pressure induced drug resistance even in anaerobes

Future prospects

The future of anaerobic microbiology should overcome the two main hurdles we face in anaerobic infections namely early detection & identification along with faster detection of drug resistance mechanisms. This seems possible with molecular methods of identification on direct specimens like sequencing or Matrix Assisted Laser De-sorption ionization- Time of Flight (MALDI -TOF). Another strategy would be the syndromic approach-based testing in clinical microbiology laboratories to include anaerobes along with other organisms for syndromic testing and diagnosis of various organisms causing infection in single specimens. Molecular methods to detect drug resistance and genes mediating resistance will greatly help in optimizing antimicrobial therapy in critically ill patients. These advances in anaerobic microbiology will further support our antimicrobial stewardship programs which is the need of the hour.

References

A.De, Varaiya A, Mathur M. (2002). Anaerobes In Pleuropulmonary Infections. *Indian J Med Microbiol*, 20(3): 150- 152

Adler A, Miller-Roll T, Bradenstein R, Block C, Mendelson B PM, Paitan Y, Schwartz D, Peled N CY. (2015). A national survey of the molecular epidemiology of Clostridium difficile in Israel: the dissemination of the ribotype 027 strain with reduced susceptibility to vancomycin and metronidazole. *Diagn Microbiol Infect Dis*. 2015;83:21–4.

Alauzet C, Marchandin H, Lozniewski A. (2010). New insights into Prevotella diversity and medical microbiology. *Futur Microbiol*. 5(11):1695–718.

Aregawi A, Tadele G. (2018). Prevalence and Associated Factors of Thinness among Adolescent Girls Attending Governmental Schools in Aksum Town, Northern Ethiopia. *Med J*. 582–5.

Assadian O, Assadian A, Senekowitsch C, Makristathis A HG. (2004). Gas gangrene due to Clostridium perfringens in two injecting drug users in Vienna, Austria. Wien Klin *Wochenschr*. 116:264–7.

Ayyagiri A, Pancholi VM, Goswami A, Malik SK (2081). Anaerobic bacteria in pleuropulmonary infections. *Indian J Med Res.* 74:809–14.

Bartlett J. (2002). Antibiotic Associated Diarrhea. *N Engl J Med.* 346(5):334–9

Bartlett J. (1992). Antibiotic-associated diarrhea. *Clin Infect Dis.* 15:573–81.

Bartlett JG , Finegold SM. (1972). Anaerobic_pleuropulmonary_infections . *Medicine.* 51(6): 413–50.

Barra-Carrasco J, Paredes-Sabja & D. (2014). Clostridium difficile spores: a major threat to the hospital environment. *Future Microbiol.* 9(14):475–86.

Bergan T. (1983). Diagnostic considerations and sample collection for anaerobic bacteria. *Scand J Gastroenterol Suppl.* 85:5–14.

Boyanova L, Kolarov R, Mitov I. (2015). Recent evolution of antibiotic resistance in the anaerobes as compared to previous decades. *Anaerobe.* 31:4–10.

Brook I . (1993). Aerobic and anaerobic microbiology of empyema/ : a retrospective review in wo military hospitals. *Chest.* 103:1502–7.

Brook I. (2004). Urinary tract and genito-urinary suppurative infections due to anaerobic bacteria. *Int J Urol.* 11(3):133–41.

Brook I. (2010). The role of anaerobic bacteria in bacteremia. *Anaerobe.* 16(3):183–9

Brook I, Wexler HM, Goldstein EJC. (2013). Antianaerobic antimicrobials: Spectrum and susceptibility testing. *Clin Microbiol Rev.* 26(3):526–46.

Brook I, Long SS. (2017). Anaerobic Bacteria: Classification, Normal Flora, and Clinical Concepts. In: Principles and Practice of Pediatric Infectious Diseases. Elsevier Inc. 987–95.

Edmiston, Jr. CE, Krepel CJ, Seabrook GR, Jochimsen WG. (2002). Anaerobic Infections in the Surgical Patient: Microbial Etiology and Therapy. *Clin Infect Dis.* 35(s1):S112–8

Elliott, D., J. A. Kufera and RAM. (2000). The microbiology of necrotizing soft tissue infections. *Am J Surg.*179:361–6.

Finegold SM. (1995). Anaerobic infections in humans: an overview. *Anaerobe.* 1(1):3–9.

Finegold S, Sussman M. (2002). Anaerobic InfectionsA Clinical Overview. In: Molecular Medical Microbiology. Elsevier; p. 1867–74.

Freeman J, Baines SD, Jabes D WM. (2005). Comparison of the efficacy of ramoplanin and vancomycin in both in vitro and *in vivo* models of clindamycin-induced Clostridium difficile infection. *J Antimicrob Chemother.* 56:717–25.

Garg R, Datta P, Gupta V, Chander J. (2017). Anaerobic Bacteriological Profile of Infected Diabetic Foot Ulcers with their Antimicrobial Susceptibility Pattern : Need of the Hour. *Natl J Lab Med.*, 6(3):9–12.

Goldstein EJC. (2002). Intra Abdominal Anaerobic Infections: Bacteriology and Therapeutic Potential of Newer Antimicrobial Carbapenem, Fluoroquinolone, and Desfluoroquinolone Therapeutic Agents. *Clin Infect Dis.* 35(s1):S106–11.

Gorbach SL, Menda KB, Thadepalli H KL. (1973). Anaerobic microflora of the cervix of healthy women. *Am J Obs Gynecol.* 117:1053–7.

Goudarzi M, Goudarzi H, Alebouyeh M, Azimi Rad M SMF, Zali MR AM. (2013). Antimicrobial susceptibility of Clostridium difficile clinical isolates in Iran. *Iran Red Crescent Med J.* 15:704 –711.

Hartmeyer GN, So′ ki J NE et al. (2012). Multidrug-resistant Bacteroides fragilis group on the rise in Europe? *J Med Microbiol.* 61:1784–8.

Hecht DW. (2006). Anaerobes: antibiotic resistance, clinical significance, and the role of susceptibility testing. *Anaerobe.* 12:115–21.

Heineman HS, Braude AI. (1963). Anaerobic infection of the brain. *Am J Med.* 35(5):682–97

Helweg-Larsen J, Astradsson A, Richhall H, Erdal J, Laursen A, *Brennum J.* (2012). Pyogenic brain abscess, a 15 year survey. BMC Infect Dis. 12.

Hill GB. (1980). Anaerobic flora of the female genital tract. DW L, Genco RJ MCK, editors. Plenum Press; 39–50 p.

Hsu HY, Lee SF, Hartstein ME HG. (2008). Clostridium perfringens keratitis leading to blinding panophthalmitis. *Cornea*. 27:1200–3.

Isenberg HD. (2007). Clinical Microbiology Procedures Handbook. 2nd editio. James I Mangels, editor. Washington DC: American Society for Microbiology (ASM) Press; 640–728 p.

Isidro J, Menezes J, Serrano M, Borges V, Paixão P, Mimoso M, et al. (2018). Genomic study of a clostridium difficile multidrug resistant outbreak-related clone reveals novel determinants of resistance. *Front Microbiol.* 9(12):1–9.

Johnson S, Driks MR, Tweten RK, Ballard J, Stevens DL, Anderson DJ et al. (1994). Clinical courses of seven survivors of Clostridium septicum infection and their immunologic responses to a toxin. *Clin Infect Dis.* 19:761–4.

Johanesen PA, Mackin KE, Hutton ML, Awad MM, Larcombe S AJ, D L. (2015). Disruption of the gut microbiome: Clostridium difficile infection and the threat of antibiotic resistance. *Genes (Basel).* 6:1347–1360.

Johansson Å, Nagy E, Sóki J. (2014). Instant screening and verification of carbapenemase activity in Bacteroides fragilis in positive blood culture, using matrix-assisted laser desorption ionization-time of flight mass spectrometry. *J Med Microbiol.* 63(PART 8):1105–10.

Kim J, Lee Y, Park Y, Kim M, Choi JY, Yong D, et al. (2016). Anaerobic Bacteremia: Impact of Inappropriate Therapy on Mortality. *Infect Chemother.* 48(2):91.

Kwon J, Olsen MA DE. (2015). The morbidity, mortality, and costs associated with Clostridium difficile infection. *Infect Dis Clin North Am.* 29(1):123–34.

Lakshmi V, Umabala P, Anuradha K, Padmaja K, Padmasree C, Rajesh A, et al. (2011). Microbiological Spectrum of Brain Abscess at a Tertiary Care Hospital in South India: 24-Year Data and Review. *Patholog Res Int.* 2011;2011(Id):1–12.

Laura J. Sherrarda, Stef J. McGratha, Leanne McIlreaveya, Joseph Hatch, Matthew C. Wolfgang, Marianne S. Muhlebache, Deirdre F. Gilpina, J. Stuart Elborna, F and MMT. (2016). Production of extended-spectrum β-lactamases and the potential indirect pathogenic role of Prevotella isolates from the cystic fibrosis respiratory microbiota. *Int J Antimicrob Agents.* 47(2):140–5.

Lassmann B, Gustafson DR, Wood CM, Rosenblatt JE. (2007). Reemergence of Anaerobic Bacteremia. *Clin Infect Dis.* 44(7):895–900.

Leffler DA LJ. (2015). Clostridium difficile infection. *N Engl J Med.* 372:1539–48.

Livermore, D. M. NW. (2000). Carbapenemases: a problem in waiting? *Curr Opin Microbiol.* 2000;3:489–95.

Löfmark S, Fang H, Hedberg M, Edlund C. (2005). Inducible metronidazole resistance and nim genes in clinical Bacteroides fragilis group isolates. *Antimicrob Agents Chemother.* 49(3): 1253–6.

Marcus G, Levy S, Salhab G, Mengesha B, Tzuman O, Shur S, et al. (2016). Intra-abdominal infections: The role of anaerobes, enterococci, fungi, and multidrug-resistant organisms. *Open Forum Infect Dis.* 3(4):1–7.

McGlone SM, Bailey RR, Zimmer SM, Popovich MJ, Tian Y UP, Muder RR LB. (2012). The economic burden of Clostridium difficile. *Clin Microbiol Infect.* 18:282–9.

Meggersee R, Abratt V. (2015). The occurrence of antibiotic resistance genes in drug resistant Bacteroides fragilis isolates from Groote Schuur Hospital, South Africa. *Anaerobe.* 32:1–6.

Merriam, C. V., H. T. Fernandez, D. M. Citron, K. L. Tyrrell YAW, Goldstein and EJ. (2003). Bacteriology of human bite wound infections. *Anaerobe.* 9:83–6.

Mikamo H, Arakawa S, Fujiwara M, Funada H, Inamatsu T, Iwata S, Kaneko A, Kato K, Kunihiro M, Kurokawa I, Kusachi S, Masaoka T, Matsumoto T, Mikasa K, Nasu M, Oguri T, Ooishi M, Ouchi K, Sakamoto H, Shiba K, Shinagawa N, Shinzato T, Suzuki K,

Takesue Y, T YT, (2011). Infections I. Chapter 2-6-1. Anaerobic infections (individual fields): Female genital infections. *J Infect Chemother*. 17:92–8.

Millard MA, McManus KA, Wispelwey B. (2016). Severe Sepsis due to Clostridium perfringens Bacteremia of Urinary Origin: A Case Report and Systematic Review. *Case Rep Infect Dis.*:1–5.

Nagy E, Urbán E. (2011). Antimicrobial susceptibility of Bacteroides fragilis group isolates in Europe: 20years of experience. *Clin Microbiol Infect*. 17(3):371–9.

Nagy E, Boyanova L, Justesen US. (2018). How to isolate, identify and determine antimicrobial susceptibility of anaerobic bacteria in routine laboratories. Vol. 24, *Clinical Microbiology and Infection*. Elsevier B.V.; 1139–48.

Nguyen MH, Yu VL, Morris AJ, McDermott L, Wagener MW, Harrell L, et al. (2002). Antimicrobial Resistance and Clinical Outcome of Bacteroides Bacteremia: Findings of a Multicenter Prospective Observational Trial. *Clin Infect Dis*. 30(6):870–6.

Nichols RL, Florman S. (2001). Clinical Presentations of Soft Tissue Infections and Surgical Site Infections. *Clin Infect Dis*. 33(s2):S84–93.

O'Donoghue C KL. (2011). Update on Clostridium difficile infection. *Curr Opin Gastrenterol.* 27(1):38–47.

Park Y, Choi JY, Yong D, Lee K, Kim JM. (2009). Clinical features and prognostic factors of anaerobic infections: a 7-year retrospective study. *Korean J Intern Med.* 24(1):13–8

Pavani Reddy BT. Mandell, (2010). Douglous & Bennett's Principles and Practice of Infectious Diseases. 7th editio. Mandell Gerald, Bennett John DR, editor. Churchill Livingstone; 3091–3095 p

Pavani Reddy BT. Mandell, Douglas & Bennett's (2010). Principles & Practice of Infectious Disesaes. 7th edition. Mandell Gerald, Bennett John DR, editor. Churchill Livingstone; 3097–3102 p.

Peng Z, Addisu A, Alrabaa S, Sun X. (2017). Antibiotic resistance and toxin production of Clostridium difficile isolates from the hospitalized patients in a large hospital in Florida. *Front Microbiol.* 8(Dec):1–8.

Percival SL, Malone M, Mayer D, Salisbury AM, Schultz G. (2018). Role of anaerobes in polymicrobial communities and biofilms complicating diabetic foot ulcers. *Int Wound J.* 15(5):776–82.

PG Bowler, , BI Duerden DA. (2001). Wound Microbiology and Associated Approaches to Wound. *Clin Microbiol Rev*. 14(2):244–69.

Rechner PM, Agger WA, Mruz K, Cogbill TH. (2002). Clinical Features of Clostridial Bacteremia: A Review from a Rural Area. *Clin Infect Dis*. 2002;33(3):349–53.

Rennie RP, Brosnikoff C, Turnbull L, Reller LB, Mirrett S, Janda W, et al. (2008). Multicenter evaluation of the Vitek 2 anaerobe and Corynebacterium identification card. *J Clin Microbiol.* 46(8):2646–51.

Rogers, M.B., A.C. Parker and CJS. (1993). Cloning and characterization of the endogenous cephalosporinase gene, cepA, from Bacteroides fragilis reveals a new subgroup of Ambler class A beta-lactamases. *Antimicrob Agents Chemother*. 37:2391–2400.

Sadarangani SP, Cunningham SA JP et al. (2015). Metronidazole and carbapenem-resistant Bacteroides thetaiotaomicron isolated in Rochester, Minnesota, in 2014. *Antimicrob Agents Chemother*. 59:4157–61.

Salonen JH, Eerola E, Meurman O. (2007). Clinical Significance and Outcome of Anaerobic Bacteremia. *Clin Infect Dis*. 26(6):1413–7.

Sasikumar K, Vijayakumar C, Jagdish S, Kadambari D, Kumar NR, Biswas R, et al. (2018). Clinico-microbiological Profile of Septic Diabetic Foot with Special Reference to Anaerobic Infection. *Cureus*. 10(3).

Shafquat Y, Jabeen K, Farooqi J, Mehmood K, Irfan S, Hasan R, et al. (2019). Antimicrobial susceptibility against metronidazole and carbapenem in clinical anaerobic isolates from Pakistan. *Antimicrob Resist Infect Control.* 8(1):1–7.

Shandera WX, Tacket CO BP. (1983). Food poisoning due to Clostridium perfringens in the United States. *J Infect Dis.* 147:167–70.

Shannon S, Kronemann D, Patel R, Schuetz AN. (2018). Routine use of MALDI-TOF MS for anaerobic bacterial identification in clinical microbiology. *Anaerobe. Dec*;54:191–6.

Sherwood JE, Fraser S CD et al. (2011). Multi-drug resistant Bacteroides fragilis recovered from blood and severe leg wounds caused by an improvised explosive device (IED) in Afghanistan. *Anaerobe.* 17:152–5.

Spigaglia P. (2016). Recent advances in the understanding of antibiotic resistance in Clostridium difficile infection. *Ther Adv Infect Dis.* 3:23–42.

Spigaglia P, Barbanti F, Mastrantonio P ESG on, (ESGCD). C difficile. (2011). Multidrug resistance in European Clostridium difficile clinical isolates. *J Antimicrob Chemother*. 66:2227–34.

Smith LDS WB. (1984). The Pathogenic anaerobic bacteria. C.C. Thomas.

Snydman D, McDermott L, Jacobus N, Thorpe C, Stone S JS, Goldstein E, Patel R, Forbes B, Mirrett S, Johnson S GD. (2015). U.S.-based national sentinel surveillance study for the epidemiology of Clostridium difficile-associated diarrheal isolates and their susceptibility to fidaxomicin. *Antimicrob Agents Chemother.* 59:6437– 6443.

Sóki J, Hedberg M, Patrick S, Bálint B, Herczeg R, Nagy I, et al. (2016). Emergence and evolution of an international cluster of MDR bacteroides fragilis isolates. *J Antimicrob Chemother.* 71(9):2441–8.

Stevens DL, Musher DM, Watson DA, Eddy H, Hamill RJ, Gyorkey F et al. (1990). Spontaneous, nontraumatic gangrene due to Clostridium septicum. *Rev Infect Dis.* 12(2):286–96.

Stevens DL, Bryant AE. (2002). The Role of Clostridial Toxins in the Pathogenesis of Gas Gangrene. *Clin Infect Dis.* 35(s1):S93–100.

Stevens DL, Aldape MJ, Bryant AE. (2012). Life-threatening clostridial infections. *Anaerobe.* 18(2):254–9.

Talan D. A., Citron D. M., Abrahamian F. M., Moran G. J. GEJC. (1999). Bacteriologic analysis of infected dog and cat bites. *N Engl J Med.* 340:85–92.

Urba′n E, Horva′th Z S ki J et al. (2015). First Hungarian case of an infection caused by multidrug-resistant Bacteroides fragilis strain. *Anaerobe.* 2015;31:55–8.

Vishwanath S, Shenoy PA, Gupta A, Menon G, Chawla K. Brain Abscess with Anaerobic Gram-Negative Bacilli: Case Series. *J Case Reports.* 6(4):467–74.

Vishwanath S, Shenoy P, Chawla K. (2019). Antimicrobial resistance profile and Nim gene detection among bacteroides fragilis group isolates in a university hospital in South India. *J Glob Infect Dis.* 11(2):59.

Wexler HM. (2007). Bacteroides: the Good, the Bad, and the Nitty-Gritty. *Clin Microbiol Rev.* 20(4):593–621.

Yiru E. Wu, Alexander Baras, Toby Cornish, Stefan Riedel and ECB. (2014). Fatal Spontaneous Clostridium septicum Gas Gangrene: A Possible Association With Iatrogenic Gastric Acid Suppression. *Arch Pathol Lab Med.* 138(6):837–41.

Zink JM, Singh-Parikshak R, Sugar A JM. (2004). Clostridium sordellii endophthalmitis after suture removal from a corneal transplant. *Cornea.* 23:522–3.

14

Anaerobes As An Emerging Probiotics

Vikas C. Ghattargi[1], *Yogesh S. Shouche*[1], *Shrikant P. Pawar*[1]

National Centre for Microbial Resources (NCMR), National Centre for Cell Science (NCCS) NCCS Complex, Ganeshkhind, Pune- 411 021 Maharashtra

Abstract

Gastrointestinal tract (GIT) harbors mixed microbial community, these are highly diversified, as revealed by molecular and phylogenetic studies. It is also estimated that microbiota of the GIT is composed of over 35,000 species and play a crucial role in human health and diseases. Thus are now of great interest to both the industrial and scientific communities. The microbes exhibit health benefits are known as probiotics. Anaerobic bacteria dominate in the GIT environment and genera like Bifidobacterium, Clostridium, *are strict anaerobes while* Lactobacillus, Lactococcus, *is facultative anaerobes implicated in the probiotic effects.*

For developing anaerobic bacterial based bio-products, it is necessary to cultivate large-scale biomass production that could be done using a modern manufacturing practice. Information on detailed understanding of the perfect growth medium and other bioprocessing parameters for the growth and viability is must. Also high cell viability is a prerequisite during downstream processing and storage of probiotic formulation. In this work we review the latest information about the history, mechanism of action, benefits of probiotic, selection criteria, sources of anaerobic probiotic bacteria, industrial cultivation medium along with biomass production, downstream processing.

Keywords: industrial, biomass production, downstream processing, manufacturing practices

Corresponding Authors: vikas.ghattargi@gmail.com

Introduction

The term probiotic has it origin from: **pro** from Latin and **bios** from Greek, which exactly means "*for life*" (Hamilton-Miller et al., 2003) and there has been no clear definition on what exactly probiotics are. In, Kollath et al. 1953 first described supplements from both organic and inorganic origin having the ability to restore the gut health. It was Ferdin and Vergin proposed the term probiotics "*Active substances that are essential for a healthy development of life*" (Kollath, 1953). The word "**Probiotic**" had several different meanings over the ages and was being used as cryptology term; therefore it is utmost important to define what probiotics are? In 1962, Lilly and Stillwell called probiotic as "*Substances secreted by one microorganism and stimulated the growth of another microorganism*". Then in 1971 Sperti described probiotics as "*tissue extracts which stimulated microbial growth*" (Fuller et al., 1992). Probiotics were also described as microbial food supplement by Parker in 1974, and he defined probiotics as "*organisms and substances, which contribute to intestinal microbial balance*". Later Fuller (1989) modified as "*live microbial feed supplement which beneficially affects the host animal by improving its intestinal microbial balance*" (Fuller et al., 1992).

In Havenaar and Huisisnt Veld (1992) have attempted to define the term probiotics "*a mono or mixed culture of live microorganisms which, applied to animal or man, affect the host beneficially by improving the properties of the indigenous microflora*". Two years after it was debated if one should include microbial stimulants such as the bifidogenic factors and even plants? Taking this into consideration the definition was modified to "*A probiotic is a preparation consisting of live microorganisms or microbial stimulants which affects the indigenous microflora of the recipient animal, plant or food in a beneficial way*". All these definitions refer to microorganisms including bacteria, yeasts, fungi and viruses (bacteriophages, plant and animal viruses). Except animal and plant viruses, others have shown to have beneficial effects when administered to animals. Microorganisms being labelled as probiotics should be non-pathogenic and mostly be the components of normal gut flora, e.g. Lactic Acid Bacteria (LAB) and some non-pathogenic variants of pathogenic species can function in the same way as traditional probiotics. For example, *Enterococcus faecium* and *Enterococcus faecalis* can also protect against infection by the respective virulent parent strain (Ozen and Dinleyici, 2015). Thus defining the word "*probiotic*" is becoming increasingly important and difficult. The latest definition stated by World Health Organisation (W.H.O.) in 2002 is "*Probiotics are live microorganisms, which when administered in adequate amounts confer a health benefit on the host*" this is now widely accepted definition all over the world (Ozen and Dinleyici, 2015).

History of Probiotic

The Indians, Egyptians, Greeks and Romans left many records showing the common usage of milk and milk products (Gogineni, 2013). By 13 B.C. nearly all the civilizations had fermented milk products for taste and health benefits without knowing the actual reason. Hippocrates, the Greek physician, considered fermented milk not only as a food product but also a medicine as a potential cure for intestinal disorders. While Plinius (Roman historian), stated that "*fermented milk products could be used for treating gastroenteritis*" (Suvarna and Boby, 2004). While countries like Japan, China and Korea were dependent on fermentation of cabbage, turnip, onion, eggplant, squash, cucumber and carrots pickle (Gogineni, 2013).

Antonie van Leeuwenhoek used a microscope to see yeast cells in fermenting beer in 1680, but he never made a correlation between yeast cells and beer, thus his work was forgotten (Suvarna and Boby, 2004). In the late 1700's Lavoisier, wrote that yeast played a chemical role rather than physical role and described the process of transformation of sugars to alcohol and carbon dioxide (Suvarna and Boby, 2004). In 1840, Thodor Schwann and Charles Cagniard-Latour suggested an association between alcoholic fermentation and the growth of yeast. To add on further, French chemist Louis Pasteur concluded that microorganisms initiated the fermentation and defined fermentation as "*respiration without air*". He also stated, "*I am of the opinion that alcoholic fermentation never occurs without simultaneous organization, development, and multiplication of cells. If asked in what consists the chemical act whereby the sugar is decomposed, I am completely ignorant of it*". Pasteur published his formative results in 1860, titled "*Mémoiresur la fermentation alcoolique*" (Fuller et al., 1992). Henry Tissier in 1899 isolated *Bifidobacteria* from the stool of healthy breastfed infant. He suggested the administration of *Bifidobacteria* to infants with diarrhoea when he found their dominance in the intestine of the healthy human gut (Malago et al., 2011).

Another leap was in 1907, a German chemist Eduard Buchner won Nobel Prize for proving that the enzymes present in yeast cells are the actual reason behind fermentation. In the same year Russian scientist Metchnikoff, based on his findings composed his work in the text "*The Prolongation of Life*" (Hamilton-Miller et al., 2003). This was the first scientific description on the improvement of human health by consuming substances that favorably alter the gastrointestinal microflora. This concept in now widely accepted as probiotics. German chemist Eduard Buchner found a long life of Bulgarian farmers with the heavy eating of yoghurt containing *Lactobacillus* species in 1908 (Hamilton-Miller et al., 2003). His contribution to immunology was on the subject of ingestion of living organism in humans for which he received the Nobel Prize in the year

1907 (Hamilton-Miller et al., 2003). The work of Eduard Buchner was further taken by Arthur Harden and Hans Euler-Chelpin and found out the enzymes responsible for fermentation. The concept of probiotic therapy had a major footstep towards reality in 1930 (Hamilton-Miller et al., 2003). The Japanese microbiologist Minoru Shirota first discovered bacterium that survives the gut after ingestion. He subsequently isolated and cultivated the bacterial strain which is now known as *Lactobacillus casei* strain shirota (Hamilton-Miller et al., 2003). This became the first bacteria-containing drink to be commercially marketed as Yakult in 1935 (Hamilton-Miller et al., 2003). Thus becoming a product that continues sold worldwide today.

Mechanisms of Probiotic Activity

Various mechanisms of action for probiotics have been proposed but the exact manner in which they exert their effects is still not fully elucidated. It includes from colonization to normalization of disturbed intestinal microbial communities (Florou-Paneri et al., 2013; Heshmati et al., 2018; Kumar et al., 2010; Ouwehand et al., 1999). In addition they produce vitamins and aid in nutrient absorption. Exclusion of pathogens, bacteriocin production (Gaspar et al., 2018; Sierra et al., 2015; Vieco-Saiz et al., 2019), enzymatic activity and generation of volatile fatty acids are another way of action of probiotics. Cell adhesion, cell antagonism, mucin production (Bilotta and Cong, 2019; Ohira et al., 2017; Plaza-Diaz et al., 2019); modulation of the immune system (Maldonado Galdeano et al., 2019; Matsuzaki and Chin, 2000; Munday, 2012); and interaction with the brain-gut axis are some more way of the mechanisms of the actions of different classes of probiotics (Liu and Zhu, 2018; Martin et al., 2018).

What can probiotics do for you?

There is increasing evidence in favor of the claims of beneficial effects attributed to probiotics, including improvement of intestinal health, enhancement of the immune response, reduction of serum cholesterol, and cancer prevention. These health properties are strain specific and are impacted by the various mechanisms mentioned above. While some of the health benefits are well documented others require additional studies in order to be established. In fact, there is substantial evidence to support probiotic use in the treatment of acute diarrhoeal diseases, prevention of antibiotic-associated diarrhoea (Gupte, 2017), *Clostridium difficile*–associated diarrhea (Guarino et al., 2015; Pattani et al., 2013a, 2013b) and improvement of lactose metabolism (Letteri et al., 1975; Oak and Jha, 2019). Further prevention of irritable bowel syndrome (IBS) (Dai, 2013; McFarland and Dublin, 2008), inflammatory bowel disease (IBD) (Saez-Lara et al., 2015), only in ulcerative colitis (Chibbar and Dieleman, 2015), and allergic disorders such as atopic dermatitis (eczema) (Chibbar and Dieleman, 2015) and allergic rhinitis (Chibbar and Dieleman, 2015) could be found.

Probiotic selection

A guided step by step approach is required for screening of probiotic microorganisms, consisting of sequences of test to find a potential probiotic candidate. The outcome would be the strain having highest functional properties without negative qualities being selected. The guidelines are stated by W.H.O and FAO and have been summarized below and in Figure 1.

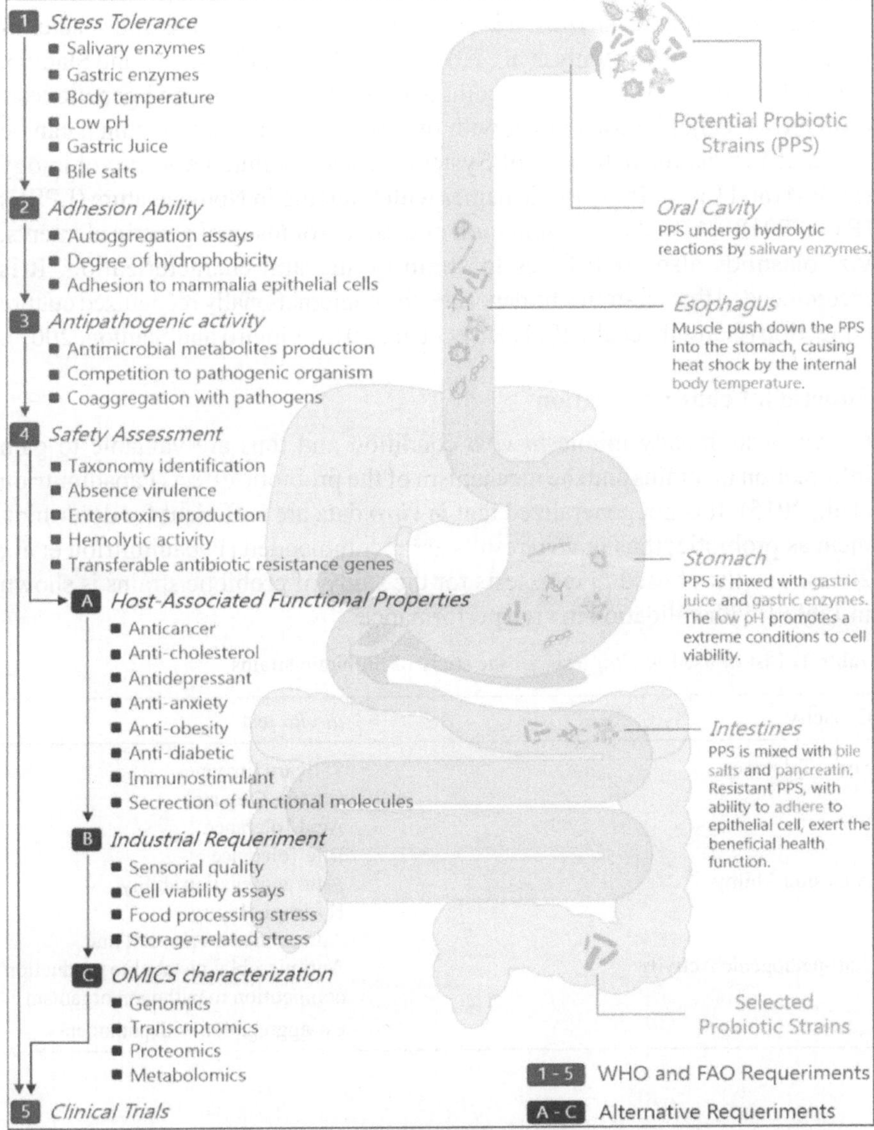

Fig. 1: Screening approaches to be used for characterization of probiotic strains according to WHO/FAO.

Strain identification by phenotypic and genotypic methods

Increasing evidence suggests that probiotics effects are strains specific and thus a property of a certain strain cannot be generalizes to the entire genus (de Melo Pereira et al., 2018; Pandey et al., 2015). Moreover, it is important to link a strain to a specific health effect as well as to enable accurate surveillance and epidemiological studies. Strain identity is important to associate a strain to a specific health effect as well as to permit correct surveillance and epidemiological studies. Thus, it is necessary to know the genus and species of the probiotic strain (Ganguly et al., 2011; Huys et al., 2013; Pineiro and Stanton, 2007). It is recommended that a combination of phenotypic and genetic tests (16S rRNA gene) be used along with the current nomenclature which can be found at International Journal of Systematic and Evolutionary Microbiology (IJSEM) and List of Prokaryotic names with Standing in Nomenclature (LPSN) (Parte, 2018). Also the determination of extra chromosomal genetic elements, viz. plasmids also contributes to strain typing and characterization. It is recommended that all strains be deposited in an internationally recognized culture collection (Ganguly et al., 2011; Huys et al., 2013; Pineiro and Stanton, 2007).

Functional characterization

in vitro tests mostly mimic *in vivo* condition and thus are valuable to gain information of strains and the mechanism of the probiotic effect (Papadimitriou et al., 2015). It is not generalized that *in vitro* data are sufficient for describing them as probiotic; thus *in vivo* results are recommended (Papadimitriou et al., 2015). The list of used *in vitro* tests for the study of probiotic strains is shown in Table 1 with validation in vivo performance.

Table 1: List of used *in vitro* tests for the study of probiotic strains

Capacity	*in-vito* test
Stress Tolerance	Salivary Enzymes Gastric Enzymes Acid Tolerance Bile Tolerance
Adhesion Ability	Auto aggregation ability Hydrophobicity Adhesion to human cell line
Anti-pathogenic Activity	Antimicrobial metabolite production competition to pathogen organism Co-aggregation with pathogens

Safety assessment

Various *in vitro* factors that are part of screening *viz*. adhesion and bile salt hydrolysis which allow survival in the gut have been also implicated in the pathogenicity. Some bacterial genera have no known virulence factors in such cases it is difficult to understand the pathogenicity. Thus screening of all known and possible virulence factors along with antibiotic resistance has to been screened (Bennedsen et al., 2011; Gueimonde et al., 2013). However, probiotics may theoretically be responsible for four types of side-effects: 1) Systemic infections, 2) Deleterious metabolic activities, 3) Excessive immune stimulation in susceptible individuals and 4) Gene transfer (Bennedsen et al., 2011; Marteau, 2001). In respect of the importance of assuring safety, even among a group of bacteria that is Generally Recognized as Safe (GRAS), it is recommended that probiotic strains be characterized for following (Ganguly et al., 2011; Huys et al., 2013; Pineiro and Stanton, 2007):

i. Determination of antibiotic resistance patterns
ii. Assessment of certain metabolic activities
iii. Assessment of side-effects during human studies
iv. Epidemiological surveillance of adverse incidents in consumers (post-market)
v. If the strain under evaluation belongs to a species that is a known mammalian toxin producer, it must be tested for toxin production.
vi. If the strain under evaluation belongs to a species with known hemolytic potential, determination of hemolytic activity is required

in vivo studies using animals and human model

Animal models are must to provide authentication of *in vitro* effects and determination of probiotic mechanism (Sorokulova, 2008). The principal outcome of efficacy studies on probiotics should be proven benefits in human trials via statistically and biologically significance and improvement in condition, symptoms and signs (Sorokulova, 2008). Each should have a proven correlation with the probiotic tested. Probiotics have been tested for an impact on a variety of clinical conditions as stated above. Standard methods for clinical evaluations are comprised of Phase 1 (safety), Phase 2 (efficacy), Phase 3 (effectiveness) and Phase 4 (surveillance) (Sorokulova, 2008). It is also recommended that human trials should be repeated by more than one Center for confirmation of results. These results should be published in peer-reviewed scientific or medical journals along with negative results. This is encouraged because it contributes to the totality of the evidence to support probiotic efficacy (Fao et al., 2002; Ganguly et al., 2011).

Labeling and proper storage conditions

In order to avoid misleading information probiotic products have to be labeled properly for the consumers and these should be checked independent by scientific experts in the field to avoid misleading information (de Simone, 2019; Jha et al., 2017; Sanders and Levy, 2011). The following information should be described on the label:

- Genus, species and strain designation.
- Minimum viable numbers of each probiotic strain at the end of the shelf-life
- The suggested serving size that deliver the effective dose of health claim
- Health claim(s)
- Proper storage conditions

Gut as anaerobic habitats

The human gut is a mixed culture system along with a large numbers of highly diversified microbes with estimated count about 10 trillion microbial cells with moreover than 35,000 species (Guinane and Cotter, 2013; Thursby and Juge, 2017). Most often these species are anaerobic in nature due to the rapid consumption of oxygen by aero-tolerant organisms (Albenberg et al., 2014; Milani et al., 2017). Intended action of probiotics is in the gut, therefore isolation of microbes from gut for probiotic purpose would give a more suitable strain as they have always been adapted to the gut conditions and mostly would be an efficient in their actions (Lebeer et al., 2008; Serra et al., 2019; Taylor et al., 2010). These probiotic microorganisms are highly diverse in their nature and belong to different eukaryotic and prokaryotic groups, including yeast and bacteria and mostly belong to *Saccharomyces, Lactobacillus, Lactococcus, Bifidobacterium, Enterococcus, Streptococcus, Pediococcus, Leuconostoc, Clostridium, Bacillus* and *Escherichia coli* (Fijan, 2014; Reid et al., 2006; Weder et al., 2014). Genera *Bifidobacterium* and *Clostridium* are strict anaerobes whiles other are facultative in nature (Lopetuso et al., 2013; Milani et al., 2016; O'Callaghan and van Sinderen, 2016).

Bifidobacterium

The genus *Bifidobacterium* is Gram-stain-positive, non-motile, anaerobic bacteria, non-filamentous, non-capsulated and non-spore forming, rod- to Y-shaped and endo-symbiotic inhabitants of gut and vagina of mammals, including humans (Nouioui et al., 2018; Sgorbati et al., 1995). It occurs in single cell multicellular chains and/or clumps in the form of branched or pleomorphic rods which gives the name of this genus (Nouioui et al., 2018; Sgorbati et al., 1995). French paediatrician, Henri Tissier in 1899 first reported the *Bifidobacteria* and described as a Y shaped (bifid) bacterium from the faeces of breast-fed infants and was names as *Bacillus bifidus communis* (Nouioui et al., 2018; Sgorbati et al., 1995). Initially it was included in the genus *Lactobacillus* and grouped within lactic acid bacteria. It was reclassified as a separate genus in the 8th Edition of Bergey's manual of determinative bacteriology as *Bifidobacteria* (Sgorbati et al., 1995). Genome show high GC content and varies from 42 to 67 mol% with the size ranging between 1.93- 2.83 Mbp and the genus is now known to include 70 species from highly diversified sources (Nouioui et al., 2018). *Bifidobacterium* intend to use a specific pathway for degradation of hexoses called as bifid shunt which differs from that of facultative anaerobic lactic acid bacteria with the involvement of key enzyme fructose-6-phosphoketolase (EC 4.1.2.2) (Felis and Dellaglio, 2007).

Probiotic activities of bifidobacterial species were first verified in 1958 and since then they have been established as probiotics because they promote desirable changes in the colon (Lee and O'Sullivan, 2010; Reid et al., 2003). As probiotics they provide beneficial effects to the body by adhesion and colonization on intestinal membrane and provide protective barrier against the invasion of pathogenic bacteria by bacteriocin production. Simultaneously they provide necessary metabolites and vitamins to the host's body (Picard et al., 2005; Quigley, 2016). Due to the benefits (therapeutic effects) they impart; they have been widely used in various food products. Breast milk and gut are the predominant source of probiotic *Bifidobacterium* (Picard et al., 2005; Quigley, 2016). It has been reported that in humans the predominant strains are *B. breve, B. parvulorum* and *B. infantis* in infants, and *B. adolescentis* and *B. longum* in adults (Reuter, 2001). With a special focus on bacteriocins, *B. infantis* produces Bifid-1 which shows inhibitory activity against *Staphylococcus, Bacillus, Salmonella and Shigella. B. bifidum* produces Bifidocin B with antimicrobial activity against *Bacillus cereus, Listeria monocytogenes and Enterococcus faecalis* (Collado et al., 2005). Bisin produced by *B. longum* active against *Streptococcus thermophiles, Bacullus subtilis* and *Serratia marcescens*; and other peptide antibiotics such as Bifilact Bb-12, Bifilact Bb-46 and Thermophilicin B67 act on a wide range of bacteria (Sarkar and Mandal, 2016).

The first *Bifidobacterium* genome was publication in 2002, since then a steady increase in the number of publicly available genome sequences have been reported (O'Callaghan and van Sinderen, 2016). The NCBI data base currently (July 2017) holds 986 publicly available Bifidobacterial genome sequences, of which 130 represent complete genome sequences. They genome sequences have been insights into the molecular mechanism of probiotic action.

Clostridium

Clostridium was first described by Pasteur and Jeu-bert (1887) as a bacterium that ferment sugars to butyric acid 'without free oxygen'(Udaondo et al., 2017). Members of the genus are Gram-positive, endospore-forming, obligate anaerobic except one species (He et al., 2004). They comprise a collection of diverse species capable of growing in a wide range of environments, including marine and freshwater sediments, wetlands, soils, human and animal gut and can be probiotic as well as pathogenic (He et al., 2004; Udaondo et al., 2017).

Of all the known species *Clostridium butyricum* a butyric acid producer commonly found in gut of healthy animals and humans (Cartman, 2011). Butyric acid (Butanoate) is a short chain fatty acid (SCFA), and plays a central metabolic role in maintaining the mucosal barrier in the gut (Chambers et al., 2018). The microbialy produced butyrate serves as a vital source of energy for the gut wall, providing up to 50% of the daily energy requirements of colonocytes (Chambers et al., 2018). Butyrate also stimulate water and ion absorption in gut, have therapeutic efficacy on patients suffering from diarrhoea and prevent remission in inflammatory bowel disease (IBD) patients (Abdel-Latif et al., 2018). Some strains of *C. butyricum* are able to survive at low pH and relatively in high bile concentrations and produce endospores having high tolerance to stress conditions (Spinler et al., 2017). These properties of *C. butyricum* provide an advantage *in vivo* over other probiotics.

The anti-inflammatory effect of *C. butyricum* was known using gnotobiotic techniques and exhibited a protective effect in an acute DSS-induced colitis model along with an increase in IL10 production in the mononuclear cells from inflamed intestine. It was also seen that probiotic mixture (*C. butyricum* and *B. infantis*) supplement is a simple and effective method to treat Antibiotic-associated diarrhea (AAD) (Abdel-Latif et al., 2018; Chambers et al., 2018; He et al., 2004; Spinler et al., 2017).

The first clostridium genome was published in 2006 was of pathogenic *C. difficile* strain, since then *C. butyricum* has been sequenced to decode the probiotic ability and mechanism of action. The NCBI data base currently (July 2017) holds 28 publicly available *C. butyricum* genome sequences, of which 8 represent complete genome sequences.

Next Generation Anaerobic Probiotics

With the development of modern culturing methodologies (culturomics), isolation of new species providing beneficial properties is now possible (Greub, 2012; Lagier et al., 2015). Furthermore, affordable genome sequencing has advanced the era in probiotic research and develop tailor made probiotics, addressing specific consumer needs and issues (Callanan, 2005). These bacteria are sometimes referred to as next-generation probiotics (NGPs) but may also be termed as live biotherapeutic products (LBPs) (Martín and Langella, 2019; O'Toole et al., 2017a). NPGs apparently follow the normal definition of a probiotic and also fit well within the US Food and Drug Administration (FDA) definition of a LBP (O'Toole et al., 2017a). Following are few of the anaerobic NGPs

Bacteroides

Few species of *Bacteroides* have capacity to act as probiotics viz. *B. fragilis*; *B. xylanisolvens; B. dorei* and *B. acidifaciens* (O'Toole et al., 2017a). Deng and colleagues isolated *Bacteroides fragilis* (strain ZY-312) from faeces of a breastfed infant and has shown health-promoting phenotypes viz. the production of microbicidal molecules and phagocytic functions in macrophages (Deng et al., 2016). *Bacteroides xylanisolvens* DSM 23964 isolated from human faeces has been found with probiotic properties moreover, heat-inactivated preparation of this organism was able to increase the levels of IgM and TFá in a manner that was dose-dependent (Brodmann et al., 2017; Ulsemer et al., 2012) *Bacteroides dorei* (strain D8) has been shown to convert cholesterol to coprostanol *in vitro* and *Bacteroides acidifaciens* has been shown to increase IgA in gnotobiotic mice (Brodmann et al., 2017).

Faecalibacterium prausnitzii

Faecalibacterium prausnitzii being one of the most abundant bacterial species found in the gut and has been reported to be depleted in individuals with inflammatory bowel disease (Dong et al., 2019; Lopez-siles et al., 2015). Thus there might be causal link between diseases given its potentially important role in promoting gut health. In animal models, evidence is available leading to/or associate with induction of anti-inflammatory cytokines or reduction of pro-inflammatory cytokines in induced models of colitis/IBD (O'Toole et al., 2017a). Thus it has been proposed that *F. prausnitzii* monitoring may therefore serve as biomarker to assist in gut diseases diagnostics (O'Toole et al., 2017a).

Akkermansia muciniphila

Akkermansia muciniphilais a bacterium of oval shape, strictly anaerobic, non-motile and Gram-stain-negative and forms no endospores (Naito et al., 2018; Zhou, 2017). It is widely distributed in human intestines, strictly anaerobic bacterium, but recently found that it can tolerate low levels of oxygen, with an oxygen reduction capacity to be 2.260.99 mU/mg total protein (van der Ark et al., 2018). This property is similar to some intestinal anaerobic colonizers such as *Bacteroides fragilis* and *Bifidobacterium adolescentis*, which could still survive after exposure to ambient air for 48 h. *Akkermansia muciniphilais* abundant in the host intestinal mucosal layer, with highest number in the caecum (Milani et al., 2017). It is found to be ubiquitous in the guts of healthy adults and infants, and accounts for 1–4% of the total gut microbiota starting from early life (van der Ark et al., 2018).

Numerous studies with *Akkermansia muciniphila* have been carried out since its discovery in 2004 and its abundance has been associated with beneficial effects in numerous animal models viz. gnotobiotic and specific immune double knock-out models (Brodmann et al., 2017; Cani and de Vos, 2017; Hemarajata and Versalovic, 2013; O'Toole et al., 2017b). It showed profound effects to improve health in metabolic disorders associated with obesity, diabetes and liver diseases (Brodmann et al., 2017; Cani and de Vos, 2017; Hemarajata and Versalovic, 2013; O'Toole et al., 2017b). Further studies have been conducted to find if the pasteurized bacteria and the bacterial constituents (small 30-kDa Amuc_1100) have been providing additional benefits (Brodmann et al., 2017; Cani and de Vos, 2017; Hemarajata and Versalovic, 2013; O'Toole et al., 2017b). Such efforts in understanding the functions and the underlying mechanisms have opened a new door to putative development of drugs.

Anaerobic Probiotic Bacteria Cultivation and Metabolism

Cultivation of anaerobic bacteria for probiotic applications is mainly carried out in submerged fermentations using both batch and fed-batch cultivation strategies and in some cases solid state fermentation (SSF) is preferred viz. *B. bifidum* and *B. longum* for the biomass production on an industrial scale (Jin et al., 2018; Othman et al., 2017). Other cultivation systems such as continuous culture and immobilized cell have also been used for very specific purpose such as co-cultivation with other bacteria (Clewer, 1983). These techniques have been proven to improve cell yield and cell stability (Clewer, 1983). However, there has been few report for the cultivation of anaerobic probiotic bacteria on a large scale but the development of optimal cultural conditions for their growth, is still limited by various factors. On priority, it is important to understand the metabolism which may help to cultivate the species.

Carbohydrate Metabolism

Most of the anaerobic bacteria like aerobes are capable of utilize a wide range of mono-, di- and oligosaccharides, thus this can considered as an advantage for the cultivation as they can retrieve energy from a various source (Gumisiriza et al., 2017; Lynd et al., 2002). In case of *Bifidobacteria*, the presence of fructose-6-phosphate phosphoketolase functions by splitting the hexose phosphate into erythrose-4-phosphate. The succeeding action of transaldolase and transketolase, leads to an increase in the formation of pentose phosphates eventually producing more ATP from carbohydrates (2.5 mol ATP/mol glucose). This is more as produced through the conventional hetero- and homofermentative pathways (Sánchez et al., 2004). The ratio between acetate and lactate production is specific to the strain. The ratio changes according to the source of the carbon used; the molar ratios between acetic and lactic acids are 0.83, 1.21, 1.18 and 1.55 when grown on FOS, fructose, glucose and skimmed milk-based medium (Doleyres et al., 2002; Margolles and Sánchez, 2012).

Growth Media

Most of the anaerobic bacteria have a very strict nutritional requirement, unlike aerobes they are not fastidious (Moore et al., 1969), and can go in a medium composed of a simple carbon source, along with few amino acids such as cysteine, glycine and tryptophan; also vitamins, nucleotides and minerals (Doleyres et al., 2002; Margolles and Sánchez, 2012). Also an additive such as L-cysteine HCl and blood from human source is also preferred (Smith, 2000). Most of the *Bifidobacteria* cultivation needs glucose but addition of oligosaccharide raffinose will promote a good growth. Some have also suggested the use of insulin (Pokusaeva et al., 2011). Besides carbon sources, addition of phosphate and mineral sources helps to improve the growth of *Bifidobacteria* (Pokusaeva et al., 2011; Smith, 2000). Additions of ammonium sulphate and yeast extract have shown to increase production of cell mass as necessary for the biomass production (Manzoor et al., 2017). For isolation of clostridia, media must include neomycin, cycloserine and kanamycin along with blood and egg yolk. This will allow selective enrichment of *Clostridium* species. For example *C. butyricum* strains will grow on a basal glucose-mineral salts medium supplemented with biotin (Popoff, 1984).

Culture Supplements

Supplement for cell growth are generally added to enhance the biomass production (Kropat et al., 2011; Manzoor et al., 2017). Along with oligosaccharide addition, medium supplemented with mupirocin, glacial acetic acid, bovine casein digest and yeast extract is helpful in cultivation (Kropat et al., 2011; Manzoor et al., 2017). Few reports on the use of natural rubber serum powder (NRSP) established the positive effect on the growth of *Bifidobacterium*. It is supposed to be rich in many kinds of amino acids, peptides and inorganic salts (Kropat et al., 2011; Manzoor et al., 2017; Popoff, 1984).

Biomass Production

Compared to aerobic bacteria and yeasts, biomass production of anaerobic species on industrial scale is very limited due to the stringent growth requirement and also anaerobic growth behavior. Productions of probiotic bacterial are protected under intellectual property rights are as trade secrets (Lynd et al., 2002).

Despite the reports on the various cultivation aspects of anaerobes batch cultivation is widely used method due to its simplicity in setup and control, but low biomass yield is the main drawback of the process (Mauerhofer et al., 2018). In-order to improve the biomass yield, fedbatch technique has also been successfully used for *Bifidobacteria* cultivation. Other techniques that are sometimes strain specific are the use of solid state fermentation (SSF), submerged fermentation systems (SMF) (Mauerhofer et al., 2018). These fermentation techniques that have been developed mostly to increase the biomass along with reduction of nutrient depletion; this could have negative effect on both cell growth and bacteriocin production. Nutrient exhaustion can be overcome by the addition of a limiting substrate during the fermentation, which can serve to increase the bacterial concentration (Gerson et al., 1988; Kovárová-Kovar and Egli, 1998). Using fed-batch strategies to produce probiotic cultures has some advantages, and by applying thermal stress to the bacteria during cultivation, the cells are able to cope with the subsequent downstream processing steps.

Downstream Processing and Stabilization

In probiotics, it's the biomass that is required in large quantity and cells are to be separated from the cultivation medium instantaneously as the fermentation is over (Fenster et al., 2019; Ranadheera et al., 2017). The next step is to do washing of cell to remove the remaining traces of the medium components that may lead to problem in downstream (Fenster et al., 2019; Ranadheera et al., 2017). Generally, probiotics are sold as dry powder in lyophilized form, due to ease in transport. Furthermore; it has been seen that lysophilized product are easy to transport and extend the shelf life of the bacteria. Freeze drying is a mild process and supports long-term stability and preservation of the microbial cells without significant loss in viability (Sahara, 1960). The process of freeze drying involves three main steps: freezing, primary drying and secondary drying. To increase cell viability during freeze drying and storage, some cell-protecting agents such as skimmed milk powder, milk whey, butter milk, glycerol, low molecular weight carbohydrates (trehalose, glucose, sucrose and lactose), dextran, polyethylene glycol, glycerol, human-like collagen (HLC), and pepsin are usually added and work as cryo-protectants (Cody et al., 2008; Fonseca et al., 2003; Prakash et al., 2013)

The cost and time consuming nature of the freeze drying method is its major limitation and use of spray drying has been considered as a potential alternative (Cody et al., 2008; Fonseca et al., 2003; Sahara, 1960). This method is widely used due to its high yield, shorter time lower capital and operating costs compared to freeze drying (Strasser et al., 2009). Other methods of fluidized bed and vacuum drying are also very well known. Fluidized bed drying is a mild drying process because cells are not exposed to extreme temperatures and the bacterial cells are granulated first and then encapsulated before drying (Strasser et al., 2009).

It is also worth noting that cell tolerance to downstream stresses, storage conditions and the harsh conditions of the GIT can be increased by provoking stress adaptations in cells by exposure to sub-lethal doses of acid or heat during different phases of the fermentation process. For example, the tolerance of *B. longum* and *B. animalis* to bile salt and low pH was increased to a certain extent when cells were exposed to short term thermal treatment at 47 C and pH 3.5 during the stationary phase. The observed increase in thermotolerance of heat treated bifidobacteria is mediated through the expression and production of heat stress proteins, as has been confirmed by proteomic studies of *B. longum* and *B. breve*. Recent research has also shown that heat shock by short time exposure to sub-lethal temperatures enhances the production and excretion of exopolysaccharides, leading to a significant increase in *B. bifidum* cell robustness and survival during freeze drying.

Quality Control and Evaluation

Quality control (QC) is accountable for the concrete testing by a wide variety of examinations of raw materials, intermediate, in-process and end-product samples. The QC lab is the one which ensure the quality of the product by vigorous testing and if the results are considered sufficiently reliable then a certificate of analysis (COA) could be given. Quality control and assurance both have the same goal of producing a quality product but they diverge in their approach. Quality assurance is accountable for keeping quality systems within the facility at check during the product development.

Probiotic are now being consumed by infants, pregnant women, people with compromised immune systems and allergic reactions, it is must that the standards and rules for maintaining QC labs and testing have to be advanced. The guidelines must be followed as mentioned by regulatory body as per the country norms. This will improve the production facilities, the production, QC processes and release a more consistent end-product. Implementation and practicing such guidelines will help in continued development in the area for process and product development.

Additionally, by adopting good laboratory practices (GLP), the cross contamination at various steps could be reduced. This also includes personal hygiene, sanitation procedures flow through the lab for both product and people. Standard operating procedure should be set in place to make equipment are sufficiently maintained and monitored both by lab members and company. Logbook records has also to created, maintained, and retained. Validation of procedures, methods along with calibration, validation and maintenance of equipment are to be kept at check to maintain the reliable results. Hazard analysis and critical control points (HACCP) is used to identify critical control points and establish acceptable protocols to minimize hazards.

Documentation of all observations along with each lot of material which is produced and these results should be recorded for all production and QC. These file should be reviewed by quality assurance to make sure all relevant paperwork is present and regulatory compliance is met. The length of time which all batch records, metrology records, etc., should be kept and how to properly dispose of all company paperwork depends on the regulatory guidance practiced and company policy.

Shortcoming of Anaerobic Probiotics

Studies on strict anaerobic bacteria, is one of the most challenging areas of research, because oxygen-free i.e. anaerobic conditions need to be created to isolate and know microbial activities and to obtain enrichments and pure culture. It is well-known that the micro/macro anaerobic environments are present everywhere on the Earth, and anaerobes comprise complex communities that play an important role in the carbon, nitrogen, and sulfur cycles on earth. Compared to aerobic bacteria and yeasts, biomass production of anaerobic species on industrial scale is very limited due to the stringent growth requirement and also anaerobic growth behavior. Also most of the anaerobic bacteria have a very strict nutritional requirement, unlike aerobes they are not fastidious (Moore et al., 1969), and can go in a medium composed of a simple carbon source, along with few amino acids such as cysteine, glycine and tryptophan; also vitamins, nucleotides and minerals (Doleyres et al., 2002; Margolles and Sánchez, 2012). Further, the cultivation of anaerobic bacteria for probiotic applications is mainly carried out in submerged fermentations using both batch and fed-batch cultivation strategies and in some cases solid state fermentation (SSF) is preferred for the biomass production on an industrial scale (Jin et al., 2018; Othman et al., 2017). These production difficulties make its application and use limited.

Conclusions and Future Perspectives

The type and number of microbial cells in the human body can influence its health status. Thus, supplementation with specific functional microbial systems, and enhancement of the growth and colonization of specific groups of beneficial microbes, could be one of the future strategies to control disease without exogenous extensive use of antibiotics. In addition, microbial cells in the human body have other functions beyond their antimicrobial effect: they provide a natural prophylactic mechanism in body homeostasis, and protection against many non-microbial related diseases.

It should be noted that just as no 2 probiotic strains can be expected to have exactly the same clinical effect, each probiotic strain, including those that have not yet been developed, would be anticipated to have a different safety profile. Perhaps more importantly, the safety of a commercially available probiotic product depends not only on the probiotic organism but on the other constituents of the product, be it a food or medicinal formulation.

References

Abdel-Latif, M.A., El-Hack, M.E.A., Swelum, A.A., Saadeldin, I.M., Elbestawy, A.R., Shewita, R.S., Ba-Awadh, H.A., Alowaimer, A.N., El-Hamid, H.S.A., 2018. Single and combined effects of Clostridium butyricum and Saccharomyces cerevisiae on growth indices, intestinal health, and immunity of broilers. *Animals 8*. https://doi.org/10.3390/ani8100184

Albenberg, L., Esipova, T. V., Judge, C.P., Bittinger, K., Chen, J., Laughlin, A., Grunberg, S., Baldassano, R.N., Lewis, J.D., Li, H., Thom, S.R., Bushman, F.D., Vinogradov, S.A., Wu, G.D., 2014. Correlation between intraluminal oxygen gradient and radial partitioning of intestinal microbiota. *Gastroenterology* 147, 1055–1063.e8. https://doi.org/10.1053/j.gastro.2014.07.020

Bennedsen, M., Stuer-Lauridsen, B., Danielsen, M., Johansen, E., 2011. Screening for antimicrobial resistance genes and virulence factors via genome sequencing. *Appl. Environ. Microbiol.* 77, 2785–2787. https://doi.org/10.1128/AEM.02493-10

Bilotta, A.J., Cong, Y., 2019. Gut microbiota metabolite regulation of host defenses at mucosal surfaces: implication in precision medicine. *Precis. Clin. Med.* 2, 110–119. https://doi.org/10.1093/pcmedi/pbz008

Brodmann, T., Endo, A., Gueimonde, M., Vinderola, G., Kneifel, W., de Vos, W.M., Salminen, S., Gómez-Gallego, C., 2017. Safety of Novel Microbes for Human Consumption: Practical Examples of Assessment in the European Union. *Front. Microbiol.* 8, 1–15. https://doi.org/10.3389/fmicb.2017.01725

Callanan, M., 2005. Mining the Probiotic Genome: Advanced Strategies, Enhanced Benefits, Perceived Obstacles. *Curr. Pharm. Des.* 11, 25–36. https://doi.org/10.2174/1381612053382377

Cani, P.D., de Vos, W.M., 2017. Next-Generation Beneficial Microbes: The Case of Akkermansia muciniphila. *Front. Microbiol.* 8, 1–8. https://doi.org/10.3389/fmicb.2017.01765

Cartman, S.T., 2011. Time to consider Clostridium probiotics? *Future Microbiol.* 6, 969–971. https://doi.org/10.2217/fmb.11.86

Chambers, E.S., Preston, T., Frost, G., Morrison, D.J., 2018. Role of Gut Microbiota-Generated Short-Chain Fatty Acids in Metabolic and Cardiovascular *Health. Curr. Nutr. Rep.* 7, 198–206. https://doi.org/10.1007/s13668-018-0248-8

Chibbar, R., Dieleman, L.A., 2015. Probiotics in the Management of Ulcerative Colitis. *J. Clin. Gastroenterol.* 49, S50–S55. https://doi.org/10.1097/MCG.0000000000000368

Clewer, J.E., 1983. Observations on the role of the orthoptist in a paediatric assessment centre. Br. *Orthopt. J.* NO. 40, 46–53.

Cody, W.L., Wilson, J.W., Hendrixson, D.R., McIver, K.S., Hagman, K.E., Ott, C.M., Nickerson, C.A., Schurr, M.J., 2008. Skim milk enhances the preservation of thawed - 80 °C bacterial stocks. *J. Microbiol. Methods* 75, 135–138. https://doi.org/10.1016/j.mimet.2008.05.006

Collado, M.C., Hernández, M., Sanz, Y., 2005. Production of bacteriocin-like inhibitory compounds by human fecal Bifidobacterium strains. *J. Food Prot.* 68, 1034–1040. https://doi.org/10.4315/0362-028X-68.5.1034

Dai, C., 2013. Probiotics and irritable bowel syndrome. *World J. Gastroenterol.* 19, 5973. https://doi.org/10.3748/wjg.v19.i36.5973

de Melo Pereira, G.V., de Oliveira Coelho, B., Magalhães Júnior, A.I., Thomaz-Soccol, V., Soccol, C.R., 2018. How to select a probiotic? A review and update of methods and criteria. *Biotechnol. Adv.* 36, 2060–2076. https://doi.org/10.1016/j.biotechadv.2018.09.003

de Simone, C., 2019. The Unregulated Probiotic Market. *Clin. Gastroenterol. Hepatol.* 17, 809–817. https://doi.org/10.1016/j.cgh.2018.01.018

Deng, H., Li, Z., Tan, Y., Guo, Z., Liu, Y., Wang, Y., Yuan, Y., Yang, R., Bi, Y., Bai, Y., Zhi, F., 2016. A novel strain of Bacteroides fragilis enhances phagocytosis and polarises M1 macrophages. *Sci. Rep.* 6, 1–11. https://doi.org/10.1038/srep29401

Doleyres, Y., Paquin, C., LeRoy, M., Lacroix, C., 2002. Bifidobacterium longum ATCC 15707 cell production during free- and immobilized-cell cultures in MRS-whey permeate medium. *Appl. Microbiol. Biotechnol.* 60, 168–173. https://doi.org/10.1007/s00253-002-1103-8

Dong, L.-N., Wang, M., Guo, J., Wang, J.-P., 2019. Role of intestinal microbiota and metabolites in inflammatory bowel disease. *Chin. Med. J.* (Engl). 132, 1610–1614. https://doi.org/10.1097/cm9.0000000000000290

Fao, J., Working, W.H.O., Report, G., Guidelines, D., London, F., 2002. Guidelines for the Evaluation of Probiotics in Food 1–11.

Felis, G.E., Dellaglio, F., 2007. Taxonomy of lactobacilli and bifidobacteria. Curr. Issues Intest. Microbiol. 8, 44–61.

Fenster, K., Freeburg, B., Hollard, C., Wong, C., Rønhave Laursen, R., Ouwehand, A., 2019. The Production and Delivery of Probiotics: A Review of a Practical Approach. *Microorganisms* 7, 83. https://doi.org/10.3390/microorganisms7030083

Fijan, S., 2014. Microorganisms with claimed probiotic properties: An overview of recent literature. *Int. J. Environ. Res. Public Health* 11, 4745–4767. https://doi.org/10.3390/ijerph110504745

Florou-Paneri, P., Christaki, E., Bonos, E., 2013. Lactic Acid Bacteria as Source of Functional Ingredients. Lact. Acid Bact. D Food, Heal. Livest. Purp. 589–614. https://doi.org/http://dx.doi.org/10.5772/47766

Fonseca, F., Béal, C., Mihoub, F., Marin, M., Corrieu, G., 2003. Improvement of cryopreservation of Lactobacillus delbrueckii subsp. bulgaricus CFL1 with additives displaying different protective effects. *Int. Dairy J.* 13, 917–926. https://doi.org/10.1016/S0958-6946 (03)00119-5

Fuller, R., Raibaud, P., Rowland, I.R., Berg, R.D., Hentges, D.J., Freter, R., Perdigon, G., Alvarez, S., Tannock, G.W., Havenaar, R., Brink, B. Ten, In't Veld, J.H.J.H., Barrow, P.A., Jonsson, E., Conway, P., Wallace, R.J., Newbold, C.J., Goldin, B.R., Gorbach, S.L., 1992. Probiotics: The Scientific Basis. https://doi.org/10.1007/978-94-011-2364-8

Ganguly, N.K., Bhattacharya, S.K., Sesikeran, B., Nair, G.B., Ramakrishna, B.S., Sachdev, H.P.S., Batish, V.K., Kanagasabapathy, A.S., Muthuswamy, V., Kathuria, S.C., Katoch, V.M., Satyanarayana, K., Toteja, G.S., Rahi, M., Rao, S., Bhan, M.K., Kapur, R., Hemalatha, R., 2011. ICMR-DBT Guidelines for evaluation of probiotics in food. *Indian J. Med. Res.* 134, 22–25.

Gaspar, C., Donders, G.G., Palmeira-de-Oliveira, R., Queiroz, J.A., Tomaz, C., Martinez-de-Oliveira, J., Palmeira-de-Oliveira, A., 2018. Bacteriocin production of the probiotic Lactobacillus acidophilus KS400. *AMB Express* 8. https://doi.org/10.1186/s13568-018-0679-z

Gerson, D.F., Kole, M.M., Ozum, B., 1988. Substrate concentration control in bioreactors. Biotechnol. *Genet. Eng. Rev*. 6, 67–105. https://doi.org/10.1080/02648725.1988.10647846

Gogineni, V.K., 2013. Probiotics: History and Evolution. *J. Anc. Dis. Prev. Remedies* 01, 1–7. https://doi.org/10.4172/2329-8731.1000107

Greub, G., 2012. Culturomics: A new approach to study the human microbiome. *Clin. Microbiol. Infect.* 18, 1157–1159. https://doi.org/10.1111/1469-0691.12032

Guarino, A., Guandalini, S., Lo Vecchio, A., 2015. Probiotics for Prevention and Treatment of Diarrhea. *J. Clin. Gastroenterol.* 49, S37–S45. https://doi.org/10.1097/MCG.0000000000000349

Gueimonde, M., Sánchez, B., de los Reyes-Gavilán, C.G., Margolles, A., 2013. Antibiotic resistance in probiotic bacteria. Front. Microbiol. 4, 1–6. https://doi.org/10.3389/fmicb.2013.00202

Guinane, C.M., Cotter, P.D., 2013. Role of the gut microbiota in health and chronic gastrointestinal disease: understanding a hidden metabolic organ. *Therap. Adv. Gastroenterol*. 6, 295–308. https://doi.org/10.1177/1756283X13482996

Gumisiriza, R., Hawumba, J.F., Okure, M., Hensel, O., 2017. Biomass waste-to-energy valorisation technologies: A review case for banana processing in Uganda. Biotechnol. *Biofuels* 10, 1–29. https://doi.org/10.1186/s13068-016-0689-5

Gupte, N., 2017. Antibiotic - Associated Diarrhoea: Pharmacotherapeutic and Preventive Aspects in Children. Gastroenterol. Hepatol. Int. J. 2, 1–6. https://doi.org/10.23880/GHIJ-16000130

Hamilton-Miller, J.M.T., Gibson, G.R., Bruck, W., 2003. Some insights into the derivation and early uses of the word "probiotic". *Br. J. Nutr.* 90, 845.

He, G.Q., Kong, Q., Ding, L.X., 2004. Response surface methodology for optimizing the fermentation medium of Clostridium butyricum. *Lett. Appl. Microbiol.* 39, 363–368. https://doi.org/10.1111/j.1472-765X.2004.01595.x

Hemarajata, P., Versalovic, J., 2013. Effects of probiotics on gut microbiota: mechanisms of intestinal immunomodulation and neuromodulation. *Therap. Adv. Gastroenterol.* 6, 39–51. https://doi.org/10.1177/1756283X12459294

Heshmati, J., Farsi, F., Shokri, F., Rezaeinejad, M., Almasi-Hashiani, A., Vesali, S., Sepidarkish, M., 2018. A systematic review and meta-analysis of the probiotics and synbiotics effects on oxidative stress. *J. Funct. Foods* 46, 66–84. https://doi.org/10.1016/j.jff.2018.04.049

Huys, G., Botteldoorn, N., Delvigne, F., De Vuyst, L., Heyndrickx, M., Pot, B., Dubois, J.-J., Daube, G., 2013. Microbial characterization of probiotics-Advisory report of the Working Group "8651 Probiotics" of the Belgian Superior Health Council (SHC). *Mol. Nutr. Food Res.* 57, 1479–1504. https://doi.org/10.1002/mnfr.201300065

Jha, S., Rollins, M.G., Fuchs, G., Procter, D.J., Hall, E.A., Cozzolino, K., Sarnow, P., Savas, J.N., Walsh, D., 2017. Trans-kingdom mimicry underlies ribosome customization by a poxvirus kinase. *Nature* 546, 651–655. https://doi.org/10.1038/nature22814

Jin, W., Yoon, C., Johnston, T., Ku, S., Ji, G., 2018. Production of Selenomethionine-Enriched Bifidobacterium bifidum BGN4 via Sodium Selenite Biocatalysis. *Molecules* 23, 2860. https://doi.org/10.3390/molecules23112860

Kollath, W., 1953. [The increase of the diseases of civilization and their prevention]. Munch. Med. Wochenschr. 95, 1260–2.

Kovárová-Kovar, K., Egli, T., 1998. Growth kinetics of suspended microbial cells: from single-substrate-controlled growth to mixed-substrate kinetics. *Microbiol. Mol. Biol. Rev*. 62, 646–66.

Kropat, J., Hong-Hermesdorf, A., Casero, D., Ent, P., Castruita, M., Pellegrini, M., Merchant, S.S., Malasarn, D., 2011. A revised mineral nutrient supplement increases biomass and growth rate in Chlamydomonas reinhardtii. *Plant J.* 66, 770–780. https://doi.org/10.1111/j.1365-313X.2011.04537.x

Kumar, M., Kumar, A., Nagpal, R., Mohania, D., Behare, P., Verma, V., Kumar, P., Poddar, D., Aggarwal, P.K., Henry, C.J.K., Jain, S., Yadav, H., 2010. Cancer-preventing attributes of probiotics: an update. *Int. J. Food Sci. Nutr.* 61, 473–496. https://doi.org/10.3109/09637480903455971

Lagier, J.C., Hugon, P., Khelaifia, S., Fournier, P.E., La Scola, B., Raoult, D., 2015. The rebirth of culture in microbiology through the example of culturomics to study human gut microbiota. *Clin. Microbiol. Rev.* 28, 237–264. https://doi.org/10.1128/CMR.00014-14

Lebeer, S., Vanderleyden, J., De Keersmaecker, S.C.J., 2008. Genes and molecules of lactobacilli supporting probiotic action. *Microbiol. Mol. Biol. Rev.* 72, 728–764. https://doi.org/10.1128/MMBR.00017-08

Lee, J.-H., O'Sullivan, D.J., 2010. Genomic Insights into Bifidobacteria. *Microbiol. Mol. Biol.* Rev. 74, 378–416. https://doi.org/10.1128/mmbr.00004-10

Letteri, J.M., Asad, S.N., Caselnova, R., Ellis, K.J., Cohn, S.H., 1975. Creatinine excretion and total body potassium in renal failure. *Clin. Nephrol.* 4, 58–61.

Liu, L., Zhu, G., 2018. Gut-brain axis and mood disorder. Front. *Psychiatry 9*, 1–8. https://doi.org/10.3389/fpsyt.2018.00223

Lopetuso, L.R., Scaldaferri, F., Petito, V., Gasbarrini, A., 2013. Commensal Clostridia: Leading players in the maintenance of gut homeostasis. *Gut Pathog.* 5, 1–8. https://doi.org/10.1186/1757-4749-5-23

Lopez-siles, M., Martinez-medina, M., Abellà, C., Busquets, D., Sabat-mir, M., Duncan, S.H., Aldeguer, X., Flint, H.J., Garcia-gil, L.J., 2015. Reduced in Patients with Inflammatory Bowel Disease. *Appl. Environ. Microbiol.* 81, 7582–7592. https://doi.org/10.1128/AEM.02006-15.Editor

Lynd, L.R., Weimer, P.J., van Zyl, W.H., Pretorius, I.S., 2002. Microbial Cellulose Utilization: Fundamentals and Biotechnology. *Microbiol. Mol. Biol. Rev.* 66, 506–577. https://doi.org/10.1128/MMBR.66.3.506-577.2002

Malago, J.J. (Joshua J., Koninkx, J.F.J.G. (Joseph F.J.G., Marinsek-Logar, R., 2011. Probiotic Bacteria and Enteric Infections. Springer Netherlands, Dordrecht. https://doi.org/10.1007/978-94-007-0386-5

Maldonado Galdeano, C., Cazorla, S.I., Lemme Dumit, J.M., Vélez, E., Perdigón, G., 2019. Beneficial effects of probiotic consumption on the immune system. *Ann. Nutr. Metab.* 74, 115–124. https://doi.org/10.1159/000496426

Manzoor, A., Qazi, J.I., Haq, I.U., Mukhtar, H., Rasool, A., 2017. Significantly enhanced biomass production of a novel bio-therapeutic strain Lactobacillus plantarum (AS-14) by developing low cost media cultivation strategy. J. Biol. Eng. 11, 1–10. https://doi.org/10.1186/s13036-017-0059-2

Margolles, A., Sánchez, B., 2012. Selection of a Bifidobacterium animalis subsp. lactis strain with a decreased ability to produce acetic acid. *Appl. Environ. Microbiol.* 78, 3338–3342. https://doi.org/10.1128/AEM.00129-12

Marteau, P., 2001. Safety aspects of probiotic products. *Scand. J. Nutr.* 45, 22–24.

Martin, C.R., Osadchiy, V., Kalani, A., Mayer, E.A., 2018. The Brain-Gut-Microbiome Axis. Cell. Mol. Gastroenterol. Hepatol. 6, 133–148. https://doi.org/10.1016/j.jcmgh.2018.04.003

Martín, R., Langella, P., 2019. Emerging Health Concepts in the Probiotics Field: Streamlining the Definitions. *Front. Microbiol.* 10. https://doi.org/10.3389/fmicb.2019.01047

Matsuzaki, T., Chin, J., 2000. Modulating immune responses with probiotic bacteria. Immunol. *Cell Biol.* 78, 67–73. https://doi.org/10.1046/j.1440-1711.2000.00887.x

Mauerhofer, L.M., Pappenreiter, P., Paulik, C., Seifert, A.H., Bernacchi, S., Rittmann, S.K.M.R., 2018. Methods for quantification of growth and productivity in anaerobic microbiology and biotechnology, Folia Microbiologica. *Folia Microbiologica.* https://doi.org/10.1007/s12223-018-0658-4

McFarland, L. V, Dublin, S., 2008. Meta-analysis of probiotics for the treatment of irritable bowel syndrome. *World J. Gastroenterol.* 14, 2650–61. https://doi.org/10.3748/wjg.14.2650

Milani, C., Duranti, S., Bottacini, F., Casey, E., Turroni, F., Mahony, J., Belzer, C., Delgado Palacio, S., Arboleya Montes, S., Mancabelli, L., Lugli, G.A., Rodriguez, J.M., Bode, L., de Vos, W., Gueimonde, M., Margolles, A., van Sinderen, D., Ventura, M., 2017. The First Microbial Colonizers of the Human Gut: Composition, Activities, and Health Implications of the Infant Gut Microbiota. *Microbiol. Mol. Biol. Rev.* 81, 1–67. https://doi.org/10.1128/mmbr.00036-17

Milani, C., Turroni, F., Duranti, S., Lugli, G.A., Mancabelli, L., Ferrario, C., Van Sinderen, D., Ventura, M., 2016. Genomics of the genus Bifidobacterium reveals species-specific adaptation to the glycan-rich gut environment. *Appl. Environ. Microbiol.* 82, 980–991. https://doi.org/10.1128/AEM.03500-15

Moore, W.E.C., Cato, E.P., Holdeman, L. V., 1969. Anaerobic bacteria of the gastrointestinal flora and their occurrence in clinical infections. *J. Infect. Dis.* 119, 641–649. https://doi.org/10.1093/infdis/119.6.641

Munday, J.L., 2012. A discourse analysis of the construction of menopause in South African newspapers. *Curr Opin Gastroenterol.* 27, 1–107. https://doi.org/10.1016/j.biotechadv.2011.08.021.Secreted

Naito, Y., Uchiyama, K., Takagi, T., 2018. A next-generation beneficial microbe: Akkermansia muciniphila. J. Clin. Biochem. Nutr. 63, 33–35. https://doi.org/10.3164/jcbn.18-57

Nouioui, I., Carro, L., García-lópez, M., Meier-kolthoff, J.P., 2018. Genome-Based Taxonomic Classification of the Phylum Actinobacteria 9, 1–119. https://doi.org/10.3389/fmicb.2018.02007

O'Callaghan, A., van Sinderen, D., 2016. Bifidobacteria and their role as members of the human gut microbiota. *Front. Microbiol.* 7. https://doi.org/10.3389/fmicb.2016.00925

O'Toole, P.W., Marchesi, J.R., Hill, C., 2017a. Next-generation probiotics: the spectrum from probiotics to live biotherapeutics. Nat. Microbiol. 2, 17057. https://doi.org/10.1038/nmicrobiol.2017.57

Oak, S.J., Jha, R., 2019. The effects of probiotics in lactose intolerance: A systematic review. *Crit. Rev. Food Sci. Nutr.* 59, 1675–1683. https://doi.org/10.1080/10408398.2018.1425977

Ohira, H., Tsutsui, W., Fujioka, Y., 2017. Are short chain fatty acids in gut microbiota defensive players for inflammation and atherosclerosis? *J. Atheroscler. Thromb.* 24, 660–672. https://doi.org/10.5551/jat.RV17006

Othman, M., Ariff, A.B., Wasoh, H., Kapri, M.R., Halim, M., 2017. Strategies for improving production performance of probiotic Pediococcus acidilactici viable cell by overcoming lactic acid inhibition. AMB Express 7, 215. https://doi.org/10.1186/s13568-017-0519-6

Ouwehand, A.C., Kirjavainen, P. V, Shortt, C., Salminen, S., 1999. Probiotics/ : mechanisms and established effects. *Int. Dairy J.* 9, 43–52.

Ozen, M., Dinleyici, E.C., 2015. The history of probiotics: The untold story. Benef. Microbes 6, 159–165. https://doi.org/10.3920/BM2014.0103

Pandey, K.R., Naik, S.R., Vakil, B. V, 2015. Probiotics, prebiotics and synbiotics- a review. *J. Food Sci. Technol.* 52, 7577–7587. https://doi.org/10.1007/s13197-015-1921-1

Papadimitriou, K., Zoumpopoulou, G, Foligné, B., Alexandraki, V., Kazou, M., Pot, B., Tsakalidou, E., 2015. Discovering probiotic microorganisms: In vitro, in vivo, genetic and omics approaches. *Front. Microbiol.* 6, 1–28. https://doi.org/10.3389/fmicb.2015.00058

Parte, A.C., 2018. LPSN – List of Prokaryotic names with Standing in Nomenclature (bacterio.net), 20 years on. *Int. J. Syst. Evol. Microbiol.* 68, 1825–1829. https://doi.org/10.1099/ijsem.0.002786

Pattani, R., Palda, V.A., Hwang, S.W., Shah, P.S., 2013b. Probiotics for the prevention of antibiotic-associated diarrhea and Clostridium difficile infection among hospitalized patients: systematic review and meta-analysis. *Open Med.* 7, e56-67.

Picard, C., Fioramonti, J., Francois, A., Robinson, T., Neant, F., Matuchansky, C., 2005. Review article: Bifidobacteria as probiotic agents - Physiological effects and clinical benefits. Aliment. Pharmacol. Ther. 22, 495–512. https://doi.org/10.1111/j.1365-2036.2005.02615.x

Pineiro, M., Stanton, C., 2007. Probiotic Bacteria: Legislative Framework— Requirements to Evidence Basis. *J. Nutr.* 137, 850S–853S. https://doi.org/10.1093/jn/137.3.850S

Plaza-Diaz, J., Ruiz-Ojeda, F.J., Gil-Campos, M., Gil, A., 2019. Mechanisms of Action of Probiotics. *Adv. Nutr.* 10, S49–S66. https://doi.org/10.1093/advances/nmy063

Pokusaeva, K., Fitzgerald, G.F., Van Sinderen, D., 2011. Carbohydrate metabolism in Bifidobacteria. *Genes Nutr.* 6, 285–306. https://doi.org/10.1007/s12263-010-0206-6

Popoff, M.R., 1984. Selective medium for isolation of Clostridium butyricum from human feces. *J. Clin. Microbiol.* 20, 417–420.

Prakash, O., Nimonkar, Y., Shouche, Y.S., 2013. Practice and prospects of microbial preservation. FEMS Microbiol. Lett. 339, 1–9. https://doi.org/10.1111/1574-6968.12034

Quigley, E.M.M., 2016. Bifidobacteria as probiotic organisms: An introduction, The Microbiota in Gastrointestinal Pathophysiology: Implications for Human Health, Prebiotics, Probiotics, and Dysbiosis. Elsevier Inc. https://doi.org/10.1016/B978-0-12-804024-9.00012-4

Ranadheera, C.S., Vidanarachchi, J.K., Rocha, R.S., Cruz, A.G., Ajlouni, S., 2017. Probiotic delivery through fermentation: Dairy vs. non-dairy beverages. *Fermentation* 3, 1–17. https://doi.org/10.3390/fermentation3040067

Reid, G., Kim, S.O., KÃ¶hler, G.A., 2006. Selecting, testing and understanding probiotic microorganisms. *FEMS Immunol. Med. Microbiol.* 46, 149–157. https://doi.org/10.1111/j.1574-695X.2005.00026.x

Reid, G., Sanders, M.E., Gaskins, H.R., Gibson, G.R., Mercenier, A., Rastall, R., Roberfroid, M., Rowland, I., Cherbut, C., Klaenhammer, T.R., 2003. New scientific paradigms for probiotics and prebiotics. *J. Clin. Gastroenterol.* 37, 105–118. https://doi.org/10.1097/00004836-200308000-00004

Reuter, G., 2001. The Lactobacillus and Bifidobacterium microflora of the human intestine: Composition and succession. *Curr. Issues Intest. Microbiol.* 2, 43–53.

Saez-Lara, M.J., Gomez-Llorente, C., Plaza-Diaz, J., Gil, A., 2015. The role of probiotic lactic acid bacteria and bifidobacteria in the prevention and treatment of inflammatory bowel disease and other related diseases: a systematic review of randomized human clinical trials. Biomed Res. Int. 2015, 505878. https://doi.org/10.1155/2015/505878

Sahara, Y., 1960. Freeze-Drying of Microorganism. Shinku/Journal Vac. Soc. Japan 3, 214–227. https://doi.org/10.3131/jvsj.3.214

Sánchez, B., Noriega, L., Ruas-Madiedo, P., De Los Reyes-Gavilán, C.G., Margolles, A., 2004. Acquired resistance to bile increases fructose-6-phosphate phosphoketolase activity in Bifidobacterium. *FEMS Microbiol. Lett.* 235, 35–41. https://doi.org/10.1016/j.femsle.2004.04.009

Sanders, M.E., Levy, D.D., 2011. The science and regulations of probiotic food and supplement product labeling. *Ann. N. Y. Acad. Sci.* 1219, 1–23. https://doi.org/10.1111/j.1749-6632.2010.05956.x

Sarkar, A., Mandal, S., 2016. Bifidobacteria—Insight into clinical outcomes and mechanisms of its probiotic action. *Microbiol. Res.* 192, 159–171. https://doi.org/10.1016/j.micres.2016.07.001

Serra, C.R., Almeida, E.M., Guerreiro, I., Santos, R., Merrifield, D.L., Tavares, F., Oliva-Teles, A., Enes, P., 2019. Selection of carbohydrate-active probiotics from the gut of carnivorous fish fed plant-based diets. *Sci. Rep.* 9, 1–15. https://doi.org/10.1038/s41598-019-42716-7

Sgorbati, B., Biavati, B., Palenzona, D., 1995. The genus Bifidobacterium, in: The Genera of Lactic Acid Bacteria. Springer US, Boston, MA, pp. 279–306. https://doi.org/10.1007/978-1-4615-5817-0_8

Sierra, H., Cordova, M., Chen, C.-S.J., Rajadhyaksha, M., 2015. Confocal Imaging–Guided Laser Ablation of Basal Cell Carcinomas: An Ex Vivo Study. *J. Invest. Dermatol.* 135, 612–615. https://doi.org/10.1038/jid.2014.371

Smith, Q.R., 2000. Transport of Glutamate and Other Amino Acids at the Blood-Brain Barrier. *J. Nutr.* 130, 1016S–1022S. https://doi.org/10.1093/jn/130.4.1016s

Sorokulova, I., 2008. Preclinical Testing in the Development of Probiotics: A Regulatory Perspective with Bacillus Strains as an Example. *Clin. Infect. Dis*. 46, S92–S95. https://doi.org/10.1086/523334

Spinler, J.K., Auchtung, J., Brown, A., Boonma, P., Oezguen, N., Ross, C.L., Luna, R.A., Runge, J., Versalovic, J., Peniche, A., Dann, S.M., Britton, R.A., Haag, A., Savidge, T.C., 2017. Next-Generation Probiotics Targeting Clostridium difficile through Precursor-Directed Antimicrobial Biosynthesis. *Infect. Immun.* 85, 1–16. https://doi.org/10.1128/IAI.00303-17

Strasser, S., Neureiter, M., Geppl, M., Braun, R., Danner, H., 2009. Influence of lyophilization, fluidized bed drying, addition of protectants, and storage on the viability of lactic acid bacteria. *J. Appl. Microbiol.* 107, 167–177. https://doi.org/10.1111/j.1365-2672.2009.04192.x

Taylor, P., Sanders, M.E., Akkermans, L.M.A., Haller, D., Hammerman, C., James, T., Hörmannsperger, G., Huys, G., Sanders, M.E., Akkermans, L.M.A., Haller, D., Hammerman, C., Heimbach, J., Hörmannsperger, G., Huys, G., Levy, D.D., Lutgendorff, F., Mack, D., Phothirath, P., 2010. Safety assessment of probiotics for human use Safety assessment of probiotics for human use. *Gut Microbes* 1, 37–41. https://doi.org/10.4161/gmic.1.3.12127

Thursby, E., Juge, N., 2017. Introduction to the human gut microbiota. *Biochem. J.* 474, 1823–1836. https://doi.org/10.1042/BCJ20160510

Udaondo, Z., Duque, E., Ramos, J.-L., 2017. The pangenome of the genus *C lostridium. Environ. Microbiol.* 19, 2588–2603. https://doi.org/10.1111/1462-2920.13732

Ulsemer, P., Toutounian, K., Kressel, G., Schmidt, J., Karsten, U., Hahn, A., Goletz, S., 2012. Safety and tolerance of Bacteroides xylanisolvens DSM 23964 in healthy adults. *Benef. Microbes* 3, 99–111. https://doi.org/10.3920/BM2011.0051

van der Ark, K.C.H., Aalvink, S., Suarez-Diez, M., Schaap, P.J., de Vos, W.M., Belzer, C., 2018. Model-driven design of a minimal medium for Akkermansia muciniphila confirms mucus adaptation. Microb. Biotechnol. 11, 476–485. https://doi.org/10.1111/1751-7915.13033

Vieco-Saiz, N., Belguesmia, Y., Raspoet, R., Auclair, E., Gancel, F., Kempf, I., Drider, D., 2019. Benefits and inputs from lactic acid bacteria and their bacteriocins as alternatives to antibiotic growth promoters during food-animal production. *Front. Microbiol.* 10, 1–17. https://doi.org/10.3389/fmicb.2019.00057

Weder, N., Zhang, H., Jensen, K., Yang, B.Z., Simen, A., Jackowski, A., Lipschitz, D., Douglas-Palumberi, H., Ge, M., Perepletchikova, F., O'Loughlin, K., Hudziak, J.J., Gelernter, J., Kaufman, J., 2014. Child abuse, depression, and methylation in genes involved with stress, neural plasticity, and brain circuitry. *J. Am. Acad. Child Adolesc. Psychiatry* 53, 417–24.e5. https://doi.org/10.1016/j.jaac.2013.12.0

Zhou, K., 2017. Strategies to promote abundance of Akkermansia muciniphila, an emerging probiotics in the gut, evidence from dietary intervention studies. *J. Funct. Foods* 33, 194–201. https://doi.org/10.1016/j.jff.2017.03.045

15

Human Gut: An Ideal Habitat for Anaerobes

*Rahul Bodkhe**, *Akshay Gaike*, *Abhijit Kulkarni* and *Yogesh S. Shouche**

The YSS lab, National Centre for Microbial Resource, National Centre for Cell Science, Pune, Maharashtra, India

Abstract

Hundreds of bacterial taxa makes up human gut microbiota and it is comprised primarily of anaerobes belonging to four main phyla which are Firmicutes, Bacteriodetes, Actinobacteria, *and* Proteobacteria. *Anaerobes colonizes the human body, including oral cavity, small intestine, vagina, skin but the majority of them are housed in the colon. The composition and distribution anaerobes varies along the length of gastrointestinal tract because of variation in oxygen gradient, anatomy of intestinal tract and other host physiological factors. Human gets colonized by anaerobes during the birth and these initial colonizers have impact on infant's health status. The colonization and development of the gut microbiome is responsive to variety of factors including the mode of delivery, feeding methods, environment etc. The human gut microbiota plays an important role in healthy status of host. It plays an important role in nutrient metabolism and absorption, synthesis of vitamins and production of short-chain fatty acids (SCFAs). In turn these metabolites have positive effect on host immunity and overall host fitness. Studying, the anaerobes using culturing method remains critical in microbiology. However, culturomics approach, combined with sequencing technology, provides new insights in studying anaerobes of human gut microbiota. In this chapter we have discussed that, how human gut is an ideal habitat for the colonization and growth of anaerobes and their impact on host health. We also tried to shed light*

**Corresponding Authors: rahulbiosciences@gmail.com; yogesh@nccs.res.in*

on latest technical advancement in sequencing culturing technology and their application in studying gut anaerobes.

Keywords: Anaerobes; Gut microbiota; vaginal microbiome; oral microbiota; small intestinal microbiota; colon microbiome; 16S rRNA gene; mucin.

Introduction

The human body is covered and bathed in microbes as are all higher organisms. This has happened over the course of evolution where the host and microbes have joined hands and developed deep symbiosis (Davenport et al. 2017). Such interactions have allowed for increased flexibility to the host to adapt to the environment and available resources and in-turn have provided shelter to the microbes. During the course of evolution, the host selectively chooses the microbes it would gain from and identifies the necessary ones form the not-so-necessary ones. The host-associated microbes have also adapted themselves to the host environment and the interactions have evolved to such high sophistication that the host development is tied to the presence of these microbes. In the case of humans, the human body is colonized by microbes during the process of birth and these microbes stay till the very end. The human gut is one of the most unique ecosystems and the diversity of microbes found in the gut is very different from the microbes present on other surfaces like the skin, oral cavity or genito-urinal tract. One of the drivers for this unique diversity is the anoxic environment in the gut. The physiological state drives the enrichment of specialized microbes that can thrive and help the host with a wide range of functions. This is probably why the gut microbiome is getting its due attention given its vast metabolic functions and is now being looked at as one of the organ systems in human biology. Microorganisms can be classified based on staining shape and nutritional requirements. Moreover, they are also classified aerobes, anaerobes, and facultative anaerobes based on their oxygen requirement for metabolism (Ravcheev and Thiele 2014). Aerobes require molecular oxygen as a terminal electron acceptor in their metabolic reactions and thus cannot grow in its absence. Anaerobes, on the other hand, utilizes other substances instead of oxygen for metabolism purposes (Moore, Cato, and Holdeman 1969). Oxygen is toxic for anaerobes; thus, they are found in anoxic conditions such as deep oceans, water sediments, and animal gut (Zhang et al. 2017). In past decades scientists have devoted much of their time in the identification and characterization of microorganisms from the animal gut especially the human gut (Bodkhe et al. 2019; Bodkhe, Balakrishnan, and Taneja 2019; Mangalam et al. 2017a). The human gut is one of the ideal ecosystems for the growth of anaerobes and harbors more than 400 bacterial species (Gerritsen et al. 2011).

Moreover, the population of anaerobes outnumbers the aerobic bacterial population (Bellali et al. 2019). The essential requirement of anaerobic bacteria is the nutrition and anoxic environment, and the human gut provides these ideal conditions for them.

In this chapter we will study the anaerobes their colonization in infants, their diversity and function and methods to study the anaerobes.

Anaerobes Colonize the Human since the Birth

Animals including humans get colonized by the bacteria since the birth. There are some reports state that the members of microbiota can transfer through placenta to fetus, however it is debatable (Perez-Muñoz et al. 2017; Rowland et al. 2018; Walker et al. 2017). Nevertheless, it is established that in case of vaginally born infants, the infant are inoculated with bacteria as they passes through the birth canal (Mueller et al. 2015; Neu and Rushing 2011). Thus neonatal gut is colonized by vagina-associated facultative and obligate anaerobes microbes such as *Staphylococcus, Lactobacillus* and *Prevotella* but within months *Bifidobacterium* and *Bacteroides* expands in the gut (Milani et al. 2017; Robertson et al. 2019; Yang et al. 2016). These initial colonizers seeds the microbiota in infants and that later transforms into a mature and stable microbiota. Over the period of days the gut gradually becomes anaerobic because of physiological changes as well as oxygen consumption by aerobes and that favors the growth and expansion of facultative and obligate anaerobic bacterial species (Friedman et al. 2018; Könönen, Asikainen, and Jousimies Somer 1992). Thus the bacterial flora in gut changes from aerobic or facultative anaerobes to strict anaerobes during the first week of life (3 to 7 days) (Liu et al. 2019). Meconium or stool samples of the infants collected after the birth are one of the best samples to study the gut colonizers in infants. Previous study by Dominguez Bello et al. (Dominguez-Bello et al. 2010) on Amerindian infants using 16S rRNA gene sequencing method found out that the guts microbiota of vaginally born infants was similar to their mothers' vaginal microbiota. Further analyses observed that the infant's guts were colonized predominantly by *Lactobacillus, Prevotella, or Sneathia spp.* Similarly, shotgun sequencing method characterized the meconium microbiome from the feces of a Chinese cohort and demonstrated that the fecal microbiota of infants had significant enrichment of obligate anaerobes such as *Bacteroides*, *Parabacteroides*, *Megamonas, Prevotella* and *Streptococcus* (Shi et al. 2018). Another study analyzed the fecal microbiome of infants living in 11 international sites (including Ethiopia, Kenya, Peru, Spain, The Gambia, Ghana, the US and Sweden) (Lackey et al. 2019). The study revealed that the microbiome in infants was different as per geographic location. Although, genera such as *Streptococcus*, *Escherichia/Shigella,* and *Veillonella,* were present in feces of all the infants irrespective of geographical location. These studies suggest that the initial colonizers might be different as per geographical locations, however are enriched with anaerobes.

Breast milk feeding is another mechanism in mammals to transfer microbiota (including anaerobes) to infants. Breast milk has long been known to contain bacterial DNA, although no full-proof available that milk contains bacterial species, also the mechanism how the gut bacteria reach to breast milk is not been established yet. Nevertheless, culture-independent techniques such as next generation sequencing technology have identified obligate and facultative anaerobes of the gut in the breast milk samples (van den Elsen et al. 2019; Pannaraj et al. 2017; Pärnänen et al. 2018).

During breast feeding infants acquire anaerobes from mother's skin and the environment. In one of the previous studies, species *Prevotella melaninogenica* was found to be the most frequently isolated anaerobe in infants, when cheeks, palate, tongue and saliva swabs samples from infants were cultured (Könönen, Asikainen, and Jousimies Somer 1992). The study also revealed occurrence of the other common anaerobic taxa such as *Fusobacterium nucleatum, Veillonella* and *Prevotella.* Undoubtedly, modes of delivery would (vaginal and cesarean sect) affect the colonization of "normal" anaerobic microbiota members. In-fact in one of the studies, the intestinal microbiome of vaginally delivered infants was more diverse than that of the infants born surgically (Yang et al. 2016). Moreover, microbiome of vaginally born infants was dominated by facultative and obligate anaerobes such as *Lactobacillus, Prevotella, Bifidobacterium* and *Bacteroides*. In contrast, the gut microbiome of a surgically born infant had reduced proportion of *Bifidobacteria* and *Bacteroides spp.* and comprised of *Staphylococcus, Corynebacteria, Propionibacterium spp.* These bacteria were probably acquired from surrounding surfaces such as mother's and others' skin surface. In addition to mode of delivery, previous studies indicated that the feeding habits also modulates the microbiota. Microbiota of formula-fed babies found be dominated with anaerobic microbial populations such as *Bacteroides* and *Clostridium*. On the other hand microbiota of breastfed infants were more commonly colonized by aerobes (Rinninella et al. 2019). These above mentioned studies suggest that anaerobic bacterial taxa colonize the human during and after the birth and expand rapidly as baby develops, however, mode of delivery and feeding affect their colonization. Differences due to mode of delivery and feeding habits changes the initial gut colonizer and that may have long-life consequences. As the first colonizers of the gut contribute to the foundation of host-microbe interactions that is essential for health in future life. Several studies suggested that these gut prokaryotic colonizers in the gut perform essential functions in infants related to the digestion and metabolism of milk and or food, the development and activation of the immune system, and the production of neurotransmitters. In order to extract these benefits from microbiota host (humans) have developed favorable environment in the gut to harbor these beneficial partners. In the next part of the chapter, we have tried to explain, the suitability of human gut for the colonization and growth of beneficial anaerobes.

Human gut is an ideal habitat for anaerobes

The human gut is anoxic thus supports the colonization and growth of anaerobic microorganism

It was thought that bacterial oxygen consumption by aerobic microflora is solely responsible for the creation of oxygen gradient and anoxic conditions. These conditions would eventually allow the colonization and growth of facultative and obligate anaerobes. However, recently, a group of scientists found that the oxygen gradient in the gut is created by the host physiological factors in addition to aerobic microbial population (Friedman et al. 2018). They used the phosphorescence quenching method in combination with tissue oxygen probe oxyphor G4 to quantify the level of oxygen in the intestinal tissue. They also correlated the measured level of oxygen to the composition of gut microbiota along the length of the gastro-intestinal tract of germ-free and conventionally raised mice. In results, the luminal oxygen levels were similar between conventionally raised and germ free mice. This observation suggests that in addition to the respiration by aerobic microbial population there present an efficient process by which oxygen is depleted in the intestinal tract. This anoxic environment is the prime requirement for the anaerobes and thus they prefer to colonize the human gut. The oxygen gradient varies along the length of the intestinal tract, and thus the abundance of anaerobic bacteria and overall microbiota composition (Figure 1) (Hillman et al. 2017). Moreover, the fluid secreted by intestinal cells also aides in shaping microbiota composition. Thus microenvironment created by oxygen and host fluids (gastric acid, bile acid, antimicrobial peptides and pancreatic fluids) and intestinal anatomy affect microbial abundance and diversity (Friedman et al. 2018; Hillman et al. 2017). Human gastrointestinal tract is lined by a layer of glycoprotein called mucin. This protein not only serve as substrate for anaerobic growth but also provide support for their growth. In the next part we will see that in details.

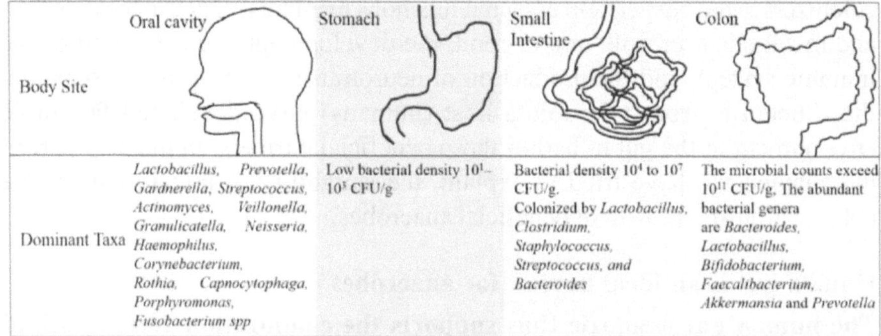

Body Site	Oral cavity	Stomach	Small Intestine	Colon
Dominant Taxa	*Lactobacillus, Prevotella, Gardnerella, Streptococcus, Actinomyces, Veillonella, Granulicatella, Neisseria, Haemophilus, Corynebacterium, Rothia, Capnocytophaga, Porphyromonas, Fusobacterium spp*	Low bacterial density 10^1–10^3 CFU/g	Bacterial density 10^4 to 10^7 CFU/g. Colonized by *Lactobacillus, Clostridium, Staphylococcus, Streptococcus, and Bacteroides*	The microbial counts exceed 10^{11} CFU/g. The abundant bacterial genera are *Bacteroides, Lactobacillus, Bifidobacterium, Faecalibacterium, Akkermansia* and *Prevotella*

Fig. 1: Distribution of facultative and strict Anaerobic Microbes across the gastrointestinal tract

Mucin acts as an anchor for the colonization and growth of anaerobes

The human gut has a protective lining of the mucus layer made up of mucins. This layer not only serves the purpose of nutrition for the anaerobes, but it also acts as an anchor for them. Mucin proteins are glycosylated, and their carbohydrate structures serve as ligands for bacterial adhesion (Corfield 2018). It has been speculated that adhesion to mucins may initiate the bacterial colonization of the intestine (Chaffanel et al. 2018; Corfield et al. 1992). To adhere to carbohydrate moiety of mucus, commensal bacteria are known to produce proteins that have an affinity to the mucus. For example, Mucus-binding proteins (MUB) have been found and studied in lactic acid bacteria (LAB) (Boekhorst et al. 2006; Derrien et al. 2010). Similarly, species belonging to *Bifidobacterium* such as *B. bifidum* express extracellular transaldolase that has shown the correlation with a mucin-binding phenotype (González-Rodríguez et al. 2012). Moreover, *B. longum subsp. infantis* possesses family-1 solute binding proteins that specifically binds to mucin (Sicard et al. 2017). Diverse bacterial taxa that are known to bind to the mucin possess the mucus-binding proteins domain, these features of the human gastrointestinal tract make it an ideal shelter to harbor anaerobes.

Human gut suffice the nutrition requirement of anaerobes

Human microbiota is rich in carbohydrate degrading enzymes and thus it can extract energy and nutrients from more complex carbohydrates. Cellulose, resistant starch and dietary fibres are considered as complex sugar compounds, and they are the part regular animal diet. Interestingly, humans do not code enzymes to degrade these complex compounds. Moreover, it is thought that during evolution, animals lost these genes as gut bacteria are capable of performing fermentation of these complex compounds. Thus, whenever animals or humans consume food containing complex compounds, it acts as substrate (energy source) for microbiota mainly for the anaerobes. In return, microbiota members produces the by-products those are essential for the host benefit. One of the important metabolites of carbohydrate breakdown is short-chain fatty acids (SCFA), like acetate, propionate, and butyrate etc. These SCFAs have been known to have positive effects on gut inflammation and immunity (Luu et al. 2018). The benefits of SCFA are but not limited to suppression of inflammatory T-cells, substrate for intestinal epithelial cells, reduction in pathogenic bacterial growth and as substrates for lipogenesis and gluconeogenesis in body organs (Abdul Rahim et al. 2019; den Besten et al. 2013; LeBlanc et al. 2013; Luu et al. 2018). Bacterial population in any ecosystem is extremely competitive for the metabolites, and preference to the same type of substrate increases this competition further. However, the human gut minimizes the competitive pressure for the substrate between the gut microbiota members. A recent study has shown that a complex diet containing a variety of dietary fibers is associated with increased bacterial diversity in the human gut (Holscher 2017). Moreover, gut microbiota evolved in such way that to avoid competitive pressures among themselves, there is an organized "food chain", by which species or group of microbiota members have specific roles to play. For example, *B. vulgatus*, cannot degrade human MUC2 protein *in vivo* but when co-cultured with mucolytic bacteria *A. muciniphila* the overall degradation of MUC2 found to be increased (Png, Lind, et al. 2010). It suggests a small food chain where *A. muciniphila* initiates the mucin degradation and it is further degraded by *B. vulgatus*.

Mucin is secreted by goblet cells and it is essential for the luminal protection of the gastrointestinal tract (Derrien et al. 2010). The main component of the mucus layer is its polysaccharide content and that is main substrate for gut bacteria. Most of the bacteria present in the outer mucus layer are possesses a mucolytic enzyme, necessary for the degradation of mucin carbohydrates (Koropatkin, Cameron, and Martens 2014; Tailford et al. 2015), For example, anaerobic bacterial species, such as *Bacteroides thetaiotaomicron, Akkermansia muciniphila, Bacteroides fragilis, Ruminococcus gnavus*, and *Ruminococcus torques* possesses mucolytic enzymes (Png, Lind, et al. 2010; Tailford et al. 2015). These bacteria make use of their mucolytic activities to degrade glycan and release mono saccharides or di-saccharides attached to the mucins. Moreover, species such as *B. thetaiotaomicron,* possess a range of different glycosidase, those are efficient for the degradation of different types of glycans (Derrien et al. 2010; Microbiology 1991; Tsai et al. 1992). However, complete the degradation of mucins, requires a combination of enzymatic activity of several microbiota members. This coordination among the gut microbiota members leads to form a food chain among them and shapes the individuals' microbiota composition.

A. muciniphila is one of the mucolytic specialists, its genome has numerous genes whose products are devoted for mucin degradation. Transcriptome and proteome studies comparing the gene expression of *A. muciniphila* grown on mucin or the non-mucin sugar also supported the mucolytic nature of this bacterium (Desai et al. 2016; Shin et al. 2019). Based on these pieces of information one can speculate that the human GI tract has a favorable oxygen gradient, an anchor for colonization and nutrition for the microbiota and makes it an ideal ecosystem for the growth of anaerobes.

Distribution of Anaerobic Microbes across the gastrointestinal tract

The oral cavity contains approximately 200 different bacterial species. It is one of the suitable habitats for a diverse microbial population including facultative and strict anaerobes. Next generation sequencing results of vaginally born infant's oral samples have found *Lactobacillus, Prevotella* and *Gardnerella* as abundant genera. Whereas, *Petrimonas, Bacteroides, Desulfovibrio, Pseudomonas, Staphylococcus, Tepidmicrobiu* and *Bifidobacterium* were dominant bacteria in the oral samples of cesarean section born infants (Li et al. 2018). These results suggests that anaerobes, colonize the human oral cavity during and after the birth and later on colonize the gut. A healthy individual harbors a unique consortium microflora colonizes the gingival sulcus in a specific site of the oral cavity and form dental plaque. Based on previous studies, a core dental microbiome is comprised of *Streptococcus, Actinomyces, Veillonella, Granulicatella, Neisseria, Haemophilus, Corynebacterium, Rothia, Prevotella, Capnocytophaga, Porphyromonas, and Fusobacterium spp.* (Figure 1) (Do, Devine, and Marsh 2013; Kolenbrander et al. 2006; Könönen, Asikainen, and Jousimies Somer 1992; Larsen and Fiehn 2017; Li et al. 2018; Marsh, Moter, and Devine 2011). Like normal gut microbiota, healthy oral microflora is also beneficial for the host. Followed by oral cavity and esophagus stomach is the part of digestive system. It was believed that its acidic secretion makes it almost sterile. However, the discovery of *H. pylori*, changed that view, stomach too supports the bacterial community. Microbial density of the stomach is 10^1–10^3 CFU/g, (Colony Forming Unit) and more than one anaerobic genera have been cultured from stomach (Hunt et al. 2015). However, there is high possibility that those cultured taxa of stomach may reflect transient passage rather than colonization.

The small intestine is also a harsh environment for the survival of microbes due to the short transit time and influx of digestive enzymes and bile (Hunt et al. 2015; Kastl et al. 2020). As a result, the bacterial populations in this region is less diverse. Compared with the colon the small intestine is elative inaccessible for sampling. Microbiome analysis of fecal samples represents only the end of the colon and leaving the small intestinal microbiome remained unexplored.

Thus, studies of the human small intestinal microbiota relies on the biopsy samples, sampled during endoscopy (Bodkhe et al. 2019). The bacterial populations increases along the length of intestinal tract, it is approximately 10^{4-5} CFU/mL in the duodenum, whereas in the distal ileum it is approximately 10^{7-8} CFU/mL. This population is mixed of facultative anaerobic and strict anaerobic, species. *Lactobacillus, Clostridium, Staphylococcus, Streptococcus,* and *Bacteroides* are among the most commonly found microbial residents of small intestine (Kastl et al. 2020). The highest number of bacteria is present in the large intestine in comparison to other parts of human body, the microbial counts exceed 10^{11} CFU/g of colonic contents in healthy individuals. The main types of bacteria in the large intestine belongs to *Bacteroidetes, Firmicutes, Proteobacteria, Verrucomicrobia* and *Actinobacteria* and comprise almost 90% of the gut species. These species are obligate anaerobes belonging to over 30 identified genera. The abundant bacterial genera in the colon are *Bacteroides, Lactobacillus, Bifidobacterium, Faecalibacterium, Akkermansia* and *Prevotella (Malmuthuge, Griebel, and Guan 2014; Rinninella et al. 2019).*

Function of anaerobic microflora in human

In past decade, a tremendous amount of research evidence has strongly suggested and demonstrated the vital role of the human gut microbiota in human health via several mechanisms. The microbiota has the potential to increase energy extraction from food mainly from dietary carbohydrates. As mentioned before human body does possess the genes for degradation of complex carbohydrates, thus organisms such as *Bacteroides, Roseburia, Bifidobacterium,* and *Fecalibacterium* ferment these complex sugars and provides SCFA to the host (Table 1) (Holscher 2017; Rivière et al. 2016). *Bacteroides thetaiotaomicron* a resident of human gut, stands the best example for the carbohydrate degradation. This bacterium codes more than 260 hydrolases, that is more than the number of genes coded human genome for carbohydrate degradation. Moreover, *B. thetaiotaomicron,* also assist in the lipid hydrolysis by up regulating expression of a colipase (Hooper et al. 2001; Zocco et al. 2007). Carbohydrate fermentation produces oxalate in the gut that in turn pose the risk of formation oxalate stone in the kidney. However, organisms such as *Oxalobacter formigenes, Lactobacillus* species, and *Bifidobacterium* degrade the oxalate and minimizes the risk of kidney stone (Table 1) (Lieske 2017).

The gut microbiota is also enriched with proteinases as well as efficient transporter system to facilitate amino acid entry from the intestinal lumen into the bacteria. Proteins such as gluten which remained undigested by the small intestine are degraded by glutenases of gut bacteria (Caminero et al. 2016; Rizzello et al. 2007). Apart from degradation, synthesis of essential metabolites such as synthesis of vitamin K and B group vitamins is another important function of the gut microbiota.. Genome hunting of 256 common gut microflora members suggested the presence of genes for the synthesis of vitamin, including biotin, cobalamin, folate, niacin, pantothenate, pyridoxine, riboflavin, and thiamin (Magnúsdóttir et al. 2015). The analysis observed that, some genomes contained all eight pathways, while others none. The most commonly synthesized vitamins by gut bacteria were riboflavin and niacin and almost all microbes belonging to Bacteroidetes, Proteobacteria and Fusobacteria possessed the potential for vitamin B biosynthesis. In addition, to nutrient metabolism a large proportion of gut microflora is involved in xenobiotic and drug metabolism.

Table 1: Function of abundant gut microbiota members

Akkermansia muciniphila Bacteroides, Roseburia, Bifidobacterium, and Fecalibacterium	Ability of this microbe to digest a broad array of complex carbohydrates provides SCFA to the host (Holscher 2017; Rivière et al. 2016). Previous studies have found a negative correlation between abundance of *Akkermansia muciniphila* and metabolic disorders such as overweight, obesity and type 2 diabetes mellitus (Depommier et al. 2019) This bacterium possesses ability to SCFA, in turn this metabolite has provides multiple benefits to human host. SCFA act as energy source for colonocytes (Den Besten et al. 2013). SCFAs, possess anti-inflammatory activity, they inhibit stimuli-induced chemokine production and expression of adhesion molecules on cells, ultimately suppress recruitment of macrophages and neutrophil (Vinolo et al. 2011).
Prevotella histicola	P. histicola possesses immune modulatory and anti-inflammatory properties. In experimental mouse model of arthritis this bacteria suppressed the inflammatory arthritis (Marietta et al. 2016). Moreover, P. histicola, also showed promising results in demyelinating disease (Mangalam et al. 2017b)
Bifidobacteria	Species belonging to *Bifidobacteria* are widely used as probiotics. Consumption of *Bifidobacteria* is effective in improvement of intestinal in metabolic problems such as constipation enteric and hepatic disorders (Mitsuoka 1990) Previous study suggested that it resist the colonization of enteric pathogens (Romond et al. 1997).
Oxalobacter formigenes, Bifidobacterium and *Lactobacillus species*	These gut microbiota members taxa degrade the oxalate and minimizes the risk of kidney stone (Abratt and Reid 2010)

The intestinal mucus serves as the first line of defense, keeps the gut microbiota separated from the gut epithelial cells (Desai et al. 2016). Earlier studies have been demonstrated that the commensals modulate gut mucus layer and the inner mucus layer in germ free mice is penetrable (Johansson et al. 2015). Moreover, intestinal lining sense the bacterial products like SCFAs, peptidoglycan to promote and strengthen mucosal barrier with the production of mucin and gastrointestinal peptide molecules (Hansson 2012; Kim, Park, and Kim 2014). These evidences indicate that gut bacteria strengthens the mucus barrier and protects its host against foreign pathogens.

The influence of gut microbiota on health through the regulation of immune system has emerged as an interesting research area of clinical importance. Interestingly, the use of germ-free and mono-colonized animal models provided us with the platform to study the role of certain gut microbiota in immunity development. For example, Recently, a study revealed that the association of polysaccharide A molecules *of Bacteroides fragilis* to induce the development of a systemic Th1 response (Mazmanian et al. 2005). Moreover, germ free animal model also helped us to understand that gut microbiota is essential for the function of macrophages. Activities such as chemotaxis, phagocytosis and microbicidal activities have been shown to be compromised in peritoneal macrophages of germ free mice (Mitsuyama et al. 1986; Mørland and Midtvedt 1984).

Gut microbiota plays an important role in development and differentiation of immunological cells. In germ free mice model significant decrease in the number of CD4$^+$ cells was observed in comparison to wild type mice (Macpherson, Martinic, and Harris 2002). Similarly, role of gut commensals have been demonstrated in the IL-22$^+$NKp46$^+$ cell differentiation (Sanos et al. 2009). External factors such as drugs, antibiotics, alcohol might lead to the dysbiotic microbiome state. In result, dysbiosis of microbiome may affect the development of intestinal as well as systemic autoimmune diseases. Indeed, microbiome dysbiosis found to be associated with several diseases (Bodkhe, Balakrishnan, and Taneja 2019; Cheng et al. 2013; Geng et al. 2013; Png, Lindén, et al. 2010; Weir et al. 2013). These pieces of evidences suggest that how important the gut microbiota members are for the development and homeostasis of the immune system.

Methods for studying anaerobic gut microbial population

Several techniques have been used to study the gut microbiota, and these methods have undergone a change or advancement over time as the advancement in the microbiological science. There are culture dependent and culture independent approached to study the gut bacteria including the anaerobic microbial population.

Culture-dependent Methods

Culturing the bacteria is the traditional, easy, reliable and cost effective technique to characterize the targeted bacteria. Culturing is followed by morphological and biochemical characteristics using several methods such as Gram's staining and biochemical tests. However to study the anaerobic microbes special efforts need to be taken, since the beginning of the experiment as obligate anaerobes, often do not survive the procedures used for sampling, or for transport. Researchers take special efforts so that the sample would be less exposed to oxygen in order to have a viable count of anaerobes. The buffer used for sampling is flushed with nitrogen to bubble out oxygen and redox indicator is also added to buffer to monitor the oxygen level (Butler, Schoonen, and Rickard 1994). Moreover, culturing, sub-culturing and growth of anaerobes requires special cabinets to create an oxygen-free environment. The development of cabinets for the growth of anaerobes was based on the roll-tube experiment. In 1944 Robert. E. Hungate developed a roll-tube approach that enabled the successful culture of *Clostridium cellobioparus* (HUNGATE 1950; Macy, Snellen, and Hungate 1972). He used tubes with rubber stopper containg boiled culture medium to remove oxygen. Further, through these tubes anoxic gas was passed to bubble-out any remaining oxygen. On the top of that he added reducing agent to scavenge residual traces of oxygen. Later, on based on the principle of roller-tube, anaerobic chamber were developed. Today, a sealed cabinets with attached gloves, filled with different anoxic gases, are routinely used. However, media and anaerobiosis are not the sole requirement for the growth of anaerobes and thus a large proportion of anaerobic gut bacteria are not been cultured yet. Hence, traditional culturing never been sufficient to study the gut anaerobes and thus microbiologist are looking for its new "omics" version known as "culturomics".

Culture Independent methods

To overcome limitations of culture techniques, several molecular approaches were developed by taking advantage of advancement in the technology to study the bacterial diversity using16S rRNA gene. The 16S rRNA gene is present in prokaryotes and it is highly conserved DNA sequence. Interestingly, it consists of conserved as well as of variable regions, the conserved region makes amplification possible using universal primer, while variable regions permits the discrimination between different bacterial taxa (Martinez-Porchas et al. 2017; Willis, Desai, and Laroche 2019). Techniques such as Pulsed field gel electrophoresis, denaturing gradient gel electrophoresis (DDGE) and temperature gradient gel electrophoresis (TGGE) were developed subsequently by taking the advantage of these conserved and variable regions of 16S rRNA gene (Fraher, O'Toole, and Quigley 2012; Sarangi, Goel, and Aggarwal 2019). However, these techniques failed to resolve the closely related bacterial groups and could not detect the rare bacterial taxa.

Advancement in sequencing technology (next generation sequencing technology) replaced these traditional molecular approached and helped to study gut anaerobes with more depth. 16S rRNA gene sequencing is commonly used for bacterial identification, its taxonomic classification and quantitation of culturable and uncultivable microbes present in the samples such as fecal sample (Figure 2) (Johnson et al. 2019). In brief, the protocol involves, PCR amplification of sequencing of 16S rRNA gene (marker gene) followed by parallel sequencing using next generation sequencing platform such Illumina (Quail et al. 2012). The DNA sequences obtained after sequencing are called as "reads". These reads are processed further using bioinformatics workflow such QIIME (Kuczynski et al. 2011), QIIME2 (Bolyen et al. 2019) or DADA2 (Callahan et al. 2016). Basically, these pipelines, discriminate and make clusters of similar reads. Further, these reads are mapped against curated databases such as The Ribosomal Database Project (Cole et al. 2014), GreenGenes (DeSantis et al. 2006), SILVA (Quast et al. 2013) or Human Intestinal 16S rRNA (Edgar 2018) database for its identification (Figure 2). Today, our knowledge about bacterial diversity in the human GI tract has increased vastly with the development of sequencing technology. In the recent past, thousands of the studies are based on this wonderful and high throughput technology. All these studies helped to understand the core human microbiome, association of microbiome with diseases and role of microbiome in the development of immune and nervous system. It is the sequencing technology that made it possible to hunt and harness novel antibiotics and novel compound from the gut microbiota.

Future Prospects

Hundreds of bacterial species make up human gut microbiota and of these, 99% are anaerobic bacteria. However, our understanding of the anaerobes present within and on the human body is limited as most of anaerobes are yet to be cultured. Culturing methods stand in need of improvement (culturomics) in order to study anaerobes in more details. Interest in the microbiome has exploded in recent years as sequencing technology and computational analysis have now made complex gut anaerobes identifiable. However, future work is warranted to understand the biological activity of these anaerobes, rather than simply creating a catalogue of microbial names. Numerous reports have been generated in recent past on the association of perturbed microbiome with diseases; however, whether the microbiome dysbiosis is the cause or the result of the disease is not established in many cases. Thus, the use of germ-free and mono-colonized mice should accompany the research in addition to showing the association between microbiome and diseases.

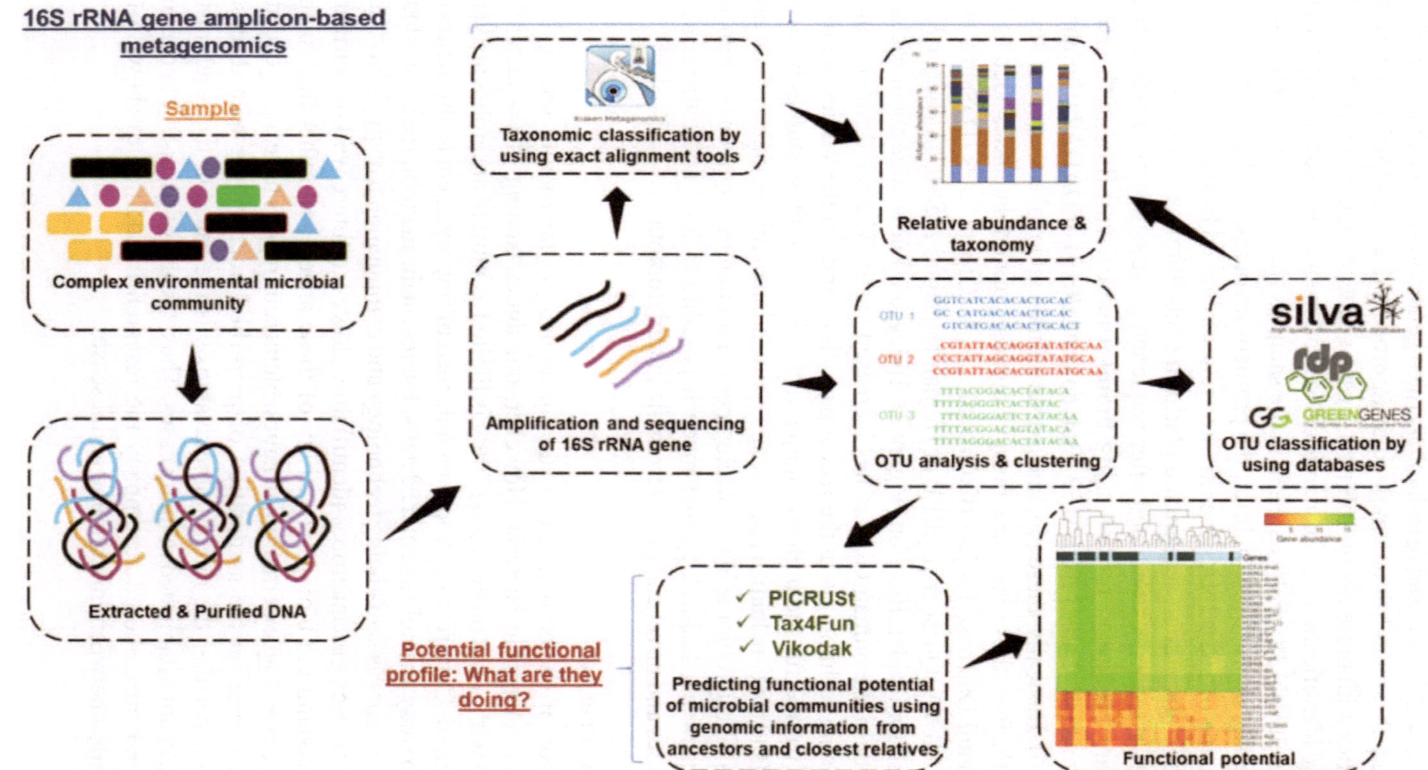

Fig. 2: Workflow for 16S rRNA gene based microbiome analysis using Next-generation sequencing (NGS) platform

Acknowledgement

We are thankful to Dr. Om Prakash Sharma (National Centre for Microbial Resource, National Centre for Cell Science, Pune, India) for his valuable suggestions.

References

———. 2017b. "Human Gut-Derived Commensal Bacteria Suppress CNS Inflammatory and Demyelinating Disease." *Cell Reports* 20(6): 1269–77.

Abdul Rahim, Mohd Badrin Hanizam et al. 2019. "Diet-Induced Metabolic Changes of the Human Gut Microbiome: Importance of Short-Chain Fatty Acids, Methylamines and Indoles." *Acta Diabetologica* 56(5): 493–500.

Abratt, Valerie R., and Sharon J. Reid. 2010. "Oxalate-Degrading Bacteria of the Human Gut as Probiotics in the Management of Kidney Stone Disease." In *Advances in Applied Microbiology*, Academic Press Inc., 63–87. http://www.ncbi.nlm.nih.gov/pubmed/20602988 (February 8, 2020).

Bellali, Sara, Jean-Christophe Lagier, Didier Raoult, and Jacques Bou Khalil. 2019. "Among Live and Dead Bacteria, the Optimization of Sample Collection and Processing Remains Essential in Recovering Gut Microbiota Components." *Frontiers in Microbiology* 10(July): 1–9.

den Besten, Gijs et al. 2013. "The Role of Short-Chain Fatty Acids in the Interplay between Diet, Gut Microbiota, and Host Energy Metabolism." *Journal of Lipid Research* 54(9): 2325–40. http://www.jlr.org/lookup/doi/10.1194/jlr.R036012.

Bodkhe, Rahul et al. 2019. "Comparison of Small Gut and Whole Gut Microbiota of First-Degree Relatives with Adult Celiac Disease Patients and Controls." *Frontiers in Microbiology* 10(feburay): 137–40.

Bodkhe, Rahul, Baskar Balakrishnan, and Veena Taneja. 2019. "The Role of Microbiome in Rheumatoid Arthritis Treatment." : 1–16.

Boekhorst, Jos, Quinta Helmer, Michiel Kleerebezem, and Roland J. Siezen. 2006. "Comparative Analysis of Proteins with a Mucus-Binding Domain Found Exclusively in Lactic Acid Bacteria." *Microbiology* 152(1): 273–80.

Bolyen, Evan et al. 2019. "Reproducible, Interactive, Scalable and Extensible Microbiome Data Science Using QIIME 2." *Nature Biotechnology* 37(8): 852–57.

Butler, Ian B., Martin A.A. Schoonen, and David T. Rickard. 1994. "Removal of Dissolved Oxygen from Water: A Comparison of Four Common Techniques." *Talanta* 41(2): 211–15.

Callahan, Benjamin J. et al. 2016. "DADA2: High-Resolution Sample Inference from Illumina Amplicon Data." *Nature Methods* 13(7): 581–83.

Caminero, Alberto et al. 2016. "Duodenal Bacteria From Patients With Celiac Disease and Healthy Subjects Distinctly Affect Gluten Breakdown and Immunogenicity." *Gastroenterology* 151(4): 670–83.

Chaffanel, Fanny et al. 2018. "Surface Proteins Involved in the Adhesion of Streptococcus Salivarius to Human Intestinal Epithelial Cells." *Applied Microbiology and Biotechnology* 102(6): 2851–65.

Cheng, Jing et al. 2013. "Duodenal Microbiota Composition and Mucosal Homeostasis in Pediatric Celiac Disease." *BMC Gastroenterology* 13(1): 113. http://bmcgastroenterol. biomedcentral.com/articles/10.1186/1471-230X-13-113 (August 6, 2017).

Cole, James R. et al. 2014. "Ribosomal Database Project: Data and Tools for High Throughput RRNA Analysis." *Nucleic Acids Research* 42(D1).

Corfield, Anthony P. 2018. "The Interaction of the Gut Microbiota with the Mucus Barrier in Health and Disease in Human." *Microorganisms* 6(3): 78.

Corfield, Anthony P et al. 1992. "Mucin Degradation in the Human Colon: Production of Sialidase, Sialate O-Acetylesterase, N-Acetylneuraminate Lyase, Arylesterase, and Glycosulfatase Activities by Strains of Fecal Bacteria." *Infection and Immunity* 60(10): 3971–78.

Davenport, Emily R. et al. 2017. "The Human Microbiome in Evolution." *BMC Biology* 15(1).

Depommier, Clara et al. 2019. "Supplementation with Akkermansia Muciniphila in Overweight and Obese Human Volunteers: A Proof-of-Concept Exploratory Study." *Nature Medicine* 25(7): 1096–1103.

Derrien, Muriel et al. 2010. "Mucin-Bacterial Interactions in the Human Oral Cavity and Digestive Tract." *Gut Microbes* 1(4): 254–68.

Desai, Mahesh S. et al. 2016. "A Dietary Fiber-Deprived Gut Microbiota Degrades the Colonic Mucus Barrier and Enhances Pathogen Susceptibility." *Cell* 167(5): 1339-1353.e21.

DeSantis, T. Z. et al. 2006. "Greengenes, a Chimera-Checked 16S RRNA Gene Database and Workbench Compatible with ARB." *Applied and Environmental Microbiology* 72(7): 5069–72.

Do, Thuy, Deirdre Devine, and Philip D. Marsh. 2013. "Oral Biofilms: Molecular Analysis, Challenges, and Future Prospects in Dental Diagnostics." *Clinical, Cosmetic and Investigational Dentistry* 5: 11–19.

Dominguez-Bello, Maria G et al. 2010. "Delivery Mode Shapes the Acquisition and Structure of the Initial Microbiota across Multiple Body Habitats in Newborns." *Proceedings of the National Academy of Sciences of the United States of America* 107(26): 11971–75.

Edgar, Robert C. 2018. "Updating the 97% Identity Threshold for 16S Ribosomal RNA OTUs" ed. Alfonso Valencia. *Bioinformatics* 34(14): 2371–75. https://academic.oup.com/bioinformatics/article/34/14/2371/4913809 (January 29, 2020).

Fraher, Marianne H., Paul W. O'Toole, and Eamonn M.M. Quigley. 2012. "Techniques Used to Characterize the Gut Microbiota: A Guide for the Clinician." *Nature Reviews Gastroenterology and Hepatology* 9(6): 312–22.

Friedman, Elliot S. et al. 2018. "Microbes vs. Chemistry in the Origin of the Anaerobic Gut Lumen." *Proceedings of the National Academy of Sciences of the United States of America* 115(16): 4170–75.

Geng, Jiawei et al. 2013. "Diversified Pattern of the Human Colorectal Cancer Microbiome." *Gut Pathogens* 5(1): 1. Gut Pathogens.

Gerritsen, Jacoline, Hauke Smidt, Ger T. Rijkers, and Willem M. De Vos. 2011. "Intestinal Microbiota in Human Health and Disease: The Impact of Probiotics." *Genes and Nutrition* 6(3): 209–40.

González-Rodríguez, Irene et al. 2012. "Role of Extracellular Transaldolase from Bifidobacterium Bifidum in Mucin Adhesion and Aggregation." *Applied and Environmental Microbiology* 78(11): 3992–98.

Hansson, Gunnar C. 2012. "Role of Mucus Layers in Gut Infection and Inflammation." *Current Opinion in Microbiology* 15(1): 57–62.

Hillman, Ethan T., Hang Lu, Tianming Yao, and Cindy H. Nakatsu. 2017. "Microbial Ecology along the Gastrointestinal Tract." *Microbes and Environments* 32(4): 300–313.

Holscher, Hannah D. 2017. "Dietary Fiber and Prebiotics and the Gastrointestinal Microbiota." *Gut Microbes* 8(2): 172–84.

Hooper, L. V. et al. 2001. "Molecular Analysis of Commensal Host-Microbial Relationships in the Intestine." *Science* 291(5505): 881–84.

Hungate, R. E. 1950. "The Anaerobic Mesophilic Cellulolytic Bacteria." *Bacteriological reviews* 14(1): 1–49.

Hunt, R. H. et al. 2015. "The Stomach in Health and Disease." *Gut* 64(10): 1650–68.

Johansson, Malin E.V. et al. 2015. "Normalization of Host Intestinal Mucus Layers Requires Long-Term Microbial Colonization." *Cell Host and Microbe* 18(5): 582–92.

Johnson, Jethro S. et al. 2019. "Evaluation of 16S RRNA Gene Sequencing for Species and Strain-Level Microbiome Analysis." *Nature Communications* 10(1).

Kastl, Arthur J., Natalie A. Terry, Gary D. Wu, and Lindsey G. Albenberg. 2020. "The Structure and Function of the Human Small Intestinal Microbiota: Current Understanding and Future Directions." *CMGH* 9(1): 33–45.

Kim, Chang H., Jeongho Park, and Myunghoo Kim. 2014. "Gut Microbiota-Derived Short-Chain Fatty Acids, T Cells, and Inflammation." *Immune Network* 14(6): 277.

Kolenbrander, Paul E. et al. 2006. "Bacterial Interactions and Successions during Plaque Development." *Periodontology 2000* 42(1): 47–79.

Könönen, E., S. Asikainen, and H. Jousimies Somer. 1992. "The Early Colonization of Gram negative Anaerobic Bacteria in Edentulous Infants." *Oral Microbiology and Immunology* 7(1): 28–31. http://doi.wiley.com/10.1111/j.1399-302X.1992.tb00016.x (January 27, 2020).

Koropatkin, Nicole M, Elizabeth a Cameron, and Eric C Martens. 2014. "How Glycans Shape." 10(5): 323–35.

Kuczynski, Justin et al. 2011. "Using QIIME to Analyze 16S Rrna Gene Sequences from Microbial Communities." *Current Protocols in Bioinformatics* (SUPPL.36).

Lackey, Kimberly A. et al. 2019. "What's Normal? Microbiomes in Human Milk and Infant Feces Are Related to Each Other but Vary Geographically: The Inspire Study." *Frontiers in Nutrition* 6.

Larsen, Tove, and Nils Erik Fiehn. 2017. "Dental Biofilm Infections – an Update." *APMIS* 125(4): 376–84.

LeBlanc, Jean Guy et al. 2013. "Bacteria as Vitamin Suppliers to Their Host: A Gut Microbiota Perspective." *Current Opinion in Biotechnology* 24(2): 160–68.

Li, Hongping et al. 2018. "The Impacts of Delivery Mode on Infant's Oral Microflora." *Scientific Reports* 8(1).

Lieske, John C. 2017. "Probiotics for Prevention of Urinary Stones." *Annals of Translational Medicine* 5(2).

Liu, Yu et al. 2019. "The Perturbation of Infant Gut Microbiota Caused by Cesarean Delivery Is Partially Restored by Exclusive Breastfeeding." *Frontiers in Microbiology* 10(MAR). https://www.frontiersin.org/article/10.3389/fmicb.2019.00598/full (January 27, 2020).

Luu, Maik et al. 2018. "Regulation of the Effector Function of CD8+ T Cells by Gut Microbiota-Derived Metabolite Butyrate." *Scientific Reports* 8(1): 1–10. http://dx.doi.org/10.1038/s41598-018-32860-x.

Macpherson, A. J., M. M. Martinic, and N. Harris. 2002. "The Functions of Mucosal T Cells in Containing the Indigenous Commensal Flora of the Intestine." *Cellular and Molecular Life Sciences* 59(12): 2088–96.

Macy, J. M., J. E. Snellen, and R. E. Hungate. 1972. "Use of Syringe Methods for Anaerobiosis." *American Journal of Clinical Nutrition* 25(12): 1318–23.

Magnúsdóttir, Stefanía, Dmitry Ravcheev, Valérie De Crécy-Lagard, and Ines Thiele. 2015. "Systematic Genome Assessment of B-Vitamin Biosynthesis Suggests Cooperation among Gut Microbes." *Frontiers in Genetics* 6(MAR).

Malmuthuge, Nilusha, Philip J. Griebel, and Le Luo Guan. 2014. "Taxonomic Identification of Commensal Bacteria Associated with the Mucosa and Digesta throughout the Gastrointestinal Tracts of Preweaned Calves." *Applied and Environmental Microbiology* 80(6): 2021–28.

Mangalam, A., Shahi, S. K., Luckey, D., Karau, M., Marietta, E., Luo, N., & Rodriguez, M. 2017. Human gut-derived commensal bacteria suppress CNS inflammatory and demyelinating disease. *Cell reports*, *20*(6), 1269-1277.

Marietta, Eric V. et al. 2016. "Suppression of Inflammatory Arthritis by Human Gut-Derived Prevotella Histicola in Humanized Mice." *Arthritis and Rheumatology* 68(12): 2878–88.

Marsh, Philip D., Annette Moter, and Deirdre A. Devine. 2011. "Dental Plaque Biofilms: Communities, Conflict and Control." *Periodontology 2000* 55(1): 16–35.

Martinez-Porchas, Marcel, Enrique Villalpando-Canchola, Luis Enrique Ortiz Suarez, and Francisco Vargas-Albores. 2017. "How Conserved Are the Conserved 16S-RRNA Regions?" *PeerJ* 2017(2).

Mazmanian, Sarkis K., Hua Liu Cui, Arthur O. Tzianabos, and Dennis L. Kasper. 2005. "An Immunomodulatory Molecule of Symbiotic Bacteria Directs Maturation of the Host Immune System." *Cell* 122(1): 107–18.

Microbiology, Medical. 1991. "Production of Mucin Degrading Sulphatase and Glycosidases by Bacteroides Thetaiotaomicron." : 97–101.

Milani, Christian et al. 2017. "The First Microbial Colonizers of the Human Gut: Composition, Activities, and Health Implications of the Infant Gut Microbiota." *Microbiology and Molecular Biology Reviews* 81(4).

Mitsuoka, Tomotari. 1990. "Bifidobacteria and Their Role in Human Health." *Journal of Industrial Microbiology* 6(4): 263–67.

Mitsuyama, M et al. 1986. "Ontogeny of Macrophage Function to Release Superoxide Anion in Conventional and Germfree Mice." *Infection and immunity* 52(1): 236–39. http://www.ncbi.nlm.nih.gov/pubmed/3007361 (January 28, 2020).

Moore, W. E.C., Elizabeth P. Cato, and Lillian V. Holdeman. 1969. "Anaerobic Bacteria of the Gastrointestinal Flora and Their Occurrence in Clinical Infections." *Journal of Infectious Diseases* 119(6): 641–49. https://www.jstor.org/stable/30102345 (January 27, 2020).

Mørland, B, and T Midtvedt. 1984. "Phagocytosis, Peritoneal Influx, and Enzyme Activities in Peritoneal Macrophages from Germfree, Conventional, and Ex-Germfree Mice." *Infection and immunity* 44(3): 750–52. http://www.ncbi.nlm.nih.gov/pubmed/6233226 (January 28, 2020).

Mueller, Noel T. et al. 2015. "The Infant Microbiome Development: Mom Matters." *Trends in Molecular Medicine* 21(2): 109–17.

Neu, Josef, and Jona Rushing. 2011. "Cesarean Versus Vaginal Delivery: Long-Term Infant Outcomes and the Hygiene Hypothesis." *Clinics in Perinatology* 38(2): 321–31.

Pannaraj, Pia S. et al. 2017. "Association between Breast Milk Bacterial Communities and Establishment and Development of the Infant Gut Microbiome." *JAMA Pediatrics* 171(7): 647–54.

Pärnänen, Katariina et al. 2018. "Maternal Gut and Breast Milk Microbiota Affect Infant Gut Antibiotic Resistome and Mobile Genetic Elements." *Nature Communications* 9(1).

Perez-Muñoz, Maria Elisa, Marie Claire Arrieta, Amanda E. Ramer-Tait, and Jens Walter. 2017. "A Critical Assessment of the 'Sterile Womb' and 'in Utero Colonization' Hypotheses: Implications for Research on the Pioneer Infant Microbiome." *Microbiome* 5(1).

Png, Chin Wen, Sara K. Lindén, et al. 2010. "Mucolytic Bacteria with Increased Prevalence in IBD Mucosa Augment in Vitro Utilization of Mucin by Other Bacteria." *American Journal of Gastroenterology* 105(11): 2420–28.

Png, Chin Wen, Sara K Lind, et al. 2010. "Mucolytic Bacteria With Increased Prevalence in IBD Mucosa Augment In Vitro Utilization of Mucin by Other Bacteria." *Am J Gastroenterol 2010;* 105(June): 2420–28.

Quail, Michael A. et al. 2012. "A Tale of Three next Generation Sequencing Platforms: Comparison of Ion Torrent, Pacific Biosciences and Illumina MiSeq Sequencers." *BMC Genomics* 13(1): 341. http://bmcgenomics.biomedcentral.com/articles/10.1186/1471-2164-13-341 (January 29, 2020).

Quast, Christian et al. 2013. "The SILVA Ribosomal RNA Gene Database Project: Improved Data Processing and Web-Based Tools." *Nucleic Acids Research* 41(D1).

Ravcheev, Dmitry A, and Ines Thiele. 2014. "Systematic Genomic Analysis Reveals the Complementary Aerobic and Anaerobic Respiration Capacities of the Human Gut Microbiota." 5(December): 1–14.

Rinninella, Emanuele et al. 2019. "What Is the Healthy Gut Microbiota Composition? A Changing Ecosystem across Age, Environment, Diet, and Diseases." *Microorganisms* 7(1): 14.

Rivière, Audrey et al. 2016. "Bifidobacteria and Butyrate-Producing Colon Bacteria: Importance and Strategies for Their Stimulation in the Human Gut." *Frontiers in Microbiology* 7(JUN).

Rizzello, Carlo G. et al. 2007. "Highly Efficient Gluten Degradation by Lactobacilli and Fungal Proteases during Food Processing: New Perspectives for Celiac Disease." *Applied and Environmental Microbiology* 73(14): 4499–4507.

Robertson, Ruairi C., Amee R. Manges, B. Brett Finlay, and Andrew J. Prendergast. 2019. "The Human Microbiome and Child Growth – First 1000 Days and Beyond." *Trends in Microbiology* 27(2): 131–47.

Romond, M. B., Z. Haddou, C. Mielcareck, and C. Romond. 1997. "Bifidobacteria and Human Health: Regulatory Effect of Indigenous Bifidobacteria on *Escherichia Coli* Intestinal Colonization." In *Anaerobe*, Academic Press, 131–36.

Rowland, Ian et al. 2018. "Gut Microbiota Functions: Metabolism of Nutrients and Other Food Components." *European Journal of Nutrition* 57(1).

Sanos, Stephanie L. et al. 2009. "RORãt and Commensal Microflora Are Required for the Differentiation of Mucosal Interleukin 22-Producing NKp46+ Cells." *Nature Immunology* 10(1): 83–91.

Sarangi, Aditya N., Amit Goel, and Rakesh Aggarwal. 2019. "Methods for Studying Gut Microbiota: A Primer for Physicians." *Journal of Clinical and Experimental Hepatology* 9(1): 62–73.

Shi, Yi Chao et al. 2018. "Initial Meconium Microbiome in Chinese Neonates Delivered Naturally or by Cesarean Section." *Scientific Reports* 8(1).

Shin, Woojung et al. 2019. "A Robust Longitudinal Co-Culture of Obligate Anaerobic Gut Microbiome with Human Intestinal Epithelium in an Anoxic-Oxic Interface-on-a-Chip." *Frontiers in Bioengineering and Biotechnology* 7(FEB): 1–13.

Sicard, Jean Félix et al. 2017. "Interactions of Intestinal Bacteria with Components of the Intestinal Mucus." *Frontiers in Cellular and Infection Microbiology* 7(SEP).

Tailford, Louise E, Emmanuelle H Crost, Devon Kavanaugh, and Nathalie Juge. 2015. "Mucin Glycan Foraging in the Human Gut Microbiome." 6(March).

Tsai, H H, D Sunderland, A Hart, and J M Rhodes. 1992. "A Novel Mucin Sulphatase From Human Faeces:Its Identification, Purification and Characterization." *Clinical Science* 82: 447–54. http://www.clinsci.org/content/ppclinsci/82/4/447.full.pdf.

van den Elsen, Lieke W.J., Johan Garssen, Remy Burcelin, and Valerie Verhasselt. 2019. "Shaping the Gut Microbiota by Breastfeeding: The Gateway to Allergy Prevention?" *Frontiers in Pediatrics* 7(FEB).

Vinolo, Marco A.R., Hosana G. Rodrigues, Renato T. Nachbar, and Rui Curi. 2011. "Regulation of Inflammation by Short Chain Fatty Acids." *Nutrients* 3(10): 858–76.

Walker, R. W., J. C. Clemente, I. Peter, and R. J.F. Loos. 2017. "The Prenatal Gut Microbiome: Are We Colonized with Bacteria in Utero?" *Pediatric Obesity* 12: 3–17. http://doi.wiley.com/10.1111/ijpo.12217 (January 27, 2020).

Weir, Tiffany L et al. 2013. "Stool Microbiome and Metabolome Differences between Colorectal Cancer Patients and Healthy Adults." *PLoS ONE* 8(8).

Willis, Ciara, Dhwani Desai, and Julie Laroche. 2019. "Influence of 16S RRNA Variable Region on Perceived Diversity of Marine Microbial Communities of the Northern North Atlantic." *FEMS Microbiology Letters* 366(13).

Yang, Irene et al. 2016. "The Infant Microbiome: Implications for Infant Health and Neurocognitive Development." *Nursing Research* 65(1): 76–88.

Zhang, Na et al. 2017. "Diversity and Characterization of Bacteria Associated with the Deep-Sea Hydrothermal Vent Crab Austinograea Sp . Comparing with Those of Two Shallow-Water Crabs by 16S Ribosomal DNA Analysis." : 1–17.

Zocco, M. A., M. E. Ainora, G. Gasbarrini, and A. Gasbarrini. 2007. "Bacteroides Thetaiotaomicron in the Gut: Molecular Aspects of Their Interaction." *Digestive and Liver Disease* 39(8): 707–12.

16

Biotechnological Applications of Anaerobes

Charu Dogra Rawat[1*] and Manisha Arora Pandit[2]

[1]*Department of Zoology, Ramjas College, University of Delhi, Delhi*
[2]*Department of Zoology, Kalindi College, University of Delhi, Delhi*

Abstract

Anaerobic microorganisms occur in oxygen-free environments as they possess biomolecules and metabolic processes that have adapted to sustain them in the stressed environments. These biomolecules and metabolic processes are harnessed and employed in various biotechnological applications. Anaerobes are abundant in various ecological niches such as freshwater or marine sediments, subsurface aquifers, deep thermal vents etc. where, in the absence or limitation of oxygen, they utilise reduced ions such as nitrate, sulphate, ferric, carbonate and certain organic compounds like degrading organic matter, fatty acids or alcohols as electron acceptors during respiration/fermentation to produce more or fewer reduced products such as, alcohols, ammonia, organic acids, hydrogen and carbon dioxide as the oxidised product. Diversity of substrates that can be fermented and the wide range of products that can be formed make anaerobic processes useful in biotechnological applications. Their substrate versatility and flexibility, for example, they can utilize simple sugars as well as many of them can utilise complex renewable biomass make their biotechnological applications sustainable and environment friendly. Anaerobes have been employed in food, chemical and material industries for the production of food items, preservative, additives, chemical intermediates, solvents, catalysts and polymers. They find widespread usage in energy sector for generation of various kinds of biofuels that can mitigate our reliance on non-renewable sources for fuel production. An emerging aspect of anaerobic biotechnology is waste valorization that involves treatment of any kind of

**Corresponding Author: cdrawat@ramjas.du.ac.in*

waste (industrial, municipal, agricultural, pharmaceutical), where organic fraction of the waste is converted into valuable products. Anaerobes play key role in maintaining the global cycles of carbon, nitrogen, and sulfur, and also facilitate bioremediation by breakdown of persistent compounds. Lately, role of anaerobes in medicine and health care sector is widely explored as they are found to produce several bacteriocins, antibiotics, immunomodulators, vitamins and anti-tumor compounds. Some of the anaerobes have also been employed as probiotics. In the present chapter, we discussed these and many more biotechnological applications of anaerobes in white, green, red and blue biotechnology.

Keywords: Anaerobes, Fermentation, Anaerobic digestion, Bioreactors, Biotechnological applications

Introduction

The existence of anaerobic bacteria was first shown by Antonie van Leeuwenhoek in 1680 when he described an experiment with "anaerobic or semi-anaerobic bacteria". He noted that even in a tightly stoppered vessel containing some moisture or meat these organisms or 'animalcules' as he called them could continue to grow in the absence of air (Goldstein 1995). Two hundred years later, in 1861, Louis Pasteur named these microorganisms as anaerobes and demonstrated that fermentation was carried out by microorganisms and was not a process based on spontaneous chemical reactions.

The process of fermentation has been known to man as early as 7000 B.C. and was utilized to produce alcohol from sugar, to make cheese, to leaven bread using yeast and to preserve milk using lactic acid bacteria (LAB) (Demain 1981). Pasteur not only demonstrated the existence of anaerobes but also discovered the bacteria, *Clostridium butyricum* that produced butyric acid in the absence of air (Goldstein 1995). Later during World War- I, Chaim Weizmann successfully carried out large scale acetone-butanol fermentations of starch with *Clostridium* (Demain 1981). It was thus clear that fermentation technology involved anaerobic microorganisms and this was probably the first demonstration of biotechnological application of anaerobes. With greater advances in science it became possible to study the products of these fermentations, understand the physiology of the microbes involved in fermentation by using pure cultures and use this knowledge to bring about industrial level production of fermentation products in both food and chemical industries (Wolfe 2019).

There are two main types of anaerobic bacteria, aerotolerant and obligate. The aerotolerant anaerobes are unable to respire oxygen (O_2) but their growth is not affected by its presence (at least for a limited time) whereas the obligate anaerobes are incapable of aerobic metabolism and are variably tolerant to oxygen; both carry out fermentation for their energy needs and the terminal electron acceptor is an organic compound instead of oxygen. The obligate anaerobes can be further subdivided into 2 types based on the percentage of oxygen that can prove toxic to them. The first category is the strict obligate anaerobes that do not survive in more than 0.5% oxygen in their environment. The second category of obligates are the moderate obligate anaerobes, that can still grow in a 2% to 8% oxygen environment (Noor & Khetarpal 2019). These bacteria are found in almost all environments like the compact layers of soil, water body sediments and even the gastrointestinal tracts of animals. Obligate anaerobes do well in oxygen deprived environments due to their ability to use other electron acceptors than oxygen. These electron acceptors are mostly reduced compounds like nitrate (NO_3^-), ferric (Fe^{3+}), sulfate (SO_4^{2-}), carbonate (CO_3^{2-}) and certain organic compounds like degrading organic matter, fatty acids or alcohols (Table 1). Some anaerobes can utilize CO, H_2, and CO_2 gases as substrates or products in biotechnological conversions. (Mauerhofer et al. 2019, Aragao 2013). When organic compounds are used as electron donors and acceptors, production of reduced compounds, like alcohols, organic acids, ammonia and hydrogen takes place and carbon dioxide is produced as the oxidized product (Aragao 2013).

Even though the anaerobic bacteria produce much less energy in comparison to the aerobic ones (only 2 ATP molecules as compared to the 36-38 ATP molecules formed by aerobic respiration of 1 mole glucose) this is in fact, an advantage in terms of their application in biotechnology as less carbon is used for cell biomass generation and maintenance and more carbon is now directed towards product formation leading to a higher product yield (Aragao 2013). Anaerobic bacteria are found in nature as part of microbial communities or consortia where they survive by interactions with member species and are capable of modifying their composition and character in accordance to the changes in the environment such as, availability of the terminal oxidant. It is therefore of prime importance to have complete knowledge of their physiology and biochemistry in order to utilize these organisms to their full biotechnological potential. (Morris 1994).

Anaerobic bacteria can utilize variety of substrates and produce wide range of products. The formation of diverse products can negatively affect the downstream production steps, however, with use of metabolic engineering shifting the production towards the required product(s) can be achieved (Aragao 2013). Metabolic engineering modulations keep all the metabolic routes active in an organism as well as is beneficial for the same strain can be used to produce the products of interest by shifting the conditions of the process.

Table 1: Characteristics of different types of anaerobic respiration as against aerobic respiration. (Hatti-Kaul & Mattiasson 2016)

Organisms	Respiration Type	Electron Acceptor	Product	Examples
Obligate & Facultative Aerobes	Aerobic Respiration	O_2	H_2O	Aerobic Prokaryotes and Eukaryotes
Obligate & Facultative Anaerobes	Iron Reduction	Fe(III)	Fe(II)	*Geobacter sp.*, *Shewanella sp.*,
	Dehalogenation	Halogenated aliphatic or aromatic organic compounds, R-X	Halide ions, X^- and dehalogenated compound,R-H	*Trichlorobacter*, *Dehalobacter*
	Sulfur Reduction	S	HS^-	*Desulfuromonadales, Pyrodictium*
Obligate Anaerobes	Sulfate reduction	SO_4^{2-}	HS^-	*Desulfovibrio desulfuricans, Desulfobacter*
Methanogenic archaea	Methanogenesis (carbonate respiration)	CO_2	CH_4	*Methanothrix thermophile, Methanobacterium, Methanococcus*
Homoacetogenic bacteria	Acetogenesis (carbonate respiration)	CO_2	CH_3COO^-	*Acetobacterium woodii. Clostridium aceticum*
Facultative Anaerobes	Nitrate reduction (denitrification)	NO_3^-	NO_2^-, N_2O, N_2	*Paracoccus denitrificans, Pseudomonas stutzeri, Escherichia coli*
	Fumarate respiration	Fumarate	Succinate	*Escherichia coli, Wolinella succinogenes*

This ability to produce many products by anaerobic processes provides a definitive advantage over aerobic processes particularly in the production of fuels and chemicals when one considers broad substrate utilization, yields and productivity, the cost of investment in equipment, energy supply, implementation and other running costs (Yazdani & Gonzalez 2007, Weusthuis et al. 2011). This can be beneficial in the development of the bio-refineries particularly in developing nations due to lower implementation costs (Davis et al. 2013).

Moreover, most anaerobic bacteria demonstrate a substrate flexibility that is absent in their aerobic counterparts. Like the ability to not only ferment sugars, but also degrade complex structural biomass like cellulose, hemicellulose etc. Due to this ability, they offer the potential to hydrolyse and ferment the complete plant biomass into an integrated process, termed Consolidated Bio-Processing (CBP). Further simplification of various processing steps could provide greater economic advantages on an industrial scale tipping the scales in favour of anaerobic bacteria (Mauerhofer et al. 2018, Podolsky et al. 2019).

Anaerobes, thus, find widespread applications in all sectors of biotechnology (Figure 1) coded majorly by four colours – white or grey biotechnology that indicate applications in industrial sector, green biotechnology that involves applications in agriculture and environment, red biotechnology that deals with applications in health and medicine including diseases and diagnostics and blue biotechnology that includes applications in aquaculture and marine environment (DaSilva & Edgar 2005). In this chapter, we describe some major applications of anaerobes in different biotechnological sectors.

Applications in Industrial (White) Biotechnology

Industrial biotechnology includes sustainable processing and production of chemicals, materials and fuels. Anaerobes have been crucial in the rise of industrial biotechnology. In 1920 the first industrial fermentation process for the production of acetone-butanol-ethanol (ABE) was done using *Clostridium acetobutylicum* by Chaim Weizmann (Wolfe 2019). Later these processes were used for solvent production in the chemical and automobile industries (Kumar & Gayen 2011). These biotechnological processes were later replaced by oil refining that could supply these solvents in much larger quantities.

With the world facing a fossil fuel crisis at the beginning of the 21st century, the focus has again moved to replacing fossil fuel derived products with renewable options. This has brought about a renewed interest in biological processes and initiated advances in industrial biotechnology in recent times. A good example is utilizing the fermentation of corn starch as the source of maximum amount of bioethanol production in the U.S. (Lawson 2019).

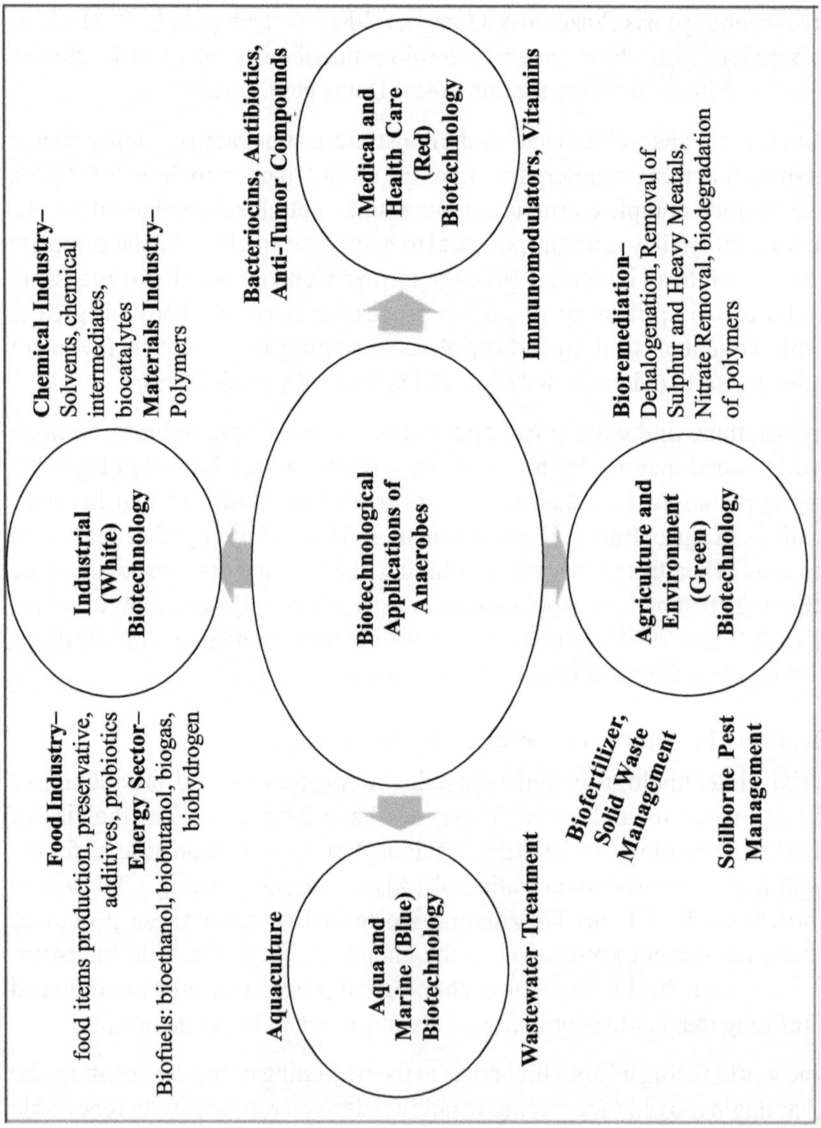

Fig. 1: Biotechnological Applications of Anaerobes

The chemical industry now comes armed with new advances utilizing genetically modified strains of anaerobic microbes and a better understanding of their metabolic routes so as to maximise production by the industry. One such example is the large-scale production of 1, 3-propanediol (Bio-PDO™) produced commercially from a genetically modified strain (Weusthuis et al. 2011).

Focus is also on the use of traditional food-fermenting microorganisms for the chemical industry like the use of lactic acid from *Lactobacillus bulgaricus* for cosmetic and pharmaceutical formulations and also in polymer production (Datta et al. 1995). Moreover, anaerobic consortia are also a rich source of both microorganisms and genes for the production of varied products in the future.

In depth studies of anaerobes living in extreme environments like oil and gas pipelines are also being carried out in order to control biocorrosion in the pipelines caused by reservoir microorganisms that results in an annual loss of billions of dollars. Emphasis is on identifying the causative organisms from the consortia and on their biochemical processes so as better mitigate such deleterious microbial processes (Johnson et al. 2010).

A lot of organic matter in landfills comes from discarded food waste. This food waste when subjected to anaerobic digestion results in the formation of biogas which is ~60% methane and a liquid sludge called digestate. This technology is widely being adopted as a waste management tool. It also reduces the cost of taking the food waste to the landfills while on the other hand it generates revenue by producing renewable energy from biogas and the use of digestate as fertilizer (Johnson et al. 2010).

Some of these industrial applications of anaerobic bacteria/treatment are elaborated below with a mention of the recent advances towards making the applications cost-effective, sustainable, cleaner, greener and more productive.

Applications in Food industry – Production of food items, preservatives additives and probiotics

For many years anaerobic microorganisms have been used in fermentation processes for production of various food items (Goldstein 1995). Many important industrial fermentations, such as ethanol production by yeasts, propionic acid, lactic acid and acetic acid production by bacteria are used to produce food items - beer, wine, bread, cheese, vinegar (Müller 2008). Ethanol fermentation by yeasts is an ancient process to produce alcoholic beverages. Ethanol is also the byproduct of many fermentations carried out by many lactic acid bacteria, enterobacteria and clostridia. Beer (produced from malted grains) and wine (produced from fruits) are the most important alcohol beverages. Distillation of the alcohol produces various spirits. For example, distillation of malt brews yields whisky and distillation of fermented grain or potato yield vodka (Bathgate 2016; Wiœniewska et al. 2015). Homofermentative lactic acid bacteria (LAB), such as *Lactobacillus delbrueckii*, produce lactic acid as the end product of fermentation (Bernardeau et al. 2006). Due to an acid being produced, the pH of their environment is lowered, which inhibits growth of other organisms, making lactic acid fermentation to be employed in the preservation of food. LAB are also employed in the production of cheese, pickles, sour dough and some types of sausages. *Propionibacterium spp.* can produce propionate from lactate and employed in the production of Swiss cheese (Rabah et al. 2017). These *Propionibacterium* species have also been used to enhance flavour in other fermented foods (Falentin et al. 2010). Further, they have been supplemented for vitamin B12 fortification and enhanced preservation through fatty acid production. Dairy propionibacteria are also gaining attention as potent probiotics –"a live microorganism which, when administrated in adequate amount, confers a health benefit on the host" (FAO, 2006). *Bifidobacterium* spp. live cultures have also been developed as whole cell probiotics and anaerobic bacteria formulations have been broadly commercialized as probiotic additives for animal feeds (Forssten et al. 2011).

Applications in Chemical & Materials Industry – Production of solvents chemical intermediates, catalysts and biopolymers

Anaerobic fermentation is the basis of production of alcohols (1, 3 – propanediol - 1,3-PD, ethanol and butanol), ketones (acetone) and organic acids (acetic acid, lactic acid, butyric acid) that are used as solvents or chemical feedstocks in the chemical industries. Naturally, solventogenic clostridia, such as *C. acetobutylicum, C. beijerinckii* and *C. pasteurianum* ferment monosaccharides, including hexose (for example, glucose) and pentose (for example, xylose) to produce butanol, acetone, and ethanol (with a mass ratio of 6:3:1) (Gu et al. 2011). *Clostridium butyricum* can produce 1, 3-PD via the fermentation of glycerol (Willke & Vorlop, 2008). Ethanol is formed as by-product by *C. butyricum* when grown in glycerol as sole carbon source. The ABE (acetone-butanol-ethanol) fermentation typically includes two phases: formation of acids (acidogenic phase) and their re-assimilation for further solvent formation during the solventogenic phase (Straathof, 2013). Anaerobic digestion of the biomass resource results in the production of biogas consisting of methane, which, in addition to its use as a fuel, is an established source of downstream chemical products (Clark et al. 2015). Methane, by steam reforming, splits into syngas, comprising of carbon monoxide and hydrogen, and provide a valuable platform for generation of many other chemicals that serve as solvents or many other chemical intermediates (Subramani et al. 2010; Clark et al. 2015).

Obligate anaerobes have a great variety of enzymes that can be used as catalysts of reactions that would be difficult to achieve by chemical means. They have naturally occurring hydrolytic enzymes that aid in degrading polymeric molecules. In pathogenic species they are used to penetrate the host. In biotechnological respect, the enzymes possessed by non-pathogenic and thermophilic species carry importance. These include amylases, pullulanases, glucosidases, cellulases, pectinases, chitinases, proteases, and lipases. It is these enzymes that provide the obligate anaerobes with the ability to carry out a wide range of diverse fermentations and in turn biotransformations (Morris 1994). These enzymes like the NADP-linked secondary alcohol-ketone oxidoreductase from *Thermoanaerobacter brockii* (Lamed & Zeikus 1981), lactate dehydrogenase of *Thermotoga maritima* (Wrba et al. 1990) etc. possess the qualities that could make them commercially viable catalysts.

For the cost-effective production of chemicals from renewable resources, there's always a need for innovative fermentation processes and new strategies for the production of alcohols and solvents (Sheldon, 2011; Luterbacher et al. 2014; Farmer & Mascal 2015). One of the methods employed is to identify or engineer microorganisms with more efficient conversion and a broader substrate uptake, including complex polymeric carbohydrates that are advantageous for the improvement of bioprocessing in the production of such compounds (Cheng et al. 2019).

Lactic acid produced through fermentation of starch and other easily available polysaccharides such as, those from corn, sugar cane, potatoes and other biomasses is the building block for the polymer poly-lactic acid (PLA) (Klotz et al., 2016). PLA due to its high strength antimicrobial, antioxidant properties and biodegradability is a prominent material used in the packaging industry, particularly for packaging of food (Jamshidian et al. 2010). Industrial production of lactic acid is by the homofermentative bacteria belonging to genus *Lactobacillus*, thermotolerant *Bacillus coagulans* and yeast-like *Candida utilis* (Mehta et al. 2005, Taskila & Ojamo 2013). PLA can then be synthesized from lactic acid by direct polycondensation reaction or ring-opening polymerization of the lactide monomer (Pretula et al. 2016). Apart from food packaging, PLA finds multiple application such as in geotextile, agricultural film, 3D printing and in the making of biomedical materials such as, absorbable sutures, and prosthetic devices (Brodin et al. 2017). Another important bio-based polymer is polyhydroxyalkanoates (PHAs). Poly (3-hydroxybutyrate) (PHB) is one of the PHAs produced on large scale for commercial consumption. PHB has been shown to be produced in *Clostridium botulinum*, *C. acetireducens* and recently by the metabolically engineered autotrophic clostridia, *C. coskatii* and *C. ljungdahlii* that utilise syngas as substrate (Emeruwa &

Hawirko, 1973, Girbal et al. 1997b Flüchter et al. 2019). PHA and its copolymers are employed in bulk packaging, advanced medical applications such as sutures, therapeutic devices, cardiovascular stents, depots for controlled drug release or implants and others (Junyu et al. 2018).

Applications in Energy Sector – Production of Biofuels: Bioethanol Biobutanol, Biogas, Biohydrogen.

Independence from using fossil fuels requires a shift towards biomass for bulk production of energy carriers and chemicals. The use of technology to convert biomass into the product of interest requires that the biosynthetic process gives high yield and productivity. Fermentation and anaerobic digestion (mentioned in the previous section) are efficient biochemical conversion processes that hold promise for global bio-energy production. Sugars (chiefly glucose) derived from starch of corn, wheat, and milo or simple sugars from sugarcane, sugar beet and oils derived primarily from soybean have been the initial resources (feedstocks) for the bio-production of industrial fuels (Yang & Yu, 2013). These resources form food and feed and, thus, their use is not sustainable. Emphasis is thereby laid on utilization of agricultural residues (leaves and vegetable wastes), solid wastes generated from industries and households, oil wastes and animal fat produced from food processing industry, edible oil industry, slaughterhouses, dairy industry and olive oil mills, sewage sludge generated during wastewater treatment and other emerging feedstocks such as perennial grass, wood and algae (Appels et al. 2011, García 2016). Waste valorization is a key component of a biorefinery – a set up/facility that integrates biomass conversion equipment and processes to produce fuel, power and value-added chemicals (Cherubini 2010, Hughes et al. 2013).

Bioethanol

Ethanol can be produced from any sugar-containing materials including sugarcane, corn, sugar beet, rice straw, sweet sorghum, etc. Extraction of fermentable sugar is either done directly as from sugarcane, or by enzymatic hydrolysis as from corn, or by thermal, acidic or enzymatic hydrolysis as from cellulosic biomass. Industrial production of ethanol uses baker's yeast, *Saccharomyces cereviseae*. Anaerobic thermophilic genera *Caldicellulosiruptor*, *Thermoanaerobacter* and *Caloramator* have shown great potential to produce bioethanol from lignocellulosic-derived material (Scully & Orlygsson 2018). Bioethanol production can be done by three types of processes: batch, fed-batch, and continuous fermentation. In batch fermentation, feedstock, microbe, nutrients, and other ingredients are added together at the beginning of the fermentation process and ethanol is recovered. In fed-batch, one or more component is added during the process of fermentation. Continuous fermentation involves continuous addition of ingredients and removal of the product (Gnansounou & Dauriat 2005). Kinetics of the used microbe and the nature of the feedstocks govern the choice of selection of the most suitable mode of fermentation. Fermentation parameters (temperature, duration etc.) also depend on various factors such as, cell density, microbe employed, composition of hydrolysate etc. 8-14% ethanol is usually expected to be produced from the biomass. The product is then recovered and concentrated by distillation. For the fermentative production of ethanol from biomass to be commercially successful, organism that is both highly ethanologenic and cellulolytic need to be considered for consolidated bioprocessing (CBP) or co-culture of organisms meeting either criteria can utilised in a simultaneous saccharification and fermentation (SSF) setup (Scully & Orlygsson 2018).

Biobutanol

Bioethanol has limitations to be used as universal fuel or as an additive in the fuel blends, for which biobutanol is gaining attention (Kolesinska et al. 2019). Higher energy content (29.2 MJ/L as compared to ethanol which has energy content of 19.6 MJ/L), low volatility, higher octane number, better miscibility with gasoline and diesel, lower miscibility with water, better storage in humid conditions and non-corrosiveness are some of the properties that make butanol a suitable fuel-material (Kolesinska et al. 2019). Few Clostridia like *C. acetobutylicum*, *C. aurantibutylicum*, *C. beijerinckii*, and *C. tetanomorphum* can produce butanol with high fermentation yields (Gottwald et al. 1984, George et al. 1983). They are able to utilize a wide variety of substrates – pentoses, hexoses, syngas, and glycerol. Yields of 16–17 g/L can be achieved in the conventional clostridial fermentation and, thus, various improvement strategies have been adopted to achieve higher yields. Isolation of strain adapted to higher butanol production, enhancement of their osmotolerance or metabolic and genetic engineering at the strain level (for example inactivation of phosphotransacetylase (*pta*) and butyrate-kinase (*buk*) genes with simultaneous overexpression of aldehyde/alcohol dehydrogenase (adhE) gene contributes to an increase by 60 % of the final butanol titer (18.9 g/L) and 154 % of the yield (0.29 g/g), co-culturing of two butanol producing strains, simultaneous saccharification and fermentation (SSF), cells immobilization (2.7-fold increase of butanol production as compared to free cells), reduction of inhibitor content, recycling of biomass etc. at the process level (Tsvetanova et al. 2018, Kolesinska et al. 2019) are some of the approaches used for improved butanol production .

Biogas

Biogas consisting of ca. 65% methane (CH_4), 35% carbon dioxide (CO_2) and trace gases such as hydrogen sulphide (H_2S), hydrogen (H_2) and nitrogen (N_2)

is produced by the microbial, anaerobic digestion of biomass. For generation of combined heat and power (in a CHP installation) produced biogas is, generally, introduced directly in power gas engines. However, for applications as vehicle fuel and fuel cells or in the natural gas grid upgraded biogas is needed. Various types of biomass and waste, such as municipal solid waste, lipid rich waste, energy crops, agricultural residues, manure and sewage sludge are used in anaerobic digestion for production of biogas, and efficiency of co-digestion process considered superior (Appels et al. 2011). Biogas production involves four steps and a consortium of acidogenic bacteria, acetogenic bacteria and methanogenic bacteria mediate the production (Tezel et al. 2011). Complex substrates first hydrolyzed by fermentative bacteria into low molecular weight and water soluble organic intermediates (glucose, fatty acids, and amino acids). During acidogenesis, volatile fatty acids (VFAs) are derived and by-products including H_2, NH_3, CO_2, and H_2S are generated. These VFAs are then converted via acetogenesis into acetate, more CO_2, H_2, etc. Acetate is utilised by methanogenic Archaea for production of biomethane in the final step (Kiran et al. 2016). As for other biofuels, improvement strategies such as optimization of production parameters, pretreatments, additives, co-digestion, bioreactor design and optimization, and genetic engineering are employed for enhancing biogas production (Zhang et al. 2019).

Biohydrogen

Higher energy yield (122 kJ/g per unit mass) and clean combustion product (water vapour only) makes hydrogen a suitable alternative to conventional fossil fuel energy resources. Over the years, anaerobic digestion has been primarily used for the production of biogas; however, technological advances have enabled it to be used for the production of biohydrogen from biomass (Cheng & Zhu 2013, Intanoo et al. 2016). The most promising method of biohydrogen production through biomass conversion is dark fermentation (fermentation in the absence of light). The net energy ratio in the production process is 1.9 (Manish & Banerjee 2008). Acidogenesis is the hydrogen production phase in anaerobic digestion of biomass. Acidogenic fermentation can be butyrate, butanol and mixed-acid (acetic and formic acid) fermentations based on the final products and depending on the microbial proportions. The butyrate butanol fermentation carried out by members of the genus *Clostridium* produces butyric acid, acetic acid, butanol, hydrogen and carbon dioxide including other compounds, such as acetone, propanol, ethanol etc. as the final products (Lukajtis et al. 2018). Mixed-acid fermentation is the characteristic of *Enterobacter* and *Bacillus* which yields formic acid, acetic acid, ethanol, hydrogen, carbon dioxide, lactic acid, succinic acid, glycerol, acetoin, and butanediol as end products (Lukajtis et al. 2018). As for the production of other biofuels, a variety of raw materials or substrates can be used for biohydrogen production. The best starting material is the one that requires simple or no pretreatment and contain large amount of readily available carbohydrate for conversion. Second generation renewable resources (rich in lignocellulose or starch biomass) as well as organic waste of various origins, glycerol, olive mill wastewater, manure, sewage sludge etc. serve as substrates for dark fermentation. Pure culture of microbes or mixed consortia can be used for biohydrogen production. Monoculture production of biohydrogen has been demonstrated mainly with Gram-positive *Clostridium* and Gram-negative *Enterobacteriaceae* (Masset et al. 2012, Patela et al. 2014). Among *Enterobacteriaceae*, *Citrobacter*, *Klebsiella* and *Enterobacter* are prominent H_2 producers. Mixed bacterial cultures capable of hydrogen production abundantly occur in municipal sewage, composts and organic waste from which they can be isolated (Kleerebezem & van Loosdrecht 2007). Apart from microbial population used and pretreatment of the inoculum, many other factors such as temperature, pH, hydraulic retention time, partial pressure of hydrogen etc. affects the fermentation and thus the yield of production. Biohydrogen production on industrial scale requires use of continuous reactors with an efficient start-up strategy (Bakonyi et al. 2014). Commonly used bioreactor configurations for biohydrogen production are continuous stirred-tank reactors (CSTR), upflow anaerobic sludge blanket reactor (UASB), anaerobic fluidized bed reactor (AFBR) and membrane bioreactor (MBR) (Lukajtis et al. 2018).

Applications in Agriculture and Environmental (Green) Biotechnology

Diverse microorganisms have been studied and exploited for improvement in agricultural production and mitigating environmental pollutions. This involves adopting new practices and processes, including microbes, for crop production and protection as well as for soil health improvement and bio-treatment of wastes.

Biofertilizer and solid waste management

While anaerobic processes dominate in flooded or poorly drained soils, anoxic or hypoxic conditions are restricted to limited zones or time periods in aerobic soils. Direct application of anaerobes in enhancing the agricultural production is thus restricted. However, application of anaerobic digestate has been reported to affect the performance of plants and modulate the soil nutrient dynamics (Kleerebezem et al. 2015). The process of anaerobic digestion for the production of biogas, an alternate source of energy has been discussed earlier. This process is a useful part in environmental biotechnology for 1) the 'digestate' – material left after anaerobic digestion is rich in nutrients and utilised as a biofertilizer to improve soil health, crop productivity and yield (Nkoa 2014). A new application is to transform the digestate into biochar, which can be used as soil enhancer or an adsorbent for purification of wastewater or fuel gas (Inyang et al. 2010). It offers a more sustainable method of solid waste management as anaerobic digesters feedstocks include agriculture waste, food waste and sewage sludge (Khalid et al. 2011). During last 20 years, production of methane-rich biogas from agricultural or municipal sludge waste *via* anaerobic digestion has expanded dramatically in China and Europe; consortia of chiefly obligate anaerobic bacteria, *Bacteroides*, *Bifidobacterium*, and *Clostridium*, mediate digestion of organic waste. Biogas is harnessed for energy production and the solid digestate is further processed into nitrogen-rich biofertilizer (Al Seadi et al. 2013). Use of anaerobic digestion technology for treatment of wastewater (low solid content) feedstocks (sewage, agro- and chemical industry wastewaters etc.).

Soilborne pest management

Anaerobic soil disinfestation (ASD) is an environmental biotechnology technique that involves stimulation of indigenous anaerobic microorganisms that utilize the organic amendment added to the soil and lead to the production and accumulation of toxic by-products, such as acetic and butyric acids. These and other volatile compounds serve to decrease the soilborne pathogens providing an alternate means of pest management rather than using harmful chemical fumigants (Blok et al. 2000, Strauss & Kluepfel 2015). ASD has been reported to suppress a wide range of pathogens such as *Agrobacterium tumefaciens* (Crown gall disease), *Fusarium oxysporum* (Fusarium wilt), *Pythium* spp. (Pythium wilt), *Ralstonia solanacearum* (bacterial wilt), *Sclerotium rolfsii* (Southern blight), *Verticillium dahlia* (Verticillum wilt) and Meloidogyne *incognita* (root-knot nematodes) (Butler et al. 2012, Meng et al. 2018, Messiha et al. 2007, SerranoPérez et al. 2017, Strauss and Kluepfel 2015).

Bioremediation

One major aspect of environmental biotechnology is to utilize scientific know-how and employ microorganisms so as to prevent or treat environmental pollutions. The process is termed bioremediation and include many facets such as, bioaugmentation (addition of microbe), biostimulation (stimulating indigenous microbes), phytoremediation (use of plants), etc. and combinations thereof for *in-situ* bioremediation (treating pollutants at the site of remediation) and biopile, bioreactor, windrow etc. for *ex-situ* remediation (excavating pollutants from the contaminated site and transporting them to another site for treatment) (Azubuike 2016). Through agriculture, industry or even through our daily routines tonnes of toxic contaminants enter different strata (soil, air and water) of the environment and result in destructive consequences on ecosystems, causing severe damage to humans and other organisms. Anaerobic microorganisms can utilize some of these contaminants as electron acceptors and thus present strong potential for anaerobic bioremediation.

The use of anaerobes is of particular importance in bioremediation as most of the areas requiring bioremediation are usually underground like aquifers where aerobic organisms cannot grow and function. In these environments the anoxic abilities of anaerobes provides a way forward in controlling contamination, particularly of ground water. For example, in areas where contamination with petroleum is common the main contaminant benzene is removed from the ground water sources by using anaerobic bacteria (Lovely 2000).

Some of the fermentative anaerobes that are widely in use for biodegradation are *Clostridium thermoaceticum* which is capable of reducing carboxylate groups, oxidizing aldehyde groups, decarboxylating aromatic acids, and hydroxylating the benzene ring (Lux et al. 1990), *Acetobacterium woodii* possesses the ability to grow on methoxylated aromatic acids (Bache & Pfennig 1981), *Syntrophococcus sucromutans* is able to utilize sugars as electron donors with a variety of electron sinks including methoxyaromatics (DeWeerd 1988), *Eubacterium limosum* that grows on a number of methoxy-substituted aromatics (Lovley 2000) and many more fermentative anaerobes like *Eubacterium oxidoreducens*, *Pelobacter acidigallici*, *Geobacter metallireducens* etc. Moreover, new species are constantly being discovered at a fast rate. It is to be noted that the utilization of anaerobes for biodegradation depends on a number of factors like the anaerobic community being used and its composition, the availability of the electron donors for maximal requisite results and the oxidants that will function as electron sinks.

Dehalogenation of Halogenated Compounds

Addition of halogen atoms in place of hydrogen atoms in many compounds alters their physical and chemical properties conferring improvements that are boon for their value as commercial products but bane for their degradation. Their persistence in the environment is a serious threat and necessitates development of technologies to tackle them. Many of these halogenated compounds, both aliphatic and aromatic, are susceptible to degradation by anaerobic microorganisms. Tetrachlorethene (TCE), for example, is completely resistant to aerobic metabolism but can be completely dechlorinated by anaerobic microorganism to non-toxic ethene which can be readily decomposed by methanogens (Gerritse 1995). Halo-respiring microbes, capable of dehalogenating halogenated hydrocarbons by various fermentative, oxidative, or reductive pathways, belong to the genera *Desulfitobacterium*, *Desulfuromonas*, *Desulfovibrio*, *Desulfomonile*, and *Trichlorobacter* and are versatile in their choice of electron acceptors and donors. Of these *Dehalococcoides* spp. convert tri- and tetrachloroethylene to ethane while isolates of *Dehalobacter* spp. convert dichloromethane to acetate and formate and use hydrogen as the electron acceptor (Damborský 1999, Justicia-Leon et al. 2012). Similarly, anaerobes belonging to genera *Pelobacter*, *Syntrophus*, *Desulfitobacterium*, *Dehalococcoides* remove chlorine atoms from chlorinated benzenes, used as solvents and pesticides; pentachlorophenol, used in fungicides and herbicides; and polychlorinated biphenyls (PCBs), once widely used in electrical transformers and capacitors rendering them susceptible to aerobic microbes attack (Sahm et al. 1986, Adrian & Görisch 2002, Field & Sierra-Alvarez 2008).

Removal of Sulphate and Heavy Metals

Heavy metals such as, aluminium, chromium, manganese, iron, cobalt, nickel, copper, zinc, cadmium, lead and various compound forms of sulfates are other pollutants that are contaminating our environment and is a significant cause of concern. Microorganisms can eliminate several heavy metal pollutants from the polluted soils by reducing them to a lower redox state (Lovley 1995). Dissimilatory metal-reducing bacteria can catalyse such reducing reactions anaerobically exploiting metals as terminal electron acceptors, though most of them use Fe^{3+} and S (sulphur) as terminal electron acceptors (Jing et al. 2007). One of the approaches to tackle heavy metals problem is to precipitate them as metal ions in the form of their respective sulphides (Hussain et al. 2016). Metal sulphide precipitation is brought about by obligate anaerobes sulfur-reducing bacteria (SRB) by generating hydrogen sulphide gas (H_2S) that reacts vigorously with metal pollutants. SRB utilize sulfates as their terminal electron acceptors during respiration to generate H_2S (Hussain et al. 2016).

Metals and metalloids like uranium, technetium and selenium that contaminate ground water are also toxic and soluble in water and can be reduced by anaerobic fermentations that convert them into insoluble forms and prevent their spread (Lloyd & Lovley 2001). The addition of organic material helps promote anoxic conditions and increases the activity of these organisms. Bioremediation by these anaerobes can be made into feasible routine operations by incorporating information about the geochemical and hydrological data into designing the fermentation processes.

Nitrate Removal

Nitrate contamination due to excessive use of fertilizers in agriculture, concentrated animal feeding operations and release of semi-treated sewage disperse into groundwater poses another persistent problem. Biological nitrate removal, brought about by denitrifying bacteria, involves conversion of nitrate to non-toxic dinitrogen (N_2) gas or other gaseous nitrogen compounds in the presence of carbon and in *bloc* or sequence of reactions, which includes nitrite ($NO2^-$), nitric oxide (NO), and nitrous oxide (N_2O) as intermediates (Zumft, 1997). Denitrifying anaerobic methane oxidation (damo) bioprocesses remove nitrate using methane as the electron donor. First damo consortia was enriched in a lab-scale bioreactor (Raghoebarsing et al. 2006). Long-term stability of the damo system for nitrate removal in the presence of methane has been demonstrated in sequencing batch reactors (SBRs) and an engineering acceptable nitrogen removal rate (NRR) has been obtained. The microbial community consisted of damo archaea and bacteria (Li et al. 2018). In a single SBR for nitrogen removal from a typical medium-age landfill leachate an anaerobic/aerobic/anoxic (AOA) process has been suggested (Li et al. 2014). The process involves, in the anoxic phase, complete denitrification of the residual nitrate and nitrite were by utilizing residual PHAs and glycogen as electron donors. Denitrifying glycogen accumulating organisms (GAOs) were considered to be playing the major role in the process (Li et al. 2014). Nitrate-dependent anaerobic ferrous oxidation (NAFO) is another valuable bioprocess for nitrate removal (Zhang et al. 2015). Several NAFO, for example, *Citrobacter freundii* strain PXL1, *Pseudomonas sp.* SZF15, Strain 2002 etc. have been isolated from lakes and sediments and applied in the remediation of nitrate contaminated groundwater and wastewater treatment (Li et al., 2014; Su et al. 2015, Weber et al, 2006, Zhang et al, 2014). A distinct process from denitrification is the dissimilatory nitrate reduction to ammonia (DNRA). It is strictly an anaerobic process occurring in nitrate-rich environments such as, anaerobic marine sediments, sulphide-rich thermal vents and the human gastrointestinal tract (Giblin et al. 2013, Slobodkina et al. 2017, Tiso & Schechter 2015). Obligate anaerobes belong to genera *Clostridium* and facultative anaerobes like *Citrobacter*, *Enterobacter*, *Erwinia*, *Escherichia* and *Klebsiella* can perform DNRA (Tiedje 1988). DNRA is considered to be a beneficial process in agriculture as it preserves nitrogen fertilizers within soil as the ammonium ion is retained in soils and sediments by absorption, whereas the nitrate ion is easily lost due to leaching. However application of DNRA in a biological wastewater treatment system is undesirable due to production of ammonia which is a waste product generally removed by nitrification-denitrification during wastewater treatment process (Silver et al. 2005).

Biodegradation of Polymers

Extensive industrial and domestic usage of polymer products mainly plastics is a significant threat to the environment as plastic materials are recalcitrant to biodegradation. Plastic waste management along with the production of bioplastics thus becomes an important strategy to combat this situation. Current technologies available for polymer degradation include various chemical, thermal, and biological techniques. Biodegradation employs microbes to convert polymers to small molecular weight fragments that can be further degraded to carbon dioxide and water (Kale et al. 2015; Ahmed et al. 2018). Buried polymeric materials of sediments and landfills subjected to anaerobic degradation in the presence of alternate electron acceptors such as nitrate, sulfate, or methanogenic conditions. During biodegradation polymers are first depolymerised i.e. broken down into monomers followed by mineralization. It should be noted here that all bio-based polymers (synthesized from biomass or renewable resources) are not biodegradable and some non-biopolymers can be biodegraded. For example, polycaprolactone (PCL), and poly (butylene succinate) (PBS) are petroleum-based but biodegradable and poly(hydroxybutyrate) (PHB) and poly(lactide) (PLA) are bio-based and also biodegradable. Polyethylene (PE) and Nylon 11 (NY11) can be produced from biomass or renewable resources but they are non-biodegradable (Tokiwa et al. 2009). Table 2 lists some anaerobic microorganisms with ability to degrade certain polymers.

Table 2: Examples of some polymer-degrading anaerobes.

Polymer	Degrading Anaerobes	References
Polycaprolactone (PCL)	*Bacteroides* sp. strain CDB 8-2 and *Clostridium sporogenes* (JCM 1416) *Alcanivorax thereius* *CLostridum* spp.	Haruo & Yutaka. 1994. Yagi et al. 2014 Abou-Zeid et al. 2001
Poly(lactide) (PLA)	*Mesorhizobium* sp. and *Xanthomonadaceae* bacteria.	Yagi et al. 2014
Polyethyleglycol (PEG)	*Pelobacter venetianus* and *Bacteroides strain* PG1	Frings et al. 1992
Poly(hydroxybutyrate) (PHB)	*Illyobacter delafieldi C Lostridum* spp.	Stieb & Schink 1984 Abou-Zeid et al. 2001

Applications in Medicine and Health (Red) Biotechnology

Anaerobes such as *Clostridium difficile, Clostridium perfringens* and *Clostridium septicum, Actinobacterium, Propionibacterium, Bifidobacterium, Lactobacillus, Peptococcus* and *Peptostreptococcusm, Bacteroides, Fusobacterium, Campylobacter, Prevotella* and *Veillonella* are causative agents of many human infections (Noor & Khetarpal 2019). They secrete toxins or different types of enzymes that are capable of cell destruction or that could inactivate or help the microbe to evade antibiotics. Whereas, on one hand, anaerobic infections are fairly common and should be managed effectively, on the other hand, anaerobes themselves produce a variety of bioactive compounds that are used for improvement of human health (Mamo 2016).

Major sources of medicinal compounds for the treatment of diseases and improvement of human health have been aerobic microorganisms. A variety of drugs have been developed from these compounds and employed in the prevention or treatment of diseases. Limited number of studies has also established the potential of anaerobes in promoting human health. Bioactive compounds such as bacteriocins, other active peptides, antibiotics, antitumor compounds etc. produced by anaerobes as well as whole cell applications of anaerobes as probiotics or in bacteriotherapy reveal their role in medicine and health care sectors.

Production of Bacteriocins

Bacteriocins are unique amino acid sequences (peptides) produced by bacteria and some fungi and possess antimicrobial activities against various groups of microorganisms (Riley & Wertz 2002, Walton et al. 2010, Chikindas et al. 2017). Anaerobes from various samples like faecal, environmental and foods have been reported to produce bacteriocins. (Table 3). Most studied bacteriocins are those from aerotolerant and GRAS (generally recognized as safe) anaerobe, lactic acid bacteria (LAB). These bacteriocins are widely applied in food preservation as they present antimicrobial activity against food spoilage bacteria and food-borne pathogens (Camargo et al. 2018, Cotter et al. 2005). Other applications of bacteriocins include cosmetic, beverage preservation and their antimicrobial activity make them useful as an alternative to antibiotics against pathogens (Behnken & Hertweck 2012).

Table 3: Some examples of anaerobes producing bacteriocins.

Anaerobes	Bacteriocins	References
Lactobacillus lactis subsp. lactis	Nisin, lactocin	Sobrino-Lopez & Martin-Belloso 2008,
Enterococcus faecium	Lactococcin G, enterocin X, Bac FL31, Enterolysin A	
Pediococcus acidilactici.	Pediocin	Jie et al. 2005.
Lactobacillus helveticus	Helveticin J	Oppegård et al. 2007,
Lactobacillus plantarum	Plantaricin A, Lysostaphin	Hu et al. 2010, Chakchouk-Mtibaa et al. 2014,
Bifidobacterium lactis	Bacteriocin like inhibitory substances (BLIS)	Khan et al. 2013. Pérez et al. 2005. Vaughan et al. 1992, Sand et al. 2010, Liu et al. 2011 Balciunas et al. 2015.

Production of Antibiotics

Microorganisms synthesize low molecular weight secondary metabolites with bactericidal or bacteriostatic properties, the antibiotics. Although aerobes produce majority of antibiotics, anaerobes have also been shown to produce similar compounds (Cotter et al. 2005, Ezaki et al. 2008, Lincke et al. 2010, Troy & Kasper 2010, Chung et al 2008). Facultative anaerobes are the major producer of anaerobe antibiotics, however some obligate aerobes have also been shown to produce antimicrobial compounds, for example, obligate anaerobe *Sporotalea propionica*, bacterium isolated from termite gut, produces naphthalecin antibiotic that is active against many Gram positive bacteria or strict anaerobe *Clostridium cellulolyticum* produces closthioamide, a metabolite that exhibits strong activity against Gram-positive pathogens, including the methicillin-resistant *Staphylococcus aureus* (MRSA) and vancomycin-resistant *Enterococci* (VRE) strains (Ezaki et al. 2008, Lincke et al. 2010). Discoveries of clostrubins (Pidot et al. 2014), barnesin (Rischer et al. 2018) and acyloins (Schieferdecker et al. 2019) have established the potential of anaerobes as valuable source of novel antibiotics.

Other bioactive compounds

Anaerobes aid in human wellbeing by means of many other bioactive compounds, such as, immune system regulators, antitumor compounds and vitamins (Challinor & Bode 2015, Piwowarek et al. 2018). *Bacteroides fragilis*, a human commensal anaerobe, influences the development of the human immune system (Troy & Kasper 2010). Other anaerobic strains belonging to *Clostridium, Bifidobacterium, Propionibacterium, Butyrivibrio,* and *Lactobacillus* have been shown to positively contribute to the immune system (Chung et al. 2008, Jiang et al. 1998, Piwowarek et al. 2018). In addition to producing immunomodulators, many anaerobes produce compounds such as equol, enterodiol, enterolactone and urolithins that are believed to have many pro-health effects such as, improving bone health, minimizing the risk of breast cancer, osteoporosis, endometrial and ovarian cancer etc. (Tousen et al 2011). Members of the genus *Eubacterium* and *Bacteroides* produce vitamins, the important organic substances that serve as vital coenzymes to regulate many cellular reactions in the human body (Hill 1997, Jayashree et al. 2010).

Whole cell applications

Strains of *Lactobacillus*, *Propionibacteria* and *Bifidobacteria* are ingested as probiotics and have been shown to improve human health by preventing and treating pathogen-induced diarrhoea and chronic inflammatory diseases, maintaining normal gut microflora and managing atopic and autoimmune diseases

(El Enshasy et al. 2016, Prabhurajeshwar & Chandrakanth, 2017, Rabah et al. 2017). Commercial probiotics are relatively simple and contain known culture(s) of live anaerobic bacteria. Patients suffering from severe *Clostridium difficile* syndromes such as diarrhoea and colitis are treated with entire faecal flora containing a vast array of microbes including anaerobes. Faecal bacteriotherapy as it is called is now widely applied to treat many human diseases. Anaerobic bacteria have also been used as anticancer agents for human tumours present hypoxic microenvironments where anaerobes can thrive well. Bacteria belonging to the genera *Clostridium*, *Salmonella*, *Bifidobacterium* and *Listeria* have been shown to localize and proliferate selectively in tumours (Lukasiewicz & Fol 2018).

Applications in Aqua & Marine (Blue) Biotehnology

Aquaculture

Microorganisms are major role players in pond culture and are particularly important with respect to its productivity, nutrient cycling, nutrition of the cultured animals, maintenance of proper water quality, disease control in the cultured population and environmental impact of the effluent. Increased demand for fishes and other seafood has exhausted natural methods of its procurement and the way forward is its cultivation in fish farms and marine cultures. In order to produce the requisite quantity and quality of these organisms and keeping up with the ever increasing demand, special focus is required on the maintenance of the microbial flora in these systems. Proper management of the activities of microorganisms in pond food chains, food webs and in nutrient cycling is required so as to maximise production (Moriarty 1997).

Bacteria, micro algae and protozoans all interact with one another and have major roles to play in aquatic ecosystems. Different microorganisms have different zones of dominance in a pond ecosystem. While the activity of aerobic bacteria and algae including blue green algae or cyanobacteria are predominant in the photic zone of the pond, the bottom of the pond where oxygen diffusion is limited and it is rapidly depleted, the anaerobic bacteria gain dominance and release reduced acids, alcohols, carbon dioxide and hydrogen which are then used by methanogenic bacteria in freshwater or sulphate-reducing bacteria in marine ponds (Moriarty 1997).

Anaerobic bacteria are also important in the utilization of detritus. Detritus is generally used to denote non-living organic matter and it includes both particulate as well as dissolved matter. Anaerobic bacteria release low molecular weight compounds while other bacteria and organisms also secrete extracellular macromolecules and organic matter released after cell lysis such as proteins,

DNA, and lipids also become bound to detritus. This makes the composition of detritus quite variable and an important nutrition source for aquatic animals. Anaerobic processes are significant in detritus decomposition in pond sediments because anaerobic fermenting bacteria do not convert a great proportion of organic matter directly to CO, and thus prevent the loss of organic matter in the form of detritus from the food web unlike their aerobic counterparts. This makes the detritus available as a food source for other organisms (Schroeder 1987).

An approach that combines high intensity rearing with less water consumption and impact on the surrounding environment is the closed recirculating aquaculture systems (RAS). Vital in this technology is microbial reconditioning of the rearing water following biofilteration using biofilters which are large surface area structures hosting microbial biofilms. Anaerobes are incorporated in the design of a synthetic biofilm community (at lower, anoxic layer of the biofilm) in a biological aerated filter (BAF) in RAS (Bentzon-Tilia 2016). Denitrification by these anaerobes will remove $NO3^{-}$, which the accumulating evidence suggests is associated with swimming disabilities and health issues in some fish species (Davidson et al. 2014).

Wastewater Treatment

A major biotechnological application of anaerobes/anaerobic system is in wastewater treatment and thus, we discuss it separately in this section.

Wastewater is the used water. It includes substances such as human/animal waste, food scraps, agricultural waste, oil, chemicals etc. emanating from domestic or commercial use, for example from industries. The major aim of wastewater treatment is to eliminate as much of the suspended solids as possible before the remaining water, called effluent, is discharged back to the environment. Anaerobic digestion plays an increasingly important role in facilitating this and is done not only for public sanitation and environmental concerns (many components of untreated wastewaters are pollutants) but also for nutrient and energy recovery (McCarty 2011). It aids in residual waste valorisation and generation of energy for downstream processing. During the waste valorisation process, anaerobic digestion allows almost complete recovery of the inherent chemical energy during its low cost conversion to methane, a transportable natural gas energy source (McCarty 2011). The advantages of anaerobic wastewater treatment include excellent removal of organic matter, lower energy requirement, less production of sludge, higher loading rates, and considerable production of biogas.

The use of technologies such as low emission anaerobic membrane processes has enabled the generation of energy and release of nutrients by anaerobic digestion even of the domestic wastewater (Batstone & Virdis 2014). Two leading processes available for the use of anaerobic digestion for wastewater treatment are: Low energy mainline (LEM), in which organics are removed and energy recovered through a low-strength anaerobic process such as an Up-flow Anaerobic Sludge Blanket (UASB) or Anaerobic Membrane Bioreactor (AnMBR) that allows net energy recovery, full phosphorous recovery, and low energy nitrogen removal (Kong et al. 2019). In UASB, a dense sludge bed is established in the bottom of the reactor on which the degradation of the organic matter relies on. The anaerobic granular sludge serves as the main component of USAB formed along with the biogas production by the up-flow velocity of the influent. UASB derivatives and upgrades, such as the expanded granular sludge bed (EGSB) reactor and internal circulation (IC) reactor have been developed and adopted for improved efficiency of the process. The AnMBR is a combination of anaerobic digester and a series of advanced membrane modules, which uses a membrane as a physical barrier to facilitate parameters that ensure the sufficient metabolism of microorganisms and the degradation of organic matters for increased production of biogas (Kong et al. 2019b).

The second process available for the use of anaerobic digestion for wastewater treatment is Partition-Release-Recover (PRR), in which nutrients and organics are concentrated to discharge limits by assimilation into organic solids by growth and accumulation in a primary contact reactor leading to the recovery of phosphorus and nitrogen and energy is recovered from the wastewater as methane by effectively capturing it using sweep gas or vacuum extraction (Batstone et al. 2015).

Anaerobic ammonium oxidation (anammox) bacteria, affiliated with phylum *Planctomycetes*, have been commercialized for ammonium removal from industrial and municipal wastewater (Lackner et al. 2014; Oshiki et al. 2016; Cao et al. 2017). Ammonium nitrogen is converted to nitrogen gas using nitrite as an electron acceptor. Newly discovered microbial processes nitrite/nitrate-dependent methane oxidation (N-damo) combined with anammox can make wastewater treatment systems more sustainable (van Kessel et al. 2018).

Conclusion

Biotechnological applications of anaerobes are widespread. Anaerobes are employed in all major biotechnological sectors industrial, agriculture, environment, medicine, health care, aqua- and marine systems for the production of variety of useful products such as, biofuels, food items, preservatives, additives, chemical -intermediates, solvents, antibiotics, bacteriocins etc. and are an integral part of many beneficial processes such as bioremediation, waste valorization, etc. We have discussed some of these applications highlighting the advances that are made in making anaerobes and anaerobic processes more efficient, sustainable and environment friendly. Prevalence of anaerobic conditions in diverse niches from freshwater to marine sediments, in soils, subsurface aquifers, deep thermal vents, even in oxygen-free pockets of human body (gut, interior of plaques, gums, skin); discovery and identification of anaerobes with novel adaptations to sustain in these conditions and advancements in biological (genomics, metagenomics) and biochemical engineering techniques are set to further widen the scope of biotechnological applications of anaerobes.

Abbreviations

1,3-PD – 1,3 – Propane Diol

ABE – Acetone-Butanol-Ethanol

AFBR - Anaerobic Fluidized Bed Reactor

Anammox - Anaerobic ammonium oxidation

AnMBR - Anaerobic Membrane Bioreactor

ASD - Anaerobic Soil Disinfestation

ATP – Adenosine Tri-Phosphate

CBP – Consolidated Bio-Processing

CHP – Combined Heat Power Installation

CSTR - Continuous Stirred-Tank Reactors

Damo - Denitrifying anaerobic methane oxidation

DNRA - Dissimilatory nitrate reduction to ammonia

EGSB - Expanded Granular Sludge Bed

GAO - Glycogen Accumulating Organisms

LAB – Lactic Acid Bacteria

MBR - Membrane Bioreactor

MJ/L - Megajoules per liter

MRSA - Methicillin-resistant *Staphylococcus aureus*

NAFO - Nitrate-dependent anaerobic ferrous oxidation

PBS - Poly(butylene succinate)

PCBs - Polychlorinated Biphenyls

PCL – Polycaprolactone

PE- Polyethylene

PHA – Polyhydroxyalkanoates

PHB - Poly(hydroxybutyrate)

PLA - Poly Lactic Acid

PRR- Partition-Release-Recover

SBRs - Sequencing Batch Reactors

SRB - Sulfur-Reducing Bacteria

SSF - Simultaneous Saccharification and Fermentation

TCE – Tetrachlorethene

UASB – Up-flow Anaerobic Sludge Blanket Reactor

VFA – Volatile Fatty Acid

VRE - Vancomycin-resistant *Enterococci*

Acknowledgements

Charu Dogra Rawat and Mainisha Arora Pandit are grateful to their College authorities for providing conducive environment for the completion of this work.

References

Abou-Zeid DM, Muller RJ, Deckwer WD. 2001. Degradation of natural and synthetic polyesters under anaerobic conditions. *J. Biotech.* 86, 113–126.

Adrian L. Görisch H. 2002. Microbial transformation of chlorinated benzenes under anaerobic conditions. *Res. Microbiol.* 153, 131-7.

Ahmed T, Shahid M, Azeem F, Rasul I, Shah AA, Noman M, Hameed A, Manzoor N, Manzoor I, Muhammad S. 2018. Biodegradation of plastics: Current scenario and future prospects for environmental safety. *Environ Sc Pol Res* 8, 7287-7298.

Al Seadi T, Drosg B, Fuchs W, Rutz D, Janssen R. 2013. Biogas digestate quality and utilization. In The Biogas Handbook, ed. A Wellinger, J Murphy, D Baxter, 267–301. Cambridge, UK.

Appels L, Lauwers J, Degrève J, Helsen L, Lievens B, Willems K, Impea JV, Dewil, R. 2011. Anaerobic digestion in global bio-energy production: Potential and research challenges. *Renew Sust Energ Rev*. 15(9), 4295–4301.

Aragao R. 2013. Exploring Anaerobic Bacteria for Industrial Biotechnology - Diversity Studies, Screening and Biorefinery Applications. 170 p. Doctoral Thesis, Lund University.

Azubuike CC, Chikere CB, Okpokwasili GC. 2016. Bioremediation techniques-classification based on site of application: principles, advantages, limitations and prospects. *World J. Microbiol Biotechnol* 32 (11), 180.

Bache R, Pfennig N. 1981. Selective isolation of Acetobacterium woodii on methoxylated aromatic acids and determination of growth yields. *Arch Microbiol* 130(3), 255-261.

Balciunas EM, Al Arni S, Converti A, Leblanc JG, Oliveira RPDS. 2015. Production of bacteriocin like inhibitory substances (BLIS) by Bifidobacterium lactis using whey as a substrate. Int *J Dairy Technol* 68, 1–6.

Bakonyi P, Nemestóthy N, Simon V, Bélafi-Bakó K. 2014. Review on the start-up experiences of continuous fermentative hydrogen producing bioreactors. *Renew Sust Energ Rev.* 40, 806-813.

Bathgate G. 2016. A review of malting and malt processing for whisky distillation: Malting and malt processing for whisky distillation. *J. Inst. Brew.* 122. 197-211.

Batstone D, Virdis B. 2014. The role of anaerobic digestion in the emerging energy economy. *Curr Opin Biotechnol* 27, 142-149.

Batstone DJ, Hülsen T, Mehta CM, Keller J. 2015. Platforms for energy and nutrient recovery from domestic wastewater: A review, *Chemosphere.*140, 2-11.

Behnken S, Hertweck C. 2012. Anaerobic bacteria as producers of antibiotics. *Appl Microbiol Biotechnol.* 96(1), 61-7.

Bentzon-Tilia M, Sonnenschein EC, Gram L. 2016. Monitoring and managing microbes in aquaculture - Towards a sustainable industry. *Microb Biotechnol.* 9(5),576-84.

Bernardeau M, Guguen M, Vernoux, JP. 2006. Beneficial lactobacilli in food and feed: long-term use, biodiversity and proposals for specific and realistic safety assessments, FEMS *Microbiol Rev.* 30 (4) 487–513.

Blok WJ, Lamers JG, Termorshuizen AJ, Bollen GJ. 2000. Control of soil-borne plant pathogens by incorporating fresh organic amendments followed by tarping. *Phytopathology* 90,253–259.

Brodin M, Vallejos M, Opedal MT, Area MC, Chinga-Carrasco G. 2017. Lignocellulosics as sustainable resources for production of bioplastics-a review. *J Clean Prod.* 162,646-664.

Butler DM, Kokalis-Burelle N, Muramoto J, Shennan C, McCollum TG, Rosskopf EN. 2012a. Impact of anaerobic soil disinfestation combined with soil solarization on plantparasitic nematodes and introduced inoculum of Soilborne plant pathogens in raised-bed vegetable production. *Crop Prot*, 39, 33–40.

Camargo AC, Todorov SD, Chihib NE, Drider D, Nero LA. 2018. Lactic acid bacteria (lab) and their bacteriocins as alternative biotechnological tools to control Listeria monocytogenes biofilms in food processing facilities. *Mol Biotechnol.* 60(9), 712-726.

Cao Y, van Loosdrecht MCM, Daigger GT. 2017. Mainstream partial nitritation-anammox in municipal wastewater treatment: status, bottlenecks, and further studies *Appl. Microb Biotechnol.* 101, 1365-1383.

Chakchouk-Mtibaa A, Elleuch L, Smaoui S, Najah S, Sellem I, Abdelkafi S, Mellouli L. 2014. An antilisterial bacteriocin BacFL31 produced by Enterococcus faecium FL31 with a novel structure containing hydroxyproline residues. *Anaerobe* 27,1–6.

Challinor VL, Bode HB. 2015. Bioactive natural products from novel microbial sources. Ann N Y *Acad Sci.* 1354, 82-97.

Cheng C, Bao T, Yang ST. 2019. Engineering Clostridium for improved solvent production: recent progress and perspective. *Appl Microbiol Biotechnol.* 103, 5549–5566.

Cheng J, Zhu M. 2013. A novel anaerobic co-culture system for bio-hydrogen production from sugarcane bagasse. *Bioresour Technol.* 144, 623-31.

Cherubini F. 2010. The biorefinery concept: using biomass instead of oil for producing energy and chemicals. *Energ Convers Manage.* 51(7), 11412-1421.

Chikindas ML, Weeks R, Drider D, Chistyakov VA, Dicks LM. 2017. Functions and emerging applications of bacteriocins. *Curr Opin Biotechnol.* 49, 23–28.

Chung SH, Kim IH, Park HG, Kang HS, Yoon CS, Jeong HY, Choi NJ, Kwon EG, Kim YJ. 2008. Synthesis of conjugated linoleic acid by human-derived Bifidobacterium breve LMC 017: utilization as a functional starter culture for milk fermentation. *J Agric Food Chem.* 56, 3311–3316.

Clark JH, Farmer TJ, Hunt AJ, Sherwood J. 2015. Opportunities for Bio-Based Solvents Created as Petrochemical and Fuel Products Transition towards Renewable Resources. *Int. J. Mol. Sci.* 16, 17101-17159.

Cotter PD, Hill C, Ross RP. 2005. Bacteriocins: developing innate immunity for food. *Nat Rev Microbiol.* 3(10),777–88.

Damborský J 1999. Tetrachloroethene-dehalogenating bacteria. *Folia Microbiol.* 44, 247 -262.

DaSilva, Edgar. 2005. Editorial - The Colours of Biotechnology: Science, Development and Humankind. *Electron J Biotechn* (ISSN: 0717-3458) Vol 7 Num 3.

Datta R, Tsai S, Bonsignore P, Moon S, Frank J. 1995. Technological and economic potential of poly(lactic acid) and lactic acid derivatives. *FEMS Microbiol Rev.* 16(2-3), 221-231.

Davidson J., Good C., Welsh C., and Summerfelt S.T. 2014. Comparing the effects of high vs. low nitrate on the health, performance, and welfare of juvenile rainbow trout Oncorhynchus mykiss within water recirculating aquaculture systems. *Aquac Eng* 59, 30–40.

Davis R, Biddy M, Tan E, Tao L, Jones, S. 2013). Biological conversion of sugars to hydrocarbons technology pathway. Pacific Northwest National Laboratory (PNNL), Richland, WA (US).

Demain A. 1981. Industrial microbiology. *Science.* 214 (4524), 987-995.

DeWeerd J, Saxena A, Nagle D, Sulflita J. 1988. Metabolism of the /sup 18/O-methoxy substituent of 3-methoxybenzoic acid and other unlabeled methoxybenzoic acids by anaerobic bacteria. *Appl Environ Microbiol.* 54(5),1237-42.

El Enshasy H, Malik K, Malek RA, Othman NZ, Elsayed EA, Wadaan M. 2016. Anaerobic Probiotics: The Key Microbes for Human Health. *Adv Biochem Eng Biotechnol.* 156, 397-431.

Emeruwa AC, Hawirko RZ. 1973. Poly-β-hydroxybutyrate metabolism during growth and sporulation of Clostridium botulinum. *J. Bacteriol.* 116, 989H993.

Ezaki M, Muramatsu H, Takase S, Hashimoto M, Nagai K. 2008. Naphthalecin, a novel antibiotic produced by the anaerobic bacterium, Sporotalea colonica sp. nov. *J Antibiot.* 61, 207–212.

Falentin H, Deutsch S-M, Jan G,, Loux V, Thierry A, Parayre S, Maillard M-B Dherbécourt J, Cousin, FJ, Jardin, J, Siguier P, Couloux A, Barbe V, Vacherie B, Wincker P, Gibrat J-F Gaillardin C, Lortal, S. 2010. The complete genome of Propionibacterium freudenreichii CIRM-BIA1, a hardy Actinobacterium with food and probiotic applications. *PLoS ONE.* 5(7),e11748.

Farmer TJ, Mascal M. 2015. Platform molecules. In Introduction to Chemicals from Biomass, 2nd ed.; Deswarte, C., Ed.; John Wiley and Sons: Chichester, UK, pp. 89–156.

Field JA, Sierra-Alvarez R. 2008. Microbial degradation of chlorinated benzenes Biodegradation. 19, 463-480.

Flüchter S, Follonier S, Schiel-Bengelsdorf B, Bengelsdorf FR, Zinn M, Dürre P. 2019. Anaerobic Production of Poly(3-hydroxybutyrate) and Its Precursor 3-Hydroxybutyrate from Synthesis Gas by Autotrophic Clostridia. *Biomacromolecules.* 20, 9, 3271-3282.

Food and Agriculture Organization. WHO Probiotics in Food: Health and Nutritional Properties and Guidelines for Evaluation. FAO; Rome, Italy: 2006.

Forssten SD, Sindelar CW, Ouwehand AC. 2011. Probiotics from an industrial perspective. *Anaerobe* 17(6), 410–13.

Frings J, Schramm E, Schink B. 1992. Enzymes involved in anaerobic polyethylene glycol degradation by Pelobacter venetianus and Bacteroides strain PG1. *Appl Environ Microbiol* 58(7): 2164–2167.

García IL. 2016. Feedstocks and challenges to biofuel development. In Handbook of Biofuels Production (Second Edition), Woodhead Publishing, 85-118.

George HA, Johnson JL, Moore WEC, Holdeman LV, Chen JS. 1983. Acetone, isopropanol, and butanol production by Clostridium beijerinckii (syn. *Clostridium butylicum*) and Clostridium aurantibutylicum. *Appl. Environ. Microbiol.* 45, 1160–1163.

Gerritse J, Renard V, Gottschal JC, Visser J. 1995. Complete degradation of tetrachloroethene by combining anaerobic dechlorinating and aerobic methanotrophic enrichment cultures. *Appl Microbiol Biotechnol.* 43, 920-928.

Giblin AE, Tobias CR, Song B, Weston N, Banta GT, Rivera-Monroy VH. 2013. The importance of dissimilatory nitrate reduction to ammonium (DNRA) in the nitrogen cycle of coastal ecosystems. *Oceanography*. 26,124–131.

Girbal L, Orlygsson J, Reinders BJ, Gottschal JC. 1997. Why does Clostridium acetireducens not use interspecies hydrogen transfer for growth on leucine? *Curr. Microbiol.* 1997, 35, 155H160.

Gnansounou E, Dauriat A. 2005. Ethanol fuel from biomass: a review. *J Sci Ind Res.* 64 (11), 809–821, 2005.

Goldstein E. 1995. Anaerobes under assault: from cottage industry to industrialization of medicine and microbiology. *Clin Infect Dis.* 20 (Supplement_2), S112-S116.

Gottwald M, Hippe H, Gottschalk G. 1984. Formation of n-butanol from D-glucose by strains of Clostridium tetanomorphum group. *Appl. Environ. Microbiol.* 48, 573–576.

Gu Y, Jiang Y, Wu H, Liu X, Li Z, Li J, Xiao H, Shen Z, Dong H, Yang Y, Li Y, Jiang W, Yang S. 2011. Economical challenges to microbial producers of butanol: feedstock, butanol ratio and titer. *Biotechnol J.* 11,1348–67.

Haruo N, Yutaka T. 1994. Confirmation of anaerobic poly(2-oxepanone) degrading microorganisms in environments. *Chem. Lett.* 23 (7), 1293-1296.

Hatti-Kaul R, Mattiasson B. Anaerobes in Industrial and Environmental Biotechnology. *Adv Biochem Eng Biotechnol.* 2016;156:1-33.

Hill MJ. 1997. Intestinal flora and endogenous vitamin synthesis. *Eur J Cancer Prev* 6 (1), S43-5.

Hu CB, Malaphan W, Zendo T, Nakayama J, Sonomoto K. 2010. Enterocin, X, a novel two-peptide bacteriocin from Enterococcus faecium KU-B5, has an antibacterial spectrum entirely different from those of its component peptides. *Appl. Environ. Microbiol.* 76, 4542–4545.

Hughes SR, Gibbons WR, Moser BR, Rich JO. 2013. Chapter 9: Sustainable Multipurpose Biorefineries for Third-Generation Biofuels and ValueAdded Co-Products. In Biofuels - Economy, Environment and Sustainability, *Prof. Zhen Fang* (Ed.), ISBN: 978-953-51-0950-1, InTech.

Hussain A, Hasan A, Javid A, Qazi JI. 2016. Exploited application of sulfate-reducing bacteria for concomitant treatment of metallic and non-metallic wastes: A Mini Review. 3 Biotech. 6, 119.

Intanoo P, Chaimongkol P, Chavadej S. 2016. Hydrogen and methane production from cassava wastewater using two-stage upflow anaerobic sludge blanket reactors (UASB) with an emphasis on maximum hydrogen production, *Int. J. Hydrogen Energ.* 41, 6107-6114.

Inyang M, Gao B, Pullammanappallil P, Ding W, Zimmerman AR. 2010. Biochar from anaerobically digested sugarcane bagasse. *Bioresour Technol.* 101, 8868–72.

Jamshidian M, Tehrany EA, Imran M, Jacquot M and Desobry S. 2010. Poly-Lactic Acid: production, applications, nanocomposites, and release studies. *Compr Rev Food Sci F.* 9 (5), 552–571.

Jayashree S, Jayaraman K, Kalaichelvan G. 2010. Isolation, screening and characterization of riboflavin producing lactic acid bacteria from Katpadi, Vellore district. *Recent Res Sci Technol.* 2,83–88.

Jiang J, Bjorck L, Fonden R. 1998. Production of conjugated linoleic acid by dairy starter cultures. *J Appl Microbiol* 85,95–102.

Jie Li, Alla A. Aroutcheva, Sebastian Faro, and Michael L. Chikindas, 2005. Mode of action of Lactocin 160, a bacteriocin from vaginal Lactobacillus rhamnosus, *Infect Dis Obstet Gynecol.*13 (3), 135-140.

Jing Y, He Z, Yang X. 2007. Role of soil rhizobacteria in phytoremediation of heavy metal contaminated soils *J. Zhejiang Univ. Sci.* B, 8;192-207.

Johnson C, Stevenson B, Drilling H, Lawson P. 2010. Molecular- and cultivation-based investigations of organisms associated with microbially influenced corrosion (MIC) of oil reservoir infrastructure on the north slope of Alaska. In Ananerobe 2010. Philadelphia.

Johnson J, Johnson S, Eames J. 2010. An economic analysis of anaerobic digestion to remove food waste on a mid-sized university campus. In Anaerobe 2010. Philadelphia.

Junyu Z, Shishatskaya EI, Volova TG, da Silva LF, Chen GQ. 2018. Polyhydroxyalkanoates (PHA) for therapeutic applications. *Mater. Sci. Eng. C.* 86, 144-150.

Justicia-Leon SD, Ritalahti KM, Mack EE, Löffler FE. 2012. Dichloromethane Fermentation by a Dehalobacter sp. in an Enrichment Culture Derived from Pristine River Sediment. *Appl Environ Microbiol.* 78 (4) 1288-1291.

Kale SK, Deshmukh AG, Dudhare MS, Patil VB. 2015. Microbial degradation of plastic: A review. *J Biochem Technol.* 6(2),952-961.

Khalid A, Arshad M. Anjum M, Mahmood T, Dawson L. 2011. The anaerobic digestion of solid organic waste. *Waste manage.* 31(8), 1737-1744.

Khan H, Flint SH, Yu PL. 2013. Determination of the mode of action of enterolysin A, produced by Enterococcus faecalis B9510. *J. Appl. Microbiol.* 115(2), 484-94.

Kiran EU, Stamatelatou K, Antonopoulou G, Lyberatos G. 2016. Production of biogas via anaerobic digestion. In Handbook of Biofuels Production: Processes and Technologies, Woodhead Publishing, 259-301.

Kleerebezem R, Joosse B, Rozendal R., Van Loosdrecht Mark CM. 2015. Anaerobic digestion without biogas? *Rev Environ Sci Biotechnol* 14, 787-801.

Kleerebezem R, van Loosdrecht MCM. 2007. Mixed culture biotechnology for bioenergy production. *Curr Opin Biotech.* 18,207–12.

Klotz S, Kaufmann N, Kuenz A, Prüße U. 2016. Biotechnological Production of enantiomerically pure d-Lactic Acid. *Appl Microbiol Biotechnol* 100, 9423-9437.

Kolesinska B, Fraczyk J, Binczarski M, Modelska M, Berlowska J, Dziugan P, Antolak H, Kaminski ZJ, Witonska IA, Kregiel D. 2019. Butanol Synthesis Routes for Biofuel Production: Trends and Perspectives. *Materials.* 23;12(3), E350.

Kong Z, Li L, Kurihara R, Zhang T, Li Y-Y, 2019b. Anaerobic treatment of N,Ndimethylformamide-containing high-strength wastewater by submerged anaerobic membrane bioreactor with a co-cultured inoculum. *Sci. Total Environ.* 663, 696-708.

Kong Z, Li L, Xue Y, Yang M, Li Y-Y. 2019. Challenges and prospects for the anaerobic treatment of chemical-industrial organic wastewater: *A review. J Clean Prod.* 231, 913-927.

Kumar M, Gayen K. 2011. Developments in biobutanol production: New insights. *Appl Energy* 88(6), 1999-2012.

Lackner S, Gilbert E, Vlaeminck S, Joss A, Horn H, van Loosdrecht M. 2014. Full-scale partial nitritation/anammox experiences – An application survey. *Water Res* 55, 292-303.

Lamed R, Zeikus J. 1981. Novel NADP-linked alcohol–aldehyde/ketone oxidoreductase in thermophilic ethanologenic bacteria. *Biochemical J.*, 195(1), 183-190.

Lawson P, Allen T, Caldwell M, Tanner R. 2019. Anaerobes: a piece in the puzzle for alternative biofuels. In: https://www.cabdirect.org/cabdirect/abstract/20113295302.

Li B, Tian C, Zhang D, Pan X. 2014. Anaerobic nitrate-dependent iron (II) oxidation by a novel autotrophic bacterium, Citrobacter freundii strain PXL1 *Geomicrobiol. J.* 3, 138- 144.

Li W, Lu P, Chai F, Zhang L, Han X, Zhang D. 2018. Long-term nitrate removal through methane-dependent denitrification microorganisms in sequencing batch reactors fed with only nitrate and methane. AMB Express. 8, 108.

Li Z, Wang S, Zhang W, Miao L, Cao T, Peng Y. 2014. Nitrogen removal from medium-age landfill leachate via post-denitrification driven by PHAs and glycogen in a single sequencing batch reactor. *Bioresour Technol*, 169, 773-777.

Lincke T, Behnken S, Ishida K, Roth M, Hertweck C. 2010. Closthioamide: an unprecedented polythioamide antibiotic from the strictly anaerobic bacterium Clostridium cellulolyticum. *Angew Chem Int Ed*. 122, 2055–2057.

Liu H, Gao Y, Yu LR, Jones RC, Elkins CA, Hart ME. 2011. Inhibition of Staphylococcus aureus by lysostaphin-expressing Lactobacillus plantarum WCFS1 in a modified genital tract secretion medium. *Appl Environ Microbiol.* 77(24), 8500-8.

Lloyd J, Lovley D. 2001. Microbial detoxification of metals and radionuclides. *Curr. Opin. Biotechnol.* 12(3), 248-253.

Lovley D. 2000. Anaerobic benzene degradation. *Biodegradation*, 11(2/3), 107-116.

Lovley DR. 1995. Bioremediation of organic and metal contaminants with dissimilatory metal reduction. *J. Ind. Microbiol.* 14; 85-93.

Lukajtis, R.; Holowacz, I.; Kucharska, K.; Glinka, M.; Rybarczyk, P. 2018. Hydrogen production from biomass using dark fermentation. Renew. Sustain. Energy Rev. 91, 665–694.

Lukasiewicz K, Fol M. 2018. Microorganisms in the treatment of cancer: advantages and limitations. *J. Immunol. Res. Article* ID 239780.

Lux M, Keith E, Hsu T, Drake H. 1990. Biotransformations of aromatic aldehydes by acetogenic bacteria. *FEMS Microbiol. Lett.*, 67(1-2), 73-78.

Mamo G. 2016. Anaerobes as Sources of Bioactive Compounds and Health Promoting Tools. *Adv. Biochem. Eng. Biotechnol.* 156, 433-464.

Manish S, Banerjee R. 2008. Comparison of biohydrogen production processes. *Int J Hydrogen Energ*. 33,279–86.

Masset J, Calusinska M, Hamilton C, Hiligsmann S, Joris B, Wilmotte A, Thonart P. 2012. Fermentative hydrogen production from glucose and starch using pure strains and artificial co-cultures of Clostridium spp. *Biotechnol Biofuels*. 5 (1),35.

Mauerhofer L, Pappenreiter P, Paulik C, Seifert A, Bernacchi S, Rittmann S. 2019. Methods for quantification of growth and productivity in anaerobic microbiology and biotechnology. *Folia Microbiologica* 64(3), 321-360.

McCarty PL, Bae J, Kim J. 2011. Domestic wastewater treatment as a net energy producer—can this be achieved? *Environ. Sci. Techno*l. 45,7100-7106.

Mehta R, Kumar V, Bhunia H, Upadhyay SN. 2005. Synthesis of poly(lactic acid): A review. *J Macromol Sci Polymer Rev*. 45 (4)325–349.

Meng T, Yang Y, Cai Z, Ma Y. 2018. The control of Fusarium oxysporum in soil treated with organic material under anaerobic condition is affected by liming and sulfate content. *Biol Fertil Soils* 54,29.

Messiha NAS, van Diepeningen AD, Wenneker M, van Beuningen AR, Janse JD, Coenen TGC, Termorshuizen AJ, van Bruggen AHC, Blok WJ. 2007. Biological soil disinfestation (BSD), a new control method for potato brown rot, caused by *Ralstonia solanacearum race 3 biovar 2*. *Eur J Plant Pathol* 117:403–415.

Moriarty D. 1997. The role of microorganisms in aquaculture ponds. *Aquaculture*. 151(1-4), 333-349.

Morris J. 1994. Obligately anaerobic bacteria in biotechnology. *Appl Biochem Biotechnol*. 48(2), 75-106.

Müller V. 2008. Bacterial fermentation. In Encyclopedia of Life Sciences (eLS). John Wiley & Sons, Ltd.

Nkoa R. 2014. Agricultural benefits and environmental risks of soil fertilization with anaerobic digestates: a review. *Agron Sustain Dev.* 34, 473–492.

Noor A, Khetarpal S. 2019. Anaerobic Infections. In: StatPearls [Internet]. Treasure Island (FL): StatPearls Publishing; 2019 Jan-.

Oppegård C, Fimland G, Thorbaek L, Nissen-Meyer J. 2007. Analysis of the two-peptide bacteriocins lactococcin G and enterocin 1071 by site-directed mutagenesis. *Appl Environ Microbiol.* 73(9), 2931-8.

Oshiki M, Satoh H, Okabe S. 2016. Ecology and physiology of anaerobic ammonium oxidizing bacteria. *Environ. Microbiol.* 18 (9), 2784-2796.

Patela SKS, Kumara P, Mehariyaa S, Purohit HJ, Leed J-K, Kalia VC. 2014. Enhancement in hydrogen production by co-cultures of Bacillus and Enterobacter. *Int J Hydrogen Energ.* 39,14663–8.

Pérez Guerra N, Bernárdez PF, Agrasar AT, López Macías C, Castro LP. 2005. Fed-batch pediocin production by *Pediococcus acidilactici* NRRL B-5627 on whey. *Biotechnol Appl Biochem.* 42, 17-23.

Pidot SJ, Ishida K, Cyrulies M, Hertweck C. 2014. Discovery of clostrubin, an exceptional polyphenolic polyketide antibiotic from a strictly anaerobic bacterium. Angew. *Chem., Int. Ed.* 53, 7856"7859.

Piwowarek K, Lipinska E, Haæ-Szymañczuk E, Kieliszek M, Œcibisz I. 2018. Propionibacterium spp.—source of propionic acid, vitamin B12, and other metabolites important for the industry. *Appl. Microbiol. Biotechnol.* 102(2),515–38.

Podolsky IA, Seppälä S, Lankiewicz TS, Brown JL, Swift CL, O'Malley MA. 2019. Harnessing nature's anaerobes for biotechnology and bioprocessing. *Annu Rev Chem Biomol,* 10(1).

Prabhurajeshwar C, Chandrakanth RK. 2017. Probiotic potential of Lactobacilli with antagonistic activity against pathogenic strains: An in vitro validation for the production of inhibitory substances. *Biomed. J.* 40(5),270-283.

Pretula J, Slomkowski S, Penczek S. 2016. Polylactides-methods of synthesis and characterization. *Adv Drug Deliv Rev.* 107,3-16.

Rabah H, Rosa do Carmo FL, Jan G. 2017. Dairy Propionibacteria: Versatile Probiotics. *Microorganisms*. 5(2),24.

Raghoebarsing AA, Pol A, van de Pas-Schoonen KT, Smolders AJP, Ettwig KF, Rijpstra WIC, Schouten S, Damste JSS, Op den Camp HJM, Jetten MSM, Strous M. 2006. A microbial consortium couples anaerobic methane oxidation to denitrification. *Nature* 440(7086), 918–921

Riley MA, Wertz JE. 2002. Bacteriocins: evolution, ecology, and application. *Annu Rev Microbiol.* 56, 117–137.

Rischer M, Raguz L, Guo H, Keiff F, Diekert G, Goris T, Beemelmanns C. 2018. Biosynthesis, synthesis, and activities of Barnesin A, a NRPS-PKS Hybrid Produced by an Anaerobic Epsilonproteobacterium. *ACS Chem. Biol.* 13, 1990H1995.

Sahm H, Brunner M, Schoberth SM. 1986. Anaerobic Degradation of Halogenated Aromatic Compounds. *Microb Ecol.* 12,147-153.

Sand SL, Oppegård C, Ohara S, Iijima T, Naderi S, Blomhoff HK, Nissen-Meyer J, Sand O. 2010. Plantaricin A, a peptide pheromone produced by Lactobacillus plantarum, permeabilizes the cell membrane of both normal and cancerous lymphocytes and neuronal cells. *Peptides.* 31(7), 1237-44.

Schieferdecker S, Shabuer G, Letzel A-C, Urbansky B, Ishida-Ito, M, Ishida K, Cyrulies M, Dahse HM, Pidot SJ, Hertweck C. 2019. Biosynthesis of diverse antimicrobial and antiproliferative acyloins in anaerobic bacteria. *ACS Chem. Biol.* 14, 1490H1497.

Schroeder G. 1987. Carbon pathways in aquatic detrital systems. In 13. Intemational Center for Living Aquatic Resources Management (pp. 217-236). Manila: D.J.W. Moriarty and R.S.V. Pullin, (Editors). Detritus and Microbial Ecology in Aquaculture, ICLARM Conference Proceedings.

Scully SM, Orlygsson J. 2018. Progress in Second Generation Ethanol Production with Thermophilic Bacteria, Fuel Ethanol Production from Sugarcane, Thalita Peixoto Basso and Luiz Carlos Basso, IntechOpen.

Serrano-Pérez P, Rosskopf E, De Santiago A, Rodríguez-Molina MC. 2017. Anaerobic soil disinfestation reduces survival and infectivity of Phytophthora nicotianae chlamydospores in pepper. *Sci Hort* 215,38–48.

Sheldon RA. 2011. Utilisation of biomass for sustainable fuels and chemicals: Molecules, methods and metrics. Catal. Today 167, 3–13. Luterbacher JS, Alonso DM, Dumesic JA. 2014. Targeted chemical upgrading of lignocellulosic biomass to platform molecules. *Green Chem.* 16, 4816–4838.

Silver WL, Thompson AW, Reich A, Ewel JJ, Firestone MK. 2005. Nitrogen cycling in tropical plantation forests: potential controls on nitrogen retention. *Ecol Appl.* 15:1604–1614.

Slobodkina GB, Mardanov AV, Ravin NV, Frolova AA, Chernyh NA, Bonch-Osmolovskaya EA, Slobodkin AI. 2017. Respiratory Ammonification of Nitrate Coupled to Anaerobic Oxidation of Elemental Sulfur in Deep-Sea Autotrophic Thermophilic Bacteria. *Front Microbiol.* 8, 87.

Sobrino-Lopez A, Martin-Belloso O. 2008. Use of nisin and other bacteriocins for preservation of dairy products. *Int Dairy J.* 18, 329–343.

Stieb M, Schink B. 1984. A new 3-hydroxybutyrate fermenting anaerobe, Ilyobacter polytropus, gen. nov. sp. nov., possessing various fermentation pathways. *Arch. Microbiol.* 140, 139–146.

Straathof, AJ. 2013. Transformation of biomass into commodity chemicals using enzymes or cells. *Chem. Rev.* 114 (3) 1871-1908.

Strauss SL, Kluepfel DA. 2015. Anaerobic soil disinfestation: a chemical-independent approach to pre-plant control of plant pathogens. *J Integ Agric* 14,2309–2318.

Su JF, Shao SC, Huang TL, Ma F, Yang SF, Zhou ZM, Zheng SC. 2015. Anaerobic nitrate-dependent iron(II) oxidation by a novel autotrophic bacterium, Pseudomonas sp. SZF15

Subramani V, Sharma P, Zhang L, Liu K. 2010. Catalytic steam reforming technology for the production of hydrogen and syngas. In Hydrogen and Syngas Production and Purification Technologies. John Wiley and Sons: Hoboken, NY, USA, pp. 14–126.

Taskila S, Ojamo H. 2013. The current status and future expectations in industrial production of lactic acid by lactic acid bacteria. In Lactic Acid Bacteria-R & D for Food, Health And Livestock Purposes, InTech.

Tezel U, Tandukar, M, Pavlostathis SG. 2011. Anaerobic biotreatment of municipal sewage sludge. In Comprehensive Biotechnology (second edition), *Environmental Biotechnology and Safety,* 6, 447-461.

Tiedje JM. 1988. Ecology of denitrification and dissimilatory nitrate reduction to ammonium. In Biology of anaerobic microorganisms. 179-244, Edited by AJ. Zehnder, Wiley-Interscience.

Tiso M, Schechter AN. 2015. Nitrate reduction to nitrite, nitric oxide and ammonia by gut bacteria under physiological conditions. *PLoS ONE* 10, 3.

Tokiwa Y, Calabia BP, Ugwu CU, Aiba S. 2009. Biodegradability of plastics. *Int J Mol Sci.* 10(9),3722-42.

Tousen Y, Ezaki J, Fujii Y, Ueno T, Nishimuta M, Ishimi Y. 2011. Natural S-equol decreases bone resorption in postmenopausal, non-equol-producing Japanese women: a pilot randomized, placebo-controlled trial. *Menopause.* 18,563–574.

Troy EB, Kasper DL. 2010. Beneficial effects of Bacteroides fragilis polysaccharides on the immune system. *Front Biosci.* 15, 25–34.

Tsvetanova F, Petrova P, Petrov K. 2018. Microbial production of 1-Butanol – Recent advances and future prospects. *J. Chem Technol Metall* 53 (4) 683-696.

van Kessel MAHJ, Stultiens K, Slegers MFW, Cruz SG, Jetten MSM, Kartal B, den Camp HJMO Current perspectives on the application of N-damo and anammox in wastewater treatment, *Curr. Opin. Biotechnol.* 50 222-227.

Vaughan EE, Daly C, Fitzgerald GF. 1992. Identification and characterization of helveticin V-1829, a bacteriocin produced by *Lactobacillus helveticus* 1829. *J Appl Bacteriol.* 73(4), 299-308.

Walton JD, Hallen-Adams HE, Luo H. 2010. Ribosomal biosynthesis of the cyclic peptide toxins of Amanita mushrooms. *Biopolymers* 94, 659–664.

Weber KA, Pollock J, Cole KA, O'Connor SM, Achenbach LA, Coates JD. 2006. Anaerobic nitrate-dependent iron(II) bio-oxidation by a novel lithoautotrophic betaproteobacterium, strain 2002. *Appl. Environ. Microbiol.* 72, 686.

Weusthuis R, Lamot I, van der Oost J, Sanders J. 2011. Microbial production of bulk chemicals: development of anaerobic processes. *Trends Biotechnol.* 29(4), 153-158.

Willke T, Vorlop K. 2008. Biotransformation of glycerol into 1, 3 propanediol. *Eur J Lipid Sci Technol.* 110(9), 831-840.

Wiœniewska P, Œliwiñska M, Dymerski T, Wardencki W, Namieœnik J. 2015. The analysis of vodka: a review paper *Food Anal. Methods.* 8, 2000-2010.

Wolfe R. 2019. Anaerobic life-a centennial view. Retrieved 19 November 2019, from http://europepmc.org/articles/PMC93796.

Wrba A, Jaenicke R, Huber R, Stetter K. 1990. Lactate dehydrogenase from the extreme thermophile Thermotoga maritima. *Eur. J. Biochem.* 188(1), 195-201.

Yagi H, Ninomiya F, Funabashi M, Kunioka M. 2014. Mesophilic anaerobic biodegradation test and analysis of eubacteria and archaea involved in anaerobic biodegradation of four specified biodegradable polyesters. *Poly Degrad Stab* 110,278–283.

Yang ST, Yu M. 2013. Integrated biorefinery for sustainable production of fuels, chemicals, and polymers. In Bioprocessing Technologies in Biorefinery for Sustainable Production of Fuels, Chemicals, and Polymers, 1-26.

Yazdani, SS, Gonzalez R. 2007. Anaerobic fermentation of glycerol: a path to economic viability for the biofuels industry. *Curr. Opin. Biotechnol.* 18,213-2197.

Zhang L, Loh K-C, Zhang J. 2019. Enhanced biogas production from anaerobic digestion of solid organic wastes: Current status and prospects. *Bioresour Technol Rep*, 5, 280-296.

Zhang M, Zheng P, Li W, Wang R, Ding S, Abbas G. 2015. Performance of nitrate-dependent anaerobic ferrous oxidizing (NAFO) process: A novel prospective technology for autotrophic denitrification. *Bioresour Technol.* 179, 543-548.

Zhang M, Zheng P, Wang R, Li W, Lu H, Zhang J. 2014. Nitrate-dependent anaerobic ferrous oxidation (NAFO) by denitrifying bacteria: a perspective autotrophic nitrogen pollution control technology. *Chemosphere.* 117, 604-609.

Zumft WG. 1997. Cell biology and molecular basis of denitrification. *Microbiol. Mol. Biol. Rev.* 61 533–616.